Floods

Floods
Physical Processes and Human Impacts

Keith Smith
Emeritus Professor of Environmental Science, University of Stirling, UK
and
Roy Ward
Emeritus Professor of Geography, University of Hull, UK

JOHN WILEY & SONS
Chichester · New York · Weinheim · Brisbane · Singapore · Toronto

Copyright © 1998 by John Wiley

Published in 1998 by John Wiley & Sons Ltd
Baffins Lane, Chichester,
West Sussex PO19 1UD, England

National 01243 779777
International (+44) 1243 779777
e-mail (for orders and customer service enquiries): cs-books@wiley.co.uk
Visit our Home Page on http://www.wiley.co.uk
or http://www.wiley.com

All Rights Reserved. No part of this publication may be reproduced, stored in a retrieval system, or transmitted, in any form or by any means, electronic, mechanical, photocopying, recording, scanning or otherwise, except under the terms of the Copyright, Designs and Patents Act 1988 or under the terms of a licence issued by the Copyright Licensing Agency, 90 Tottenham Court Road, London W1P 9HE, UK, without the permission in writing of the Publisher and the copyright owner.

John Wiley & Sons Inc., 605 Third Avenue, New York,
NY 10158-0012, USA

WILEY-VCH Verlag GmbH, Pappelallee 3,
D-69469 Weinheim, Germany

Jacaranda Wiley Ltd, 33 Park Road, Milton,
Queensland 4064, Australia

John Wiley & Sons (Asia) Pte Ltd, Clementi Loop #02-01,
Jin Xing Distripark, Singapore 129809

John Wiley & Sons (Canada) Ltd, 22 Worcester Road,
Rexdale, Ontario M9W 1L1, Canada

British Library Cataloguing in Publication Data
A catalogue record for this book is available from the British Library

ISBN 0 471 95248 6
Typeset in 10/12pt Times from the authors' disks by Dobbie Typesetting Limited, Tavistock, Devon
Printed and bound in Great Britain by Bookcraft (Bath) Ltd
This book is printed on acid-free paper responsibly manufactured from sustainable forestation, for which at least two trees are planted for each one used for paper production

Contents

Preface ix

Acronyms and Abbreviations xi

SECTION ONE   THE FLOOD HAZARD IN CONTEXT

Chapter 1   Floods: Physical Events and Natural Hazards 3

1.1   Introduction and definitions 3
1.2   Floods as physical features 9
1.3   Floods as hazards 19

Chapter 2   Impacts and Interpretations of Flood Hazard 34

2.1   Impacts of flood hazard 34
2.2   The benefits of floods 35
2.3   Estimating the losses from floods 38
2.4   Direct losses from floods 45
2.5   Indirect losses from floods 51
2.6   Interpretations of flood hazard 54

SECTION TWO   PROCESSES OF FLOODING

Chapter 3   River Floods: Geophysical Processes 61

3.1   Introduction 61
3.2   The raw materials of flooding 61
3.3   The flood hydrograph 65
3.4   Flood-producing rainfalls 71
3.5   Snowmelt and icemelt 76
3.6   Time trends in flood production 79

Chapter 4   River Floods: Spatial Characteristics 96

4.1   Introduction 96
4.2   Spatial characteristics of river flooding 97
4.3   Regionalisation 97
4.4   Flood climate zones 103
4.5   Floods in cold climates 104

4.6	Floods in hot climates	111
4.7	Floods in temperate climates	131

Chapter 5 Coastal Floods 143

5.1	Introduction	143
5.2	Flood-producing processes in coastal and estuarine areas	145
5.3	Storm surges	148
5.4	Tsunamis	164
5.5	Sea-level change and coastal flooding	170
5.6	Floods and coastal geomorphology	175

Chapter 6 Flood Estimation 178

6.1	Introduction	178

RIVER FLOODS

6.2	Estimating river floods	179
6.3	Enhancing the floods database	192
6.4	Contemporary approaches to river flood estimation	194
6.5	Meltwater floods	198

COASTAL FLOODS

6.6	Estimating coastal floods	199

SECTION THREE   RESPONSES TO THE FLOOD HAZARD

Chapter 7 Flood Defence 205

7.1	Introduction to flood defence	205
7.2	River flood engineering	208
7.3	River flood abatement	220
7.4	Coastal flood engineering	226
7.5	Coastal flood abatement	231
7.6	Flood proofing	233

Chapter 8 Flood Forecasting and Warning 238

8.1	Introduction	239

RIVER FLOOD FORECASTING

8.2	Hydrological components	241
8.3	Hydrograph routing	247
8.4	Real-time flood forecasting systems	249

FLOOD WARNING

8.5	The nature of flood warning	264
8.6	Prior assessment for flood warnings	268
8.7	Dissemination of flood warnings	270
8.8	Response to flood warnings	274
8.9	Effectiveness of flood warning schemes	275

CONTENTS   vii

      COASTAL FLOODS
8.10 Introduction 280
8.11 The components of coastal flood forecasts 280
8.12 Storm-surge forecasting and warning systems 283
8.13 Tsunami warning systems 290

Chapter 9 Mitigating and Managing Flood Losses 293
9.1 Introduction 293
9.2 Disaster aid 295
9.3 Insurance 300
9.4 Emergency planning and disaster management 308
9.5 Land-use management 312
9.6 Living with floods 319
Appendix 9.1 Subdivision of Special Flood Hazard Areas 326

Chapter 10 Outlook 328
10.1 Introduction 328
10.2 Physical processes: Problems 329
10.3 Physical processes: Some possible solutions 331
10.4 Human impacts: Problems 334
10.5 Human impacts: Some possible solutions 336

Appendix   Metric Conversion Tables 341

References 343

Index 377

Preface

Throughout history floods have been part of human destiny. They are widely discussed today as a result of increased public awareness and greater research activity on the part of both physical and social scientists. The media headlines often present flood events as rare and random caprices of a hostile world inflicted on unsuspecting victims. In more developed countries, especially, a prevailing view has been that science and technology are bringing these unpredictable excesses under safe control.

The reality is very different. Floods are recurring phenomena which form a necessary and enduring feature of all river basin and lowland coastal systems. In an average year they bring important benefits to millions of people who depend on these environments. But major floods are the largest cause of economic losses from natural disasters, mainly in more developed countries, and they are also a major cause of disaster-related deaths, mainly in the less developed countries. Despite recent advances in the understanding of the relevant climatological, fluvial and marine mechanisms, and a greater investment in flood reduction measures, floods take a larger number of lives and damage more properties each year, mainly because of unwise land management practices and growing human vulnerability.

For decades we have not coped well with floods. Part of the reason for this lies in the complex nature of floods and the varied responses to them. Earlier reliance on engineering structures, geared to the control of floodwaters, gave way after the 1960s to a broader approach involving a combination of structural and non-structural measures designed to reduce human exposure to floods. More recently, concern about the economic and technical performance of some flood control measures, together with a heightened awareness of environmental issues and the relationships between flooding and human vulnerability in the Third World, has led to a new policy era. Now the emphasis is on living with floods rather than fighting them, despite the fact that, in future decades, the frequency and magnitude of floods can be expected to increase along many of the coastlines and major rivers of the world as a result of climate change (including the effects of sea-level rise in coastal areas), land-use changes and further encroachment on to floodplains and coastal lowlands.

The current understanding of floods represents a plateau of knowledge which, as the 1990s International Decade for Natural Disaster Reduction draws to a close, is well worth summarising. Our intention in this book is to produce a comprehensive and readable survey of this important topic. The book presents a balanced overview of all aspects of flooding arising from the varying mix of natural processes, technology applications and societal responses in the modern world. Throughout we illustrate common problems and policies using case studies drawn from a wide range of global experience.

The book is intended to serve both as an undergraduate teaching aid and as a key source of reference for postgraduates and professionals, including practitioners and policy makers in agencies, consultancy and government departments. But we hope that it will also appeal to everyone who is concerned about flooding and the flood hazard and who wishes to increase their understanding of the relevant issues.

All the diagrams were prepared by Bill Jamieson and David Aitchison (Department of Environmental Science, University of Stirling), to whom we express our gratitude and admiration. In addition, both of us would wish to thank the many friends and colleagues who have generously provided data and information, helped with the selection of diagrams and photographs, and commented on sections of the draft manuscript. Particular thanks are due to Heinz Engel (Bundesanstalt für Gewässerkunde), Bob Moore and Duncan Reed (Institute of Hydrology), Edmund Penning-Rowsell (Middlesex University), Ian Reid (University of Loughborough), and Richard Yates (Environment Agency). Their contributions have done much to improve this book; for the failings that remain, however, we take full responsibility. We also thank our wives, Muriel and Kay, for their continued help and encouragement in the preparation and indexing of this book.

Keith Smith and Roy Ward
October, 1997

Acronyms and Abbreviations

AVHRR	Advanced Very High Resolution Radiometer
DTM	Digital Terrain Model
EA	Environment Agency (UK)
ENSO	El Niño–Southern Oscillation
ERS	Earth Resources Satellite
FEH	*Flood Estimation Handbook* (UK)
FEMA	Federal Emergency Management Agency
FRIEND	Flow Regimes from International Experimental Data
FRONTIERS	Forecasting Rain Optimised using New Techniques of Interactively Enhanced Radar and Satellite
FSR	*Flood Studies Report* (UK)
GCM	General Circulation Model
GDP	Gross Domestic Product
GIS	Geographical Information System
GLOCOPH	Global Continental Palaeohydrology [Project]
GPS	Global Positioning System
HEC	Hydrologic Engineering Center (US Corps of Engineers)
HYRAD	Hydrological Radar
IDNDR	International Decade for Natural Disaster Reduction
IPCC	Intergovernmental Panel on Climate Change
IRF	Intermediate Regional Flood
LDC	Less developed country
MAFF	Ministry of Agriculture, Fisheries and Food (UK)
MCC	Mesoscale convective complex
MDC	More developed country
MORECS	Meteorological Office Rainfall and Evaporation Calculation System (UK)
NERC	Natural Environment Research Council (UK)
NFIP	National Flood Insurance Program (USA)
NGO	Non-Governmental Organisation
NOAA	National Oceanographic and Atmospheric Administration
NWS	National Weather Service (USA)
OFDA	Office of Foreign Disaster Assistance (USA)
PMF	Probable Maximum Flood
PMP	Probable Maximum Precipitation
PSMSL	Permanent Service for Mean Sea Level
RFFS	River Flow Forecasting System

SEPA	Scottish Environment Protection Agency
SFHA	Special Flood Hazard Area
SMD	Soil moisture deficit
SPF	Standard Project Flood
SSSI	Site of Special Scientific Interest (UK)
SST	Sea surface temperature
STWS	Storm Tide Warning Service
UNDRO	UN Disaster Relief Organisation
USGS	United States Geological Survey

SECTION ONE
The Flood Hazard in Context

CHAPTER 1

Floods: Physical Events and Natural Hazards

CONTENTS

1.1 Introduction and definitions 3
    1.1.1 An ever-present hazard 4
    1.1.2 An historical perspective 5
    1.1.3 Definitions 8
1.2 Floods as physical features 9
    1.2.1 The flood process 10
        1.2.1.1 Types of flood 10
        1.2.1.2 Causes of floods 10
        1.2.1.3 Flood-intensifying factors 12
    1.2.2 The river flood hydrograph 14
    1.2.3 The dimensions of flooding 16
1.3 Floods as hazards 19
    1.3.1 Flooding as a natural hazard 19
    1.3.2 World patterns and trends 21
    1.3.3 Human causes of flood hazard 26
    1.3.4 Human responses to flood hazard 32

1.1 INTRODUCTION AND DEFINITIONS

The terms 'floods', 'flooding' and 'flood hazard' cover a very wide range of phenomena, not all of which are treated with equal emphasis in the following discussions. The main focus in this book is on river floods, but the low-lying deltas and estuaries of many of the world's major rivers are also exposed to the hazards of flooding from the sea by storm surges and tsunamis, as indeed is all low-lying coastal land. In Britain, for example, coastal flooding poses a major risk and dominates British flood history (Parker, 1985). Accordingly, it would have been difficult, and arguably irrational, to have ignored the phenomenon of coastal flooding which therefore forms an important but secondary focus.

    It is perhaps initially surprising, even disappointing, that in an age of virtual reality and multimedia communications, events as simple to comprehend as floods should remain sufficient of a 'problem' to merit a new book. In fact, floods continue to be not only a problem but in some respects an *increasing* problem, catching individuals and communities by surprise in a repetitively exasperating way, and causing disruption, damage and even death. Although many of the reasons for this are explored in what follows, one important underlying factor which may be noted at the outset is the inadequacy of the database for both physical and socio-economic analyses of the flood hazard. Reliable and comprehensive hydrological data are still not universally collected and in some countries have not been collected until quite recently. Where data on river flow and rainfall *are* collected they may be less than adequate for detailed flood studies, e.g. flood peaks or significant periods of intense rainfall may be missed by periodic rather than continuous recording (Wain, 1994). Even where continuous data on river flow are collected, the largest floods are seldom recorded since they normally destroy the recorders and control sections involved.

1.1.1 An Ever-Present Hazard

Floods are rarely out of the local, national, or international news on both TV and in the newspapers. Most of the news items relate to fairly minor events which cause little damage and are soon forgotten except by those most directly affected. Some, however, reflect major disasters involving loss of life and the destruction of property, and bring in their wake hardship, suffering, disease and famine.

In the 1980s there were at least 60 major flood disasters, each involving the loss of more than 100 lives and including 10 where more than 1000 lives were lost. These occurred in 17 countries, mainly in Asia and South America (UNEP, 1991a). A similar catalogue occurred in the 1990s. In April 1990 floods in the Australian outback covered an area the size of Western Europe and three years later the state of Victoria suffered its worst flooding in more than half a century. In July 1991 China experienced the worst floods of the century in which more than 1400 people died; more than 200 died two years later when a dam burst in Qinghai province; and in July 1996 the heaviest rains in 50 years caused flooding in which more than 250 people died in south-east China. Between 2000 and 3000 people died during flash flooding in the Philippines in November 1991 and a similar number in Pakistan in September 1992. In July 1993 a tsunami, approaching 30 m in height in places, crashed onto the shores of Okushiri Island in the Sea of Japan killing almost 300 people. In the same month the Mid-West of the USA experienced record-breaking flooding with further calamitous flooding in May 1995 and March 1997. Europe was severely affected in December 1993 and January 1994 when major flooding occurred in southern England, southern France, Spain and the countries bordering the lower Rhine. The Rhine countries were again flooded in early 1995 as a result of the wettest January for nearly 50 years. This prompted the largest evacuation of people in the Netherlands (about 250 000) since the great North Sea flood of 1953. And in July 1997 ageing dikes on the River Oder collapsed in many places after prolonged heavy rainfall, resulting in the worst floods for about 200 years in adjacent low-lying areas of the Czech, Polish and German Republics. Many people died and losses of livestock, homes and industrial plant exceeded US$ 3 billion.

The repeated occurrence of major flooding, as in China, or the Indian sub-continent, or in North America where catastrophic flooding has occurred in every year since 1993, and the use of phrases like 'the worst in half a century' and 'the worst in living memory', give the impression that floods are getting worse. Although human memory is notoriously transient (on average, 'living memory' has a duration of less than 10 years), there is some evidence that flood conditions may be deteriorating. Climate change, irrespective of global warming, is undeniable. Small changes in atmospheric circulation and subtle changes in the fine interrelationships between sea surface temperatures and the overlying atmosphere cause local alterations in the frequency of intense or prolonged precipitation, and thereby in the size and frequency of river flood events, although the evidence is still largely circumstantial and difficult to interpret. Land-use changes, such as urbanisation, forest clearance, agricultural under-drainage and the ploughing up of former extensive grasslands, have increased the flood potential of some river basins. In addition, the character of river flooding may have changed over many centuries of development in countries like Britain. In early historical times rivers probably flooded more frequently but because of the primitive, poorly drained character of river basins and floodplains the floodwaters would have risen gradually, peaked, and then drained equally gradually back into the channels. Over the centuries land-use changes and improved drainage have speeded the movement of water into stream channels which, as a result, have been gradually dredged, straightened and embanked so as to accommodate a greater flow of water. Now, minor floods are contained but when larger floods overtop the embankments the resulting inundation takes place rapidly and the floodwaters are unable to drain back easily into the channels.

There is even more evidence that the flood problem is getting worse in terms of the damage caused by flooding. Despite massive expenditure on flood defence, flood damage losses continue to rise in many countries, and it is now accepted that protective measures are often counterproductive and that, by engendering a false sense of security and encouraging more intensive development, they may give rise to even higher damages than would otherwise have occurred.

This is an important point which underlines a major theme of this book, namely that *although most floods are more or less natural phenomena* (albeit intensified by human action such as land-use change), *the flood hazard is largely of human origin*. Most floods result from moderate-to-large events (rainfall, snowmelt, high tides) occurring within the expected range of streamflow or tidal conditions. In the case of river floods, for example, fluvial channels can carry only a fraction of the flows occurring during floods so that the remainder must spill on to the floodplain. In flood conditions, therefore, channels and their adjacent floodplains are complementary and *together* form the proper conveyance for the transmission of floodwaters. In many cases even major floods simply spill their waters on to unoccupied floodplains or 'washlands' where they do little damage and may even be beneficial, as in arid zones, where irrigation and soil fertilisation may depend on the natural flooding of rivers. Floods constitute a 'hazard' only where human encroachment into flood-prone areas has occurred.

1.1.2 An Historical Perspective

The generally successful long-term utilisation of flood-prone areas has been interrupted frequently by major tragedies and recent history has seen many disastrous flood events. Few can match the sheer devastation caused in China by the 1887 and 1931 flooding on the Huang Ho which may have claimed 900 000 and 3.7 million lives respectively, although such estimates are notoriously unreliable (Burton, 1989), and the 1911 flooding on the Ch'ang Chiang (Yangtze) in which 100 000 died. In China, where millions live at risk of flooding, records go back in some districts for some 2200 years and reveal more than 1000 occurrences of severe flooding (Yang, 1989). Chinese history is full of tales of the struggles against the notorious Huang Ho, often called 'China's Sorrow', which not only floods frequently but also tends to change its course, bringing unexpected disaster to new areas.

Some of the maximum floods on record are shown in Table 1.1 and Figure 1.1 for river basins ranging in size from about 1000 km² to more than 4.5×10^6 km². Although, for reasons discussed earlier, the precision of the flood estimates shown is relatively poor, it is clear that absolute flood magnitude generally increases and specific magnitude (i.e. discharge per unit area) generally decreases with basin size. Accordingly, neither is necessarily a satisfactory index of whether one flood is larger than another. For such comparisons Rodier and Roche (1984) preferred the K-index, proposed by Francou and Rodier (1969), where

$$K = 10\,[(1 - (\log(Q) - 6))/(\log(A) - 8)] \tag{1.1}$$

when Q is the largest flood in m³ s⁻¹ (cumecs) and A is the catchment area in km². On a log–log plot the envelope to all the largest floods in the world from catchments larger than 100 km² is a straight line corresponding approximately to a K-value of 6.00 (Rodier and Roche, 1984). All the flood events shown in Table 1.1 and Figure 1.1 have K-values greater than 5.75 and the largest have K-values well in excess of 6.00.

Throughout history and indeed pre-history, floods have focused attention on the interactions between people and environment and epic stories of flood disasters are found in the myths, poems and narratives of many civilisations. Some of these legends have been at least partially verified by subsequent archaeological and anthropological evidence. Probably the best-known flood epic is the biblical story of Noah, saved from

Table 1.1 Selected maximum floods

Country	River	Basin area (km²)	Maximum discharge (m³ s⁻¹)	Specific discharge (m³ s⁻¹ km⁻²)	K-value*	Year
New Zealand	Haast	1020	7690	7.54	5.76	1979
Mexico	Cithuatian	1370	13500	9.85	6.16	1959
Australia	Pioneer	1490	9840	6.60	5.84	1918
USA (Texas)	West Nueces	1800	15600	8.67	6.16	1959
Taiwan	Tam Shui	2110	16700	7.91	6.20	1963
Japan	Shingu Oga	2350	19025	8.10	6.29	1959
North Korea	Daeryong Gang	3020	13500	4.47	5.83	1975
Philippines	Cagayan	4244	17550	4.14	5.98	1959
USA (California)	Eel Scotia	8060	21300	2.64	5.92	1964
USA (Texas)	Pecos	9100	26800	2.95	6.11	1954
Madagascar	Betsiboka	11800	22000	1.86	5.78	1927
North Korea	Toedong Gang	12175	29000	2.38	6.06	1967
South Korea	Han Koan	23880	37000	1.55	6.05	1925
Pakistan	Jhelum	29000	31100	1.07	5.74	1929
China	Hanjiang	41400	40000	0.97	5.87	1583
Madagascar	Mangoky	50000	38000	0.76	5.70	1933
India	Narmada	88000	69400	0.79	6.21	1970
China	Ch'ang Chiang	1010000	110000	0.11	5.20	1870
Russia	Lena	2430000	189000	0.08	5.52	1967
Brazil	Amazonas	4640000	370000	0.08	6.76	1953

*See text for explanation
Source: From data in Rodier and Roche (1984)

FLOODS: PHYSICAL EVENTS AND NATURAL HAZARDS 7

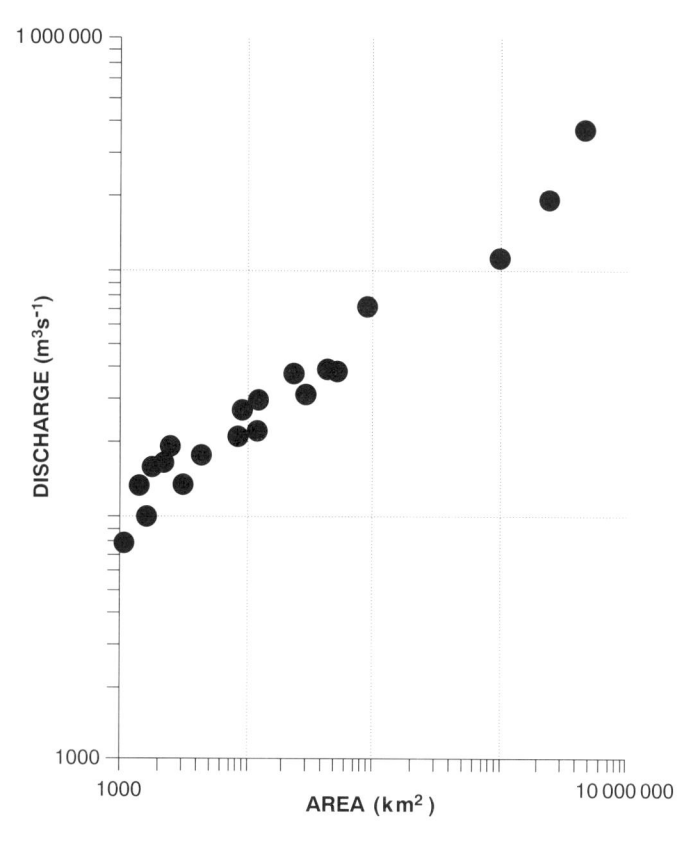

Figure 1.1 Selected maximum flood discharges plotted logarithmically against catchment area. (Based on the data in Table 1.1)

the floodwaters with his family, birds and animals by building and entering the Ark. Whether Noah's flood was an historical fact, or merely the stuff of myth and legend, is unclear. One view is that the story reflects the break-through of the Mediterranean into the Black Sea basin (Ahuja, 1996). Sedimentological and stratigraphical evidence appears to support the view that rising sea level during the emergence from the last Ice Age resulted in the Mediterranean spilling through the Bosphorus about 7500 years ago flooding the inhabited plain that is now the Black Sea. Alternatively, Hebrew and Babylonian traditions may have chronicled the long-term struggle with the wayward rivers of the Tigris and Euphrates valleys, the legends spreading later to Syria and Palestine (Lambert and Millard, 1969). Certainly, archaeological evidence of flood deposits separating the strata of different Sumerian and Babylonian civilisations indicates that floods did destroy existing cultures, although it is not clear whether the tradition of the flood as an historical time-reference reflects one severe flood or the memory of many disasters. In the Babylonian flood saga, described by Lambert and Millard, King Atra-hasis was given seven days by the god Enki to pull down his reed house and build a boat with the materials. The boat was loaded with his possessions, animals and birds and as soon as they were aboard the flood came and, apart from

Atra-hasis and his passengers, the entire human race was destroyed. In Babylonian tradition the flood lasted seven days and seven nights.

Clearly, Noah appears under various aliases, not only in the Middle East, but in many other parts of the world including India, China, the East Indies, Polynesia and particularly the Americas. Milne (1986) noted that from the Canadian North-West, through North and Central America and into the southern continents, virtually every tribe and ethnic group has a legend referring to a worldwide flood.

Universally, flood legends repeat the common themes of arks, mountains and godly retribution and, as Milne (1986) observed, they are often couched in terms of a new beginning rather than of the end of the world and may have diffused from a relatively small number of source areas. Some seem clearly to relate to coastal flooding from tsunamis, e.g. the tales of the Araucanian Indians of Chile; others were undoubtedly the embellished account of a local flood or violent storm.

By comparison, our scientific interest in floods, based on programmes of intensive and organised data collection, is largely very recent. Flood databases as long as those in Egypt or China are very rare. In the USA it was not until the mid-nineteenth century that the Federal Government first actively participated in flood protection and organised surveys of the Mississippi system. In fact it was not until 1928 that a modern approach to flood studies began when Congress authorised the expansion of the Mississippi flood control project—an action which was broadened by subsequent legislation, including the significant Flood Control Act of 1936 (Chow, 1956). In Britain, the Institution of Civil Engineers' *Interim Report of the Committee on Floods in Relation to Reservoir Practice*, published in 1933, was the basic guide to engineering practice until the publication in 1975 of the NERC Flood Studies Team's *Flood Studies Report*. Together with subsequent amendments, this has since formed the basis for flood hydrology and flood estimation in Britain and in many other areas as well. It is due to be succeeded before the end of the century by the NERC *Flood Estimation Handbook*.

1.1.3 Definitions

For present purposes a meaningful definition of floods should not only incorporate the notions of *inundation* and *damage*, but also move beyond the restrictive definitions of river floods given, for example, by Chow (1956)

> A flood is a relatively high flow which overtaxes the natural channel provided for the runoff

or by Rostvedt et al. (1968)

> A flood is any high streamflow which overtops natural or artificial banks of a stream

to encompass the additional processes involved in coastal flooding. The definition given by Ward (1978) that

> A flood is a body of water which rises to overflow land which is not normally submerged

FLOODS: PHYSICAL EVENTS AND NATURAL HAZARDS

usefully reflects the approach to flooding which is adopted in this book. However, although explicitly including all types of *surface inundation, damage* is implied only in its final three words. Nor does the definition necessarily illuminate either the great range of scale at which inundation and damage occur, e.g. from the flooding of a small suburban shopping centre to the flooding of an entire nation, or the full spectrum of outpourings during a flood event. These may range from the nearly clear water generated by a melting glacier to, on the one hand, the slush avalanche of water-lubricated snow and ice crystals cascading from a collapsing snowpack, and on the other hand, the mudflow or 'mud rush' moving rapidly down a hillslope or channel and comprising water so full of mud, volcanic ash or other suspended solids that its specific gravity is closer to 2.0 or 2.5 than to 1.0. The definition does, however, cover the main elements of river and coastal flooding as well as flooding in shallow depressions which is caused by water-table rise, rainwater flooding on level surfaces and sheetwash flooding on low-gradient slopes, both resulting solely from torrential rainfall, and flooding caused by the backing-up or overflow of artificial drainage systems, especially in urban areas.

In some major urban areas, especially those having complex underground transport networks and services, the threatened *subsurface flooding* of these facilities as a result of rising groundwater levels now constitutes a major problem. Under central London, for example, the water-table rose 36 m between the 1970s and the mid-1990s and is predicted to rise a further 25–30 m by the year 2005 (Midgley, 1995). Similar rises have occurred under Liverpool, where the underground rail system has been disrupted from time to time, and under Birmingham, Glasgow, Nottingham and elsewhere in the UK (e.g. Greswell et al., 1994; Lerner and Barrett, 1996). Subsurface flooding is not dealt with further in this book and may, in any case, be counteracted by increasing the rate of groundwater abstraction for public water supply or other purposes in threatened areas.

1.2 FLOODS AS PHYSICAL FEATURES

The definitions discussed in Section 1.1.3 imply a wide variety of events and processes leading to inundations of water carrying varying concentrations either of solids in suspension (i.e. ranging from nearly clear water to mudflows or sewage) or of salts in solution (i.e. ranging from nearly clean freshwater to seawater). These inundations may be deep or shallow and may comprise water which is either virtually static or which is a raging torrent. The impact of floodwaters through erosion and deposition, or through social and economic loss, depends largely on the combination of water quality, depth and velocity. However, it also depends on other factors, such as seasonality and frequency, the type of flooding and the shape of the flood hydrograph.

Apart from the geomorphological role of floods, for a treatment of which the reader is referred to specialist texts such as Baker et al. (1988) and Beven and Carling (1989), these and related issues will form recurrent themes throughout this book. They are introduced briefly, in this discussion of floods as physical features, under the headings of the flood process, the river flood hydrograph, and the dimensions of flooding.

1.2.1 The Flood Process

1.2.1.1 Types of Flood

The main types of flood, mentioned briefly in Section 1.1.3, are listed in more detail here.

River floods

- Floods in river valleys occur mostly on floodplains or washlands as a result of flow exceeding the capacity of the stream channels and *overspilling* the natural banks or artificial embankments.
- Sometimes inundation of the floodplain, or of other flat areas, occurs in wet conditions when an already shallow water-table rises above the level of the ground surface. This type of *water-table flooding* is often an immediate precursor of overspill flooding from the stream channels.
- In very dry conditions, when the ground surface is baked hard or becomes crusted, extensive flat areas may be flooded by heavy *rainfall ponding* on the surface. This *rainwater flooding* is typical of arid and semi-arid environments but is also experienced much more widely.
- Also typical of arid and semi-arid areas is the situation where there are no clearly defined channels and where *sheetwash flooding* occurs by the unimpeded lateral spread of water moving down a previously dry or near-dry valley bottom or alluvial fan.
- In urban areas flooding often results from overspilling or surface ponding, as described above, but may also occur when *urban stormwater drains* become surcharged and overflow. This is a growing problem in Britain where ageing and inadequate combined sewage and stormwater systems give rise to frequent foulwater flooding from combined sewer overflows in urban areas (Parker and Penning-Rowsell, 1983b; Murray, 1995).

Coastal floods

- Floods in low-lying coastal areas, including estuaries and deltas, involve the inundation of land by brackish or saline water. *Brackish-water floods* result when river water overspills embankments in coastal reaches as flow into the sea is impeded by high-tide conditions. Overspill is exacerbated when high-tide levels are increased above normal by storm-surge conditions or when large freshwater flood flows are moving down an estuary.
- Direct inundation by *saline water floods* may occur when exceptionally large wind-generated waves are driven into semi-enclosed bays during severe storm or storm-surge conditions, or when so-called 'tidal waves', generated by tectonic activity, move into shallow coastal waters.

1.2.1.2 Causes of Floods

The causes of flooding, which are discussed in more detail in later chapters, are summarised in the upper part of Figure 1.2. Most *river floods* result directly or indirectly from climatological events such as excessively heavy and/or excessively prolonged rainfall. In cold-winter areas, where snowfall accumulates, substantial flooding usually

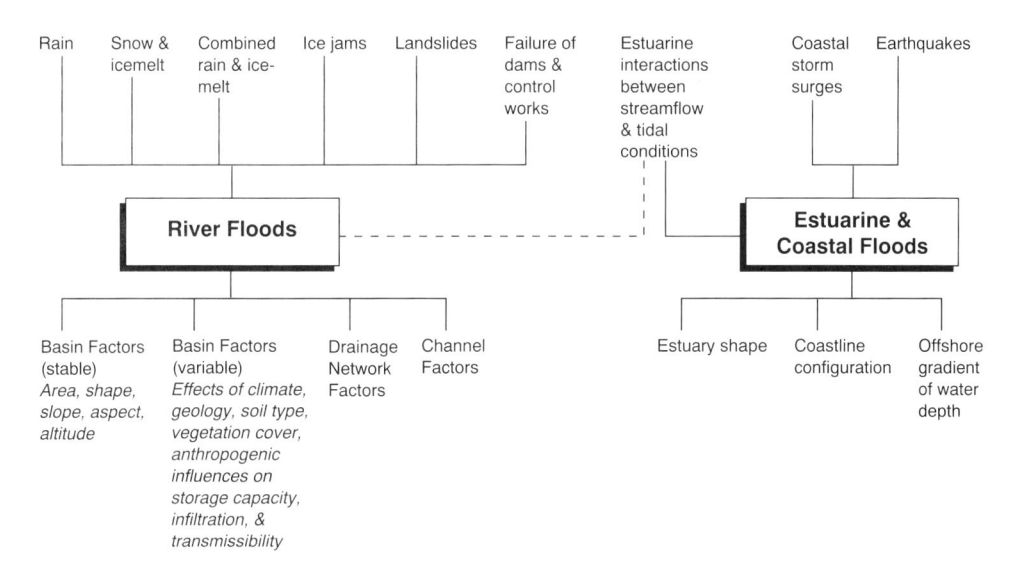

CAUSES OF FLOODS

```
Rain   Snow &   Combined   Ice jams   Landslides   Failure of   Estuarine         Coastal    Earthquakes
       icemelt  rain & ice-                        dams &       interactions     storm
                melt                               control      between          surges
                                                   works        streamflow
                                                                & tidal
                                                                conditions
```

River Floods — — — — — — — — — — — — — — — **Estuarine & Coastal Floods**

Basin Factors	Basin Factors	Drainage	Channel		Estuary shape	Coastline	Offshore
(stable)	(variable)	Network	Factors			configuration	gradient
Area, shape,	Effects of climate,	Factors					of water
slope, aspect,	geology, soil type,						depth
altitude	vegetation cover,						
	anthropogenic						
	influences on						
	storage capacity,						
	infiltration, &						
	transmissibility						

FLOOD-INTENSIFYING FACTORS

Figure 1.2 Causes of floods and flood-intensifying factors

occurs during the period of snowmelt and icemelt in spring and early summer, particularly when melt rates are high. Floods may also result from the effects of rain falling on an already decaying and melting snowpack. An additional cause of flooding in cold-winter areas is the sudden collapse of ice jams, formed during the break-up of river ice.

Major landslides, such as 'La Josefina' in the Paute valley near Cuenca, Ecuador, in 1993, may cause flooding in two ways. First, ponding occurs behind the debris dam across the valley causing upstream flooding; then, as the debris dam is overtopped, erosion or collapse delivers massive flows downstream. In the case of 'La Josefina' outflows were in the range 7000–10 000 m^3 s^{-1} (United Nations, 1993). River floods may also result when landslides fall directly into upstream lakes or reservoirs causing a sudden rise in water level which overspills the outlet or dam. Similar catastrophic flooding normally results from the failure of natural or constructed dams and control works, or of canals. In 1976, for example, the newly built Elbe canal burst its floor at a viaduct near Luneburg, Germany, and in Yorkshire, England, there is concern that coal-mining in the Selby area will cause the collapse of river flood embankments. It is recognised, however, that the breaching or overtopping of embankments and other control works, following mining, erosion or inadequate maintenance or construction, is not normally a primary cause of *flooding* although it is often a direct cause of flood *disasters*, especially in the countries of Southern and South-Eastern Asia (e.g. Watanabe, 1989).

Estuarine and coastal floods are usually caused by a combination of high tides and the elevated sea level and large waves associated with storm surges which result from severe cyclonic weather systems. In estuarine and coastal areas protected by walls or embankments inundation may result from either *overflow*, when the water level exceeds the level of the crest of the defence, or from *overtopping*, when the combined effect of waves and water level results in waves running up and breaking over the defence, or from *structural failure* of the defence (Burgess and Reeve, 1994). When a storm surge coincides in an estuary with large inflows of river water, flood conditions in the estuary may become especially severe and may also magnify flooding in the lowland reaches of the rivers draining into the estuary (as shown by the broken line in Figure 1.2). Estuarine and coastal flooding may also be caused by *tsunamis* which result from the great increase in amplitude of seismically generated long ocean waves when these enter shallow coastal waters.

1.2.1.3 Flood-Intensifying Factors

As the lower part of Figure 1.2 shows, floods may be intensified by a number of factors. In the case of estuarine and coastal floods intensification results largely from the funnelling effects of estuary shape and coastline configuration, e.g. the Bay of Bengal and the southern part of the North Sea, which causes an overall increase in sea level as water is driven into the narrower section. For the same reason the overflow of sea defences is less common than overtopping on the open coast but is an important aspect of flooding in estuarine areas. The height of tsunami waves at their point of impact upon the coastline is dependent upon the offshore slope of the seabed, i.e. upon the rate of shallowing of water depth encountered as the tsunami moves towards the shore.

River floods may be intensified by factors associated either with the catchment itself or with the drainage network and stream channels. Most of these operate to speed up the movement of water within the catchment, although some work in a very different way, as when dense summer growths of water buttercup on the chalk streams of southern England displace a sufficient volume of water to cause flooding of adjacent land (Mann, 1988). Few of these factors, however, operate either unidirectionally or independently. Area, for example, is clearly important in the sense that the larger the catchment, the larger is the flood produced from a catchment-wide rainfall event. However, when the rainfall event covers only part of the catchment, the attenuation of the resulting flood hydrograph, as it moves through the channel network to the outlet, is likely to be greater in a large catchment than in a small one. Again, basin shape and the pattern of the drainage network combine to influence the size and shape of flood peaks at the basin outlet as shown in Figure 1.3A. Some of the most complex relationships, those between the variable basin factors, have a significant influence on three important hydrological variables, i.e. water storage, infiltration and transmissibility.

Water storage in the soil and deeper subsurface layers may affect both the timing and magnitude of flood response to precipitation, with low storage often resulting in rapid and intensified flooding. High infiltration values allow much of the precipitation to be absorbed by the soil surface, thereby reducing catchment flood response, and low infiltration values encourage the normally swifter over-the-surface movement of water leading to rapid increases in channel flow. In basins where most precipitation infiltrates the soil surface, flood response may be greatly modified by subsurface transmissibility, i.e. the ease with which water can move through the subsurface materials.

FLOODS: PHYSICAL EVENTS AND NATURAL HAZARDS 13

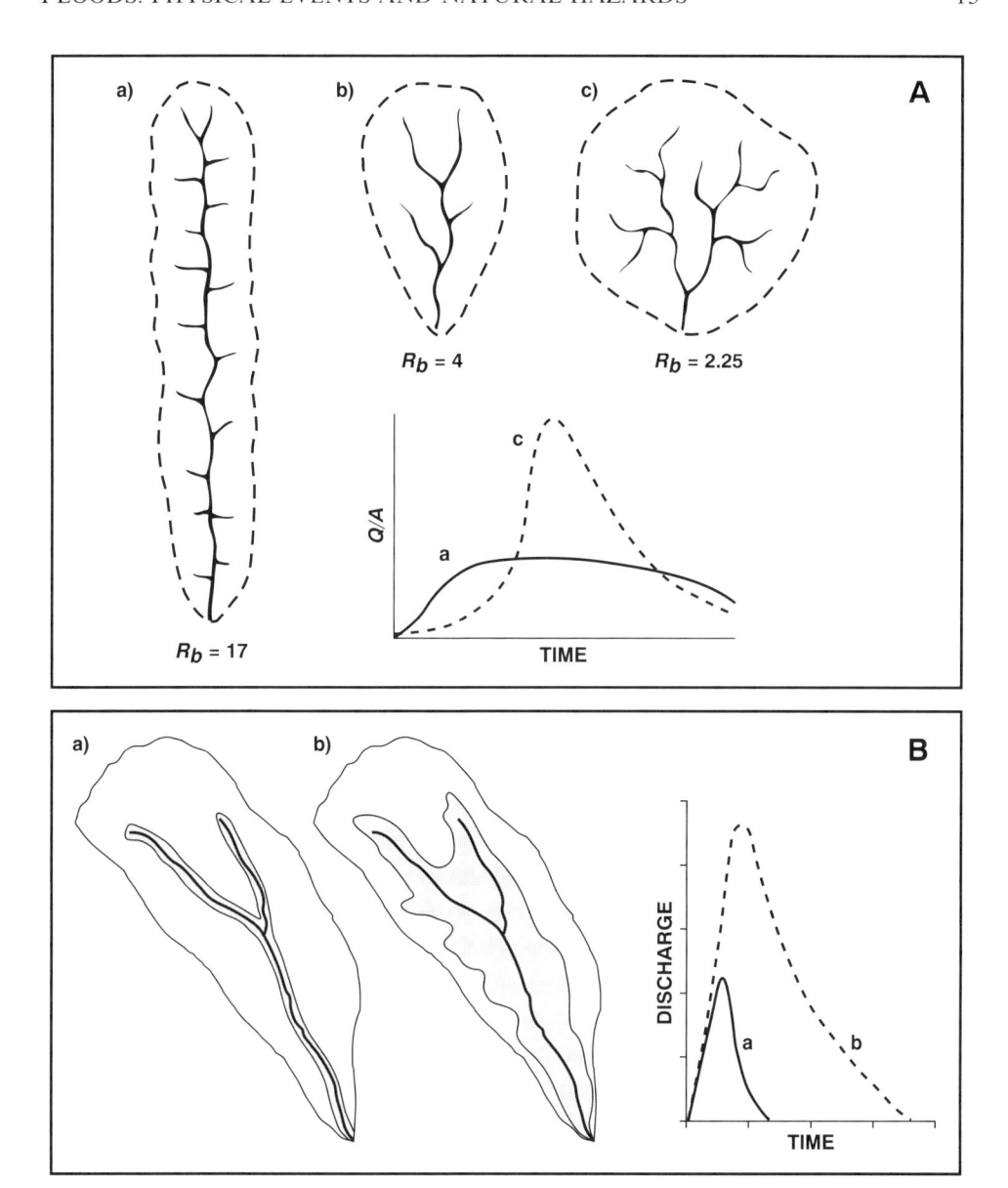

Figure 1.3 Examples of river flood intensification: (A) relations between basin shape, bifurcation ratio (R_b) and flood peaks; (B) variable source-area effects on flood peaks
Source: (B) is based on a diagram in Strahler (1964) with permission of The McGraw-Hill Companies

Apart from drainage pattern, channel and network factors within a drainage basin are essentially dynamic and their effects on flooding may change noticeably within a few hours. Of major importance is the proportion of the catchment area having interconnected water and waterlogged surfaces where the effective infiltration capacity is zero, so that all rain falling on such surfaces contributes directly and rapidly to streamflow. At the onset of rainfall these source areas for quickflow (Chapter 3) may be restricted to the water surfaces of the channel network but as rainfall continues the source areas expand (see Figure 1.3B) causing a major increase in the volume of rapid runoff reaching the stream channels.

Human activity frequently acts as a flood-intensifying factor by modifying key hydrological variables such as water storage, infiltration and transmissibility. The effects on flooding of anthropogenic influences associated with urbanisation, agriculture, forestry and dam construction are discussed in Section 3.6.

1.2.2 The River Flood Hydrograph

A river flood hydrograph is the plot of discharge or water-surface elevation against time during a flood event and defines the shape of the flood wave at the location of the gauging station. Hydrographs at successive locations can be used to define the changing shape of the flood wave as it moves downvalley. For coastal floods the flood hydrograph is defined by the water-surface elevation corresponding to the tidal curve or, more accurately, to the departure of the 'observed' from the 'expected' tidal curve.

The hydrographs of river floods caused by rainfall and/or melt have certain common features which are illustrated in Figure 1.4. During dry periods streamflow decreases exponentially (AX). At such times discharge consists solely of baseflow which is largely sustained by throughflow in small, steep, impervious catchments, by groundwater flow in flat, permeable catchments or by a combination of throughflow and groundwater flow in intermediate catchments. Rainfall begins at time X and the rapid increase of discharge between X and Y results from the generation of quickflow from source areas which expand as water is added to them by infiltration and throughflow (see Section 3.2 for more detailed discussion of quickflow generation). The hydrograph peak occurs at time Y, shortly after rainfall ends, and thereafter discharge is determined largely by the amount of water held in storage on and under the catchment surface. The rate at which this storage is exhausted will be reflected in the shape of the hydrograph's recession limb (YB). The early part of the recession is sustained largely by saturated throughflow, while the later stages (ZB and AX) will be sustained by the combination of unsaturated throughflow and groundwater flow, depending on catchment conditions.

Figure 1.4 illustrates that during a flood event the river flood hydrograph is dominated by the quickflow component, with a time base XZ and a peak Y. In most cases flood conditions do not begin immediately discharge increases at time X, since the initial increase in discharge is either contained within the channel or results in increased water levels which are still below the chosen arbitrary flood level. In other words the flood hydrograph actually begins and ends when the pre-selected flood level or discharge is equalled on the rising and falling limbs respectively. In Figure 1.4, therefore, the time base of the flood hydrograph is actually X_1Z_1.

The shape of the river flood hydrograph will depend not only on the nature and intensity of the climatological input (rainfall intensity, snowmelt, etc.) but also on variations in flood intensifying conditions (see Section 1.2.1 and Chapter 3) and may vary from the low peak and long time base of a sluggish stream to the high peak and short time base of a flashy stream (Figure 1.4 inset). In the latter case the *time of rise* or *basin lag* is very much shorter than in the former and means that the occupants of downstream flood-prone areas may receive a shorter warning of an incipient flood event. Lag depends on the interaction between precipitation or melt characteristics and

FLOODS: PHYSICAL EVENTS AND NATURAL HAZARDS

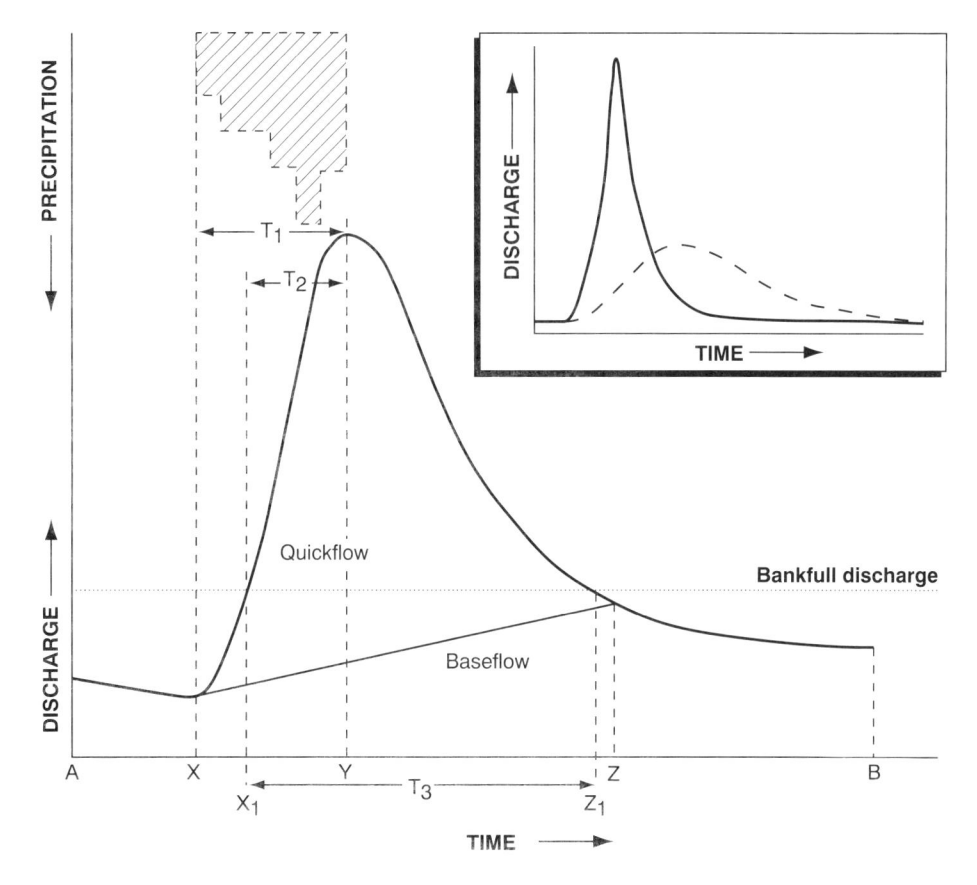

Figure 1.4 River flood hydrograph. Inset hydrographs on flashy and sluggish streams
Source: Hoyt and Langbein (1955). ©1955 Princetown University Press

the flood-intensifying conditions shown in Figure 1.2. Two measures of lag are shown in Figure 1.4, i.e. T_1—the time from the beginning of rainfall to the peak discharge, and T_2—the time from the onset of actual flood conditions, such as bankfull discharge, to the peak discharge. T_2 is often considerably shorter than T_1 and because people tend to become concerned about flooding only when actual flood conditions have arrived, represents more closely the time available for adjustment to the oncoming flood (Sheaffer, 1961).

In summary, the speed with which flood conditions develop after a rainfall or melt event is determined by a number of influences including rainfall or melt intensity, catchment characteristics and the scale effects associated with catchment size. Singly, or in combination, these influences determine whether the resulting flood is a slow-onset event or a sudden-onset event. One extreme is illustrated by the flood wave moving steadily down a major river system, such as that of the Mississippi in the USA or the Darling in Australia, where the inhabitants of downstream flood-prone settlements may have warning of the impending flood weeks or even months ahead. At the other extreme are the so-called *flash floods* which often strike without warning and pass just as quickly and which are a major cause of flood-related deaths. The term 'flash flood' is ill-defined and imprecisely used and is associated by many only with extreme events in arid or semi-arid environments. In reality, flash floods may occur anywhere that heavy rainfall, catastrophic melting or dam collapse occurs (Archer, 1994) but are especially severe in areas of steep slopes, narrow valleys and low density of population, and especially dangerous in the tourist season in wilderness areas when canyon floors are used as campgrounds. Classic examples were provided in Britain by the Dolgarrog dam collapse of 1925 and the Lynmouth rainfall flood of 1952 and in the USA by the Big Thompson, Colorado, flood of 1976.

1.2.3 The Dimensions of Flooding

The hydrograph is an important guide to flood characteristics. In the case of river flooding, hydrographs may have Y-axis units of either discharge or water surface elevation (sometimes referred to as 'stage'), although only the latter are used for coastal flood hydrographs. Hydrograph analysis can yield information on peak flood discharge and water level, the duration of flooding, the time taken to attain peak conditions, and the total volume of floodwater.

The rate of *discharge* provides the basis for most methods of predicting river flood magnitude and frequency (see Chapter 6) and is normally derived indirectly from the measurement of water level via a stage–discharge rating curve. Of particular importance are the water level and discharge at the peak of the flood (time Y in Figure 1.4) since this is normally when greatest inundation and damage occur. However, information on peak discharge is usually available only for gauging stations having automatic water-level recorders but even these may be overtopped in severe conditions. Supplementary measurements may be made using simple peak velocity meters and maximum water-level gauges; theoretical estimates of out-of-channel flow have improved as a result of recent research but still depend heavily on the accuracy of estimated channel and floodplain roughness.

The hydrograph may also be defined by the raw data on water surface *elevation* which is of immediate relevance, in both fluvial and coastal flood situations, if flood-retaining structures are likely to be overtopped and which is an equally valid measure for both fluvial and coastal flooding. The *volume* of discharge, i.e. the total quantity of water flowing past a given point in the time period X_1Z_1, is a significant factor in the optimum design and operation of several river flood protection strategies, including reservoirs intended to alleviate downstream flooding.

There are, however, other dimensions of flooding which are of equal importance in the analysis and understanding of floods whether fluvial or coastal. For instance, the *velocity* of floodwaters, together with water depth and the area of inundation, is a crucial determinant of flood damage. *Seasonality* of flooding is another important dimension. Although river floods may occur at any time of year, they tend to be most numerous in the wet season (either the rainy season or, in areas of uniform precipitation, the season of lowest evaporation when soil moisture values are highest) and in the melt season in the case of high-altitude and high-latitude rivers. Large rivers, which normally show marked seasonal characteristics, invariably attain their peak flows during these seasons. Where flooding is clearly seasonal in character, adaptation to the flood hazard is normally more easily achieved. Some other types of flooding also have a seasonal rhythm. This might be true, for example, of flooding resulting from convectional storms or of coastal flooding resulting from storm surges, but not of coastal flooding resulting from earthquake-induced tsunamis.

Flood frequency is a statistical measure of the probable occurrence of a flood of a given magnitude. Large floods occur relatively infrequently and have a long average return period or recurrence interval (T) of perhaps several hundreds of years. For example, the Pecos River flood shown in Table 1.1 is believed to have had a recurrence interval of about 2000 years and most of the other floods listed are likely to have had T-values of at least 100 years (Rodier and Roche, 1984). Small floods occur frequently, perhaps annually, and therefore have a very small return period or recurrence interval. The five-year flood (Q_5) and the 100-year flood (Q_{100}) can occur at any time, although the former is more likely, having a 20% probability of occurring in a given year, compared with a 1% probability of occurrence for the 100-year flood (see also Section 6.2.2). This has important repercussions for the encroachment of human activity into flood-prone areas since the floods most likely to have been experienced are the more frequent minor ones whose impact is generally small. Flood frequency is an important concept in the design of many flood-alleviation measures which are constructed to withstand the effects of floods up to and including the 'design flood'. For instance, minor dams and urban stormwater systems may be built to cope with a design flood having a return period of only 20 to 30 years, while for major earth dams the design flood may have a return period of 1000 years. In the Netherlands, where catastrophic loss of life is likely to result from overtopping of the flood defences, the major river flood dikes are designed against the 1250-year flood and the sea dikes are intended to cope with events having a return period of 10 000 years.

The extent of inundation associated with a given water-surface elevation defines another dimension of flooding, i.e. the *flood outline*, which expands as water level increases. Floods of given return periods each have a unique outline which, when mapped, represents a valuable planning tool. Some of the earliest flood inundation maps were in US Geological Survey flood reports and *Water-Supply Papers*. In 1959 the USGS began a new series of flood outline maps, published as *Hydrologic Investigations Atlases*, which were used to establish zoning regulations and to minimise encroachment. Burkham (1978) compared the accuracy of the methods used by the USGS. Flood outline mapping by the US Army Corps of Engineers concentrated largely on showing the outlines of floods of selected frequencies, e.g. the Standard Project Flood (SPF), with a return period of about 100 years, and smaller floods (Intermediate Regional Floods) with return periods of between 5 and 50 years (Figure 1.5). These maps were an integral part of the *Flood Plain Information Reports*, prepared by the Corps of Engineers under the provisions of the amended Flood Control Act of 1960. They were intended to provide useful information on floods and flood hazards so as to encourage optimum use of flood-prone land and thereby minimise future expenditure on flood alleviation resulting from the improper use of floodplains (Molinaroli, 1965). Flood outline mapping in the USA was given a further stimulus by the Flood Disaster Protection Act of 1973.

In England and Wales floodplain delineation and some flood outline mapping for specified return periods were carried out by the Regional Water Authorities under the terms of Section 24(5) of the Water Act 1973 and the location of the principal 'washlands' will be mapped in the Catchment Management Plans of the former

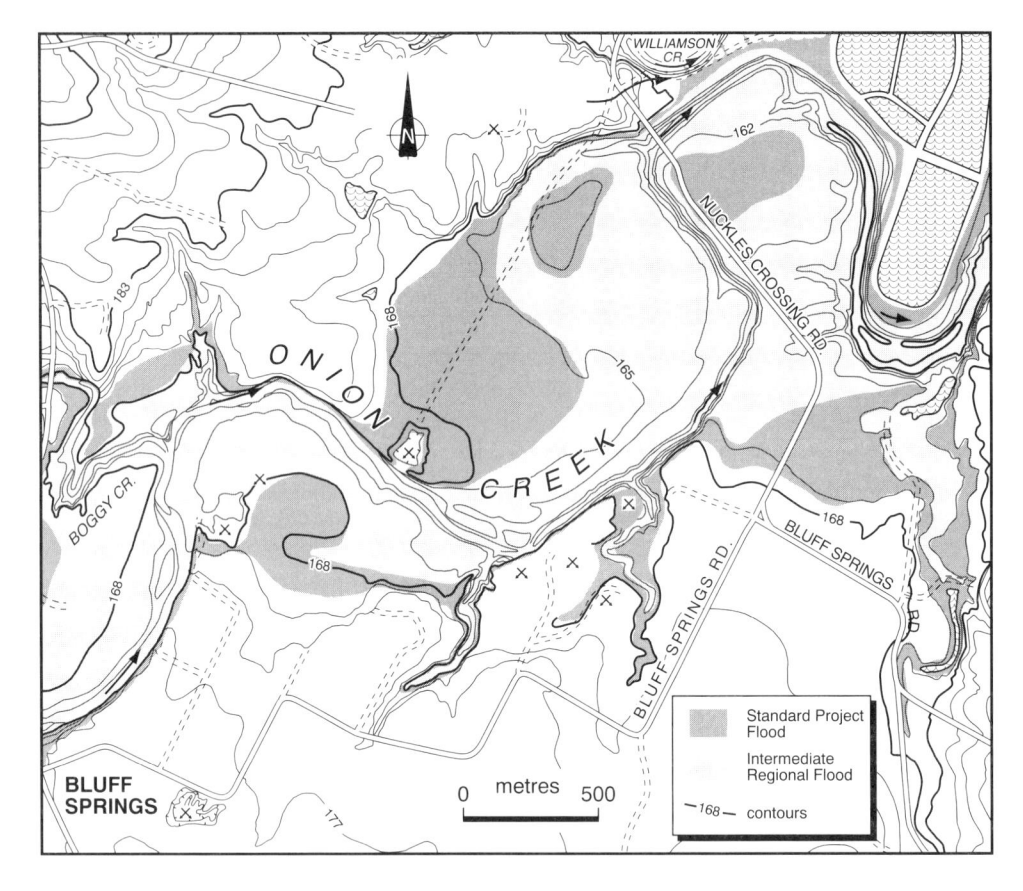

Figure 1.5 Flood outlines for the Intermediate Regional Flood and the Standard Project Flood on Onion Creek near Austin, Texas
Source: US Army Corps of Engineers (1973)

National Rivers Authority regions. The Institute of Hydrology was commissioned by MAFF to estimate the total area of England and Wales that, in the absence of any flood defences, would be inundated by river floods having an approximate 100-year return period (MAFF, 1995b; Morris and Flavin, 1995). This work became feasible only after the development of the IH Digital Terrain Model (IHDTM) for England and Wales and the completion of digital maps of river networks and appropriate digital spatial datasets. The analysis was conducted on a 50 m square grid, which is the horizontal resolution of the IHDTM, and the 100-year flood depth was computed for every point where the catchment area exceeded 10 km^2. The areal extent of inundation is determined by taking each flooded river point in turn and using the IHDTM to identify contiguous areas of its catchment that are lower than or equal to the flood surface elevation. Such maps were intended to give regional rather than local indications of flood extent and showed that 7.1% of the land area of England and Wales (10 683 km^2, of which 611 km^2 is urbanised) would be flooded by the 100-year event (Morris and Flavin, 1995).

From early in its development satellite imagery has offered considerable scope for both one-off and also repetitive mapping of major fluvial and coastal floods (e.g. Brown et al., 1987; Blyth and Biggin, 1993; Koopmans et al., 1995). Flood outline mapping could also benefit from the development of a Global Positioning System (GPS), in which fixes on three satellites permit a selected location on the ground surface to be determined routinely to within a few metres and if necessary to within a few millimetres (Cornelius et al., 1994). This could lead eventually to the definition of outlines for floods having specifed return periods, at least for large rivers and for flood-prone coastal areas.

Detailed flood-outline and flood-risk mapping of floodplain and coastal areas, at scales larger than 1:50 000, may become more readily available by using computer-digitised three-dimensional aerial photography. Such a system has been developed by the British Geological Survey, to produce large-scale digital elevation models with a vertical precision of better than one metre. This information could make a major contribution in the setting of flood insurance premiums and in the improved design of flood warning and evacuation schemes.

1.3 FLOODS AS HAZARDS

1.3.1 Flooding as a Natural Hazard

So-called *natural hazards* result from the potential for extreme geophysical events, such as floods, to create an unexpected threat to human life and property (Smith, 1996). When severe floods occur in areas occupied by humans, they can create *natural disasters* which involve the loss of human life and property plus serious disruption to the ongoing activities of large urban and rural communities. Although the terms 'natural hazards' and 'natural disasters' emphasise the role of the geophysical processes involved, these extreme events are increasingly recognised primarily as the 'triggers' of disaster, which often have more complex origins including many social and economic factors.

A flood in a remote, unpopulated region is an extreme physical event, of interest only to hydrologists. Entirely natural floodplains can be drastically changed—but not damaged—by the events which create them. Indeed, most floodplain ecosystems are geared to periodic inundation. Terms such as *flood risk* and *flood losses* are, therefore, essentially human interpretations of the negative economic and social consequences of natural events. As with other human value-judgements, different groups of people have been found to differ markedly in their selection and definition of the risks from flooding (Green et al., 1991). In addition, the flood risk in any given locality may be increased by human activity, such as unwise land-use practices related to deforestation or urban development. Equally, the flood risk may be reduced by flood control structures or by effective emergency planning. The real risk from floods stems from the likelihood that a major hazardous event will occur unexpectedly and that it will impact negatively on people and their welfare.

Flood hazards result from a combination of *physical exposure* and *human vulnerability* to geophysical processes. Physical exposure reflects the type of flood events that can occur, and their statistical pattern, at a particular site, whilst human vulnerability reflects key socio-economic factors such as the numbers of people at risk on the floodplain or low-lying coastal zone, the extent of any flood defence works and

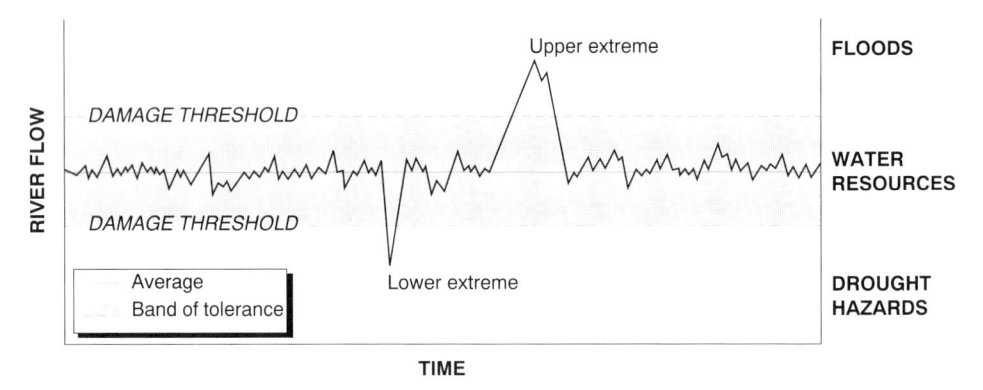

Figure 1.6 Sensitivity to flood hazard expressed in relation to the variability of river discharge and the degree of socio-economic tolerance at a site
Source: Modified from Hewitt and Burton (1971)

the ability of the population to anticipate and cope with hazard. It is the balance between these two elements, rather than the physical event itself, which defines natural hazard and determines the outcome of a natural disaster. In Figure 1.6, variations of river stage through time are plotted in relation to the band of social and economic tolerance available at a hypothetical location. As long as the river flows close to the average, or expected level, there is no hazard and the discharge will be perceived as a resource because it supplies water for useful purposes, such as irrigation or water transport. However, when the riverflow exceeds some predetermined threshold of local significance and extends outside the band of tolerance, it will cease to be beneficial and be perceived as a hazard. Thus, very low or very high flows will be considered to create a drought hazard or a flood hazard respectively. The impact of the hazard will, in part, be determined by the magnitude of the event (expressed by the peak deviation beyond the damage threshold on the vertical scale) and the duration of the event (expressed by the length of time the threshold is exceeded on the horizontal scale). But the true significance of flood disaster will depend primarily on the vulnerability of the local community. Rivers often overflow their banks without creating a significant hazard and such hydrologically defined 'flood flows' may create little economic damage and produce no response from the emergency services. Indeed, like many other natural hazards, low-magnitude–high-frequency floods provide gains as well as losses to the community at risk (see Section 2.2).

The relationship between physical exposure and human vulnerability is highly dynamic and can change through time. Figure 1.7 shows various possibilities giving rise to increased flood risk. Figure 1.7A represents the effect of a constant band of socio-economic tolerance and constant variability of flows but a trend to higher mean values leading to more frequent breaches of the tolerance threshold, perhaps due to channel constrictions which limit the capacity of the river banks to contain specified flood flows. Figure 1.7B represents a constant band of tolerance and constant mean value of flow but an enhanced risk arising from increased variability of flows. This might be due to a shift in climate leading to more intense rainstorms. In Figure 1.7C the flow regime of the river does not change but the socio-economic band of tolerance narrows, perhaps because of floodplain invasion placing more people and property at risk.

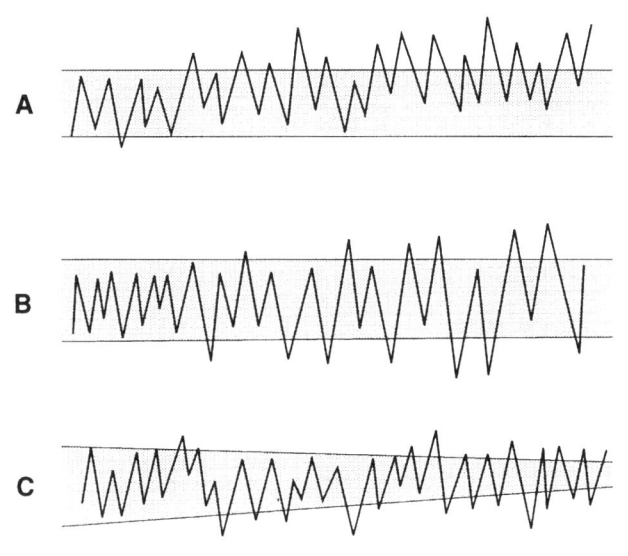

Figure 1.7 A schematic illustration of increases in human vulnerability to flood hazard caused by changes in the distribution of flood events and a decrease in socio-economic tolerance at a site.
See text for details
Source: After De Vries (1985)

Another important attribute of flood risk is the relative unpredictability of the event. Unexpectedness, combined with the difficulty of issuing precise warnings of location and timing, is a major cause of flood disaster, especially with flash floods. On the other hand, many rivers exhibit regular floods which, especially in large drainage basins, will rise slowly and predictably in a seasonal 'flood pulse', thereby offering an opportunity for an efficient loss-reducing response. Most important of all from the standpoint of effective loss mitigation, floods recur in well-identified topographical settings which can be accurately mapped and can either be defended against by engineering works or mitigated by other response strategies. In most countries the greatest threat comes from fluvial flooding but coastal flooding poses the major hazard in countries as diverse as Bangladesh and the UK.

1.3.2 World Patterns and Trends

Flooding along low-lying plains adjacent to rivers and coasts is commonly regarded as the most frequent and widespread natural hazard in the world. It is often asserted that floods regularly account for about one-third of all global disasters arising from geophysical hazards and adversely affect more people than any other natural hazard, apart from drought. This dominance is usually explained by the fact that the overtopping of the natural or artificial boundaries of a watercourse, together with the submergence of coastal zones, is a frequent occurrence compared with the incidence of other hazards such as damaging earthquakes or major volcanic eruptions. In addition, the density of much floodplain and coastal settlement places large numbers of people at risk. It is also believed that increased numbers of flood disasters are being recorded worldwide. This may reflect a variety of quite different factors including a genuine

Table 1.2 The population estimated to be at risk from flooding in England and Wales

Flood type	Number of properties	Estimated population	Percentage of total population
River floodplains			
Unprotected	101 600	274 200	0.54
Protected	231 700	625 590	1.25
Coastal flood zones			
Unprotected	56 000	150 000	0.30
Protected	500 000	1 350 000	2.70
Totals	889 300	2 399 790	4.79

Note: Excludes those at risk from sewer flooding and from dam failure
Source: After Parker (1992)

increase in storm rainfall due to changed hydrometeorological activity, better reporting of more minor flood events or the continuing invasion of flood-prone land placing progressively more people and property at risk.

There are great regional variations in flood hazard. In the more developed countries (MDCs), the scale of the flood risk from rivers is often determined by the extent of urban development on relatively restricted floodplains rather than by the overall population density. For example, there are some 5–6000 urban flood locations in England and Wales (Parker and Penning-Rowsell, 1983b). The greatest concentration is in London where over 25 000 houses alone were threatened by a combination of high tides and storm-surge conditions before the construction of the Thames tidal barrier. Over the country as a whole, a lot of pre-war development of residential property took place before the introduction of more effective planning controls in 1947 (Penning-Rowsell, 1976). But, in total, less than 5% of the population in England and Wales lives in flood-prone areas—over half in low-lying coastal areas protected by sea defences (Table 1.2)—whereas almost 70% of urban centres with populations exceeding 20 000 are flood-prone in more thinly populated New Zealand (Ericksen, 1986). About 10% of people in the USA live on the 1:100-year floodplain and floods account for around two-thirds of all the federally-declared disasters. In comparison, certain less developed countries (LDCs) have a much greater exposure to floods. Some 80% of the area of Bangladesh is composed of floodplain land and it is estimated that 20–30% of the national territory is regularly flooded by monsoon rainfall and tropical storm surges. Vietnam is also widely exposed to major floods. Individual events can leave more than half a million people homeless and, each year, the country loses some 300 000 tonnes of food from floods (Wickramanayake, 1994). Throughout Asia, river floods alone adversely affect the lives of about 20 million people every year.

Coastal areas contain a significant amount of population and economic assets. Many of the most serious flood problems exist in developed areas adjacent to deltas and estuaries, especially when there is a combined risk from fresh- and salt-water flooding. About 60% of the world's population, nearly three billion people, already lives within 60 km of the coast. By the year 2100, these figures are expected to rise to 75% and 11 billion respectively. In England and Wales alone, over 5% of the population, and over half of all Grade 1 agricultural land, lies below the 5 m contour and is dependent on either drainage or flood defence to maintain its productivity (MAFF, 1993). In addition, flood-prone coastal zones are often areas with high ecological value and amenity value. The ecological benefits of coasts are based on high biological productivity, and the diversity of animal and plant life which they sustain, e.g. as a habitat for migrating birds and waders. Many low-lying coasts are also important for cultural and recreational reasons and about one-third of the coastline of England and Wales has been designated as Heritage Coast in recognition of its scenic qualities.

Any estimate of flood hazard is plagued by the problem of defining basic terms (e.g. 'flood disaster' or 'flood-prone') and the related inadequacy of recorded data on flood events and losses. In the absence of any agreed international database of flood disasters, it is difficult to authenticate claims about flood losses worldwide and any related trends. The evidence is weakest for the LDCs, which often lack firm statistical information on direct flood losses, and which also suffer badly from many indirect impacts, such as the lagged mortality from water-borne diseases and flood-induced famine. In addition, the LDCs have relatively high agricultural losses, which are usually harder to quantify than urban losses. Even in the MDCs discrepancies in data may occur. Official estimates made by various organisations after the 1974 Brisbane floods in Australia placed fatalities at 3, 12 and 15 respectively, whilst damage estimates ranged from A$ 178 million to A$ 450 million.

Any reliable database must specify the minimum reporting threshold, in terms of human fatalities or economic damage, for the definition of a flood disaster. It also has to distinguish flood losses from those caused by associated natural disasters, such as 'cyclones', 'storms' and 'landslides'. The database compiled since 1964 by the US Office of Foreign Disaster Assistance (OFDA, 1996) includes all flood disasters recorded worldwide, other than in the USA and its territories, where the number of people killed and injured totals at least 50, or at least 1000 people are homeless or affected, or damage is estimated at a minimum of US$ 1 million. A similar database was assembled by Glickman et al. (1992). This dataset includes all natural disasters with 25 or more fatalities between 1945 and 1986. When a tropical cyclone or other disaster resulted in a flood, the event was counted as a flood only if the flooding was reported to have caused most of the deaths. This clearly causes some under-reporting of flood disasters. For example, coastal flooding associated with a storm surge is attributed to 'storm' rather than 'flood'. On the other hand, the database does list each country affected by the same major flood event on an international river, e.g. a flood passing down the Ganges–Brahmaputra–Meghna drainage basin and affecting both India and Bangladesh.

Because of differing definitions and more general reporting difficulties, all the resulting figures should be treated with caution. However, according to Glickman et al. (1992), floods were responsible for 31% of the natural disasters which claimed 25 or more lives worldwide between 1945 and 1986. These data also show that, between 1945 and 1986, the average annual number of floods causing 25 or more deaths more than trebled, a trend which is confirmed by the OFDA series from 1964 to 1996 (Figure 1.8). The OFDA figures indicate that, since 1964, 835 floods that occurred outside the USA have killed over 130 000 people, have made about 70 million people homeless and adversely affected well over one billion humans. The direct economic cost of such flood disasters was estimated at over US$ 91 billion. The burden of flood disasters bears most heavily on the impoverished countries of Asia. Table 1.3 shows that, whilst less than half of all flood disasters occurred in Asia, over 80% of people killed, affected or homeless were located in this continent. Although economic losses were also

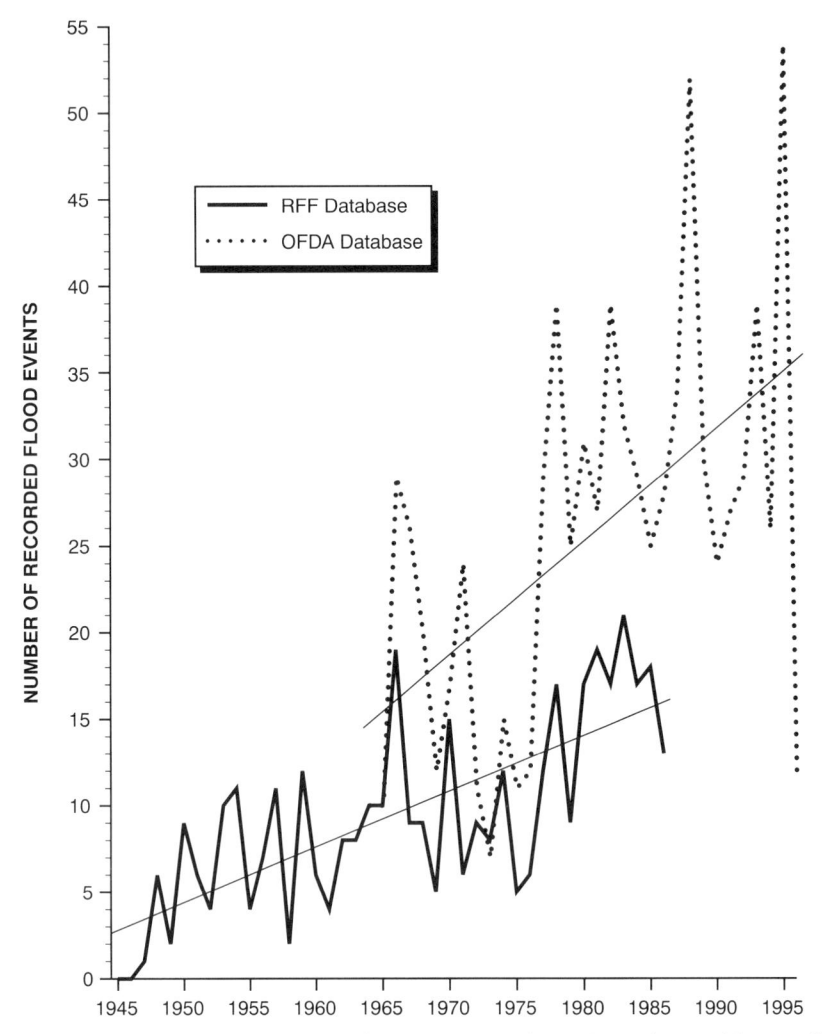

Figure 1.8 The increase in recorded flood disasters throughout the world according to two databases from 1945 to 1986 and 1964 to 1995
Source: Data from Glickman et al. (1992) and OFDA (1996)

disproportionately high in Asia, the highest absolute direct costs of flooding are experienced in relatively wealthy regions, such as Europe.

Generally speaking, the most reliable information on flood losses is available for some individual countries in the developed world. Figure 1.9 shows flood loss data for the USA over the period 1925–1994. Despite large inter-annual variations, there is a clear upward trend, especially for economic damage to property which has risen about thirty-fold (at constant prices) over the 70-year period. The 1993 peak for economic losses due to the Mid-West floods is well in excess of the previous high in 1972. However, 1972 remains the year for maximum deaths, due to widespread flooding in New York State and Pennsylvania associated with hurricane Agnes, together with the localised flash flood in Rapid City, South Dakota. The latter event, partly caused by a

FLOODS: PHYSICAL EVENTS AND NATURAL HAZARDS

Table 1.3 Percentage of all recorded flood disasters and associated flood impacts 1964–96 by continental area (excluding USA)

Area	Events	Killed	Affected	Homeless	Damage
Africa	16.0	4.5	1.0	5.1	1.9
Asia	41.1	82.2	95.3	85.1	65.6
C. America	7.1	1.7	0.1	0.5	1.3
Caribbean	3.2	1.2	0.3	0.2	0.2
Europe	8.7	2.1	0.6	0.5	20.1
Near East	4.1	1.5	0.2	1.0	0.7
Pacific	1.2	0.1	0.0	0.0	0.2
S. America	18.6	6.7	2.5	7.6	10.0
Totals	100.0	100.0	100.0	100.0	100.0

Source: OFDA (1996)

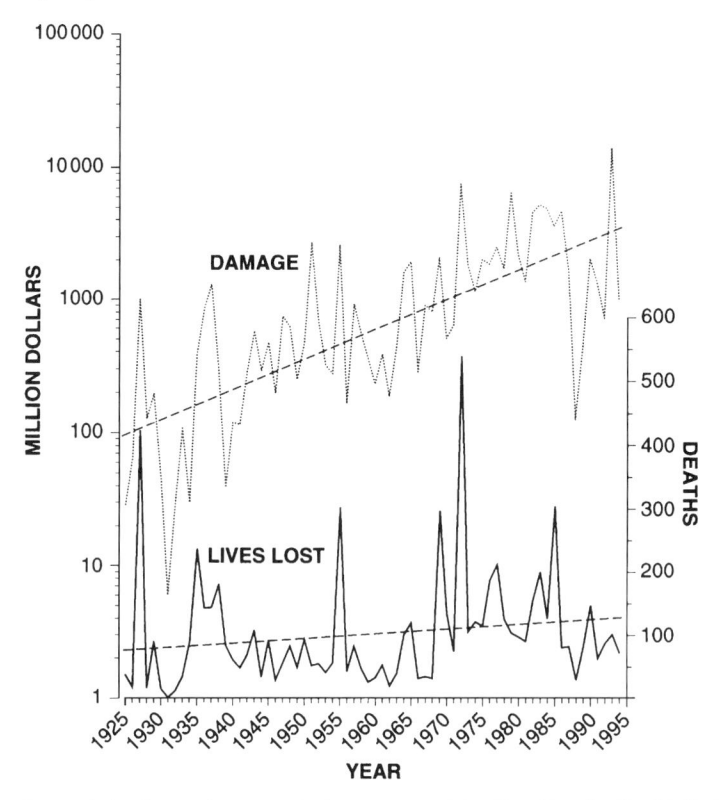

Figure 1.9 Annual deaths and economic losses caused by flooding in the United States 1925–1994. Damages are in millions of US$ adjusted to 1990 values
Source: Data from US Department of Commerce

dam collapse, claimed 238 lives, which is the highest number of deaths so far recorded from a single flood in the United States. Flash floods are now the main cause of weather-related deaths in the USA and in many other countries. This is because of their rapid-onset characteristics, which limit warning procedures and emergency actions, plus the high velocity of the flood flows and the associated debris load.

1.3.3 Human Causes of Flood Hazard

Flood hazards are created by countless individual locational decisions which encourage the settlement and economic development of floodplains and flood-prone coastal areas. Historically, the process is well-established. For example, river flooding has been a problem from the time of the first European settlement in Australia and the dangers were recognised by one state governor as early as 1819 (Douglas, 1980). Despite a recognition of the threat, many towns and cities—such as Kansas City, USA (Driever and Vaughn, 1988)—underwent largely unregulated expansion into flood-prone areas throughout much of the nineteenth and early twentieth centuries. The intrinsic land-use attractions of major floodplains and coasts have made these zones some of the most densely populated—and hazardous—settings in the world. Because flood hazard is essentially created by human decisions and actions, it follows that the social and political context for such actions is an important element in the selection of any subsequent flood mitigation schemes.

Floodplain invasion in the MDCs has been driven mainly by prolonged economic development together with the growth and redistribution of population during the twentieth century. In many densely populated countries, like England, a shortage of alternative building land has also been an important factor. As previously implied, there are major differences in flood exposure levels within, as well as between, individual countries. For example, perhaps 4% of all buildings in Australia have a flood risk but in Queensland nearly 15% are at risk from river flooding alone. In some instances the greatest hazards have been borne voluntarily by the richer members of society seeking river views and waterfront facilities for recreational activities. It has been claimed that throughout most of Britain, a river, lake or coastal view will add 15–25% to the average market value of a house (Taylor, 1995; Webb, 1996). Houses with direct access to a waterfront may cost 30–50% more than average, with a private berth at the bottom of the garden adding a further 15%. In very many cases of flood-zone occupation, the real costs of such occupation have not been borne by the beneficiaries, for example when subsequent flood control schemes designed to protect such sites are funded out of national taxation rather than from local resources.

Throughout the twentieth century there has been a strong, and well-documented, migration towards flood-prone areas in the MDCs, first identified by Burton and Kates (1964) over 30 years ago. Burton et al. (1968) estimated that there were over 2000 towns and cities in the USA already situated on floodplains. A few years later, White (1975) calculated a current 1.5–2.5% annual rate of additional expansion onto floodplains and, as early as 1960, the US Army Corps of Engineers admitted that existing flood-control expenditure was unlikely to keep pace with the rising damages from floodplain invasion (Corps of Engineers, 1960). In a study of 26 American cities, Schneider and Goddard (1974) reported that the average proportion of the floodplain occupied by urban development nationally was almost 53%. Similarly, Goddard (1976) found that one-sixth of all urban land was within the 1:100-year floodplain and more than half of all available floodplain land had been developed by 1974. By the late 1980s, about seven million buildings were located within the 100-year floodplain in the USA. Even with low rates of floodplain invasion, there can still be increased floodplain investment—and the potential for greater losses—as rising prosperity and property prices increase the value of houses and their contents, often at a level above general inflation.

Much floodplain encroachment has resulted from the expansion of existing urban settlements. According to Platt (1987), the main growth in flood hazard in the US has been on tributary creeks and streams in developing metropolitan areas. These streams typically lack federal flood control schemes, such as dams, and the small watersheds are particularly susceptible to increased flood discharges resulting from the hydrological effects of the urbanisation process. In the fastest growing communities, the municipal authorities have often been lax in curbing new investment in flood-prone areas by private property developers. This was confirmed by Burby et al. (1988) from a study of 10 American cities seeking to reduce floodplain invasion through land-use management between 1976 and 1985. It was found that, although floodplain land-use policies did steer some potential new development to alternative locations, it proved incapable of stopping urban development in the floodplain.

Where economic prosperity and population growth occur in combination, the demand for new building sites is usually high. One such area has been the lower Thames floodplain west of London where some 5500 properties, 90% residential, are now at risk in various settlements. The greatest pressures tend to be experienced where developable land outside the floodplain is scarce, as illustrated by the floodplain invasion process at Datchet (Figure 1.10) where, in the decade following 1974, over 400 new houses were constructed despite the absence of any flood protection works. In this case, restrictions on alternative building land had been imposed by the construction of a water supply reservoir in the 1970s and the fact that the settlement is entirely surrounded by designated green belt, which is land protected from new building by planning legislation (Parker, 1992). Such pressures increase the flood loss potential, not only through the construction of new floodplain properties, but also through the higher investment value associated with house extensions and redevelopment and the threat of stormwater overflows from sewage treatment plants which cannot now cope with the extra drainage created by the spread of housing.

There are many similarities between river and coastal flood zone invasion. Two-thirds of all cities with populations over 2.5 million throughout the world are situated near estuaries and some important urban centres, such as Venice and London, have long been under a storm-surge threat, which has increased in recent decades as a result of a relative rise in sea level. These patterns are mirrored in individual countries. For example, in the USA large-scale migration to the Sun Belt states has led to the coastal population of Florida almost doubling within 20 years and, on present trends, over 75% of the US population will be concentrated within 100 km of the coast by the year 2000. Many of these people are at risk from flooding both along the coast and on barrier islands during coastal storms because of the land reclamation projects which accompany development, especially when construction involves the removal of protective vegetation, such as mangrove swamps (to improve ocean views) and coral reefs and sand dunes (to provide building material). Such destruction of the coastal ecosystems reduces the natural ability of the shoreline to dissipate the energy of storm-surge waves.

Much of this development has been in pursuit of leisure activities, as along the Gold Coast of Queensland, Australia (Hobbs and Lawson, 1982), and it has frequently led to high economic investment values in water-front facilities such as high-rise residential accommodation and marinas. In addition, such areas often contain large proportions of older and retired people plus the prospect of difficult emergency evacuation procedures

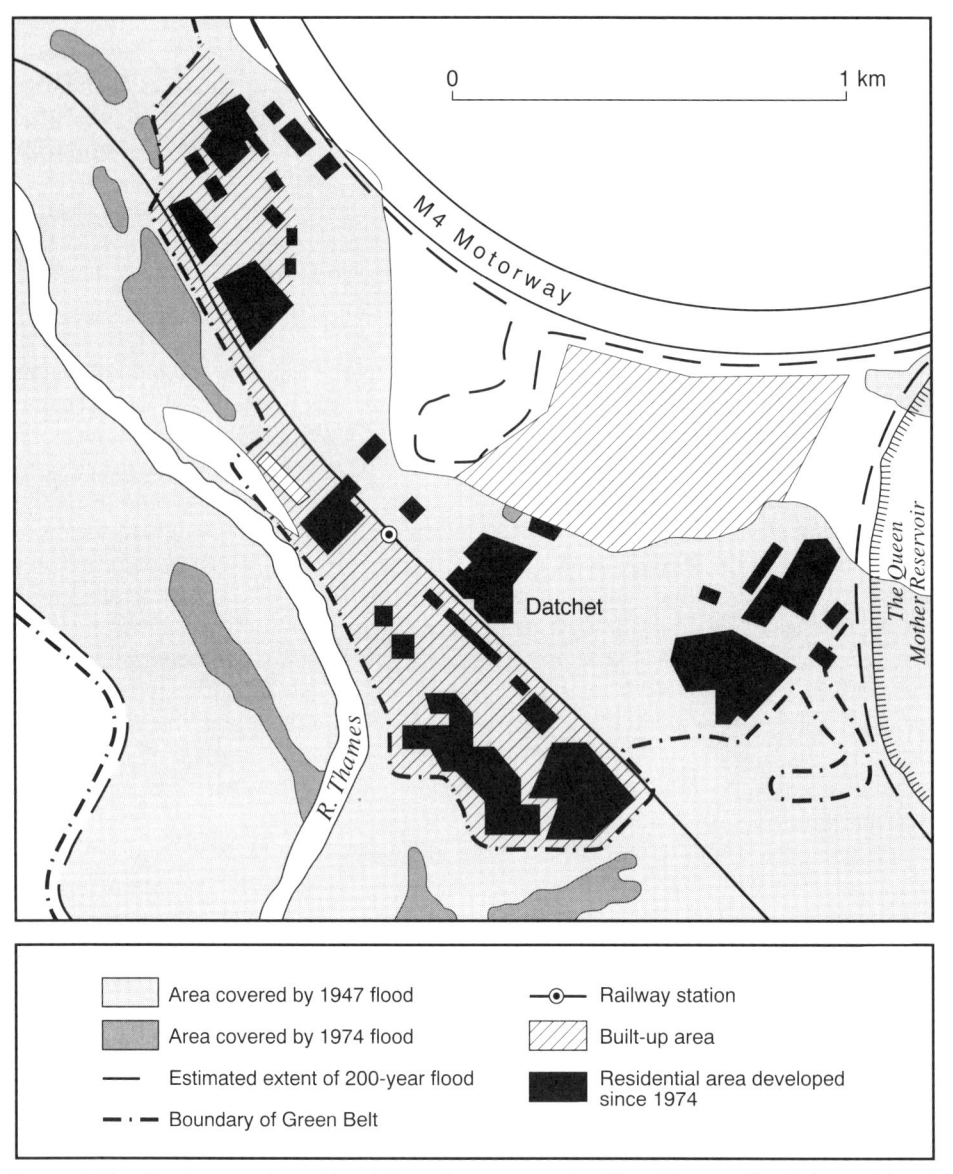

Figure 1.10 The increase in residential development on the River Thames floodplain at Datchet, England, since 1974
Source: After Parker (1992)

along restricted coastal highways. Some highly developed coasts also face threats from erosion resulting from unwise human activities. For example, at Miami Beach, Florida, much of the beach area has been lost with an increasing vulnerability to storm damage over several decades, mainly as a result of dredging the inlets for recreational boating (Carter, 1988). California has nearly 1800 km of coastline, over 80% of which is actively eroding. Since about three-quarters of the state's 30 million people lives within 50 km of

FLOODS: PHYSICAL EVENTS AND NATURAL HAZARDS 29

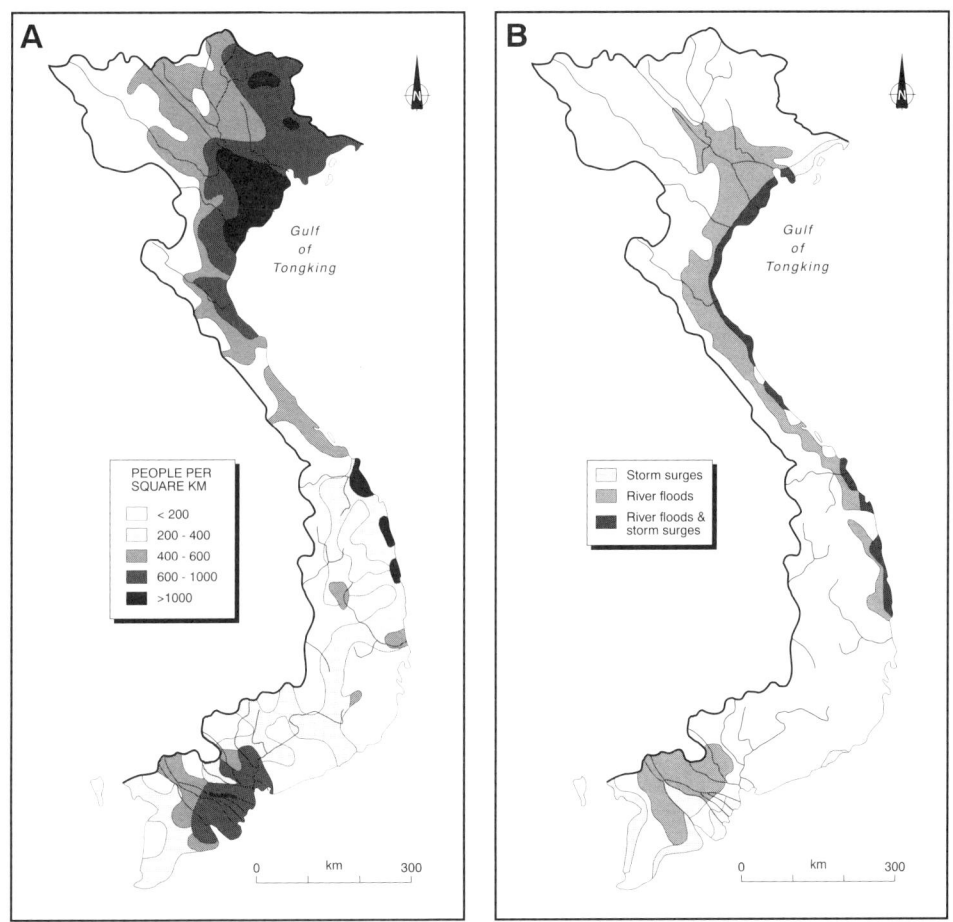

Figure 1.11 Vulnerability to flooding in Vietnam shown by (A) distribution of population and (B) areas subject to different types of flooding
Source: After UN Department of Human Affairs (1994)

the shore, the conflict between geological instability and accelerating ocean-front development is growing. Over the past 15 years, the California coast has suffered nearly US$ 5000 million in storm damage. In addition, defending the coast and repairing infrastructure costs about US$ 100 million annually.

In the LDCs, the high population densities on river floodplains and in coastal areas often reflect the more basic requirements of traditional agriculture and fishing, together with the lack of alternative land for settlement by poor people. For example, in Vietnam the alluvial river deltas and low-lying coastal plains have been intensively settled for wet-rice cultivation. As shown in Figure 1.11, these areas are subject to both river floods and storm surges with the result that at least 70% of the 71 million population, and almost all of the economic infrastructure of the country, is at risk from flooding (Department of Humanitarian Affairs, 1994). Bangladesh, arguably the most flood-prone country in the world, has more than 115 million people living in an area the size of Wisconsin and this leads to very high rural population densities. More than 50% of these inhabitants are functionally landless and more than 80% live below the poverty line (Haque and Blair, 1992). Floods in 1988 inundated 46% of the land and killed an estimated 1500 people. In Guayaquil, Ecuador, when the population was already over one million in 1975, some 60% of the population lived in squatter communities built over coastal swampland (Wijkman and Timberlake, 1984). These people were housed in bamboo and timber huts erected on poles above mud and polluted water, over 40 minutes' walk away from dry land.

Despite an association which exists between affluence and a voluntary migration to floodplains and seashores in some parts of some MDCs, there is little doubt that the strongest links between vulnerability to floods and human lifestyles worldwide involve poverty. Even in the MDCs, regularly flooded communities are often of low-income status, with a large proportion of households containing elderly people or young families living in rented housing on state support. For example, a survey conducted after the 1993 Mid-West floods found that many homes had a market value of less than US$ 20 000 (Inter-Agency Floodplain Management Review Committee, 1994). In the LDCs it is also the very poor who are at greatest risk because rapid population growth and scarce resources have encouraged the spread of informal (shanty) housing into flood-prone areas both within the cities and in the agriculturally-subsistent rural areas. These unplanned 'squatter' settlements have placed at risk millions of people who live in flimsy, self-constructed buildings because of a chronic shortage of suitable building materials and construction skills. These houses are vulnerable to flood forces which have often been exacerbated locally by poor watershed practices and inadequate coastal zone management.

As with other natural hazards, many apparently short-term flood events have their roots in wider economic and social movements and a history of environmental degradation. The world's poor are increasing in number faster than the earth's population as a whole, with a disproportionate number of children. In some LDCs as much as 50% of the population is under 15 years of age and vulnerable to economic exploitation and disease. For example, Chan (1995) demonstrated that poverty, combined with low educational attainment and poor job mobility, are the main reasons why people continue to occupy hazardous floodplains in Peninsular Malaysia. Rural–urban migration brought about by land pressure may drive such people into unsafe towns which have expanded onto steep slopes and flood-prone areas. The prime factor tends to be the managerial and financial weakness of the appropriate policy-making and regulatory bodies at all levels. This is particularly true in the LDCs where adverse environmental change and land degradation (urbanisation, deforestation, land clearance, soil erosion) increase the physical frequency and magnitude of flooding in many floodplain and coastal areas. Inadequate municipal planning, budgeting and technical support then leads to weak policy analysis and a lack of coordination between agencies who lack properly qualified staff. The unplanned expansion of squatter housing onto flood-prone land soon follows, complete with inadequate drainage systems. Finally, an absence of basic infrastructure, unenforced legislation and the poor maintenance of services means that such settlements cannot cope with the increasing pressure of population (Kreimer and Preece, 1991).

Many of these Third World cities suffer flooding routinely. The poorly built and maintained drainage systems are easily blocked with sediment and uncollected solid wastes, so uncontrolled waste water ends up in the streets. For example, Taiz, Yemen Arab Republic, has a population of more than 150 000 people and is growing at more than 15% annually. Floods, which cause moderate property damage and the disruption of traffic for periods of two or three hours by the deposition of sediment and refuse in the streets, occur 5–10 times per year. Nearly 30% of all houses are flooded each year (Kreimer and Preece, 1991). The prime reason is imprudent development which involves wadis used for roads, houses built on floodplains and natural drainage channels used as the main streets of residential areas. High-income housing and roads are already extending onto steep slopes and increasing soil erosion. The municipal authorities are using sanitary sewers to discharge stormwater which increases health hazards as sewers clog with sediment and refuse, thus allowing water contaminated by human wastes to overflow onto the streets. In Benin City, Nigeria, over two-thirds of the drains are disfunctional and fail to remove water from the streets (Odemerho, 1993). In these Third World cities the poor are often both the perpetrators and the victims of flood hazard.

The very poor are equally vulnerable in rural settings. Blaikie et al. (1994) showed that Third World disaster impact is a function of complex and powerful underlying causes and that certain groups in society, such as urban squatters and landless peasants, suffer most. The problem is not poverty alone, but specific household-scale issues relating to occupation, social class, ethnicity, age and gender, all of which impinge on the ability of people to sustain their livelihood and resist flood disaster. For example, people with disabilities or chronic malnourishment suffer more from water-related diseases common in floods, such as dysentery.

In some LDCs farmers pay one-third of their crop income to absentee landlords and have few material reserves with which to recover from a flood. During the 1987 floods in Bangladesh, the main losses were to house structures and crops (Chowdhury, 1988). The largest absolute losses were sustained by the landowning households, mainly due to the damage to standing crops. Since the landless possessed little or no arable land, their agricultural losses were relatively low although they suffered equal damage costs to their house structures. This was mainly attributed to the fact that the poor people's structures were more dilapidated, with walls of clay and straw, allowing them to be damaged more easily by floodwaters. The houses of richer people tended to be relatively strong with brick and corrugated iron sheets making them more flood resistant. Overall, the relative impact was greatest for the poor families who suffered losses amounting to nearly 50% of average annual income. In the 1991 Bangladesh cyclone only 3% of the houses were strong enough to withstand the tidal surge, although no-one died in the few two-storied buildings occupied by more wealthy people which were subjected to the floodwaters. In another study, Chowdhury et al. (1993) showed the clear effect of age and gender on mortality in the 1991 Bangladesh cyclone. Nearly three-quarters of all deaths occurred among children under 15 years of age. Overall deaths were lowest in the 15–49 age group but females died more than males in all age groups with the gender differences being most pronounced amongst the very young and the very old.

The socio-economic gradient of flood vulnerability can be very steep. Winchester (1992) provided a vivid illustration of the differential effects of a cyclone in south-east India on a rich and poor family living only 100 m apart. The wealthy household, headed by a man owning a small grain business, lived in a brick house, had six draught animals and owned 1.2 ha of good paddy land. In the poor family, both husband and wife

```
                    FLOOD HAZARD ADJUSTMENTS
                   /                        \
         STRUCTURAL                          NON-STRUCTURAL
         FLOOD DEFENCE                       FLOOD MANAGEMENT
              |                                    |
    ENGINEERING METHODS                   LOSS-SHARING METHODS
    Channel improvement                   Disaster aid
    Levees and sea walls                  Insurance
    Storage dams
    Flood proofing
              |                                    |
    ABATEMENT METHODS                     LOSS-REDUCTION METHODS
    Topographic modification              Preparedness
    Vegetation modification               Forecasting and warning
    Land-use modification                 Land-use planning
```

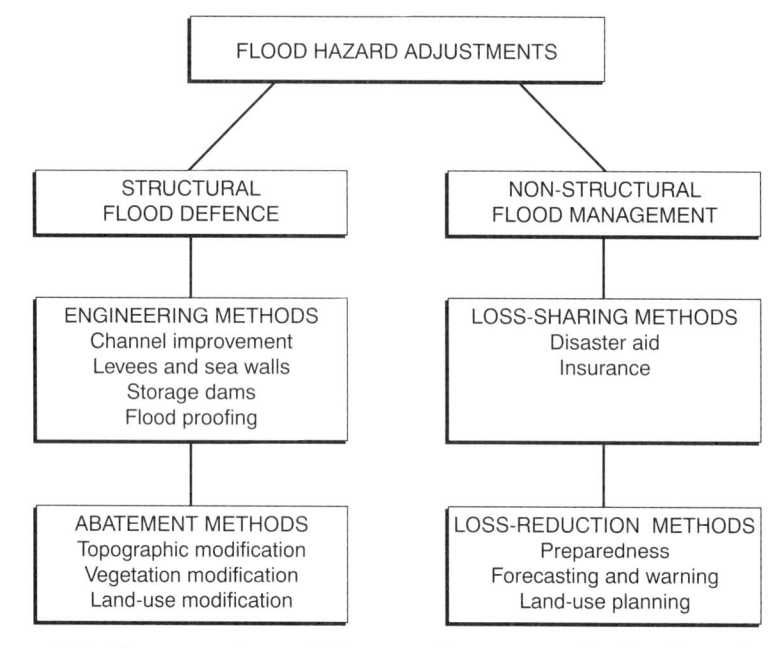

Figure 1.12 The range of potential human adjustments to the flood hazard

worked as agricultural labourers and lived in a poorly constructed thatch and pole house. They had one draught ox and one calf, 0.2 ha of poor quality, unirrigated land and share-cropping rights on another 0.1 ha. The wealthy family received a cyclone warning on their radio and were able to leave the area in their truck. The storm surge partly destroyed the house, three cattle were drowned and all the crops were flooded and destroyed. The poor family remained on site, the youngest child was drowned, together with both animals, and the house was completely destroyed along with the crops. After the flood, the wealthy family returned and were able to deploy their savings to rebuild their house within one week, replace the lost cattle and replant their fields immediately the floodwaters subsided. The poor family lost less in economic terms but had no savings. Therefore, they had to borrow money to provide shelter and to hire bullocks for ploughing their field. But, since they were unable to compete effectively in the post-flood market, where draught animals were scarce, they suffered crucial purchasing delays and could not replant their land sufficiently quickly to avoid a period of hunger that lasted for eight months after the cyclone.

1.3.4 Human Responses to Flood Hazard

Given the nature and severity of the flood hazard worldwide, it is not surprising that many attempts have been made down the centuries to reduce the negative human impacts associated with flooding. A great variety of human responses, or adjustments, to the flood hazard has been implemented in a wide range of social and economic settings. In practice, these responses can be summarised within two major groups represented by structural flood defence measures and non-structural flood management measures, as shown in Figure 1.12. These groups have very different objectives.

Although each individual measure has been in existence in some form as a flood mitigation tool from the earliest recorded times, the relative importance of each method, and the scope for combining one method with another, has often depended on local environmental and social conditions and has also evolved through time as flood hazards have become better understood (Ericksen, 1975).

Structural flood defence captures all the physically-based techniques which seek to modify the flood as a natural event. This means the construction and maintenance of the engineering works which are designed to contain and control flood flows up to a certain level of magnitude and frequency irrespective of whether the works operate on a drainage-basin scale, as is possible in the case of a storage dam, or on the scale of an individual building, as is possible in the case of flood proofing. The physical land-based characteristics of floods can also be influenced in a rather less direct, and more pre-emptive manner through abatement methods. These attempt to modify certain catchment and coastal zone characteristics so that water which might pose a threat is delayed and dissipated on its potentially destructive path through the relevant hydrosystem, whether this be down a slope towards the headwaters of a river or over a saltmarsh towards an estuary.

Non-structural flood management aims to reduce disaster by the process of adjusting people to damaging events. Loss-sharing methods are the most limited responses since their prime aim is to spread the financial burden of flood disasters beyond the immediate victims through emergency aid and insurance programmes. Increasingly, however, disaster aid and insurance are being linked to future loss reduction. More traditional loss reduction is achieved mainly by preparedness programmes and the installation of flood forecasting and warning schemes, which enable threatened communities to anticipate the hazard better and, where possible, take protective or even evasive action through evacuation. Longer-term planning is an important means of halting the invasion of flood-prone land and encouraging more appropriate land-use allocations in the future.

In recent years, there has been a trend away from a strong reliance on flood defence towards a more integrated approach to flood hazard which seeks to blend structural and non-structural methods within a socio-economic context which increasingly has to recognise that complete protection from flooding is impossible and that some risk has to be accepted. This trend, together with its causes and consequences, forms a major theme of this book.

CHAPTER 2

Impacts and Interpretations of Flood Hazard

CONTENTS

2.1 Impacts of flood hazard 34
2.2 The benefits of floods 35
2.3 Estimating the losses from floods 38
2.4 Direct losses from floods 45
    2.4.1 Tangible losses 45
    2.4.2 Intangible losses 47
2.5 Indirect losses from floods 51
    2.5.1 Tangible losses 51
    2.5.2 Intangible losses 52
2.6 Interpretations of flood hazard 54
    2.6.1 The engineering paradigm 54
    2.6.2 The behavioural paradigm 55
    2.6.3 The development paradigm 56

2.1 IMPACTS OF FLOOD HAZARD

Most flood studies emphasise the negative impacts of flooding, which are summarised in Figure 2.1. In particular, attention is given to the *direct losses* which occur immediately after the event as a result of the physical contact of the floodwaters with humans and with damageable property. However, *indirect losses*, which are less easily connected to the flood disaster and often operate on long time-scales, may be equally, or even more, important. Depending on whether or not these losses are capable of assessment in monetary values, they are termed *tangible* or *intangible*. Tangible and intangible losses can also be divided into primary and secondary categories. *Primary* losses result from the event itself while *secondary* losses are at least one causal step removed from the flood.

Some of the most important direct consequences of flooding, such as the loss of human life or the consequent ill-health of the survivors—either through water-borne disease or mental stress—are intangible. Direct tangible losses tend to be relatively important in the MDCs. This is not only because of the lower incidence of fatalities and the greater monetary losses but also because there may be secondary expenses, such as the costs of extra weed control after floodwaters have contaminated good-quality agricultural land, which do not apply in the LDCs. On the other hand, the indirect and intangible consequences of flooding are probably greatest in the LDCs, especially where frequent and devastating floods create special impacts for the survivors. These include the increased vulnerability of widows and children following the death of the head of the

IMPACTS AND INTERPRETATIONS OF FLOOD HAZARD

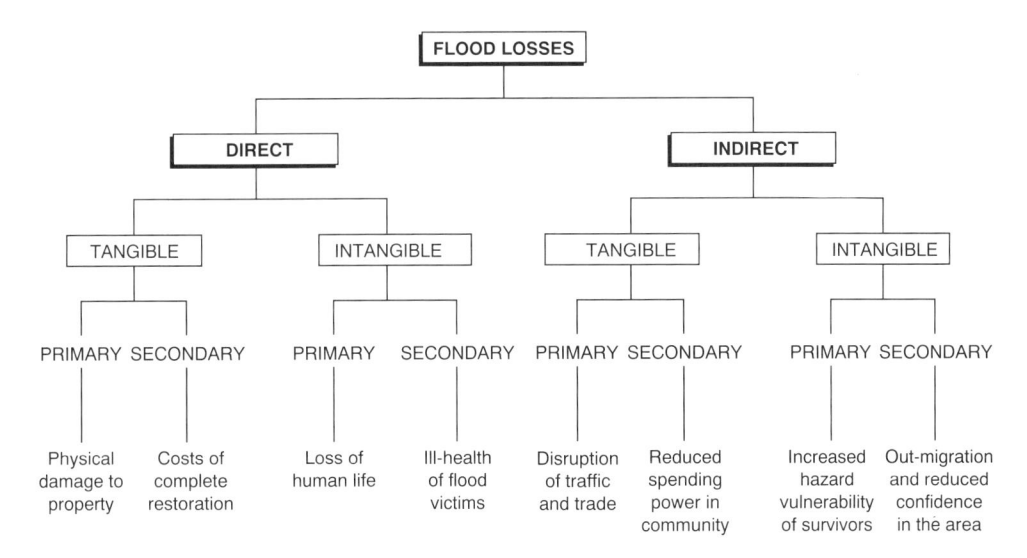

```
                            FLOOD LOSSES
                    ┌────────────┴────────────┐
                  DIRECT                   INDIRECT
              ┌─────┴─────┐             ┌─────┴─────┐
           TANGIBLE    INTANGIBLE    TANGIBLE    INTANGIBLE
           ┌──┴──┐     ┌──┴──┐       ┌──┴──┐     ┌──┴──┐
       PRIMARY SECONDARY PRIMARY SECONDARY PRIMARY SECONDARY PRIMARY SECONDARY

       Physical  Costs of   Loss of   Ill-health  Disruption  Reduced    Increased   Out-migration
       damage to complete   human life of flood   of traffic  spending   hazard      and reduced
       property  restoration          victims     and trade   power in   vulnerability confidence
                                                              community  of survivors in the area
```

Figure 2.1 The categories of flood loss potential

family and, in extreme cases, the long-term economic and social decay of an area if the more active members move away.

The gains associated with occasional flooding are often ignored, mainly because they are less understood and are more difficult to assess than the losses. For example, the flow of relief aid into a flood area is a direct and tangible benefit, which can be easily monitored, but aid is unlikely to be evenly distributed or to constitute an overall net gain for the community. On the other hand, a floodplain location offering basic natural resources, such as building land, water supplies or amenity value, is a long-term asset to a riverine community but it is also very difficult to cost either in absolute terms or in relation to episodic flood losses. Floodplains and coastal zones are some of the most productive areas of the world ecologically, agriculturally and economically, and they often act as important buffer zones between intensive lowland land-use systems and the preservation of high-quality water supplies in rivers. Without floodplain development, many communities in the MDCs would experience greater urban sprawl with the need for more extensive transport and public utility support systems. In view of these gains, the most realistic policy goal is to strive for the optimal use of floodplain land rather than to seek to eliminate the hazard through a total ban on development (Milliman, 1983). The greatest difficulty is in arriving at a determination of what is a socially and politically acceptable level of both development and hazard mitigation, bearing in mind that some residual risk will always remain, and that it is rarely clear who will bear the cost, and who will benefit, from investment in flood mitigation.

2.2 THE BENEFITS OF FLOODS

'Normal' floods, which are regular and expected, bring benefits to most riverine communities. The local economy and ecology are usually well adapted to the normal 'floodpulse', which may be defined as the slow seasonal rise and fall of the river hydrograph which is a feature of many hydroclimatic regimes. These benefits are often greater than those associated with other natural disasters, despite being largely indirect and intangible. As already indicated, they stem from the permanent resource attributes of floodplain land and from the value of frequent low-level inundations to which people are adapted through traditional activities.

Floodplains offer flat land for development combined with river water for domestic supply or irrigation. Large floodplains, such as those of the Euphrates and Tigris, have supported important hydraulic civilisations in the past. Even today, many floodplain communities are agriculturally based and dependent on crops, such as rice, which require large quantities of water. Such patterns are not confined to the LDCs. For example, it has been estimated that some 7% of the national territory of the USA is floodplain land. The floodplains of New South Wales, Australia, cover almost one-quarter of the state. They range from narrow strips of mainly rural land to vast inland plains and form the most productive agricultural land supporting cane sugar, cotton, wheat, dairy products, beef and wool. Apart from agriculture, floodplain environments provide the natural resources for many other economic activities, such as fishing, river navigation and industrial production. Floodplains also act as important corridors for road and rail communications. For example, in the Philippines the annual floods are used to float bamboo and logs downstream to markets that otherwise would be difficult to reach. In the MDCs, the scenic attractions of a river valley, or a river view, frequently provide an incentive to high-quality residential development.

The floodwaters themselves can bring advantages, particularly with respect to common property resources. Flood-borne silt builds large deltaic ecosystems, such as those of the River Nile or the Ch'ang Chiang (Yangtze), which are amongst the most fertile farming areas on earth. In Bangladesh, offshore alluvium deposited by rivers forms silt islands or 'chars' which are quickly settled by landless peasants without other means of subsistence. In other settings, coastal sedimentation processes create sand bars forming barrier islands which act as a natural defence against tropical storm surges. In most cases, the sediments can also be exploited for construction purposes or for topsoil. In the MDCs there is great aesthetic value, plus the economic potential of tourism, when floods help to retain the concept of 'wild' rivers, and sustain sports such as white-water rafting, in rural areas where the landscape quality is already high. Floods also help to preserve areas of floodplain marsh and swamp which provide storage against damaging flood flows.

Some of the most widespread benefits of floods are linked to the maintenance of high biological diversity in the floodplain ecology. Wetland vegetation provides important resting, feeding and nesting areas for many waterfowl species, partly because some major river courses are used as flyways for migrating birds. The floodplain vegetation and soils also serve as water filters removing excess nutrients, pollutants and sediments from runoff before it reaches the river, thus reducing the need for costly water treatment. Alongside lakes or estuaries, fisheries and wading birds are often the most direct winners from this process. On partially developed floodplains in the MDCs, landowner and community partnerships have helped to create green-space 'river corridors' where nature trails, wetland protection, fish and wildlife habitat improvement, woodland conservation, outdoor recreation, water quality enhancement and environmental education can all co-exist with effective stormwater management and erosion control.

Regular annual floods provide abundant water resources to replenish lakes and ponds which, in turn, support irrigation or fish farming. Common property fisheries are of great value to poor people in the LDCs for nutritional purposes, since they act as a major source of animal protein, and also as a means of income for fishers and traders. In the tropics, floodwaters carry nutrients that stimulate fish development and help to maximise productivity from temporary seasonal ponds. Standing water in the fields may help to recharge shallow aquifers, which can then be tapped by tube wells for water supply. Regular inundation by floodwaters may be vital to the carrying capacity of pastures, especially if it replenishes soil moisture in low rainfall areas. Nimmo (1947) found that, in the Queensland channel country of Australia, the volume of streamflow in the flood season statistically explained over 85% of the annual variance in the stock fattening capacity of the land. In certain areas, such as California, riparian owners who benefit from the seasonal flooding of pastures have been legally compensated after flood control schemes denied them this benefit.

Many rivers carry minerals and nutrients which support the more intensive agricultural production on floodplains. Together with the increased availability of water for crops, this greatly increases crop production in normal flood years. Fresh deposits of silt are an annual event on many floodplains and are of great agricultural benefit as long as the particles remain small and do not extend to more infertile sand and gravel. Although silt-laden water reaches only a small part of the regularly flooded areas in Bangladesh, the new alluvium enriches the phosphorus and potash content of these soils (Brammer, 1982; quoted in Rogers et al., 1989). Renewed soil fertility is also associated with the flushing of salts from the surface layers and, in the rain-flooded areas, nitrogen fertility is provided by the biological activity of blue–green algae in the floodwaters. It has been estimated that these organisms provide up to 30 kg ha^{-1} of nitrogen annually in areas flooded by rainwater.

As a result, Bangladeshi peasants distinguish between the *barsha* flood, which is regular, anticipated and benevolent, and *bonna* flood years, when more extreme events cause major damage to crops and property (Paul, 1984). The *barsha* festivals celebrate the economic stimulus of the slow-rising floodpulse season and villagers take to their boats to renew trade links and family ties. It is difficult to find a relationship between direct flood losses and annual foodgrain production in Bangladesh because agricultural production may well increase in the years when major economic losses also occur. This is because damage in the *aus* or *aman* rice seasons (early and late monsoon planting respectively) is often offset by improved soil moisture status, and the farmers' compensating willingness to plant a larger area with more intensive inputs, in the dry *rabi* season (Rogers et al., 1989). For example, after the 1988 floods in Bangladesh, the forecast was for losses of up to 40% on the total harvest based, in part, on the loss of 14% of the main summer (*kharif*) rice crop compared with the previous five years. In fact, the country produced 10% more rice than normal and other staples also showed higher yields in the 1988/89 crop year. This was due to the abundant residual moisture left by the flood for intensified cultivation in the winter dry *rabi* season. Cropping in areas outside the severely flooded zone often benefits from higher than average rainfall too.

Organised flood retreat agriculture is practised widely in the tropics where the moist soil left after flood recession is planted with food crops. The significance of these seasonally inundated floodplains increases greatly in times of drought when the additional flexibility and diversity of wetland and dryland cropping is especially beneficial to the rural economy. Adams (1993) emphasised the overall ecological and economic importance of seasonal flooding across the large floodplains within semi-arid West Africa. For example, the fringing floodplains of the Senegal River and the Niger cover some 5000 and 6000 km² respectively in flood, areas which shrink to about one-tenth and one-half respectively in the dry season. The total area of floodplain land on the Senegal amounts to around one million hectares but the amount actually cultivated varies greatly with the size of the seasonal flow. These rivers maintain extensive deltaic floodplains. The Niger Inland Delta, within the Sahel region, supports an estimated 550 000 people and supplies grazing for one million cattle and two million sheep and goats in the dry season. In addition, there are about 80 000 fishermen and over 15 000 hectares of rice, which is about half the total area of rice in Mali. Kimmage and Adams (1992) also drew attention to the agricultural importance of the floodplain of the Hadejia–Jama'are rivers in northern Nigeria where both sorghum and rice are grown on seasonally flooded land. Once the rice plants have germinated and grown to about 12 cm, they tend to survive, although they are always at risk from early flooding. Despite these uncertainties, it is likely that the area under informal irrigation within the Hadejia–Jama'are floodplain is larger than the total area of controlled irrigation in the whole of Nigeria. Despite the high capital cost of such irrigation projects in northern Nigeria, the volume and economic value of the output remains trivial compared to that of the much larger areas of informal floodplain agriculture.

Efficient floodplain agriculture is only possible when wetland and dryland cropping patterns are closely integrated. Farming on the Niger Inland Delta involves both flood cropping of sorghum and rice with rainfed bullrush millet on unflooded land. Floodplain farmers minimise the risk from severe flooding by careful crop and variety selection which is often related to the variable topography and sedimentary characteristics of the floodplain. In addition, the farmers engage in other activities, such as fishing or herding, to maximise their economic flexibility. For example, on the upper Ganges floodplain north-east of Delhi, India, there are many abandoned river channels some of which become dry in the dry season and offer good grazing for animals. The wet channels or *jhils* suffer serious waterlogging but their margins can also be used for livestock and there is freshwater fishing in these permanent lakes. The easily perishable crops are harvested before the onset of the monsoon. During the monsoon months (June to September) an annual cash crop of sugar cane and fodder crops remains in the field, although the latter are easily damaged by floods.

2.3 ESTIMATING THE LOSSES FROM FLOODS

Severe floods temporarily wipe out the long-term benefits of floodplain locations and the main problem is to determine accurately the losses to be expected from floods of different magnitudes and return periods. Only when this information is available for key flood-prone areas is it possible to estimate the total risk and to draw up policies for rational flood alleviation based on cost-effective measures. In practice, the effects of flooding vary greatly between urban and rural areas, between rich and poor countries and between direct and indirect losses. Direct tangible losses in urban areas within the MDCs have been subject to the greatest scrutiny and are consequently better understood than many other impacts. One consequence of this is a danger of emphasising what can be measured rather than what is important. The priority given to direct tangible losses has been driven by the reliance of government funding agencies on cost-benefit appraisals of direct damage for the economic justification of flood alleviation schemes. Despite this, the methods employed to estimate mean annual flood damages have been rather rudimentary until recent years.

There are two main approaches to the calculation of direct economic losses from floods. The first is based on the collection of *actual* flood damage information which is reported after the event. The advantage of this method, which relies on field interviews and questionnaires, is that it deals with real events. In some cases, the methodology is well-established. For example, a nationwide system of flood loss reporting, compiled from questionnaire returns administered on a river basin basis, has been used in the USA since 1902 and is the main source of long-term damage information for that country (see Figure 1.9). In Australia, a detailed investigation of actual flood damage at Lismore, New South Wales, produced one of the most comprehensive investigations ever undertaken at a single location (Smith and Munro, 1980). One useful outcome of this work was the computer-based flood damage maps for the area, an example of which is shown in Figure 2.2 (Smith and Greenaway, 1980). The residential sector in Lismore contained some 1900 houses organised into 25 000 m² grid squares and the damage map approximates to the 1 in 100 year flood.

Since such data collection methods depend on waiting for floods to happen, the resulting estimates are only available for the range of floods recorded at the sites surveyed. As a result of more frequent flood experience, some areas will be much better documented than others and there may be problems in extrapolating the results to another river basin. Despite attempts at standardisation, the data collection methods are less than perfect. For example, the timing of the survey is important. Immediately after the flood, many people tend to over-estimate direct damage, a tendency which is encouraged because repair bills may not have been received at the time of the survey. On the other hand, some types of longer-term structural damage, such as undermining of the foundations or wet rot to floorboards, may not be detected until much later. However, if the survey is made months later, flood victims may be unable to recall important, but undocumented, details. Unless special precautions are taken, interviewer–respondent error can lead to unreliable assessments for individual households, notably by the double-counting of certain types of loss.

The second method of calculating damages involves estimating the *potential* losses expected to result from flood events of a specified severity based on generalised relationships between certain flood characteristics and physical damage. This produces synthetic, rather than actual, direct loss values but the method is regarded as more systematic than post-flood field surveys alone and the results can be more easily transferred to areas where flood experience is either non-existent or outdated. When flooding does occur, a comparison of actual and synthetic losses can be made which can help to refine the methodology. It is important to realise that unmodified synthetic losses can be higher than actual recorded losses simply because they ignore the damage-reducing actions that emergency managers and floodplain residents take in a flood event. The key to obtaining reliable synthetic data lies in the representative sampling of building types and Smith (1994) reviewed the main problems involved in deriving the resulting stage–damage curves.

Several physical flood characteristics can influence the type and the amount of economic loss. For example, the seasonality of flooding may well be critical for agricultural land whilst the rate of rise and the timing of the flood peak, either during the day or during the week, can influence the efficiency of flood warning and response. The

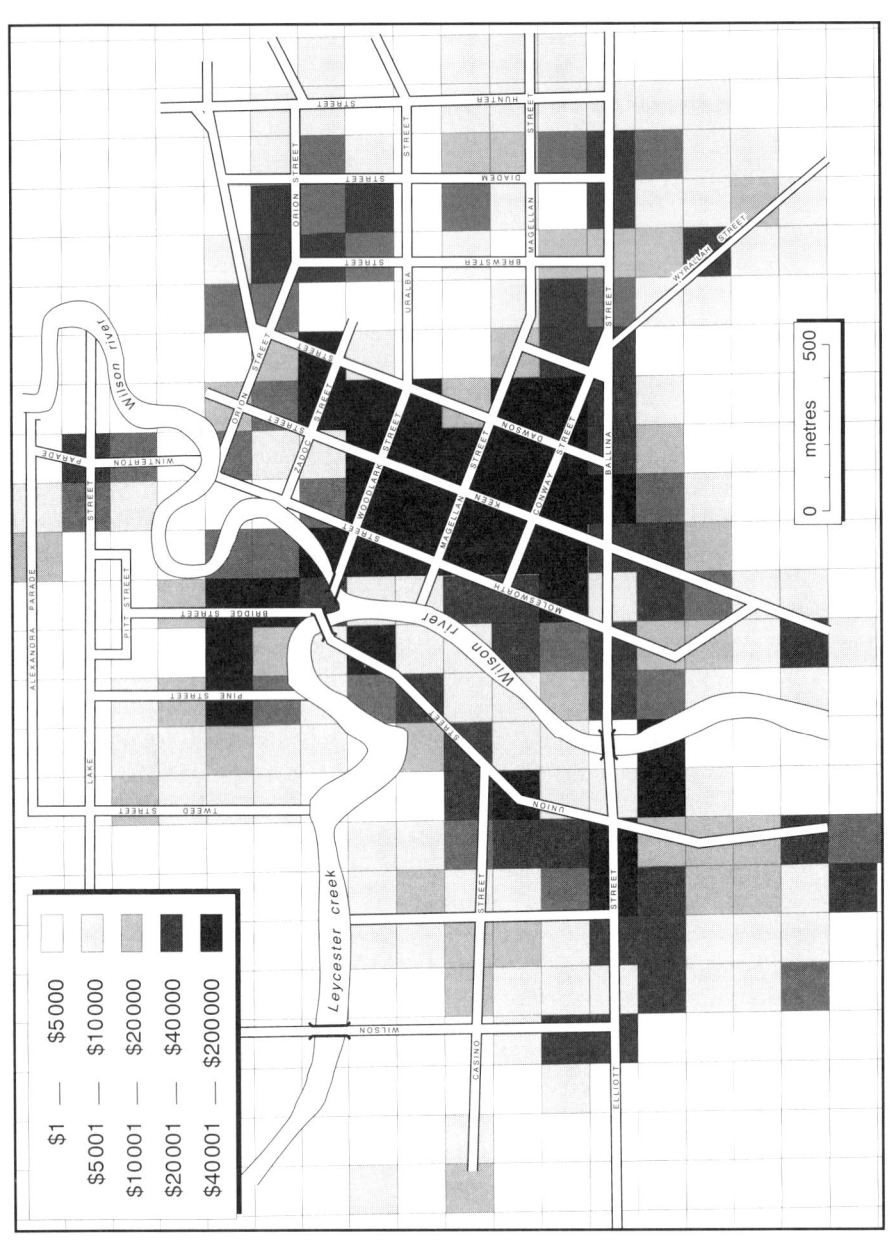

Figure 2.2 The location of flood damage areas in Lismore, Australia. Values show direct combined residential, commercial and industrial losses at 13.0 m gauge height in $A
Source: After Smith and Greenaway (1980)

nature of the flood flows, particularly high velocity, high turbulence, high sediment load and long duration, are all thought to increase flood losses. Some of these variables are interrelated (Smith and Tobin, 1979). For example, turbulence, velocity and sediment load may well be positively correlated whilst flood duration is often a function of flood depth. Because of these interrelationships, and the difficulty of making independent measurements, most synthetic methods of estimating flood losses rely on simpler relationships between flood depth, or stage, and damage. In other words, depth–damage functions beyond some base level, commonly the ground-floor level of buildings, are regarded as of primary importance. In the most sophisticated assessments, where the appropriate data are available, these functions can be modified according to secondary variables such as flood duration and water quality.

Figure 2.3 shows some early flood loss estimates based on stage–damage curves for different building types at La Follette, Tennessee, by White (1964) based on the following procedure:

1. Determine by levelling the elevation of the first floor for each type of building or establishment.
2. Determine from stream profiles for floods of three critical stages the elevation each flood would reach in the building.
3. Determine the stage at which damage would begin. Below this level zero damage is assumed.
4. Estimate the amount of damage that would result from flooding at each of the selected stages and graph a smooth curve.
5. Convert this curve to a mean annual damage estimate by assigning a frequency to each of the elevations and measuring the area under the curve.

Average losses for residential premises were derived from observations of current real estate values, the likelihood of the (mainly wooden) structures being moved off their foundations and the quality of the furnishings, all verified by independent field checks. From Figure 2.3 it can be seen that the damage assessments varied for different establishments and that the standard stage–damage curves can be extended to incorporate some design flood events, including the maximum probable flood.

More recently, attempts have been made to develop nationally applicable standard flood damage databases for more detailed property types and their contents. Actual and synthetic depth–damage datasets were constructed for Britain in the 1970s (Parker and Penning-Rowsell, 1972). Their approach was based on an improved procedure:

1. Select the land-use category for analysis.
2. Identify the relative importance of the main flood characteristics (depth, duration, velocity, load).
3. Within each land-use category, identify a limited number of significant sub-groups of building types (one or two storeys, presence of a basement, etc.).
4. Using the main characteristic of flooding identified in (2), establish a relationship between that variable and damage, e.g. derive a depth–damage curve for each land-use sub-group.
5. Secondary flood characteristics, e.g. velocity, are then used to modify the curve. For example, the stage–damage curve could have low, medium and high velocity variants.

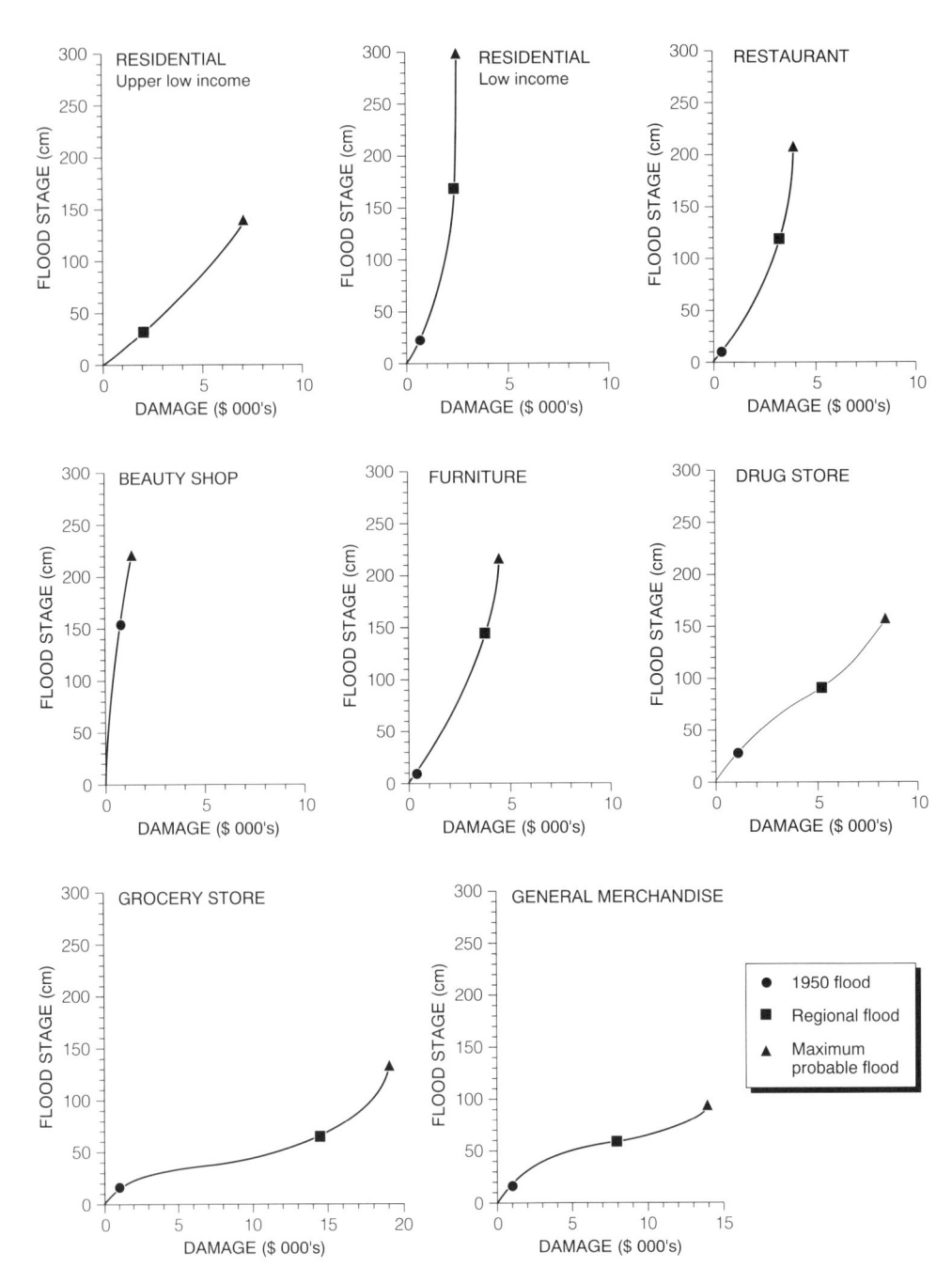

Figure 2.3 Stage–damage curves for representative properties on the floodplain at La Follette, Tennessee, USA
Source: After White (1964)

IMPACTS AND INTERPRETATIONS OF FLOOD HAZARD 43

6. Test the synthetic damage estimates so derived by comparing them with actual damage assessments as these become available.

This work was eventually collated into the *Blue Manual* which was the standard instrument used by the UK water industry for flood benefit–cost assessment from 1977 to 1987 (Penning-Rowsell and Chatterton, 1977). As shown in Figure 2.4, residential depth–damage functions were presented for several combinations of house type, age category and social class of occupants. The functions were derived by selecting a typical example of each house type and age, and a typical contents inventory for each social class. The damages incurred at each level of flooding were hypothesised using contractors' prices for repair work and susceptibility ratios for damage to particular items. Building fabric and contents damages were tabled separately for a flood duration of both less than, and greater than, 12 hours. A similar approach was adopted for deriving non-residential depth–damage functions. The result was a range of standard depth–damage relationships. Standard datasets were also developed for tidal flooding, taking account of the extra damage owing to salt contamination, and for the situations where potential flood damages may be reduced by the response to flood warnings. These methods have been computerised and are capable of regular updating (Suleman et al., 1988; N'Jai et al., 1990).

Similar work was undertaken in the USA by Grigg and Helweg (1975), who reviewed residential flood damage estimation procedures and recommended the use of depth–damage functions adopted by the Federal Insurance Administration. In order to permit international comparisons of similar methodologies which reflect different currencies and cost structures, it is necessary to non-dimensionalise the damage axis, as shown by Higgins and Robinson (1981). Damage D can be expressed as a proportion of $D1$ occurring at a specific index depth $d1$. The relationship shows d versus $D/D_{0.6}$ because all functions are similar for depths of flooding less than about one metre, which is the most common situation. Figure 2.5 shows a comparison of stage and relative damage curves for the broad datasets used by Penning-Rowsell and Chatterton and by Grigg and Helweg, and this indicates a high degree of coincidence. Such databases can be used to produce maps of different flood damages at various flood heights for each land-use sector.

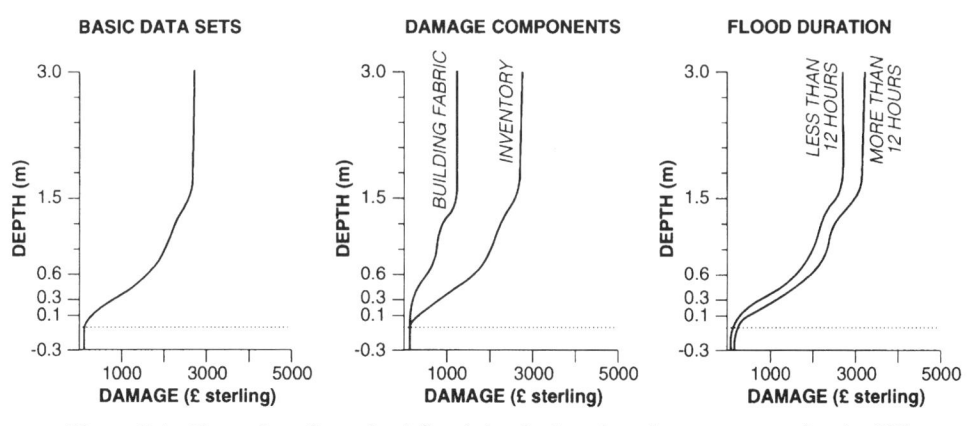

Figure 2.4 Examples of standard flood depth–duration–damage curves for the UK
Source: After Penning-Rowsell and Chatterton (1977)

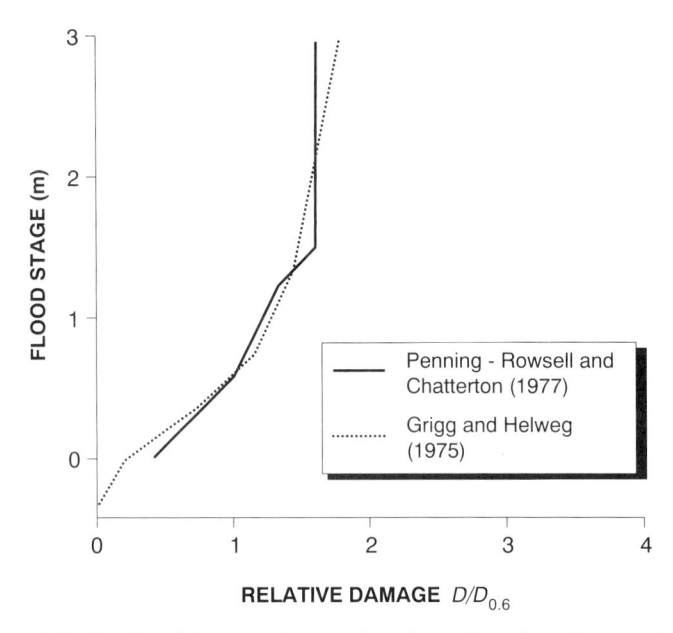

Figure 2.5 Residential stage–damage functions plotted as d versus $D/D_{0.6}$
Source: Modified after Higgins and Robinson (1981)

The estimation of indirect flood damages is more difficult. Green et al. (1983) proposed a method for estimating the indirect losses from urban flooding based on the extent to which a flood disrupts the network of linkages between goods, people and information. This method was elaborated with regard to a major urban problem, that of road traffic disruption (Green, 1983) and the general approach was eventually compiled in the *Red Manual* which provided systematic methods for estimating the disruptive effects of floods, including manufacturing industry value-added losses, traffic disruption costs and the marginal costs of flood emergency services (Parker et al., 1987). More recently, a *Yellow Manual* has been produced which extended the analytical evaluation methods for intangible benefits, such as those associated with recreation and the environment, to coastal areas subject to flooding (Penning-Rowsell et al., 1992). There are many difficulties in making precise estimates of indirect losses. For example, for retail outlets the extent to which trade is either deferred, or transferred to other businesses, is often unclear. Other data may be difficult to obtain because of commercial confidentiality and access to health records is usually restricted. When dealing with the more intangible losses, like health effects, the methodology is often dependent on controlled samples for a simple 'before and after' comparison by analysing visits to local doctors, hospital admission rates or psychiatric referrals. Investigation of the stress-related effects of flooding, such as anxiety, depression, belligerence, alcohol abuse or wider family problems, such as the loss of kinship ties, unemployment or involvement in crime, depends on resource-intensive individual and group interviews and questionnaires with victims after the event. These negative effects of floods can be operative from a few hours to over a year after the event and there is a lack of standardisation of data collection and assessment techniques in this field. One specific problem is that of making a reliable financial assessment of the loss of a human life.

Methods of flood loss estimation will continue to improve and may well need further refinement in the light of new conditions. For example, despite all the uncertainties associated with future climate change, the prospect of an increase in the frequency of extreme flood events in many countries cannot be ruled out. Smith (1993) noted that any rise in the occurrence of low probability events will create disproportionately large increases in flood damages due mainly to the greater depth and velocity of the flows. The two latter parameters are especially important if they lead to the structural collapse of certain types of building. For the town of Queanbeyan, New South Wales, it was estimated that, if a change of one standard deviation occurred in the frequency of floods, the mean annual direct damages would increase by a factor of eight, even if the loss estimates were confined to the same number of properties at risk.

2.4 DIRECT LOSSES FROM FLOODS

2.4.1 Tangible Losses

Primary losses are caused by the direct physical damage to property. Because of the relatively high levels of investment per unit area of flood-prone land, these losses tend to be highest in absolute terms on urban floodplains in the MDCs. Flood losses in the USA now exceed US$ 5 billion in individual years and initial assessments of the upper Mississippi floods of 1993 indicate a loss range between US$ 15 and 20 billion, with more than 50 000 homes destroyed or damaged and nearly 54 000 people evacuated (NOAA, 1994). Flood damages cost Australians between A$ 350 and 400 million per year in 1994.

LDCs are also vulnerable to direct flood losses. This is due to the concentration of urban development in primate cities. In 1988 Khartoum, Sudan, suffered storms leading to more rain in one day than the average rainfall for an entire year. Some 200 000 homes were destroyed or extensively damaged, about 2 million people were rendered homeless and more than 80% of schools in the Khartoum area were damaged or destroyed. Buildings in developing countries are particularly susceptible to flood damage because they may not be designed and constructed to high standards. Hughes (1982) showed significant differences in the flood-resistance of different building styles and construction materials particularly when architect-designed and engineered dwellings were compared with the more common vernacular houses, which may be little more than poorly built shelters.

Primary losses can also be high in rural areas where most of the damage is sustained by crops, livestock and the agricultural infrastructure, such as irrigation systems, levees, walls and fences. According to Ramachandran and Thakur (1974) almost 75% of direct flood damage in India is to standing crops, with the remainder involving houses, livestock and public utilities. Livestock losses are a particular concern in India where animals commonly suffer injuries and starvation, as well as death from drowning, and Sastry (1994) recommended the construction of multi-purpose livestock shelters in areas prone to floods and cyclones. The greatest hazard exists on the extensive, densely populated floodplains of Asia where river floods alone damage about 4×10^6 ha of land each year, creating average annual property damage estimated at more than US$ 3 billion (Smith, 1989). During the exceptional monsoon floods of 1987 and 1988, the estimated material cost in Bangladesh alone was close to US$ 3 billion (Rogers et al., 1989). The worst affected countries are those, like Bangladesh and Vietnam, which are exposed to both monsoon rains and coastal cyclones. Shan (1996) drew attention to the direct effect of floods on grain yields in China and, since grain prices and inflation are closely related, to the wider and less direct economic consequences.

Agricultural losses depend very much on the season of flooding and the type and state of the crop. The timing of a flood in relation to harvest is often critical with the maximum damage frequently occurring immediately before harvest or even after the harvest if the crops are still stacked in fields. The extent of loss will be reduced if there is time to plant another crop after the flood has receded. Flood depth is an important determinant of crop damage. For example, whether or not the water covers the ripening heads of grain crops is crucial. But, on extensive floodplains, this information is difficult to collect and apply because flood depths are generally small and show little variation between floods of different magnitudes. Also different floods break river banks at different points and spread to different floodplain areas depending on the height of vegetation and the condition of repair of small features like walls and fences. Flood duration is equally important. Grass is generally killed if submerged longer than three weeks due to lack of oxygen. Potatoes and other root crops can be completely destroyed if flooded for as little as 24 hours. Secondary direct losses include longer-term factors, such as reduced soil fertility due to the spread of gravel deposits or salt contamination of the land from storm surge.

Secondary losses relate to restoration of the direct damages. Thus, rebuilding costs for property may be unusually high after a flood because of shortages of labour and materials and this may exacerbate existing financial pressures on property owners, especially if they are under-insured. If complete restoration is not possible, the property will suffer a loss in market value. Even with restoration, it might be argued that the occurrence of a flood would devalue the affected properties, at least for a short time after the event. On the other hand, repairs and improvements made to damaged houses could increase the value in the longer-term, especially if the event is seen to be a once-in-a-lifetime flood, either because of its recurrence interval or because of the subsequent construction of defence works. Evidence from case studies conducted in the USA has tended to confirm that, after an initial decline, selling prices for houses tend to recover. In some communities experiencing rare flooding, values may even exceed pre-flood levels. Montz and Tobin (1988) studied post-flood residential real estate values in two Californian communities after a levee break and showed that both the list prices and the actual selling prices of houses declined immediately after the event but then slowly recovered. But the recovery in the real estate market was not even across the floodplain, indicating that the spatial inequalities in flood experience did not end after the event. The drop in property values, and the length of the recovery periods, tended to be largest for those properties flooded to the greatest depths. For example, in the Californian houses that had suffered 45 cm of water, the values recovered to near pre-flood values in less than one year, whilst houses subject to greater depths of flooding lagged behind. On the other hand, in an area of more frequent flooding, values for flooded properties rather reflected the number of times they had been flooded (Tobin and Montz, 1990). Extending this work to three communities located on the North Island of New Zealand, Montz (1992) found an even more complex situation where the influence of housing sub-markets, independent of hazardousness, was sufficiently powerful to either mask or to exaggerate the flood-related impacts. As a result, the impacts on house prices varied although, once again, any differences between flooded and non-flooded properties decreased with time and it was concluded that flood hazard was not an important long-term consideration in house purchase decisions.

2.4.2 Intangible Losses

Primary losses are the enhanced rates of death resulting from disaster, known as 'mortality'. Flood deaths occur most directly as a result of drowning and PAHO (1981) claimed a death:injury ratio of 1:6 for floods, compared to 1:3 in earthquakes. Historically, China's problems have been dominant. The Huang Ho (Yellow River) has often been termed 'China's Sorrow' and has probably killed more people than any other single natural feature on the earth's surface. As recently as 1931 over 3 500 000 people were killed in flooding on the Ch'ang Chiang (Yangtze) when some 3.5×10^6 hectares were inundated. In this case, many deaths were attributed to water-borne disease and famine. China has addressed the flood problem over recent decades, mainly with improved evacuation and follow-up support services for sheltering and feeding refugees. Although as much as 10% of the Chinese population continues to be affected by damaging floods, the risk to the population now appears to be decreasing.

To some extent, the flood risk focus has passed to other Asian countries. In Bangladesh, most of the population lives with little protection on the floodplain of the Ganges–Brahmaputra–Meghna system, widely regarded as the most aggressive river system in the world. Severe flooding from the monsoon rains may cover almost half of the country inflicting massive damage on the economy and infrastructure (Rogers et al., 1989). Not surprisingly, people here perceive floods as the worst disasters which they can suffer (Alam, 1990). For example, in September 1987 about 25 million people were affected by monsoon flooding and over 2×10^6 ha of crops and some 5 million homes were lost. But the greatest mortality rates occur as a result of the storm surges generated by tropical cyclones. In November 1970 and April 1991 respectively, some 300 000 and 140 000 people lost their lives compared with the abnormal floods of 1987 and 1988 in which fewer than 2000 people were killed. This difference in the total number of casualties is due to the fact that the river floods have a much longer onset and warning time than the cyclones. In India, perhaps the country most prone to flood disasters, the fertile agricultural tracts of the Ganges and Brahmaputra valleys, as well as the deltaic areas of Andhra Pradesh and Orissa, are the most affected areas. In contrast, the Indus Basin has a highly developed water resource system regulated with dams and canals. But the area is still vulnerable to excess monsoon rains. In 1992 there were floods in the Indus and Jhelum basins of Kashmir and Pakistan which resulted in more than 2000 deaths and economic losses estimated at over US$ 1 billion. The floods arising from the monsoon rain could not be controlled and water released from the Mangla dam inundated the city of Jhelum.

On a world scale, the mortality from flash floods is believed to be increasing, largely because of the lack of adequate forecasting and warning procedures for such events, even in the MDCs. Settlements at the foot of mountains, like Boulder, Colorado, are particularly vulnerable. The Rapid City and Big Thompson floods are other examples of flash flood disasters in the USA. One-half of all flood deaths in MDCs are vehicle related and occur mainly in darkness. Floods can wash cars away in about 0.5 m of fast-flowing water and, the faster one drives through floodwaters, the more dangerous they become.

Secondary losses are associated with various forms of physical and mental ill-health created by injury and disease after disaster, known as 'morbidity'. This is a complex subject and Legome et al. (1995) called for an international reporting scheme for injuries imposed by floods. Ill-health can arise from physical trauma in high-velocity flood flows or from contact with water highly polluted with dangerous chemicals, but the greatest problems are associated with water-borne disease. The literature that links floods to an increase in communicable diseases is largely restricted to a few LDCs and is plagued by poor quality data based on limited spatial sampling in the field, with problems of 'before and after' comparisons. But the studies show that certain endemic diseases, especially gastrointestinal diseases, rise to epidemic proportions in areas of the LDCs where sanitation standards are low. For example, in Bangladesh, where less than 1% of the rural population had sanitary latrines in 1981, diarrhoea is endemic and the greatest cause of death for children under five years of age. In Ecuador about one-third of the urban population has no access to water services or piped sewage disposal, a fraction which rises to about three-quarters for the rural population. Knowledge of public health and sanitation is often low and people may not be able to afford to boil drinking water.

Disease morbidity rises due to ecological changes and the additional exposure created by the migration of people, often already infected with disease, from rural areas to overcrowded urban shelters in the first few weeks after the flood. The problem is most severe with water-related diseases which spread easily due to the failure of sewage systems and the contamination of drinking water supplies by microbiological pollution after floods. Surface water sources, such as rivers and ponds, which may normally be used for washing and personal hygiene, become highly contaminated. One example is the marked El Niño floods of 1982–83 which affected much of the west coast of South America. This flood covered about 15% of the national territory of Ecuador, causing losses estimated at 6.4% of the 1983 GDP, and extensively damaged waste-disposal systems leading to the presence of water and sediments polluted by coliforms in the streets (Hederra, 1987). In this case the incidence of typhoid and malaria doubled above the endemic rate. The authorities had no plans for such events and the suspension of the normal programme of spraying of dwellings with insecticide aided a multiplication of mosquitoes. According to Cedeno (1986), the malarial epidemic spread to large urban areas, such as Guayaquil, which had previously been free from the disease. In many tropical countries, the incidence of malaria and yellow fever also rises because of the multiplication of insect vectors in stagnant water. A similar picture was reported from the adjacent areas of northern Peru by Gueri et al. (1986) and Russac (1986). Here there was an increase in diagnosed gastrointestinal diseases of over 150%, mainly in the 5–14 age group, together with increases in respiratory and skin diseases. During February 1983 the monthly death rate from gastrointestinal disease was about three times the equivalent 1983 figure and the floods created a 94% increase in total mortality, primarily in the 1–4 years age group. Figure 2.6 shows the increase in malaria in the Departments of Puira and Tumbes of northern Peru where it can be seen that, in the worst month, the incidence rose to about seven times the average level experienced in previous years. This outbreak was associated with a rise in temperature and humidity, a proliferation of temporary mosquito breeding sites and the exposure of individuals in cramped shelters. In Bolivia mosquito breeding grounds have been specifically linked to the receding floodwaters but increases were also noted in small mammals, especially rats, which were implicated in disease arising from their infestation of markets and food crops (Telleria, 1986).

Long-term development aid combined with short-term public health responses, such as the distribution of water purifying tablets, may not be a complete answer. A development

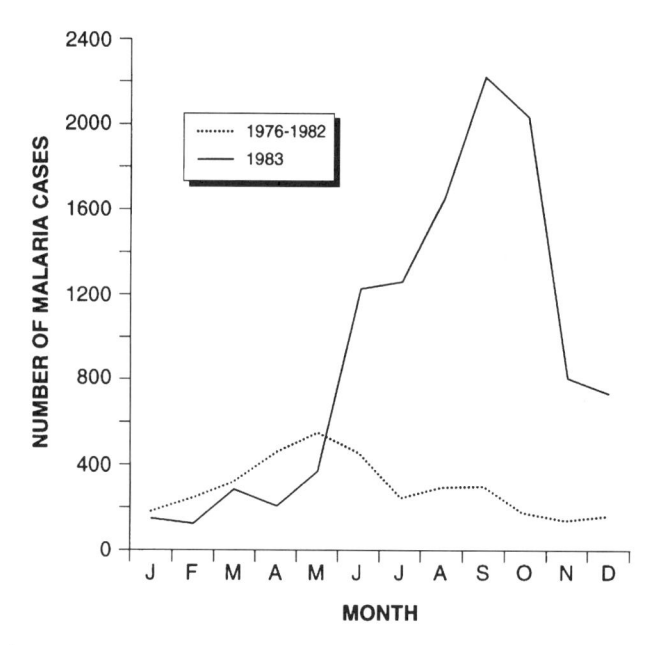

Figure 2.6 The number of reported cases of malaria after floods in 1983, compared with the average monthly incidence 1976–82, on the north coast of Peru
Source: After Russac (1986)

programme in Bangladesh which introduced pit latrines was described by Hoque et al. (1989), but after the 1987 river floods the use of these latrines by the population aged 5 years and over decreased from 88% to 78%. The main reason seems to be that about 60% of the latrines suffered flood damage necessitating considerable expenditure on repairs. In another study, Hoque et al. (1993) showed the devastating effect of the 1991 cyclone on public health. In Bangladesh 80% of the population take drinking water from tube wells and use ponds and other surface sources for bathing, washing and cooking. After the cyclone the surface sources became highly polluted with saltwater and sewage. As a result, the tube wells were used for all domestic purposes and the demand on them doubled, despite the fact that 40% of them had been damaged in the cyclone. Water then became extremely scarce with a big increase in disease and death. In such situations, the water purifying tablets may have limited benefits. Although over 60% of the affected population received tablets, they were of different types and required complex dosage instructions which could not be followed by largely illiterate people. More importantly, nearly two-thirds of the distributed tablets were of poor quality and had lost their potency. After the flood at Khartoum, Sudan, in 1988 diarrhoeal diseases and malaria accounted for the greatest number of visits to medical clinics and there was a rise in morbidity associated with both these causes. This was despite relief precautions which included the provision of potable water, immunisation, the distribution of oral rehydration salts and vitamin supplements to children (Woodruff et al., 1990).

The sudden changes in the physical environment associated with floods can cause severe emotional stress and create long-term mental health problems amongst the victims. This problem has been most studied in the MDCs. The emotional suffering is tied to the strain of economic and sentimental loss arising from the destruction of homes and treasured personal possessions, the disruption when uprooted from a community and separated from family and friends, coupled with the daunting physical effort of the clean-up operations. In the MDCs, the impact of floods on individual households may well involve the loss of some 'essential' features of family life such as a television set or aesthetic surroundings, such as wall decorations and carpets. These losses can be severe if a house is all-electric and the wiring is damaged. One-parent families are less able to clear up while continuing to earn a living and elderly or disabled people may suffer greater ill-health in damp accommodation. Such stresses have varying effects on people. Common results are manifestations of isolation, impotent anger and dismay and unresolved grief which may lead to clinical symptoms of severe anxiety or depression, leading to alcohol or drug abuse. These effects are most severe amongst the most vulnerable groups who tend to be young, elderly or mentally ill. Flood experience, especially if combined with permanent relocation, can lead to prolonged psycho-social effects which may endure for many years and may be responsible for long-term morbidity.

A pioneering study of the mental trauma associated with flood disasters was undertaken by Bennet (1970) who investigated the effects on a controlled group of flood victims after a comparatively minor event in Bristol, England, when 3000 properties were flooded in 1968. The health of the flood victims was worse than that of the control group and for older people, who were well-represented in the sample, there was an increased likelihood of death within 12 months of the event. All of the patients referred for psychiatric care were having difficulties in their lives before the flood which simply created an additional burden. Other studies have tended to validate these findings, albeit sometimes at a lower level of impact. For example, Abrahams et al. (1976) found no increase in mortality after the major Brisbane floods of 1974 but the number of flood victims seeking medical attention increased significantly in the year after the flood. In Lismore, New South Wales, which experiences frequent floods, a relatively large event in 1974 failed to increase the number of hospital admissions or deaths but there was evidence that hospital admissions were related to the extent of flood experience. Those victims whose dwellings had been flooded to a metre or more above the floor level were twice as likely to be admitted to hospital as residents of the flood-free areas (Handmer and Smith, 1983).

One of the most comprehensively studied floods was the Buffalo Creek disaster of 26 February 1972 when a poorly maintained dam burst without warning in a coal-mining valley in West Virginia, USA, killing 125 people. A total of 500 homes were destroyed, at an estimated cost of US$ 450 million, and 4000–5000 people were made homeless. The resultant mental trauma was severe because the survivors were living in a relatively isolated area and suffered the sudden destruction of their tightly knit community as a result of the perceived irresponsibility of others. The victims experienced spatial and temporal disorientation, apathy, feelings of hopelessness and separation. This led to insomnia, amnesia and eating problems and some people were unable to relate to other family members or make new relationships (Erikson, 1976). Many families became hostile and depressed when placed in over-crowded trailer parks, where there was no concern for natural community groupings and where the victims had no decision-making powers (Church, 1974). A group of 625 survivors filed negligence claims against the mining company who owned the dam when it became clear that the event was still causing apathy and withdrawal some two years after the event. Of the 615 survivors examined in connection with the legal action 18 months after the event, 570 were found to be suffering from an emotional disorder. In particular, it was claimed that many children had been emotionally impaired by the Buffalo Creek flood and rendered more vulnerable to future stresses (Newman, 1976). For the first time, a court accepted psychic impairment claims from persons not present at the scene of the disaster and an overall settlement of US$ 13.5 million was reached, of which US$ 6 million was distributed for psychological damages (Stern, 1976).

2.5 INDIRECT LOSSES FROM FLOODS

2.5.1 Tangible Losses

Primary losses relate mainly to the disruption of economic and social activities, especially in urban areas, immediately after a flood. To this extent they can be called consequential losses and are usually regarded as impacts on traffic flows as well as losses to industrial production and trade. Although these losses are theoretically tangible, they have been so under-researched compared with direct losses that it is often difficult to assign them reliable monetary values. In addition, the indirect losses can spread well beyond the flooded area. In the case of coastal flooding on the Isle of Portland, Dorset, some 48% of total maximum flood damages were indirectly caused by the disruption of communications which affected employment nearly 50 km from the inundated area (Penning-Rowsell and Parker, 1987).

The indirect losses to rural economies often relate to damage to subsistence agriculture, although increasing investment and industrial development are placing non-agricultural assets at risk. The agricultural costs of severe floods include not only the damage to a single year's production but much longer-term costs due to under-investment in the more costly inputs, such as higher-yielding varieties, fertiliser, irrigation and labour because of a reduced level of confidence. Major indirect flood losses to agricultural production are possible in the MDCs. For example, wet soils, high humidity, unseasonably low temperatures and a lack of sunlight associated with the 1993 Mid-West floods resulted in some of the worst plant disease epidemics in the area for decades (Munkvold and Yang, 1995). Livestock are of key importance in the rural economies of many LDCs. In the case of Bangladesh, cattle provide over 95% of the power for tillage, together with manure for the fields, fuel for the poorest families, foreign exchange from the hides and skins, as well as meat and milk for human consumption. A survey of 1000 households in Bangladesh after the 1984 floods showed comparatively few losses of livestock due to drowning but large losses due to disease and hunger following the destruction of animal feeds. This not only caused distress sales of the remaining livestock, plus land and other assets, but also created a tillage power problem for farmers anxious to plant their winter crops on time (Jabbar, 1990).

Despite important losses in rural areas, the levels of investment in urban infrastructure ensure that the disruptive effects of floods are at a maximum in the built-up areas of the MDCs. Road transport is an integral feature of towns and indirect losses include the greater consumption of fuel if longer journeys are necessary. With long delays, some goods (e.g. fresh fruit, newspapers) may decline in value during transit and the occupants of delayed vehicles lose the opportunity cost of the extra time involved in deliveries. These impacts are worst where few alternative routes exist and when local communications depend heavily on scheduled services, e.g. by bus or train. Unexpected flash floods, in particular, lead to unplanned delays and allow no substitution, such as the prior rescheduling of services. In industrial areas, transport disruption can have a major impact on industry and trade. For example, production can be lost if plant and machinery are damaged or if employees are unable to reach their place of work. Access to raw materials or the distribution of finished goods can be restricted. The worst effects on manufacturing industry occur where continuous production processes, with costly shut-down and start-up operations, are interrupted and where no spare capacity exists to make up production after the event. This can happen if all the production is concentrated on one flooded site. Most retail outlets are dependent on warehouse access for stock. The food retail industry depends on regular deliveries and additional problems may occur if food production cannot be restarted until public health inspections have taken place or where release of toxic chemicals creates a risk of fire or contamination of the site. Losses in the retail trade range from the short-term problems of access by employees and goods to longer-term impacts if there is reduced spending power in the local community because of unemployment. The heaviest losses are likely to be sustained if flooding occurs at peak trading periods (e.g. immediately before Christmas) and local customers either go to competing outlets or spend on replacement household items lost in the flood, rather than on regular purchases. On the other hand, non-urgent purchases of more durable goods will be much less affected.

Other indirect tangible costs of flooding include the deployment and response of the emergency services, such as police, fire and ambulance (Parker, 1988), and a reduced level of other public facilities, such as interruptions to electricity or gas supplies, failure of the telephone system and delayed postal deliveries. Public utilities suffer the greatest loss when re-routing is not possible. For example, it may be possible to deliver mail by alternative roads but a severed water main may offer no possibilities for substitution. Any loss of electricity supplies has many consequential effects, including more difficult drying-out of flooded premises, the maintenance of food supplies through refrigeration and the disruption of information flows through the loss of computers and telephone links.

Secondary tangible costs are mainly caused by the longer-term effects of floods. For example, in the MDCs the flooding of a holiday resort or caravan site in the summer season may lead to reduced tourist spending for several years. The need to protect property against floods in the future may lead to stockpiling costs for sandbags, food supplies and medicines by the appropriate agencies. Other impacts can influence spending patterns in the local community. Sikander (1983) showed how the price of basic consumables rose after floods in certain rural areas of Pakistan (Table 2.1). In areas where supplies of electricity and raw materials were reduced, food shortage tended to be acute. Thus, supplies were hoarded with the emergence of a black-market economy. Prices for some foods rose significantly and families had to make cuts in their budgets elsewhere to buy these commodities. This re-prioritisation of family expenditure can have multiplier effects throughout the local economy. For example, in the MDCs a family may be unable to afford a new car because of the cost of repairing a flood-damaged home. The car dealer and manufacturer lose a sale as a result and suffer a secondary economic loss.

2.5.2 Intangible Losses

Indirect intangible flood losses are the most difficult of all to identify. It is especially difficult to separate *primary* and *secondary* effects. This is partly because they are bound

Table 2.1 The increased cost of household consumables during the flood season in Pakistan

Commodity	Price before flood (rupees)	Price during flood (rupees)	Increase (%)
Rice	160	172	7.5
Pulses	240	280	16.0
Onions	200	275	37.0
Potatoes	150	195	30.0
Wheat flour	50	60	20.0
Meat	16/kilo	20/kilo	25.0
Kerosene oil	4/bottle	6/bottle	50.0

Souce: After Sikander (1983)

up with the traditional adjustment strategies adopted by people regularly exposed to flood hazard. These strategies can carry an ongoing burden of extra effort and reduced productivity. For example, throughout much of South-East Asia farmers routinely grow deep-water rice and raise their houses above the ground. In some flood-prone areas of Vietnam, farmers often grow only one crop per year compared with two or three in the flood-protected areas of the delta. In coastal areas where the salinity of the soil has been increased by saltwater incursions, farmers prefer to grow saline-resistant traditional rice varieties rather than high-yielding ones. The loss of morale from repeated flooding may discourage people to the extent that they do not wish to make more than minimum efforts to survive. This may well lead to lower living standards. Sikander (1983) showed how floods led to prolonged isolation and lack of recreational opportunities in Pakistan. In the area surveyed, 60% of families did not receive any post for 30-60 days, compared with the normal twice weekly deliveries, and for a similar length of time in the high flood season people were confined to their houses, in some cases the roofs, which clearly affected social behaviour as well as their working capacity.

Land erosion, combined with the re-deposition of coarse sediments, can be a major source of agricultural loss, especially where aggressive rivers excavate deep, unconsolidated material. It has been estimated that 5% of the total floodplain of Bangladesh is directly affected by riverbank erosion. This can be locally catastrophic when the average holding is only 0.5 ha (Hossain, 1993). In a study of one village between 1979 and 1989 almost 20% of the farmland was eroded which resulted in the loss of more than half the predicted crop income. Although this alluvial material would be deposited elsewhere, the individual victims of riverbank erosion are unlikely to secure the new land which is usually vigorously disputed between rival claimants. In such circumstances, it is the women, children and the poor who suffer most. These people already have a low nutritional status and, if sick or injured, cannot work with a subsequent loss of income and further deprivation. In turn, this can lead to famine and greater hazard vulnerability together with the out-migration of younger, fitter members of the community.

In total, riverbank erosion in Bangladesh renders an estimated one million people landless and homeless every year through displacement (Haque and Zaman, 1989; Zaman, 1991). About 19 million rural people are at risk from bank erosion in the Brahmaputra–Jamuna and Ganges floodplain alone. The majority of the displaced households tend to remain in the vicinity of their previous residence because they are too poor to relocate elsewhere and they also believe that their land will re-emerge soon from the rivers. Many have no land of their own to re-settle and cannot rely on material assistance from their equally poor relatives. Eventually they may migrate to the cities or remain in the *chars* as 'patron-tied' dependants destined to form a cheap labour pool. Clearly there is a need to develop better post-flood resettlement policies at all levels of government in Bangladesh.

2.6 INTERPRETATIONS OF FLOOD HAZARD

The dominant view of what creates a flood hazard, and what comprises the best strategy for the prevention of flood disaster, has evolved through three distinct phases during the twentieth century. This evolution can be linked to the wider emergence of hazard paradigms (Smith, 1996). As in other fields of science, each succeeding paradigm neither answers all the outstanding questions nor completely replaces the previous school of thought. According to Platt and McMullen (1980), the structural period of engineering flood control responses lasted until about 1960 and was replaced by an era of unified floodplain management. During this period more integrated approaches, adopting loss-sharing techniques (insurance) and flood-avoidance techniques (warning, land-use planning) designed to reduce human vulnerability to flooding, became prominent. Since 1980 it has been possible to recognise a third phase, often termed the post-flood hazard mitigation era. This period is still developing but is typified by measures such as property acquisition and relocation as part of a growing awareness of the need for floodplain communities to adopt a more ecological and sustainable relationship with their natural environment.

2.6.1 The Engineering Paradigm

This represents the earliest interpretation and was founded on the long-lasting premise that the flood problem is caused by extreme hydrological events and that the cure is to exert physical control over flood flows. This approach dates back to the earliest hydraulic civilisations but the modern expression of river control culture is closely associated with the USA, starting with the formation of the Army Corps of Engineers in 1799. The first large financial appropriation for flood control works, on the Mississippi and Sacramento rivers, was made in 1917 and the first comprehensive plan for the control of a major river, the Mississippi, was authorised in 1928. During the 1930s this approach flourished through demands for a greater development of natural resources whilst the availability of capital made large-scale engineering projects possible. The Tennessee Valley Authority was established in 1933 and in 1936 the Flood Control Act authorised numerous projects for the purpose of curbing destructive floods. An amendment to this Act in 1938 placed the entire cost of such schemes on the federal government. The only contribution required from the beneficiaries was that the state, or other local authorities, should provide the necessary lands and property rights for construction, should protect the federal government from damage claims and should maintain the works after completion.

The result of this centralised policy was the creation of one of the greatest public works programmes ever undertaken and this essentially technocratic approach rapidly spread worldwide. For example, in New Zealand the Soil Conservation and Rivers Control Act

IMPACTS AND INTERPRETATIONS OF FLOOD HAZARD 55

1941 was specifically geared to the better protection of property from damage by floods. In many countries large dams and levees were provided wherever the direct benefits, to whoever they might accrue, were assessed to exceed the direct costs. A common test of feasibility was the comparison of the direct constructional costs with the difference between the present and the expected market value of floodplain land. This paradigm has been successful in protecting life and property in key areas and has always enjoyed powerful support. The engineering approach has traditionally been favoured by those standing to lose property and income in urban areas and by large landowners who will benefit from increased land values in rural areas. It has also been advocated by members of the technical flood community, such as engineers and construction companies, whilst government ministries and aid donors have often seen the results as a symbol of achievement. Engineering structures are highly visible and their presence often indicates that some political action has been taken against floods.

2.6.2 The Behavioural Paradigm

This interpretation of flood hazard originates from the work of White (1945) who was the first person to highlight the failure of structural schemes in the USA to contain urban flood losses. White's work, subsequently expanded through the work of other geographers into what became known as the 'Chicago School', was based on criticism of two existing failings:

- The policy failure of the flood prevention authorities to consider the implementation of non-structural alternatives, such as land zoning or forecasting and warning.
- The behavioural failure of individual floodplain managers and residents to assess the full risk from floods.

This approach has exerted a worldwide influence. Within the MDCs, much criticism focused on the so-called 'levee effect' whereby the reliance on engineering structures was erroneously perceived to provide absolute freedom from floods for the structurally defended area of the floodplain. The resulting rise in land values encouraged further floodplain invasion, which placed more property at risk when major floods eventually exceeded the design capacity of the schemes. One example of the levee effect is the investment which has taken place in the water-front Docklands area of London following the erection of the Thames barrier. To counter this effect, the behavioural paradigm proposed a diversification of flood loss reduction methods whereby engineering works would be supported by forecasting and warning schemes and better land management would curb further investment in the floodplain. Such an approach still depended on the use of appropriate technology, such as advanced communication systems and accurate mapping techniques, and its successful transfer to non-industrial societies in the Third World.

At first the behavioural approach was transferred rather uncritically to other developed countries. In some New World countries, such as Canada and Australia where floodplain encroachment was a problem, the criticism of flood defence structures and institutional weakness in land-use planning struck a chord. But in some other countries, the direct implementation of such ideas was deemed less appropriate. In Britain, for example, characterised by short, rapidly rising rivers, there was already a long tradition of land-use control, very different from the intrinsic rights of American landowners to develop their own land, and investment in flood forecasting and warning systems had already taken place (Parker and Penning-Rowsell, 1983a). Therefore, non-structural measures were already to some extent in place. In addition, because of the stronger planning restrictions governing flood-prone land in the UK, the hazard perception of individual floodplain dwellers was not deemed to be such an important feature of floodplain invasion.

The chief criticism of the behavioural paradigm has been with respect to the LDCs. Despite according humans a role in both hazard creation and mitigation, the behavioural approach was attacked by Torry (1979) and other social scientists with first-hand experience of the Third World who rejected the view that it is sufficient to deal with the hazard threat alone. In particular, they criticised what they saw as 'modernisation' theory and its triumphalist outlook, which they interpreted as environmental determinism written in a subtle form (Blaikie et al., 1994). The more recent paradigm stresses the wider global forces and the institutional constraints which limit the actions of people at risk from floods, an interpretation which also has some relevance for the MDCs. For example, if a government ministry is given the task of flood reduction, the priority will be a technical approach based on engineering principles. This approach tends to be rather standard and to overlook local priorities and responses. As a result, smaller-scale, self-help, community-based measures have not usually been taken into consideration.

2.6.3 The Development Paradigm

This interpretation of flood hazard first emerged in the mid-1970s from a growing realisation that the western-style technocratic flood response, even when moderated by non-structural responses, was less applicable in the LDCs where natural disasters are a more recurrent feature. In the Third World, floods are increasingly viewed as the 'triggers' of disaster which has wider roots in civil war, foreign debt, uncontrolled urbanisation and poor building construction. Anthropologists and development specialists working in poor countries have questioned whether large-scale flood control projects are suitable for the LDCs, since they may increase the national debt for comparatively little economic return. New technology can marginalise people or give them wages and remove them from traditional, community-based flood coping responses. For example, in many countries like Bangladesh, the average land holding is less than one hectare. Land expropriated for a flood embankment could cover much of such a plot and reduce the land holding to an uneconomic size for the local cultivators. The local advocates of the development approach are concerned for the future of traditional rural societies. They tend to be the less powerful members of society, such as subsistence farmers and fishermen, supported by environmentalists and development workers who have little formal political representation.

In the post-industrial age, there is a need to consider floods in the context of sustainable development throughout the world. We are moving into an era when the capacity of large areas to support the population in the future without serious environmental and human consequences is in some doubt. Freedom from disastrous events is one important aspect of sustainable use and in the LDCs a 'living with floods' culture, based on a clearer distinction between the frequent but benign floods and the infrequent, disastrous events is ever more important (Cuny, 1991). Rivers and coasts are public resources and the public should be given the opportunity to understand and participate in the planning processes surrounding them. For this to happen, the weakest groups in all societies must gain some empowerment and a better access to resources. Previous flood mitigation programmes have been unduly single-purpose and, in some cases, they have proved counter-productive. For example, the construction of engineering works often reduces the incentive for an individual to floodproof property and the routine provision of insurance or disaster aid makes the salvaging of flood-damaged goods less likely. The challenge today is to integrate flood-loss reduction strategies with policies serving wider river and coastal zone management goals for sustainable use.

SECTION TWO
Processes of Flooding

CHAPTER 3

River Floods: Geophysical Processes

CONTENTS

3.1 Introduction...61
3.2 The raw materials of flooding 61
3.3 The flood hydrograph..65
3.4 Flood-producing rainfalls 71
    3.4.1 Empirical evidence 71
    3.4.2 Defining a flood-producing rain...........................72
    3.4.3 The role of atmospheric systems..........................74
3.5 Snowmelt and icemelt 76
    3.5.1 Factors affecting melt rate...............................76
    3.5.2 The conversion of meltwater to floods 77
3.6 Time trends in flood production..................................79
    3.6.1 Human modifications 79
        3.6.1.1 Urbanisation 80
        3.6.1.2 Forestry 84
        3.6.1.3 Agricultural drainage 85
    3.6.2 Floods and fluvial geomorphology.........................86
    3.6.3 Climate variations......................................88
        3.6.3.1 Climate variability and flood hydrology..............89
        3.6.3.2 Climate change and flood hydrology 92

3.1 INTRODUCTION

This chapter attempts to clarify the nature of flood-producing processes and the principal characteristics of the floods which result from them by considering the mechanisms which produce river floods. Treatment of this large topic is simplified by first discussing in general terms the 'raw materials' of flooding, namely rainfall and melting, and the flood hydrographs which they produce. Then follows a more detailed discussion of specific flood-producing processes, namely, flood-producing rainfalls, snowmelt, rain-on-snow and icemelt. Finally, the chapter addresses the ways in which the influence of these processes upon flood hydrology may be modified by human actions, by fluvial geomorphology, and by climate variability and change.

3.2 THE RAW MATERIALS OF FLOODING

Apart from the rare effects of landslides, ice jams or dam failures, floods in most river basins are caused almost entirely by excessively heavy and/or excessively prolonged rainfall, or in areas of snow and ice accumulation, by periods of prolonged and/or intense melt (see also Section 1.2.1). In each case the operative processes result in a large volume of *quickflow* which, as the term implies, reaches the stream channels very rapidly during and immediately after a rainfall or melt event.

Quickflow deriving directly from rain falling on a river basin originates mainly from hydrologically responsive source areas which vary in size (hence their description as *variable source areas*) depending on the interaction of rainfall and catchment conditions. The source areas have saturated ground surfaces which effectively shed, as saturation overland flow, virtually all the rain falling on them.

In most catchments these variable source areas are relatively limited in extent, but they increase in size with total storm rainfall and duration (see Figure 1.3B). The processes involved were first described by Hewlett (1961) and later refinements of Hewlett's ideas were summarised by Ward and Robinson (1990). In the early stages of a storm (Figure 3.1A) all rainfall (P) infiltrates the soil surface. Then, as a result of infiltration and throughflow (Q_t) in the soil profile, the riparian areas and the lower valley slopes become saturated as the shallow water-table rises to the ground surface (Figure 3.1B). In these surface-saturated areas infiltration capacity is zero so that all precipitation falling on them, at whatever intensity, becomes saturation overland flow ($Q_o(s)$).

Although, at the onset of rainfall, variable source areas may be restricted to the stream channels themselves and to adjacent valley-bottom areas, convergence of shallow subsurface flowpaths may also result in surface saturation and saturation overland flow in slope concavities and areas of thinner soils throughout the catchment. Continuing throughflow from upslope unsaturated areas of the catchment will result in the areal growth of source areas wherever they are located initially. Consequently, source areas which often cover less than 5% of the catchment at the onset of rainfall may expand to cover 20–25% of the catchment as a storm of several hours' duration continues, thereby resulting in a five-fold increase in the volume of quickflow generated by a given rate of rainfall. Evidence of massive source area expansion during the great Mississippi flood of 1993 was provided by satellite imagery which showed clearly that surface water spread widely over almost flat upland surfaces and that the full extent of inundated land (almost 58 000 km^2) far exceeded the combined area of the floodplains of the Mississippi, Missouri and their major tributaries (Prince, 1995). The greater the rainfall amount, the greater will be the expansion of the source areas contributing quickflow to the stream channels so that, in small catchments, after prolonged heavy rainfall most of the area may be contributing. It follows therefore that even moderate rates of rainfall may produce severe flooding provided either that the duration of current rainfall or the depth of antecedent rainfall is large.

Where rainfall is particularly intense, or where natural infiltration capacities have been reduced by anthropogenic effects such as soil compaction or overgrazing, overland flow may result from the process described by Horton (1933) and illustrated in Figure 3.1C. During those parts of a storm when rain falls at a rate that is greater than the rate at which it can be absorbed by the ground surface there will occur an excess of precipitation which will flow over the ground surface as overland flow, i.e.

$$(i - f)t = P_e = Q_o \tag{3.1}$$

where i is rainfall intensity, f is infiltration capacity, t is time, P_e is precipitation excess and Q_o is overland flow.

RIVER FLOODS: GEOPHYSICAL PROCESSES

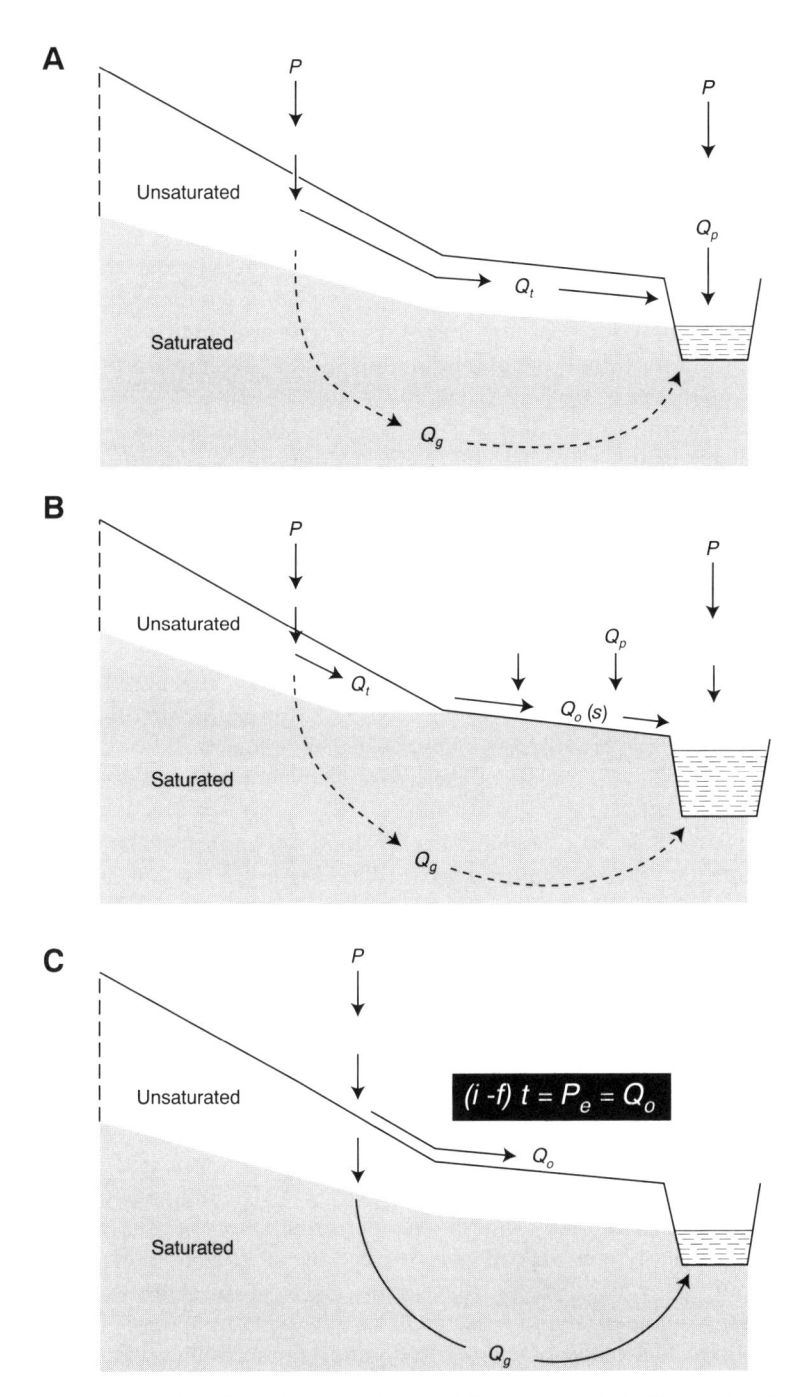

Figure 3.1 The generation of quickflow during a rainfall event: (A) the flowpaths of water at the onset of rainfall; (B) the formation of areas of saturation overland flow later in the rainfall event; (C) the formation of overland flow in areas of low infiltration capacity. Symbols: P is precipitation, Q_g is groundwater flow, Q_o is overland flow, $Q_o(s)$ is saturation overland flow, Q_p is precipitation onto stream surfaces, Q_t is throughflow

No overland flow will occur if the rainfall intensity is lower than the infiltration capacity. However, a combination of surface- and soil-profile effects normally causes a rapid reduction in infiltration capacity soon after the onset of rain so that rain falling at moderate intensity, although incapable initially of generating overland flow, may do so once the early high infiltration rate has declined. Furthermore, since infiltration capacity is likely to show a continued decrease through a sequence of closely spaced storms, it is commonly found that rain falling late in the storm sequence will generate more overland flow and therefore more severe flooding than the same amount of precipitation falling early in the storm sequence. In other words, the development of Hewlett-type saturation overland flow and Horton-type overland flow may both lead to similar intensification of flooding as rainfall and source-area growth proceed.

Irrespective of the way in which quickflow source areas develop initially, when the 'rainprint' of the storm, i.e. the area on to which rain falls simultaneously, covers all or most of the catchment, and when the rainfall duration is prolonged, most of the catchment will eventually contribute quickflow simultaneously. Then, notwithstanding initial infiltration or saturation conditions, the amount of quickflow generated for a given depth of precipitation will be about the same from a forested, ploughed or urbanised area. This has important implications for the effectiveness of catchment management strategies aimed at reducing flood runoff, e.g. afforestation, and for the unintentional effects on flooding of catchment modifications introduced to achieve a variety of economic and social aims (see Sections 3.6.1 and 7.3). Similarly, Wolman and Gerson (1978) found insignificant differences in flood discharge per unit area between distinctive physiographic provinces in the USA (coastal plain, piedmont, Appalachians) and regarded this as confirmation of the adage that 'floods are caused by too much rain'.

Where meltwater is a major component of flooding, as in high-latitude and high-altitude catchments, variable source areas and flood intensity may decline, rather than increase, with time. The main reasons for this are that first, the overland flow of meltwater at the base of the snowpack will be more efficient with a frozen than with an unfrozen ground surface. Normally frozen ground will be more extensive during the early part of the melt season and will reduce in area as melting proceeds. The complexity of the relationships between quickflow production and soil freezing near Leningrad were examined by Kapotov (1989). Second, the volume of the residual snowpack, which ultimately determines the maximum volume of meltwater that can be produced, declines as melting proceeds. And third, since melting normally proceeds from lower to higher altitude, the remnants of a melting snowpack tend to be located at an increasing distance from mainstream channels. In high-altitude catchments, source areas for quickflow may be very widespread, not only because of the extent of frozen ground during the early melt season but also because of the large areas of thin soils, steep slopes and bare rock outcrops which often characterise such catchments.

Although the accumulation and subsequent melting of snow and ice may significantly influence the flood hydrology of high-latitude and high-altitude areas, the sparsity of relevant data means that this influence is not always well understood. Essentially, the severity of meltwater flooding is determined by the water equivalent of the accumulated snowpack (i.e. the equivalent water depth upon melting) and the rate of melting. As might be expected, some of the most severe floods result when several factors coincide, e.g. abundant rainfall in a short period, high temperatures, deep snow cover having a high water equivalent, and high values of soil water. Bolt et al. (1975) described the severe flooding which may occur on rivers draining the southern Sierra Nevada in California, USA, where snow accumulation is frequently 5 to 6.5 m in depth, compared with 0.5 to 1 m in the mountain areas of Germany, and the water equivalent may exceed 2.5 m. In such conditions the water volume generated by melting reaches the equivalent of tropical rainfall although, of course, flood characteristics are determined by the rate and timing of the melt.

Icemelt normally takes place more slowly than snowmelt and by itself is rarely responsible for severe flooding. However, floods known as jökulhlaups (see Section 4.5.2) may occur when the melting of glacier ice suddenly releases large volumes of meltwater which have been ponded back within the glacier system, or when the break-up of the icepack in a river or estuary results in an ice jam which may hold back large volumes of water before suddenly giving way.

Flood production is also possible in catchments where source areas for quickflow may not develop in the ways described above, e.g. limestone, volcanic and other very permeable areas, where subsurface flowpaths dominate the movement of water through river basins. An obvious example is provided by the way in which soil freezing may reduce catchment infiltration to near zero and effectively expand the source area to cover the entire catchment. Flood hydrographs from a limestone catchment in conditions of frozen and non-frozen ground are shown in Figure 3.2A.

In cavernous carbonate rocks, such as massive limestone, concentrated subsurface water movement through macro-fissures may give rise to springs with average discharges of more than $20 \, m^3 \, s^{-1}$ (Stringfield and LeGrand, 1969). Where an impermeable caprock is present surface streams may disappear underground through swallow-holes or sinks at the limestone margin, reappearing subsequently at larger resurgences or risings. In heavy rainfall conditions in 1968 resurgent discharge in the Mendip Hills of south-west England increased by up to 16 times and resulted in the so-called 'Great Flood' of 10 July (Hanwell and Newson, 1970). Even in non-massive limestone, such as the chalk of southern and eastern England, flood flows on relatively minor intermittent streams (bournes or gypseys) are quite common. Such peak flows are almost entirely groundwater-based and often show a time lag of several months after the seasonal peak of precipitation (Figure 3.2B). Significant groundwater flooding on the larger perennial rivers draining chalk areas is much rarer. However, a dramatic example, with a return period of at least 100 years, occurred on the River Lavant at Chichester in early 1994 after 190 mm of rain in 17 days fell on the 90 km² catchment which had already been saturated by earlier heavy falls (Taylor, pers. comm., 1994; Holmes, 1995; Newman, 1995). Records show that flooding on the Lavant usually results when heavy rainfall coincides with high groundwater levels and is believed to be due to the non-uniform transmissivity properties of the upper chalk aquifer.

3.3 THE FLOOD HYDROGRAPH

The hydrograph of river flow synthesises the flood hydrology of a river basin without necessarily shedding very much light on the main flood-producing processes. Precipitation events vary widely in the size of their rainprints (Table 3.1) and since these often cover less than the entire catchment area, their location within the catchment and their direction of travel will have an important influence on hydrograph

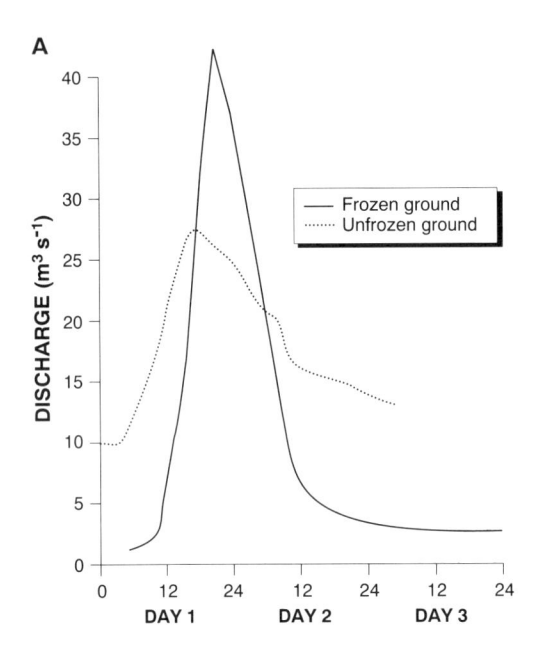

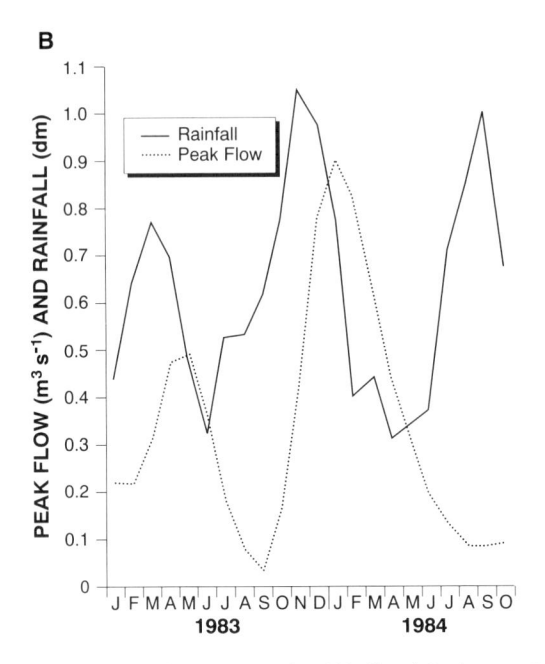

Figure 3.2 Floods in carbonate terrain: (A) flood hydrographs for a limestone catchment in conditions of frozen and non-frozen ground; (B) three-month moving averages of rainfall and peak flow in the Gypsey Race, near Bridlington, UK
Source: (A) from a diagram in White and Reich (1970). With kind permission from Elsevier Science–NL, Sara Burgerhartstraat 25, 1055 KV Amsterdam, The Netherlands; (B) based on data in IH (1986a; 1986b)

Table 3.1 Scale characteristics of some precipitation events

Scale	Example	Size (km^2)	Duration
Synoptic	Tropical storm Major front	25 000–1 000 000	Several days
Large mesoscale	Mesoscale convective complex (MCC)	> 50 000	6–24 h
Mesoscale	Multicell thunderstorm	< 2500	1–12 h
	Thunderstorm	< 750	0.5–3 h
Microscale	Convective cell	< 10	< 0.5 h

Source: Based on data in Maddox (1980), Perry (1981), Bras (1990)

shape (see Figure 3.3). Furthermore, as a flood peak moves through a river system, the effects of channel storage and travel time gradually overwhelm the effects of the quickflow-producing rainfall or melt event on the headwater hillslopes (see Figure 3.4). Burt (1989) observed that there is surprisingly little empirical evidence that can be used to relate the flood response of headwater catchments to the downstream flood hydrograph in the river basin to which they contribute their flow. Figure 3.4 illustrates that although the magnitude of the flood peak increases downstream, the specific discharge (i.e. discharge per unit area) decreases downstream and thus sheds some light on Hewlett's (1982) assertion that

> it is not the peak discharge in the headwaters that produces the downstream flood, but rather the volume of stormflow released by the headwater areas.

This clearly has potentially important implications for the effect on downstream flood magnitudes of deliberate or accidental changes of land use in headwater areas, especially where such changes are known to influence the volume of runoff produced. This issue will be discussed further in Section 3.6.

A concomitant modification, illustrated in Figure 3.4, is the extension of the time base of the flood hydrograph as it moves downstream. This is a function partly of the area of the drainage basin contributing to channel flow, partly of the attenuation of the flood hydrograph as it is routed through the channel system, and partly of the duration and severity of the flood-producing rainfall or melt event. Large rivers, draining large basins, generate flood peaks with time bases measured in weeks or months, compared with the hours or days of smaller streams. The attenuation of the flood hydrograph, which will be discussed in more detail in Chapter 8, is also illustrated in Figure 3.3A where the reduction in peak discharge and increase in time base reflect the distance of the flood-producing event from the basin outlet.

Finally, the nature of the flood-producing event may have a profound effect on hydrograph shape. Localised high-intensity rainstorms commonly produce 'flash' floods which rise extremely rapidly, often within a few hours, and fall again equally quickly. In such events the time base of the entire flood peak may occupy less than half a day. In contrast, the heavy rainfall associated with major frontal systems or stagnating tropical storms may cover many thousands of square kilometres and continue for several days (Table 3.1). The resulting flood peaks may extend over many days or even weeks. In regions marked by extreme seasonality of precipitation or, in cold areas, by seasonal melting, the flood hydrograph may have a time base of several

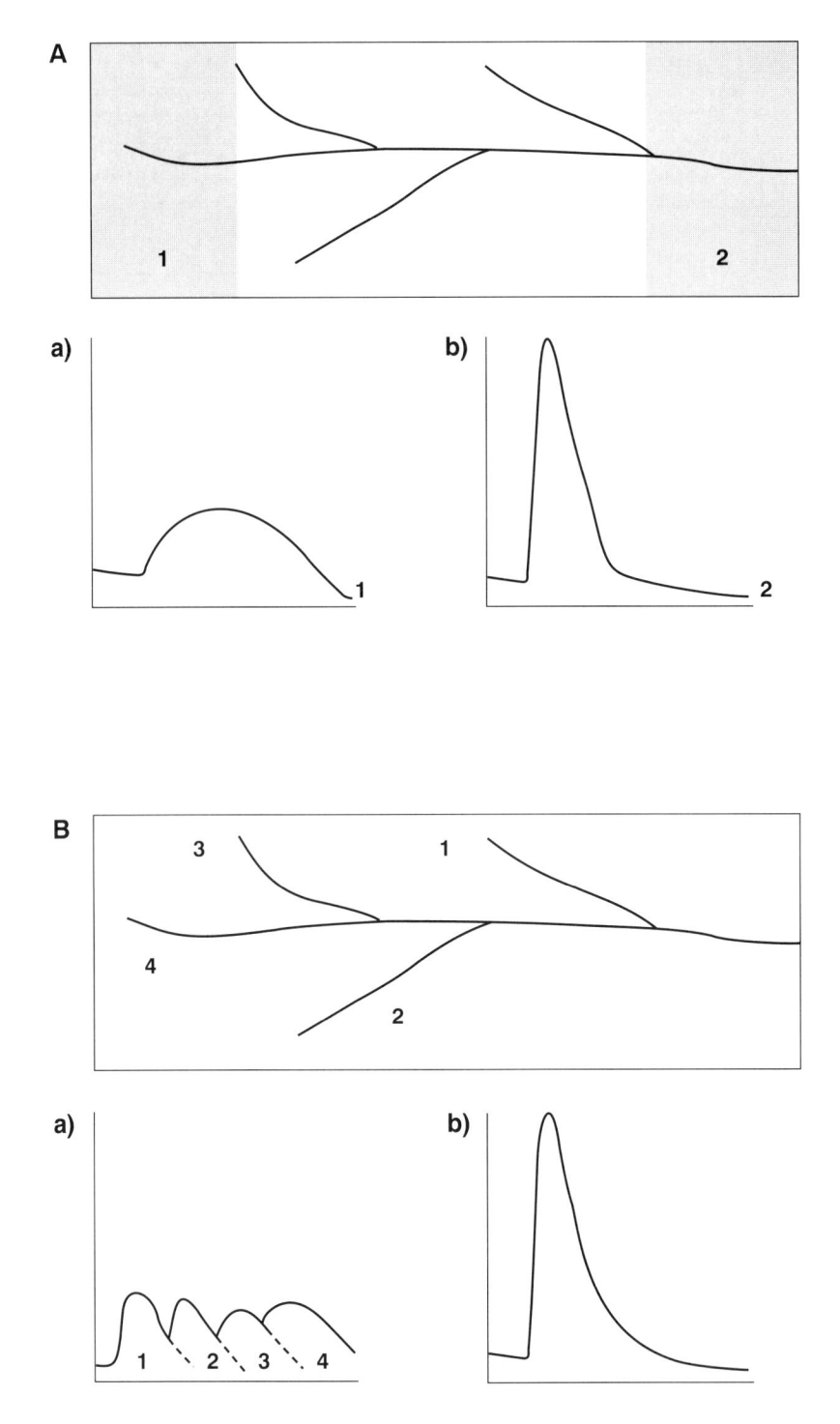

Figure 3.3 Effect of (A) storm location (rainfall area shaded) and (B) direction of storm movement (upstream (a) and downstream (b)) on the shape of the flood hydrograph

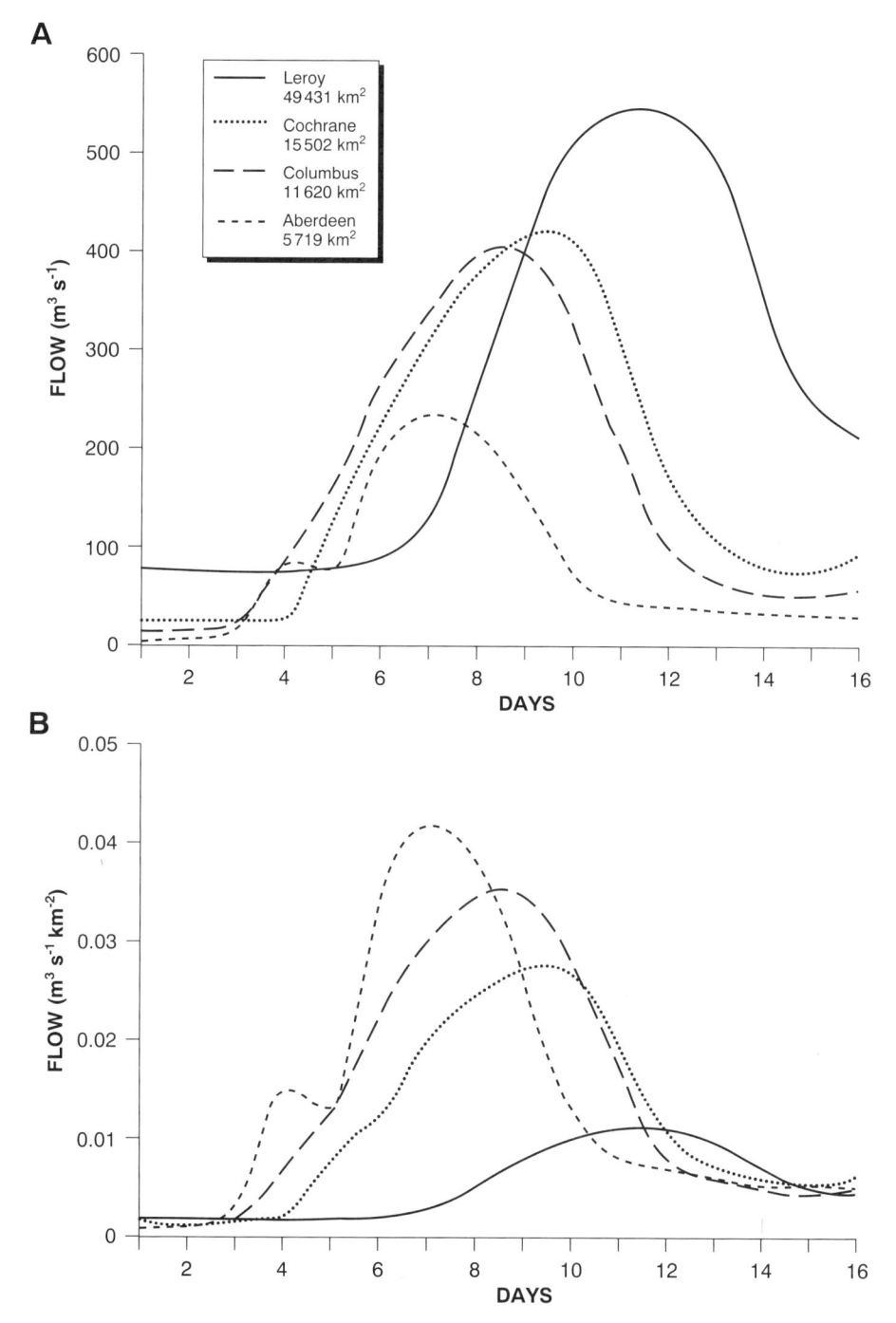

Figure 3.4 Flood hydrographs on the Tombigee River at Aberdeen and Columbus, Mississippi, and Cochrane and Leroy, Alabama, USA, plotted as (A) actual discharge and (B) specific discharge

Source: Based on data in USGS (1944)

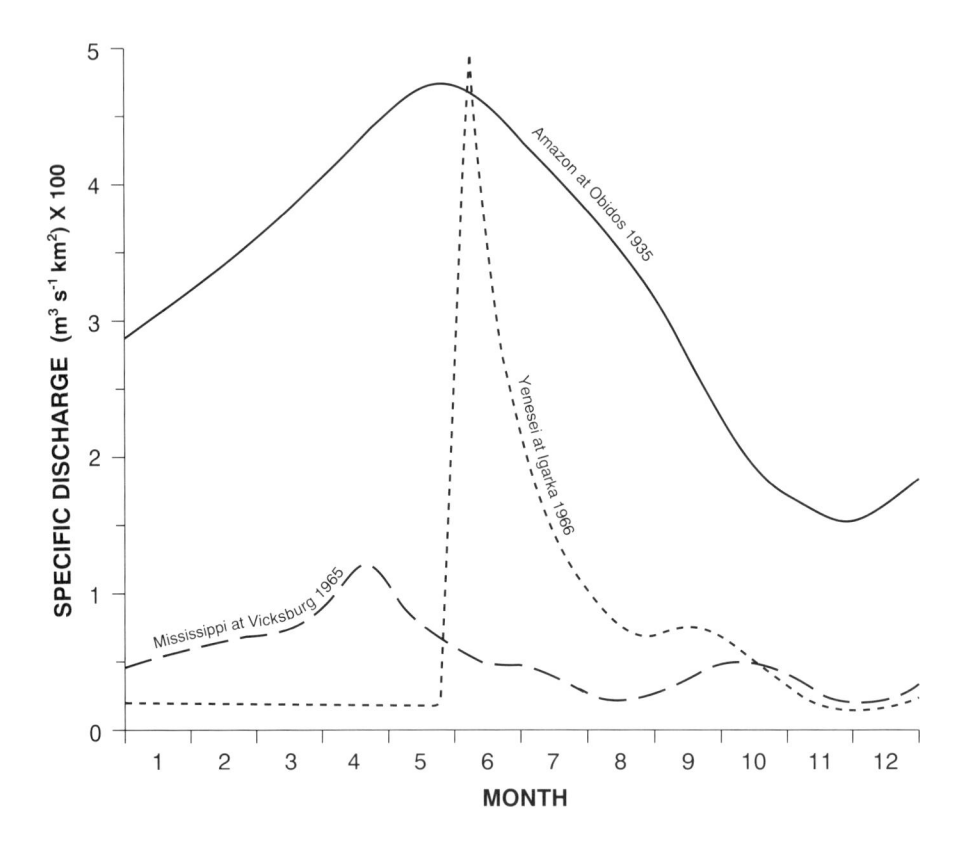

Figure 3.5 The flood regimes of the Amazon at Obidos 1935, the Mississippi at Vicksburg 1965, and the Yenesei at Igarka 1966
Source: Based on data in UNESCO (1969)

months. Figure 3.5 shows that the Amazon in South America, the greatest river in the world in terms of volume and drainage basin area, rises gradually from November to June, when flood levels may reach 15 m above low flow levels, and then falls until spring (Bolt et al., 1975).

Again, hydrograph shape may reflect differences between rainfall and melt events. Bolt et al. (1975), for example, suggested that in the southern Sierra Nevada of California snowmelt floods do not have the sharp peaks characteristic of rainfall floods, but instead are shaped by sustained large volumes of water spanning periods of weeks to several months. Similarly, Kattelman (1990b) reported that, although snowmelt in the Sierra Nevada maintains high flood discharges for several weeks each spring, levels rarely go much beyond the bankfull stage. Instead, it is the rain-on-snow events that generally account for the highest peak flows in most rivers. In the Colorado Front Range, USA, Caine (1995) found that the influence of rainfall on flood flows during a rain-on-snow event was insignificant where the areal extent of the snow cover was greatest and the melt regime continued to dominate the flood hydrograph. Where the areal snow cover was lowest the same rainfall event caused a six-fold increase in streamflow.

RIVER FLOODS: GEOPHYSICAL PROCESSES

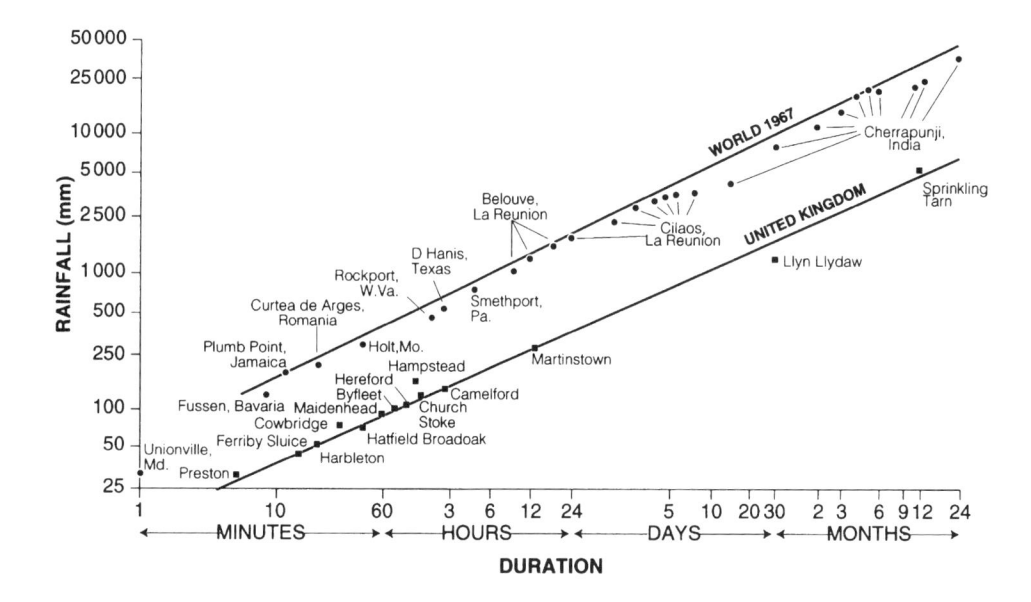

Figure 3.6 Extreme rainfalls showing envelope curves for the world and for the United Kingdom
Source: Based on a diagram in Rodda (1970) and on data from various sources

Rain-on-snow events may continue to be important in generating major flood hydrographs even late in the melt season. Naef and Bezzola (1990), for example, described an extreme flood in the Reuss valley in Switzerland which resulted from a rainfall in August 1987 of about 12 hours' duration and an average intensity of about 10 mm h^{-1}. Embedded in this event was a short shower of less than one hour duration but having an intensity of about 40 mm h^{-1}. Moreover, the higher parts of the catchment were still covered with old snow which had been melting rapidly, so that the saturation deficit of the slope materials was already low when rainfall began. The authors concluded that neither the rainfall, nor the shower, nor the snowmelt individually would have resulted in other than modest flows. It was the combination of the long snow melting period followed by the lasting rainfall of moderate intensity and the short but intense burst of rain, which led to the extremely high discharge.

The prolonged time bases of flood hydrographs on the world's largest rivers are usually the result of several of these factors operating simultaneously. Figure 3.5 shows the flood regime for specified years of the Amazon, Mississippi and Yenesei rivers, draining catchment areas of 4 688 000 km², 2 964 300 km² and 2 440 000 km² respectively.

3.4 FLOOD-PRODUCING RAINFALLS

3.4.1 Empirical Evidence

In very simple terms empirical evidence confirms that large rainfalls, i.e. events which are exceptional for their magnitude and low frequency, usually result in large floods.

For example, the July 1976 floods in Big Thompson Canyon in the Colorado Rockies, USA, resulted from 250 mm of rain falling in five hours, compared with an annual average rainfall of less than 400 mm, and on 19 August 1969 more than 630 mm of rain fell in less than eight hours as Hurricane Camille passed over Virginia, USA, causing flooding with an estimated return period of about 1000 years.

Many tropical areas experience dramatic downpours and the widespread monsoonal flooding in the Indian sub-continent results from almost unbelievable amounts of rain, such as the 12 000 mm which falls annually over the Khasi Hills of Assam. In some years twice that amount falls and from August 1860 to July 1861 some 27 000 mm were recorded from scattered gauges (Milne, 1986). Again, Mount Waialiali, Hawaii, is renowned for an annual rainfall exceeding 10 000 mm and the world record 24-hour rainfall of 1870 mm was achieved on the rugged Indian Ocean island of La Réunion on 15–16 March 1952 (Figure 3.6).

In comparison, British flood-producing rainfalls are rather modest, with many of the severe floods appearing to be associated with localised, short-duration, summer thunderstorms intensified by an orographic component. Up to 20% of the average annual precipitation may occur in one day in parts of south-east England (Perry, 1981). For some of the most severe flood events no rainfall data are available, e.g. the cloudburst which struck Langtoft, East Yorkshire, in July 1892, demolishing two cottages and entirely eroding the soil from some steeper fields, or the 'gullybuster' which drowned 21 people at Louth, Lincolnshire, in 1920 (Milne, 1986). In other cases the rainfall amounts associated with severe flooding are known, for example, the devastating Lynmouth, Devon, floods of August 1952, were caused by about 230 mm of rain falling on the Lyn catchment in less than 24 hours.

With uniform-precipitation regimes, as in Britain, it is duration rather than intensity of rainfall that is often responsible for severe flooding, especially during the winter months, when evaporation and soil moisture deficit (SMD) are at a minimum. The record flooding of 1894 in southern England, for example, resulted from a period of 26 days in October and November during which daily rainfall amounts were unexceptional. However, the number of wet days and the persistence of rainfall meant that up to 30% of the annual mean rainfall was recorded during this period resulting in flooding whose severity has still not been surpassed in this area (Brugge, 1994). In turn, seasonal variation of SMD implies that a given rainfall may produce a flood in one season but not in another. Even in a country as small as the UK, there is a marked spatial variation in the occurrence of peak river flow, with catchments in the west having flood dates early in the winter and those in the east, where early winter SMD values are much higher and the geology is more permeable, having flood dates as late as February or March (Figure 3.7).

3.4.2 Defining a Flood-Producing Rain

It follows, therefore, that the definition of flood-producing rain cannot be expressed simply in terms of the amount or intensity of rainfall but must also take into account the conditions of the catchment upon which the rain falls. However, Schulze (1980) observed that when, as is often the case, only daily rainfall data are available, many of the factors which influence flood magnitude, such as storm intensity, duration and antecedent moisture conditions, are usually unknown. For convenience, Schulze

RIVER FLOODS: GEOPHYSICAL PROCESSES

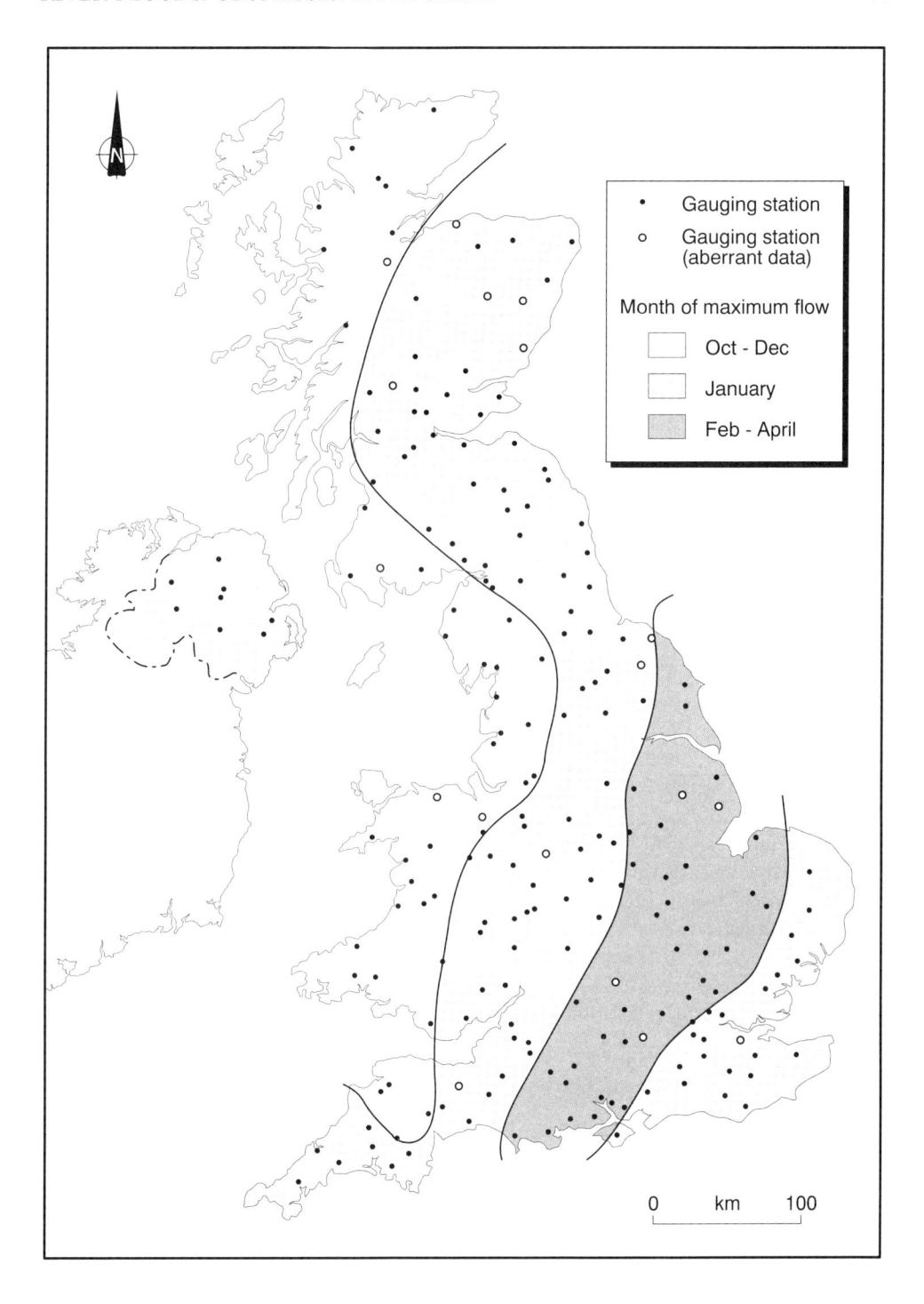

Figure 3.7 Spatial distribution of flood dates in the UK
Source: Based on data in IH (1994a)

Table 3.2 Some tropical flood-producing rainfalls

Duration	Rainfall (mm)	Location	Date	Cause
15 minutes	41	Ibadan, Nigeria	16 June 1972	
90 minutes	254	Colombo, Sri Lanka	[1907]	
24 hours	1168	Baguio, Philippines	14 July 1911	
	1870	Cilaos, Reunion	16 March 1952	
	1248	Pai Shih, Taiwan	11 September 1963	Typhoon
48 hours	1671	Funkiko, Formosa	19 July 1913	
	2789	Bowden Pen, Jamaica	22 January 1960	Frontal
96 hours	2587	Cherrapunji, India	12 June 1876	
7 days	3388	Cherrapunji, India	09 June 1876	Monsoon

Source: Based on data in Newson (1994)

defined flood-producing rains in Southern Africa as events of two or three consecutive days each with a rainfall of 25 mm, thereby echoing the earlier work of Finch (1972) who found that in the UK the threshold for extensive flooding appeared to be two consecutive days each with 30 mm or more of rainfall. The two-day rainfall was also used as a basic 'building block' in the more sophisticated analyses of the UK *Flood Studies Report* (FSR) (NERC, 1975). The two-day depth of rain having a return period of five years (two-day M5) was mapped using values from 6000 stations and was used to derive M5 rainfalls of different durations (e.g. 30 minute M5, 1 day M5) and also rainfalls of different return periods (e.g. M10, M50, M1000).

The shorter rainfall durations are likely to be more significant in flood production from small catchments than from large ones. For example, Kadoya et al. (1993) found that flood peaks in the Yamato River basin in Japan are strongly influenced by 12-hour intense rainfalls, especially when such rainfalls occur in urbanised areas along the main channel. In larger catchments most short-duration rainfall events affect only a limited area of the flood-producing catchment so that the flood response to a rainfall event may be modified by its location within the river basin and by attenuation within the channel system (see Section 3.3). Even in the context of the small river systems of Britain, the rainprints of the heaviest 24-hour rainfalls, i.e. >175 mm, are normally less than 100 km^2 in area (Perry, 1981) and therefore cover only a small part of most catchments.

3.4.3 The Role of Atmospheric Systems

It follows then that 'catastrophic' flooding, defined in the sense of having great magnitude, sudden occurrence, or extreme destructiveness (Hirschboeck, 1987), especially on larger rivers, is less likely to result from the rainfall of a single meteorological event than from the rainfall of substantial atmospheric systems. Obvious examples are the very heavy rainfalls which dominate the flood behaviour of rivers in the humid tropics, e.g. parts of south-east Asia, humid India, Madagascar, the Caribbean and northern Australia (Newson, 1994). These rainfalls are a component of monsoonal or tropical storm systems, although they are often intensified by orographic effects (see Table 3.2).

In addition, however, Hirschboeck (1987) demonstrated that both major regional flooding and also flash flooding in the USA can be tied to less obvious large-scale features of the atmospheric circulation, thereby providing independent verification of important earlier work on relations between flash floods and synoptic and mesoscale atmospheric flow patterns (e.g. Maddox et al., 1979; 1980) (see also Section 4.6.2.3). Specifically, it appears that 'blocking', a large-scale perturbation of the normal zonal (latitudinal) movement of pressure systems, especially in the mid-latitudes, is instrumental in setting up many catastrophic flood events at a wide range of scales. In the northern hemisphere, blocking highs, characterised by high-level meridional air flow, generally persist for one or two weeks and occur most frequently over the north-eastern areas of the Atlantic and Pacific Oceans, with the former having the greatest influence on flood-producing weather systems in Europe and the latter the greatest influence in the USA. As Hirschboeck observed, the role of blocking appears to corroborate earlier claims (e.g. Knox, 1984) that large floods in North America are likely to be associated with periods of meridional circulation. Of most significance perhaps is the apparent influence of large-scale circulation patterns on small-catchment flash floods which have been attributed more usually to the *random* occurrence of local thunderstorms.

Investigations of the atmospheric mechanisms responsible for the onset and prolongation of the disastrous summer floods of 1993 in the USA (Bell and Janowiak, 1995; Mo et al., 1995), showed how the development of a very strong meridional flow from the western Pacific to the eastern USA in late May served as a conduit to guide a series of intense Pacific cyclones directly into the Mid-West throughout June. The convective complexes associated with these storms resulted in major flooding. Then in July a persistent wave pattern became established over the western and central USA, with a lee trough forced by the Rocky Mountains in the presence of the strong westerly air-flow being maintained by transient eddies upstream. The continued excessive rainfall and flooding in the Mid-West were concluded to be the effect of the eddies which maintained a strong upper-level meridional flow, the Rockies which sustained a lee trough, and an associated low-level jet which brought tropical moist air into the central USA (see also Section 4.7.2).

More generally, there is increasing evidence that a disproportionately high percentage of summertime extremes of 1–12 hour convective rainfall in the central United States is produced by *mesoscale convective complexes* (MCCs). These were defined by Maddox (1980; 1983) as large, convective weather systems which often persist for periods exceeding 12 hours and whose size, depending on cloud temperature, exceeds $50\,000\,km^2$ (see also Table 3.1). Since they are likely to produce some of the more important flood-producing rainfalls, the role of MCCs may eventually be shown to be very significant in flood hydrology. Tollerud and Collander (1993) argued that future general circulation models must be capable of incorporating them, their statistical characteristics must be better understood, and it will need to be established whether the effect of climate changes on MCC precipitation extremes may be different from the effect of climate changes on other types of heavy precipitation.

In summary, flood-producing rainfalls are generated by systems of widely differing scale and complexity, many of which are linked via the macro- and mesoscale elements of the atmospheric circulation in ways which are not yet clearly understood. Accordingly, assumptions that flood-producing rainfalls are independent events

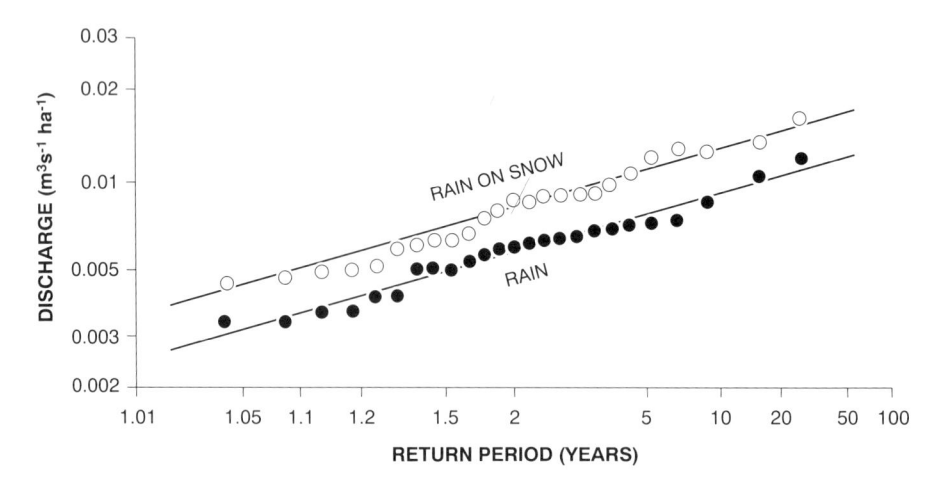

Figure 3.8 Magnitude–frequency plots for flood peaks derived from rain and from rain-on-snow, Oregon, USA
Source: From a diagram in Church (1988). Reprinted by permission of John Wiley & Sons, Inc

coming from the same statistical 'population' may not be appropriate. This is a theme to which we return elsewhere in this book (see especially Chapter 6) and whose importance has been stressed by others. Hirschboeck (1988), for example, suggested that flood frequency analysis should separate populations of flood peaks resulting from convective, frontal and orographic rainfall within the same river basin, and Church (1988) noted the contrasting magnitude–frequency characteristics of floods derived from rain and from rain-on-snow shown in Figure 3.8.

3.5 SNOWMELT AND ICEMELT

Melting plays a significant role in the flood hydrology of many high-altitude and high-latitude areas but is also important in some mid-latitude areas as well. Meltwater is generated by the melting of snowpacks and of glaciers and ice-sheets. Furthermore, related processes in areas of snow and ice may also contribute to flood formation. Important examples are channel obstruction by ice blockages during thaw and the sudden outflows of meltwater which occur when ice barriers move or when moraine or solifluction blockages fail. The nature of floods resulting from these melt and melt-related processes is discussed in Chapter 4 together with floods resulting from rain-on-snow and rain-on-ice events. In this section attention is focused on the process of melting.

3.5.1 Factors Affecting Melt Rate

Snowmelt is the result of several different processes which result in a net transfer of heat to a snowpack which is isothermal at 0 °C. The most important are solar radiation, long-wave radiation (especially at night), sensible heat transfer from the air to the snow by convection and conduction, and latent heat transfer by evaporation and condensation at the snow surface (Colbeck et al., 1979). Comparatively minor contributions are made via the heat gain from rainfall and the heat exchange with the underlying ground. Snowmelt takes place predominantly at the snowpack surface and shows a marked diurnal rhythm.

In forested areas, or in other situations where vegetation or urban structures protrude through the snow surface, the energy exchange is much more complicated than in open bare-snow conditions and the relevant components are often difficult to quantify.

Just as there has been much interest in the notion of flood-producing rainfalls (Section 3.4), so too has attention been paid to high rates of snowmelt. In general terms the upper bounds on daily snowmelt at a given latitude can be estimated from the potential net radiation. Kattelmann (1991a) noted that, although the highest daily snowmelt value reported in the literature (Price et al., 1979) is 60 mm in the Canadian subarctic, average daily maxima over a range of latitudes are normally of the order of 40 mm. Similarly, the highest hourly snowmelt found in the literature is 9 mm (Kuzmin, 1961) although more commonly reported maxima are in the range $1-3$ mm h^{-1}.

Even in the UK, maximum melt rates may be substantial. Estimates by the Meteorological Office (reported in NERC, 1975) indicated a five-year melt rate (i.e. the melt rate having a return period of five years) of 15.1 mm day^{-1}, a 25-year rate of 25 mm day^{-1}, and a 100-year rate of 33 mm day^{-1} for the Trent catchment. Subsequent estimates for the UK (NERC, 1975) showed that a rare melt rate might be as high as 42 mm day^{-1} and might be sustained for two or three days if snow depths exceeded 250 mm. The importance of temperate-zone snowmelt is illustrated in the UK where it contributes significantly in the production of some high-magnitude flood events. The major flooding of the Ouse at York in 1982 was exacerbated by melting snow; on the Trent at Nottingham at least one in three of the 35 highest discharges during the past century had significant snowmelt contributions; and the worst flood in 120 years on the Wansbeck at Morpeth in March 1963 was generated almost entirely by melting snow (Moore et al., 1996).

Snowmelt calculations normally assume that snow is pure ice with a temperature of 0 °C. This is not always the case, so that snow covers which look alike may produce different outputs of meltwater. Snowpack temperatures in winter may be well below 0 °C and an initial input of heat is needed to raise the temperature to melting point. Until that temperature is reached there will be no meltwater output. During the melt season however the snowpack may not only be isothermal at 0 °C but may also contain some liquid water in the interstices of the ice matrix up to a water content equivalent to the retention capacity of the snowpack. In this condition the snowpack is said to be 'ripe'. Liquid water in excess of the retention capacity will drain through the pack as gravity flow.

In mountainous areas there may be an altitudinal zoning of these different snowpack conditions (Figure 3.9). However, on the plains, where catastrophic meltwater floods are more likely to be generated, the distribution of ripe and unripe snowpack will be much less predictable. In either situation, when a snowpack containing liquid water melts, the liquid water is also released and the total outflow from the snowpack exceeds that derived simply from energy-balance calculations.

3.5.2 The Conversion of Meltwater to Floods

Although, over the melt season, the total volume of snowmelt runoff is determined largely by the amount of water stored in the snowpack, the individual meltwater flood

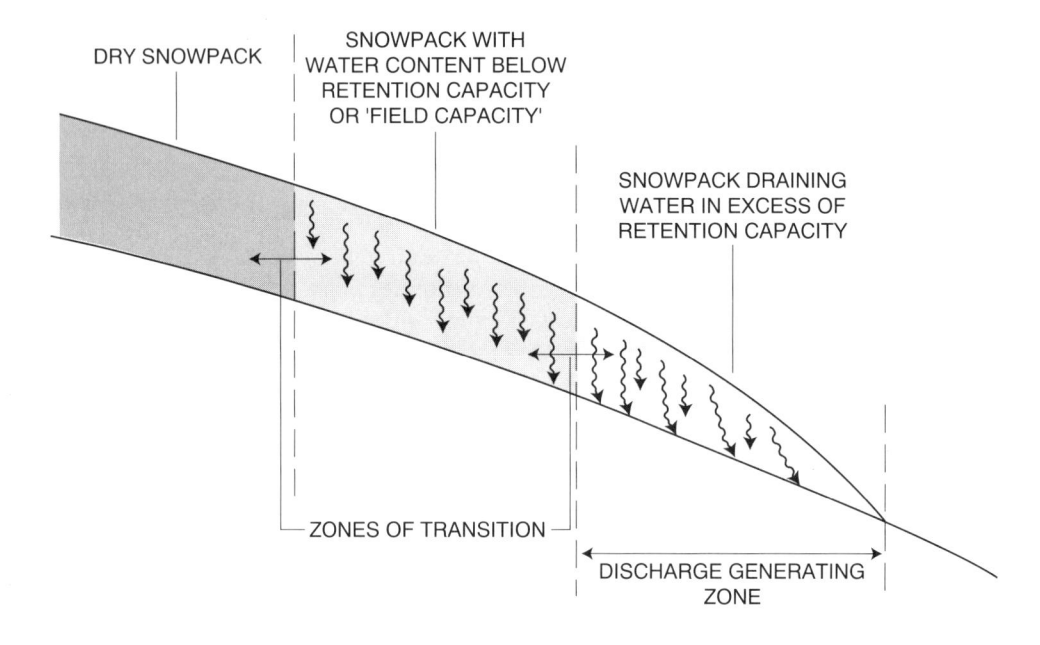

Figure 3.9 Typical distribution of snowpack conditions in a mountainous area
Source: van de Griend (1981). Reproduced by permission of Dr Adriaan van de Griend

peaks during the melt season depend on the interaction of several factors. These include not only the diurnal pattern of melt at the surface (discussed in Section 3.5.1) but also the passage of water through the snowpack, the saturated flow at the base of the snowpack (see also the discussion of quickflow in Section 3.2), and the way in which tributary contributions combine.

The speed with which meltwater generated near the surface can move through a snowpack to the underlying ground surface and then to the channel system depends on the depth, structure and stratification of the snowpack. Most snowpacks develop a layering of less permeable ice strata, within the generally permeable snow matrix, which divert percolating meltwater so that complicated flowpaths develop which delay and diffuse the outflow of meltwater.

Initially, if the retention capacity of the snowpack is not exceeded, no meltwater runoff will occur. Subsequently, the downward movement of the wetting front into a deep stratified snowpack is characterised by delays and ponding at the ice layers and the development of flow fingers which eventually penetrate to the base of the snowpack. These flow fingers, which may remain as zones of higher permeability even after snowpack ripening, therefore offer increasingly effective flowpaths for meltwater through much of the period of snowpack melting (Colbeck, 1979; Marsh and Woo, 1985). Similarly, beneath the snowpack, defined meltwater channels begin to develop. As the melt season progresses, therefore, the speed of movement of meltwater towards the stream channels increases and hydrographs become more peaked (Church, 1988).

Snowpack 'collapse' may occur when an ultraripe snowpack, comprising a blend of ice matrix, snow and the maximum possible amount of liquid water, is subjected to rapid thaw conditions, such as a sudden temperature rise and heavy warm rainfall. This may result in a slush flow or slush avalanche in which a cascade of snow, ice and water moves downslope to the channel system (e.g. Onesti, 1985). Wolf (1952) discussed this component of the Glen Cannich floods in Scotland.

Icemelt influences flood hydrology through the seasonal and diurnal melting of quasi-permanent ice-sheets and valley glaciers and through the melt and break-up of winter-frozen rivers. Icemelt alone is rarely a direct cause of severe flooding. This is mainly because the thermally driven meltwater regime is more subdued than that from melting snowpacks, producing a *periodic* variation of flow with distinctive diurnal and seasonal rhythms. Flooding in glacierised basins tends to reflect an *aperiodic* component of flow. This may consist either of extreme meteorological events, such as periods of very rapid melting over a week or more or the occurrence of intense rainfall, especially late in the afternoon when meltwater runoff is at a maximum, or of sudden releases of water previously stored on, within, or alongside the glacier or ice body. In addition, icemelt flooding may vary in the long term because of climate variability and change.

3.6 TIME TRENDS IN FLOOD PRODUCTION

Since the principal processes of flood production, i.e. heavy rainfall and melting, vary with time in both random and organised ways, it follows that the occurrence of flooding will be similarly variable over time. This is confirmed by all river flood data, although statistically significant time trends in flood occurrence may be easier to detect than to explain. Such trends usually indicate the hydrological effects of either climate variation or geomorphological changes, especially in the river channel, or human modifications to the upstream catchment. However, although it is comparatively easy to demonstrate qualitatively that variation of climate and changes of land use may affect hydrological processes, including runoff processes, it is more difficult to quantify their effect on flood production, despite many claims to the contrary. Furthermore, land-use changes may affect downstream flood flows either through their direct influence on hydrological processes, such as infiltration and evaporation, or indirectly through their alteration of sediment loads and channel morphology. The natural processes of sediment production and sediment movement in river channels both influence and are influenced by the downstream passage of successive flood waves.

It would not be feasible to provide encyclopaedic coverage of human influences on flooding. Even within the restricted scope of agricultural land-use changes in a low-intensity flood environment such as England, it can be shown that virtually every alteration in the pattern of agricultural production (e.g. the spread of sugar beet; the shift to winter cereals), modifies flood occurrence (e.g. Boardman, 1995). Instead, the examples of human modifications discussed in this section have been chosen primarily to illustrate ways in which they affect flood processes. This theme continues through the subsequent discussion of relationships between flood occurrence and channel sediment dynamics which is then followed by a consideration of the effects on flood trends of climate variability and climate change.

3.6.1 Human Modifications

The human modifications illustrated here are largely accidental by-products of a land-use change or activity which has been carried out for some purpose not directly related to its influence on streamflow. However, many human modifications of flood hydrology are more deliberate and some are at least partially 'reversible', so that they may be used for flood mitigation. It is appropriate, therefore, that such human influences on flooding and the flood hazard are discussed further in Section 7.3.

3.6.1.1 Urbanisation

Although urban areas occupy less than 3% of the earth's land surface, the effect of urbanisation on flood hydrology and flood hazard is disproportionately large. In part this reflects the large and rapidly growing urban population. Between 1970 and 1990 the share of the world's population living in urban areas rose steadily from 37% (3686 million) to 43% (5300 million) and is expected to reach 47% by the end of the century, with the greatest growth taking place in developing countries (UNEP, 1991a). In part also the hydrological importance of urbanisation reflects the wide diversity of flood-producing processes upon which it impinges. It has long been accepted, for example, that urban areas make their own climate and there is strong evidence that more rain falls over urban areas than elsewhere (e.g. Perry, 1981). It can also be shown that flood conditions within and downstream of urban areas are modified by changes in channel morphology which are either engineered deliberately to improve channel efficiency and capacity or are brought about by downstream channel adaptation to the modified outputs of water and sediment from urbanised areas. In addition, flood hydrology is influenced directly through the changed timing and volume of quickflow generated from the extensive impermeable surfaces of urban areas, by the import of clean water for domestic and industrial use, and by the generation and export of dirty water through stormwater and sewerage systems.

The extent to which flood characteristics are modified by urbanisation depends very much on the nature of the modified urban surface, on the design of the urban hydrological system, and on the climate.

The main distinguishing characteristic of *urban surfaces* is that they are less permeable than most of the surfaces which they replace. As a result they are effective source areas for quickflow and their flood hydrographs tend to have both higher and earlier peaks (Figure 3.10), reflecting the greater volume of quickflow and its rapid movement across the urban surface. Table 3.3 illustrates, however, that the permeability of urban surfaces varies considerably so that careful urban design and planning can do much to minimise the adverse hydrological effects of urbanisation. These effects also depend on the permeability contrast between an urbanised area and the pre-urban surface, so that flood conditions are exacerbated more by the urbanisation of a high infiltration, sandy area than by the urbanisation of a low infiltration clay area. As an example, much of the city of Perth, Australia, has been built on very pervious sands where infiltration rates after steady rain may remain as high as $10\,\text{mm}\,\text{h}^{-1}$; in contrast the basaltic clays common in the northern and western suburbs of Melbourne, Australia, have infiltration rates as low as $1\,\text{mm}\,\text{h}^{-1}$ when saturated (Aitken, 1975). An additional consideration is that seasonality of flooding may be less marked in urbanised than in non-urbanised catchments, since infiltration is less affected by variations of soil moisture deficit.

The ways in which flood hydrographs are modified by urbanisation reflect not only the production of more quickflow but also its routing through *urban hydrological*

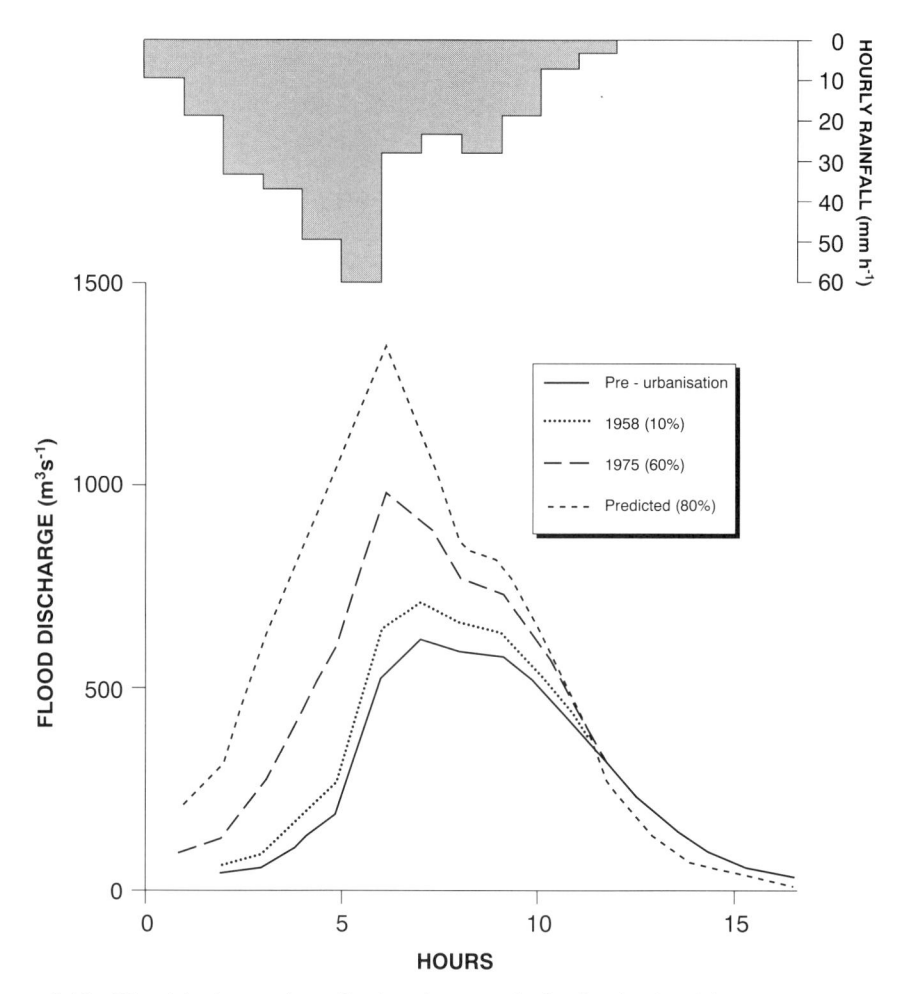

Figure 3.10 Flood hydrographs reflecting the spread of urbanisation (shown as percentage of basin urbanised) in the western part of the Tokyo megalopolis, Japan
Source: Yoshimoto and Suetsugi (1990). Reproduced by permission of IAHS Press

systems. For example, Figure 3.11 illustrates different systems commonly found in the UK and their implications for changes in flood hydrographs. Earlier systems routed stormwater and sewage through the same pipe network, with the combined discharge passing through a sewage treatment plant before entering the river system. For low-magnitude events this arrangement causes a minimum of hydrological change; indeed when the sewage treatment plant is located downstream of the urban area, flood peaks may even be lower than before urbanisation. In modern urban and suburban developments the stormwater system is kept separate from the foulwater system and is either routed directly into the drainage system, in which case the effect upon flood peaks may be severe, or is retained temporarily in ponds and lagoons. Again, if sewage treatment occurs downstream of the urban catchment, the effects of urbanisation may be somewhat ameliorated.

Table 3.3 Impermeability characteristics of typical urban surfaces

Type of surface	Impermeability (%)
Watertight roof surfaces	70–95
Asphalt paving in good order	85–90
Stone, brick and wooden block paving:	
with tightly cemented joints	75–85
with open or uncertain joints	50–70
Inferior block paving with open joints	40–50
Macadam roads and paths	25–60
Gravel roads and paths	15–30
Unpaved surfaces, railway yards, vacant sites	10–30
Parks, gardens, lawns—depending on the surface slope and character of subsoil	5–25

Source: Newson (1992). Reproduced by permission of Routledge Ltd

Since many urban stormwater systems in countries like the UK are old and are able to cope only with low-magnitude events, having return periods of up to about 25 years, stormwater surcharges are common and may lead to widespread flooding and disruption within urban areas. This problem is exacerbated in many cities, in both developed and developing countries, by ageing systems and by poor maintenance, silting and blockage of stormwater channels. Ultimately, however, the influence of urbanisation diminishes with the severity of the flood-producing event in the sense that, after prolonged and heavy rainfall, the infiltration characteristics of urban and saturated non-urban surfaces are very similar (see also Section 3.2).

It follows from the immediately preceding comment that the effect of urbanisation upon the flood hydrograph is also influenced by *climate*, especially rainfall, conditions. In Lagos, Nigeria, the 24-hour rainfall having a return period of 25 years is 175 mm and in Oxford, UK, it is 75 mm and however well designed the urban hydrological system is, it will have much more difficulty coping with the Lagos rainfall. Similarly, a five-year return period stormwater drainage system for a catchment of 3000 ha in Darwin, Australia, would be designed for rainfall intensities ranging from about 55 mm h^{-1} up to 200 mm h^{-1} but in Hobart, Australia, the design intensities would range from 20 mm h^{-1} to 60 mm h^{-1} (Aitken, 1975). A fuller discussion of the effects of rainfall severity on the flood characteristics of urbanised areas was presented by Hall (1984).

The identification of urban influences on flooding may be hampered by climate variability. Walsh et al. (1994), for example, argued that, in the Khartoum area of the Sudan, the effects on flooding of the accelerating urbanisation during the 1970s and 1980s was disguised by a simultaneous decline in rainstorm frequency until the disastrous floods produced by a 200 mm rainstorm in August 1988. A particular form of urban flooding, which is influenced by climate variability, may occasionally occur where major urban areas, such as Salt Lake City and Chicago, have developed in lakeside locations since lake levels reflect changing trends in the local or regional water balance (see also Section 3.6.3.1).

The threat of subsurface flooding has increased in major urban areas which have complex underground transport and service systems when industrial decline has resulted in reduced groundwater abstraction and rising groundwater levels (see Section 1.1.3).

RIVER FLOODS: GEOPHYSICAL PROCESSES

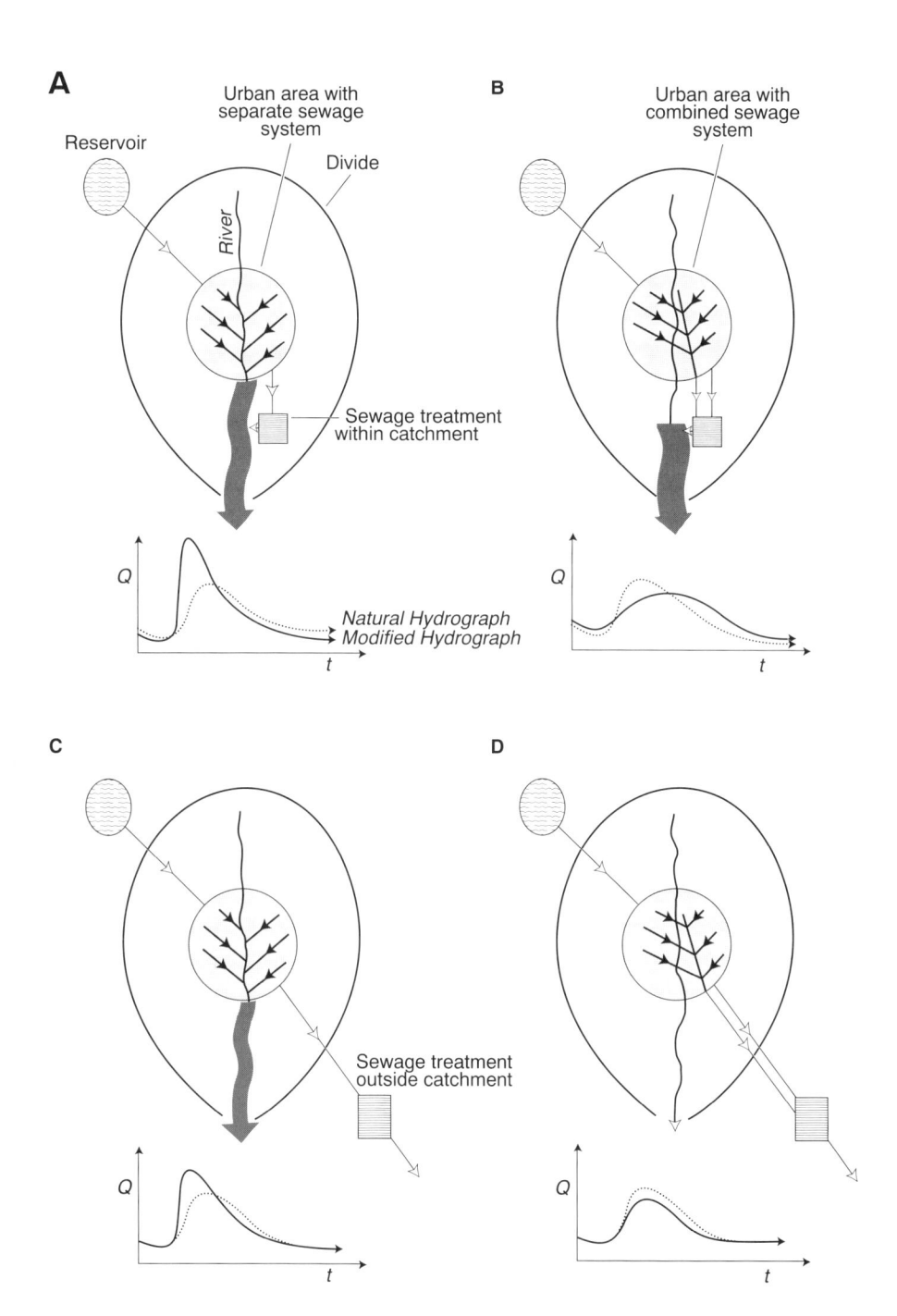

Figure 3.11 Urban hydrological systems in the UK and their impact upon flood hydrographs
Source: From a diagram in Roberts (1989). Copyright John Wiley & Sons Limited

3.6.1.2 Forestry

The role of forestry in modifying river basin flood characteristics has long been a source of vigorous debate (e.g. Anon., 1904; Bates, 1921; Zon, 1927). The arguments continued in Britain (e.g. Law, 1957), as the area of forest increased rapidly between 1945 and the early 1980s covering nearly 2×10^6 ha, or 9% of the land area, by 1984 (Acreman, 1985a). Such controversy will remain in relation, for example, to the proposed creation of ten 'community' forests close to urban and derelict areas in England and of the National Forest which would extend through Leicestershire, Derbyshire and Staffordshire in central England (MAFF, 1994), or the afforestation likely to accompany the growth of set-aside agriculture in the European Union, or the deforestation which has already occurred in the foothills of the Himalayas.

Deforestation may intensify river flooding by: adversely affecting soil structure and volume; reducing infiltration rates, either through the effects of diminished root mass or by facilitating the development of concrete as opposed to granular frost; and reducing water storage, either in the soil profile or within the canopy. However, these influences will normally be significant only during frequent low-magnitude storms. Their effect during increasingly severe flood-producing storms will diminish as prolonged heavy rainfall and/or melting fills available storage and creates widespread conditions of surface saturation and zero infiltration. Accordingly, their effects will also be insignificant where initial storage values are low, e.g. swamplands and steep slopes with shallow soils, which produce rapid quickflow whether forested or not (Hewlett, 1982). Furthermore, as was seen in Section 3.3, the effect on the flood hydrograph of an additional volume of quickflow will diminish with distance down the channel system, so that headwater deforestation will become less significant as the flood peak moves downstream.

Often the initial modification of river flood behaviour is brought about not by afforestation or deforestation *per se*, but by associated, temporary, mechanical procedures, e.g. ploughing and drainage prior to planting, and skid-road construction and more general compaction during felling. Their effects on runoff volumes and suspended solids loads may be dramatic but are normally short-lived. Robinson (1989), for example, observed that much of the land being afforested in northern Europe required draining prior to planting and that the effect of drainage was to increase peak flows for the first five years after planting. However, in Swedish forested catchments investigated by Iritz et al. (1994), peak flows tended to *decrease* after drainage works. And more generally, in the longer term, afforestation of previously non-forested areas appears to cause an average decrease of flow equivalent to 200 mm of precipitation. Although average first-year increases of streamflow after forest removal are about 300 mm, the effect diminishes rapidly with time after cutting (Hibbert, 1967).

Forest clearance for agriculture has frequently been blamed for increased flooding downstream and nowhere more contentiously than in the case of the Himalayan foothills of the Indian sub-continent. The basic hydrological sequence following prolonged forest degradation was postulated by Haigh et al. (1990, p. 423) thus:

> The reduced vegetation cover allowed more water to reach the soil more rapidly and returned less to the atmosphere. Surface runoff began to occur more frequently and, as it did so, rates of erosion began to rise. Erosion rapidly reduced the depth of the soil and thus its capacity to store water...

Flooding became more serious because of the increased volume of water in the environment, because of the increased frequency and volume of surface runoff, and because of the rising levels of affected river beds.

However, examination of several decades of hydrological data led Hofer (1993) to conclude that there is no evidence that flooding on the Gangetic plain has increased. Others (e.g. Hamilton, 1987; Bruijnzeel and Bremer, 1989; Ives and Messerli, 1989; Rogers et al., 1989; Kattelmann, 1991b) have argued that, although clearance of the lowland jungle had been a recent phenomenon, conversion of forests to agriculture and pasture in the Middle Hills of Nepal had taken place over several hundred years. Moreover, much of the 'deforestation' has involved replacing forest with 'hydrologically and erosionally benign' paddy terraces (Hamilton, 1987). Also the main influence of deforestation on flood hydrology results normally from the reduction in available soil moisture storage consequent upon the lower evaporation from the deforested surface. In the Himalayan context, however, the prolonged rains of the Asian monsoon ensure that there is little difference in soil moisture storage between forested and deforested areas since during major flood-producing events soil moisture levels approach saturation everywhere, resulting in massive extension of the source areas for quickflow (Section 3.2). Furthermore, as Kattelmann (1991b) emphasised, even if deforestation resulted in increased floodpeaks in some upstream tributaries, the downstream effect on floodpeaks in the lowlands would be negligible because of the overwhelmingly large volume and variability of runoff from elsewhere in the basin.

Another area in which deforestation has taken place on a massive scale is the Amazon basin, which contains about 50% of the world's tropical rainforest. Various attempts to model the effect of deforestation upon rainfall were reviewed by Mitchell et al. (1990). Total deforestation could reduce rainfall locally by as much as 20% and result in a more seasonal rainfall regime, with corresponding implications for the flood hydrology of the world's largest river (see also Figure 3.5).

As with other aspects of the relationship between flood hydrology and environmental change, there remain frustrating gaps and uncertainties in our understanding. Certainly, if the underlying problem in the Himalayan region is as serious as the evidence on changes in downstream sediment loads and channel morphology suggest, then it may already be too late to effect remedies which will bring about short-term or even medium-term improvements.

3.6.1.3 Agricultural Drainage

The substantial growth in the area of drained agricultural land which occurred through some four decades after 1939, especially in Europe, gave rise to a controversial and often ill-informed debate on the effects of such drainage on flood hydrology. The apparently conflicting evidence was reviewed in a major contribution by Robinson (1990) who showed that earlier interpretations had failed to recognise either the inadequacy of much of the available data or the importance of soil type. Robinson found that, in contrast to earlier views, the drainage of heavy clay soils, which are prone to prolonged surface saturation in their undrained state, generally results in a lowering of large and medium flood peaks, as drainage ameliorates their naturally 'flashy' response by greatly reducing surface saturation. However, on more permeable soils, which are less prone to surface saturation, the effect of drainage is usually to accelerate the speed at which water follows subsurface flowpaths, thereby tending to increase flood flows.

In some upland areas of northern England attempts were made over many years to improve moorland sheep pastures by digging 'grips'. These are quasi-parallel ditches which drain waterlogged peat areas. Unfortunately, their indiscriminate location and alignment has led, in some areas, to rapid runoff and to the increased 'flashiness' of some main rivers such as the Swale in Yorkshire. Flooding appears to have increased in both magnitude and frequency and is often followed by periods of very low water. Moorland 'gripping' has also caused damaging channel and hillside erosion and increased downstream sedimentation in some main rivers.

3.6.2 Floods and Fluvial Geomorphology

The passage of a flood wave through a river system constitutes a geomorphological as well as a hydrological event since floods may erode, transport and deposit large quantities of sediment as they move downstream. This causes changes to the morphology, not only of the river channel but, in the case of larger floods, of the floodplain and contributing hillslopes as well. As a result, the flood-carrying capacity of the channel and floodplain may be significantly modified and this in turn will change the subsequent relationship between river discharge and flood magnitude. This important, dynamic linkage between the water and sediment dynamics of flood events impinges fundamentally upon most aspects of river engineering and is taken up again briefly in Chapter 7.

In the short term, very high-magnitude floods may have catastrophic geomorphological effects on the fluvial environment which play a disproportionately important role in the shaping of fluvial landforms (e.g. Bathurst et al., 1990; Kale et al., 1994; and other papers on extreme floods and their geomorphological effects in Sinniger and Monbaron, 1990). Such events confirm the rarity of steady-state conditions in sediment dynamics, to which Trimble (1995) drew attention. Over very long time-scales, the reconstruction of flood palaeohydrology from sedimentological and related evidence (e.g. Baker, 1989; 1994) provides a means of reconstructing ancient floods and of extending the documentary and observed flood database (see also Chapter 6). These and similar issues are addressed in the relevant geomorphological literature (e.g. Schumm, 1965; Beven and Carling, 1989; Newson and Lewin, 1991). Here, attention is focused on ways in which, over time-scales appropriate to policy planning and decision making, i.e. a few decades to a century or so, the processes and characteristics of subsequent floods may themselves be affected by the geomorphological processes of erosion and aggradation involved in earlier events.

Whether or not channel dimensions in alluvial rivers are associated, as was long believed, with frequent floods having a return period of one or two years, it is evident that floods which are larger and less frequent than this will exceed channel capacity. This will result in overbank flow and localised erosion and deposition in the floodplain/ channel complex. Very high-magnitude events may have major morphological effects as a result of catastrophic degradation or aggradation, triggered when discharge exceeds a

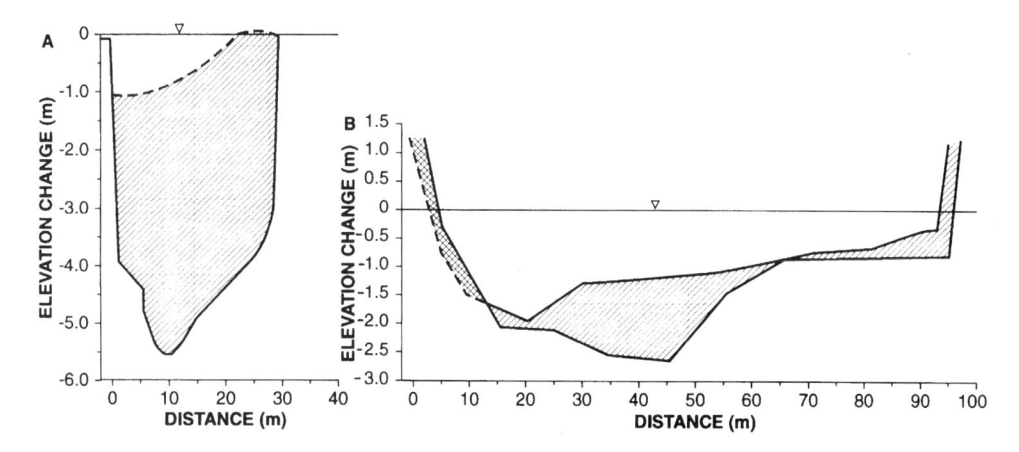

Figure 3.12 Examples of bed aggradation in mountain river channels, northern Italy: (A) Mallero River at the Piazza Vecchia Bridge; (B) Adda River at Fuentes
Source: Based on diagrams in Di Silvio (1994)

particular threshold, and recovery from which may be slow and prolonged. The channel/floodplain systems of some alluvial rivers may thus follow an irregular cycle whereby sudden change, occurring over days or weeks, is followed by a long period of recovery, measured in decades, centuries or even millennia (Nanson, 1986; Lewin, 1989). The flood characteristics of a given river reach expressed, for example, in terms of the extent or depth of inundation for a given discharge of water will reflect this cyclical pattern, changing imperceptibly during periods of recovery but dramatically when a threshold aggradational or depositional event has been triggered. Deliberate flooding of the Grand Canyon of the Colorado River in the spring of 1996 was carried out to counter three decades of sediment starvation caused by the trapping effect of the upstream Glen Canyon dam and resulted in a 53% increase in sand bars (Sapsted, 1996).

Headwater systems of mountain rivers are particularly susceptible to bed aggradation during catastrophic floods (see Figure 3.12) and often maximum water elevation is determined more by bed height than by the depth of flow (Di Silvio, 1994). In the lower reaches of major alluvial rivers, low gradients and high sediment loads can lead to rapid bed aggradation. In Bangladesh, for example, from the border with India to its confluence with the Brahmaputra, the Ganges has aggraded its bed by as much as 5–7 m in recent years (Alexander, 1989) and the Old Brahmaputra, which was navigable for steamers until the 1960s, had become an abandoned channel by the early 1990s (Khalequzzaman, 1994). Sudden aggradation on this scale is likely to modify flooding through the mechanism of channel avulsion whereby the river channel is catastrophically relocated following a major flood. Outstanding examples are provided by the episodic Holocene relocation of Mississippi delta lobes (Lewin, 1989), and by the disastrous historical changes to the course of the Huang Ho in China.

Although accelerated hillslope erosion, including soil-wash, gullying and landslipping, is often caused by heavy rainfalls or severe melt events, this influences the magnitude of contemporary or subsequent flood events in a largely indirect way through the mechanisms of channel and floodplain sediment dynamics previously discussed. However, in areas of recent volcanic activity, which are often characterised by steep slopes and a deep layer of unconsolidated slope materials, including volcanic ash, heavy rainfall may mobilise very large volumes of slope material and so add considerably to the volume and destructiveness of the ensuing flood wave (Pierson, 1989; Tayag and Punongbayan, 1994). Dramatic examples occurred in the Mount St Helens area of the USA in 1980 and in the Philippines during September and October 1995, when tropical storms Nina and Sybil struck the Mt Pinatubo area and triggered disastrous mudflows (known as 'lahars'), several metres in depth. Such mudflows tend to recur annually during the rainy season and, in the case of the Mt Pinatubo area, will continue for at least a decade until the volcanic debris from the 1991 eruption has been washed away. Again, volcanic activity can contribute importantly to flooding in mountain regions. Lahars may also be associated with meltwater, especially along the high volcanic chain of the northern Andes. An outstanding example occurred during an eruption of Cotopaxi in Ecuador in 1877 when the volume of snow and ice which was melted caused enormous lahars, 160 km long, to discharge simultaneously to the Pacific and Atlantic drainage basins (Smith, 1996).

When floodwaters erode and redistribute mine tailings, spoil and other waste heaps, significant and damaging quantities of pollutants may be released. This is a widespread problem in areas of mineral extraction but is especially severe in areas of high rainfall such as Malaysia (e.g. Chan and Parker, 1996). In recent years flooding has also been responsible for the downstream spread of radioactive contamination from the sites of nuclear accidents. Undoubtedly the most dramatic example is provided by the Chernobyl, Ukraine, disaster of April 1986 which deposited 380×10^{12} becquerels of strontium and plutonium on the Pripyat River floodplain. In the succeeding decade floods washed radioactivity downstream and by 1996 fish in Lake Kojanovskoe 250 km away contained levels of radioactivity 60 times higher than the European Union safety limit (Edwards, 1996).

3.6.3 Climate Variations

Variation of climate is undoubtedly another main cause of significant trends in the occurrence of flooding over time, although one that poses problems whose interpretation and explanation are less advanced and less convincing than might be expected from its continuing high-profile coverage in the media and in scientific publications. As in the preceding discussion of floods and geomorphology, this large and enormously complex topic is considered within the narrow but essentially relevant context and time-scale of the policy planner and decision maker concerned with problems of flooding and the flood hazard. This means that the main focus of the discussion has a time-frame of, at most, a hundred or so years although, inevitably, some consideration of climate variation over longer time-scales is unavoidable.

However fiercely the debate rages about climate variability and climate change, there seems to be general agreement on the following points:

- both climate variability and climate change are a reality;
- significant short-term variations of climate occur, some of them associated with El Niño–Southern Oscillation (ENSO) events;

- there is clear evidence that, over prolonged periods of time, substantial areas of the earth have been both significantly warmer and significantly colder than at present;
- such past variations were far greater than anything predicted to result from anthropogenically induced 'global warming';
- the same will probably be true of future climate changes;
- climate can change quite dramatically over comparatively short periods of time, e.g. trends in the 1940s and 1950s indicated global cooling and even as recently as 1970 Danish scientists were predicting the onset of a new mini-'ice age' (Anon., 1970). However, trends over the past century are now interpreted mostly as indicating the onset of global warming.

Variations of climate occur on many different time-scales, ranging from the clustering of a few wet years (or dry years), i.e. persistence, to the broad trends of precipitation or temperature which are detectable over centuries or millennia, i.e. climate change. On whatever time-scale they occur, climate variations challenge the assumption, once commonly held in hydrology, that the past provides a reasonable model for the future. The broad patterns of climate change are often difficult to determine from climatological and hydrological records but are clearly evident in the much longer geological record.

Depositional sequences and the dimensions of palaeochannels are just two of the indicators showing that floodwater volumes have varied enormously through geological time (e.g. Baker, 1989; 1994) and one of the aims of the Global Continental Palaeohydrology (GLOCOPH) Project is to investigate how studies of palaeohydrology can shed light on past and future interrelationships beween climate and hydrology.

In the remainder of this section attention is focused, not on the mechanisms and causes of climate variability and change, but on their hydrological implications and especially their implications for the occurrence of flooding, which remains one of the most important unanswered questions in flood hydrology. Climate variability is considered first, with particular reference to ENSO events. Then follows a discussion of the hydrological implications of climate change, with particular reference to the implications for river flooding of various levels of global warming. Although this inevitably involves some consideration of the influence of sea-level rise on flooding in the lowland reaches of major rivers, the predicted effects of sea-level rise resulting from global warming are more fully addressed in the context of coastal flooding in Chapter 5.

3.6.3.1 Climate Variability and Flood Hydrology

There are numerous examples of medium- to short-term variations in the patterns of rainfall and temperature operating at the scale of a decade or so. 'Persistence' is the term often used to describe the way in which wet years or dry years tend to group together with the result that periods of higher rainfall, and a greater intensity of local or regional flooding, alternate with periods when annual rainfalls are below average. Climate polarisation, or 'partitioning', describes the situation whereby precipitation gradients become steeper, i.e. wet areas get wetter and dry areas get drier. Partitioning may be identified over distances as short as that between Scotland and southern England (IH, 1993b). In western Scotland, for example, the mean annual flow of the River Clyde increased by 60% over the 25 years from 1969 to 1993 in response to a substantial increase in annual rainfall, although evidence for an increase in flooding is less persuasive, and previous experience suggests that this trend may eventually be reversed (Curran, 1994). Decadal-scale fluctuations in flood frequency on the River Tyne in northern England over three centuries were associated by Rumbsby and Macklin (1994) with short-term fluctuations in upper atmosphere circulation patterns, the highest frequencies of major floods occurring during periods of meridional circulation.

Shorter-term shifts, such as those triggered by sea surface temperature and atmospheric pressure variations associated with ENSO events, appear to influence even more dramatically the magnitude and frequency of flooding, both at a regional scale (e.g. Waylen and Caviedes, 1987; Ely et al., 1994) and also at a global scale. El Niño events are extended periods of unusually warm sea surface temperature that occur periodically off the coast of South America and which are now recognised as part of a global pattern of anomalies in both the atmosphere and the ocean (Cane, 1983; Barnett et al., 1988a; Bergeron, 1996). Such events, lasting one or two years and very rarely, as with the El Niño which ended in June 1995, up to five years, occur as part of a natural oscillation of winds and ocean currents that repeats itself on average about once every four to seven years, although the interval has been as short as two years and as long as 10 years. These events are accompanied not only by a warming of oceans around the world, but also by associated changes in weather patterns resulting in torrential rain over deserts in South America, drought in Australia and Indonesia, and possibly also the failure of the monsoon in India and East Africa (Pearce, 1988) and fewer hurricanes and tropical storms in the Pacific and Atlantic Oceans. Indeed, perceptions of global warming may partly reflect the succession of record warm years brought about by the El Niños of the 1980s. In contrast, the trough in the El Niño cycle, known as anti-El Niño or La Niña, causes normal winds and currents across the Pacific to intensify and ocean temperatures around the world to fall. La Niña is typically accompanied by above-average rainfall in India, Australia and Africa and in 1988, for example, was blamed for catastrophic flooding in Sudan and Bangladesh (Pearce, 1988).

That changes of sea surface temperature play an important role in modifying climatological and hydrological processes at a global scale is not in doubt. The storage of energy in the oceans, and the way it is transferred into the overlying air, exerts a major control on the atmospheric system. Less clearly understood, however, are the reasons for such temperature changes or the way in which the processes relating to, say, the events of the ENSO cycle are linked to other, longer-term, variations in global sea surface temperature (SST) anomalies which are generally thought to result from variations of ocean circulation and vertical mixing processes over time-scales of decades or longer. Gray (1990), for example, demonstrated that SST anomalies appeared to trigger both flood-producing rainfalls in the Sahel region of West Africa and intense hurricane activity in the eastern USA and argued that hurricane activity could be forecast from observation of West African rainfall patterns.

A better understanding of the linkages between spatially widespread atmospheric phenomena and their effect on flood hydrology is clearly fundamental if our understanding of floods is to progress significantly. Barnett et al. (1988b), for example, showed how variations in springtime snow depths over the Eurasian continent are linked to the subsequent strength of the Asian summer monsoon and to

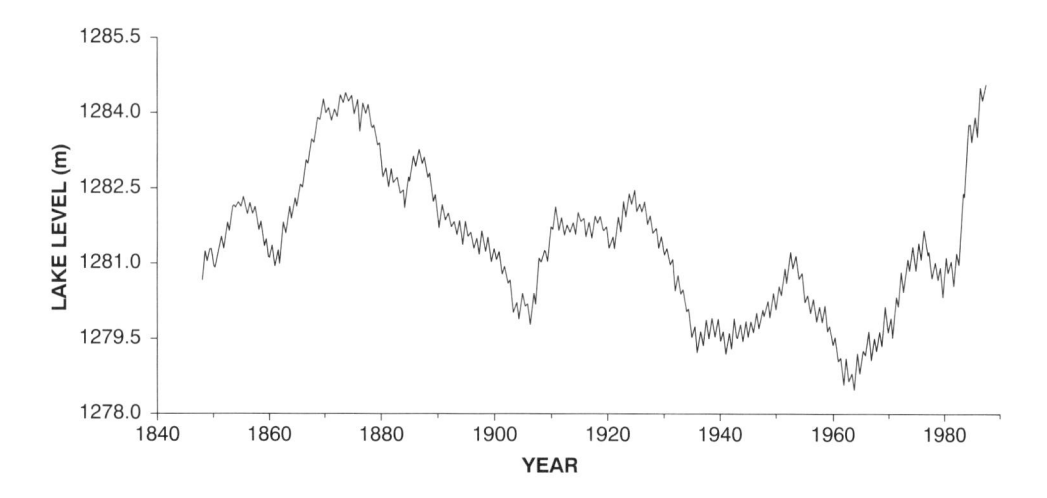

Figure 3.13 Variations of lake level, Great Salt Lake, Utah, USA
Source: Based on a diagram by Morrisette (1988)

ENSO phenomena. Further research is expected not only to clarify our understanding of the role of ocean temperatures in modifying flood hydrology at a global and regional scale but also to define more exactly the relative importance of SST changes and land surface temperature changes. Not until the complex interrelationships between hydrological, climatological and ocean circulation processes are better understood will it be possible to explain more satisfactorily than at present the occurrence, magnitude and distribution of both river and coastal flooding.

Freshwater lakes and inland seas change in volume, and therefore also in depth and areal extent, in response to shifts in the water balance of their catchment areas, growing and deepening during wet periods and shrinking during dry periods (Figure 3.13). Climate variability therefore impacts very directly upon lake levels and upon the flood risk to lakeside areas. This is particularly important where major urban development has taken place during periods of relatively stable lake levels. For example, increased flooding in Salt Lake City, Utah, during the 1980s, resulted as the level of the Great Salt Lake rose in response to the higher snowfall which had occurred on the surrounding mountains after 1960. By the beginning of 1988 the level of the lake had risen by more than 6 m from the record low level reached in 1963, prompting a massive pumping scheme to evacuate surplus water into a desert depression west of the lake (Morrisette, 1988). As Figure 3.13 indicates, similar, though smaller, secular variations have occurred during the period of record. Geomorphological evidence shows that the still-water level of the Great Salt Lake can be expected to rise to near, or slightly above, the historic maximum about once every 100 years (Atwood, 1994). Similarly, in northern Tanzania, a long dry period prior to the mid-1960s appears to have caused the shrinkage of Lake Babati to about one-quarter of its normal size (Stromquist, 1992). Since then flooding has increased, culminating in 1990 when a surge from the lake destroyed buildings in the centre of Babati town.

On lakes where the wave fetch is very long, the effect of increased water levels on flooding may be intensified by shoreline erosion and wave overtopping. Even on the Great Salt Lake wind action extends flooding in some locations to elevations 2 m above the still-water elevation of the lake (Atwood, 1994). And in Chicago and elsewhere at the southern end of Lake Michigan in the mid- to late-1980s, four decades of cooler and cloudier weather and 15 years of above-average precipitation led to a 0.7 m increase in lake level between 1980 and 1986 (Changnon, 1987). Such high water levels decline only slowly even when drier conditions return because the outflow of Lake Michigan, and the other Great Lakes, is small in comparison to lake volume.

3.6.3.2 Climate Change and Flood Hydrology

Although some of the causes of climate change, such as continental drift, major mountain-building episodes, and oscillations of the earth's orbit around the sun, are of little direct relevance to flooding at the planning and policy-making time-scale adopted here, others certainly are. Large meteorite impacts and major volcanic eruptions, for example, have long been known to affect patterns of precipitation and evaporation for several years or even decades after the initial event. In addition, sunspot activity was linked to the occurrence of flooding in a study of flood data for 141 rivers in Asia, Europe, Africa, Oceania, North America and South America by Wang (1993). Floods exceeding $10\,000\,m^3\,s^{-1}$ appeared to occur mostly in the decreasing phase of the sunspot cycle, with very few occurring in the increasing phase of the cycle. The implications for flood prediction and flood hazard amelioration are considerable and verification from further studies is awaited with interest.

Contemporary discussions of the relations between climate change and flooding are:

- largely concerned with the middle and higher latitudes;
- conducted in the context of the continuing emergence of the middle and higher latitudes from the atypically cold climate of the Pleistocene;
- strongly influenced by concerns about the possibility of global warming resulting mainly from anthropogenic causes;
- increasingly reflective of the spatial complexity of climate change, even over an area as small as the UK (e.g. Petts, 1988). This makes it difficult to interpret the evidence, although climate change appears to have been more important than land-use change in explaining recent variations in flood hydrology in the UK (Lawler, 1987);
- increasingly willing to acknowledge that the climate change 'signal' is weak in comparison with the 'noise' of the flood-producing process. This problem is exacerbated by the short period for which flow data are available but will be partly alleviated by attempts to pool flow data for large areas, e.g. the FRIEND Water Archive which covers a large area of Northern and Western Europe (e.g. Arnell, 1989; 1994).

Although the hydroclimatic effects of some of the factors referred to above are reasonably clear, e.g. increased cloudiness and rainfall and reduced evaporation for long periods following major meteorite strikes or volcanic eruptions, there is great uncertainty about whether the events themselves will occur during, say, the next 100 years. For example, the major eruptions of Laki, Iceland, in 1783, Krakatau, Indonesia, in 1883 and Agung, Bali, in 1963 were quite evenly spaced in time but recently a similar number of major eruptions occurred in little more than a decade, e.g. Mount St Helens, USA, in 1980, El Chichón, Mexico, in 1982 and Pinatubo, Philippines in 1991. In other cases the event may be more predictable, e.g. the sunspot cycle, but its effect on flooding is still not fully demonstrated. In the worst case, the event *and* its effects on flooding are uncertain.

Unfortunately, global warming falls into this category to the extent that its development with time, its spatial distribution, and its effect on flooding are all unclear, especially in the mid-latitudes, where the interaction of tropical and polar airmasses means that climate is very sensitive to shifts in wind systems. The important international work of the Intergovernmental Panel on Climate Change (IPCC) (Houghton et al., 1990; 1992) predicted future warming at rates of about 0.3 °C per decade (range 0.2 to 0.5 °C per decade) over the next century under the business-as-usual scenario. Subsequent work at the UK Meteorological Office on the complexity of feedback processes between different sorts of pollutants, suggested that sulphate aerosols, common in industrial areas, may offset, by as much as 50%, the warming effects of greenhouse gases such as carbon dioxide, thereby reducing the rate of warming to 0.15 °C per decade. Some American scientists believe that the offset could be 100%, meaning no rise at all in global average temperature. Revisions of the primary database, including satellite microwave radiometer data, indicate either no significant warming or a slow upward drift of temperature at only about a quarter of the figure predicted by global climate models (Burroughs, 1995). Other recent results, particularly those from the UK Natural Environment Research Council's Terrestrial Initiative in Global Environmental Research (TIGER), have emphasised the spatial heterogeneity of warming, so that some areas, including Britain, may become cooler than originally predicted while other areas, including the Caribbean coast of South America and large parts of Australia and South-East Asia, become much warmer.

The effect of global warming on river flooding will depend on the nature of precipitation changes, in terms of both precipitation amounts and intensities, and also whether precipitation falls largely as rain or snow. Individual flood events are likely to be more important than average conditions. This was illustrated by Knox (1993) who showed that major changes of flood magnitude and frequency on the Mississippi River over a period of about 7000 years appear to have been associated with changes of only 1–2 °C in mean annual temperature. The sensitivity of river basins to precipitation change depends crucially on the proportion of rainfall which becomes runoff; even in a small country like Britain this varies widely from more than 75% in the wetter areas to less than 25% in the drier areas (Ward, 1981).

Projected hydrological scenarios for future climates vary widely although on physical and empirical grounds it would be expected that global warming would result in increased precipitation and in an increased frequency of extreme events (Fowler and Hennessy, 1995). However, there is less agreement about the spatial distribution of that precipitation, or about the relative importance of rain and snow. Major ice ages through geological history appear to have been preceded by atmospheric warming and increased precipitation, which in high-latitude areas fell as snow, leading to the increased storage of water as ice and to a consequent *decline* of mean global sea level.

The main effects of global warming on river flooding are likely to result from changes in the variability and frequency of short-period, extreme events, especially mid-latitude and tropical storms. The number of tropical storms, cyclones and hurricanes is likely to increase, though with little change in their average structure or intensity (Houghton et al., 1992). Summer convective storms in the mid-latitudes are likely to increase in frequency, partly as a result of the development of severe storms from the remains of tropical hurricanes and cyclones, and partly as a result of the poleward spread of warmer water in the northern-hemisphere oceans which would lead to greater evaporation, a poleward shift of the westerlies and an initial steepening of latitudinal temperature gradients as warmer water and snow- and ice-covered surfaces move into closer proximity.

The generally increased convective activity in the tropics and mid-latitudes implies more intense local rain at the expense of gentler but more persistent rainfall events (Houghton et al., 1992). This could mean fewer rainfall events in total and this is indeed projected, especially for mid-latitudes, in some General Circulation Model (GCM) analyses (Fowler and Hennessy, 1995). Partial confirmation of a trend towards greater precipitation was provided by the analyses of Karl et al. (1995) which showed that the proportion of total warm-season precipitation derived from extreme precipitation events has increased substantially during the twentieth century in the USA and in north-eastern Australia. Increased rainfall intensity could result in increased flooding, even in dry summer conditions on hydrophobic soils.

Reviewing the likely situation in Britain, Newson and Lewin (1991) argued that if global warming became an analogue of the emergence, after the sixteenth century, from the 'Little Ice Age', the location and efficiency of cyclonicity could well produce an increase in the westerlies and therefore sustained and intense flood activity, particularly in the north and west, with a correspondingly drier rainshadow area in the east. There is, in fact, evidence that increased westerly airstreams across Scotland since 1970 have already led to significantly increased river flows (Smith and Bennett, 1994; Smith, 1995). Arnell (1992) suggested that higher temperatures might mean that less precipitation would fall as snow. If true, this could increase rainfall event floods in the winter and reduce the severity of snowmelt flooding in the spring. However, even with a trend towards milder winters, the occasional dominance of prolonged continental conditions, accompanied by substantial snow accumulation, could cause very high melt rates once 'normal' mild, wet conditions returned (Newson and Lewin, 1991). Such increased variability is likely to be most pronounced in climatically 'marginal' areas of the UK, such as Scotland, and may lead to a greater frequency of events like the floods of 1993 on the River Tay, whose flow of almost $2000\,m^3\,s^{-1}$ at Ballathie surpassed the previous highest flow in the UK national river flow archive (IH, 1993b). Generally, in the UK, the risk of river flooding is likely to increase partly because of increased winter rainfall (CCIRG, 1996) and partly because flash floods in summer will result from an increased number of intense summer rainfalls. IH (1994b) noted, however, that specific increases are very uncertain and are likely to vary greatly between catchments.

Global warming is also likely to have a significant, though complex, influence on flood hydrology in areas of snow and ice. Martinec and Rango (1989) modelled global warming influences in three mountainous areas in the Rocky Mountains of Canada and the USA and in the Swiss Alps and suggested that the main modification of flooding would result, not from changes in precipitation, which would be variable, but from earlier melting in the spring. Earlier melting would occur when runoff losses were smaller and would result in an increased seasonal runoff volume of the order of 10%.

Accelerated melting of glaciers resulting from an increase in air temperature of 2 °C was modelled by Fukushima et al. (1991) for a catchment in the central Nepal Himalaya. Summer flood peaks were expected to more than double, given no change in glacier-covered area, and to increase by about 30% assuming a concomitant reduction in the glacier-covered area.

The effects on flood hydrology of past and predicted climate change are even more difficult to determine for low-latitude areas. It appears that while the northern mid-latitudes were experiencing glacial periods, intertropical areas experienced periods of greater aridity, and that a pluvial period, when monsoonal rains were restored, occurred at the end of the last glaciation. During the past century temperatures in the tropics seem to have undergone little change although rainfall decreased over much of the tropics, excluding monsoon Asia, by about 30% (Barry and Chorley, 1982). The predicted effect of global warming is very uncertain and for some tropical areas is further confused by ambiguity about the hydrological influences of large-scale deforestation.

CHAPTER 4

River Floods: Spatial Characteristics

CONTENTS

4.1	Introduction	96
4.2	Spatial characteristics of river flooding	97
4.3	Regionalisation	97
4.4	Flood climate zones	103
4.5	Floods in cold climates	104
	4.5.1 Floods resulting from hydrometeorological factors	105
	4.5.2 Floods resulting from channel obstruction by ice and snow	108
4.6	Floods in hot climates	111
	4.6.1 Humid environments	112
	4.6.1.1 The 1987 and 1988 Bangladesh floods	116
	4.6.2 Arid and semi-arid environments	119
	4.6.2.1 Hydrological characteristics	121
	4.6.2.2 Implications for flooding	122
	4.6.2.3 Flash floods	125
4.7	Floods in temperate climates	131
	4.7.1 Susquehanna River floods, 1972	132
	4.7.2 Mississippi River basin: The great flood of 1993	133
	4.7.2.1 The climatological context	134
	4.7.2.2 The role of human interference	136
	4.7.2.3 Description of the floods	137
	4.7.3 River Rhine: The Christmas floods of 1993/94	139
	4.7.3.1 The climatological context	140
	4.7.3.2 The role of human interference	140
	4.7.3.3 Description of the floods	142

4.1 INTRODUCTION

This chapter examines the extent to which the operation of the flood-producing processes discussed in Chapter 3 may result in explicable spatial patterns of river flooding. Clearly, if such patterns exist they may be helpful both in understanding better the nature of river flooding and also in forecasting and predicting flood characteristics at a specific location.

    The topic is approached by considering first the internal consistency of flow characteristics within river basins, and particularly the relations between flow in headwater catchments and that at downstream locations. Consideration is then given to the extent to which neighbouring river basins may be grouped to define hydrologically homogeneous areas within which flood behaviour is recognisably similar. Next, the extent to which flood characteristics may reflect the broad zones of global climate is examined. Then, finally, the nature and form of river floods are illustrated by reference to selected examples drawn from cold, hot and temperate climates.

4.2 SPATIAL CHARACTERISTICS OF RIVER FLOODING

Floods at any location in a river system are a function of the floods generated in the catchment upstream of that location. As was shown in Section 3.3, however, the relationship between flood behaviour in headwater catchments and the flood behaviour of the entire river basin is often complex and sometimes imperfectly understood. The downstream flood hydrograph differs from the upstream flood hydrograph, partly because of lag and routing effects, partly because of the changing nature of the basin geology, physiography and climate from headwaters to outlet, and partly because of scale effects.

Scale effects are important in relation to both catchment conditions and precipitation inputs and frequently restrict our ability to generalise from existing flood data and to predict flood occurrence and distribution. The fact that flood peak discharges tend to increase downstream when measured absolutely (i.e. $m^3 s^{-1}$) but decrease downstream when expressed as specific discharge (i.e. $m^3 s^{-1} km^{-2}$) in part reflects a mismatch between the scales of catchment and precipitation event. In other words, large catchments normally have a lower specific discharge than small catchments partly because the flood-producing precipitation event is smaller than the total catchment area. In small catchments the precipitation event is often at least as large as the area of the catchment, thereby encouraging high specific flood discharge. It follows, therefore, that a large basin subjected to a macroscale precipitation event, such as a major tropical storm (see Table 3.1), should generate a higher specific discharge than when it is subjected to a meso- or microscale precipitation event.

The extent to which, despite the multiplicity of influencing factors, it is possible to discern some consistency in flood behaviour within river basins, e.g. from headwater catchments to the entire basin, is clearly fundamental to attempts to generalise about flood characteristics over even larger areas. This issue attained prominence in relation to the 'representativeness' of the small experimental catchments used in the 1960s to investigate hydrological processes. Ward (1971) reviewed a large body of indirect evidence and concluded that the general level of representativeness appeared to be low. Although, almost a decade later, Baron et al. (1980) argued that there was still relatively little information available that related *directly* to the hydrological relations of small and large catchments, their own analysis of Australian flood data made a valuable contribution. They investigated the effects of catchment area on flood response by using mainstream length and stream slope as surrogates for area and were able to demonstrate consistent relationships for 52 catchments in Queensland, New South Wales and Tasmania which ranged in area from 0.05 to 15 000 km^2.

4.3 REGIONALISATION

From the discussion in Section 4.2 and in much of Chapter 3, one might expect the flood-producing potential of each river basin and sub-catchment to be distinctively different. However, there is some evidence of a spatial dimension to river flooding on a scale larger than that of a river basin. In Britain, for example, the best estimate of mean

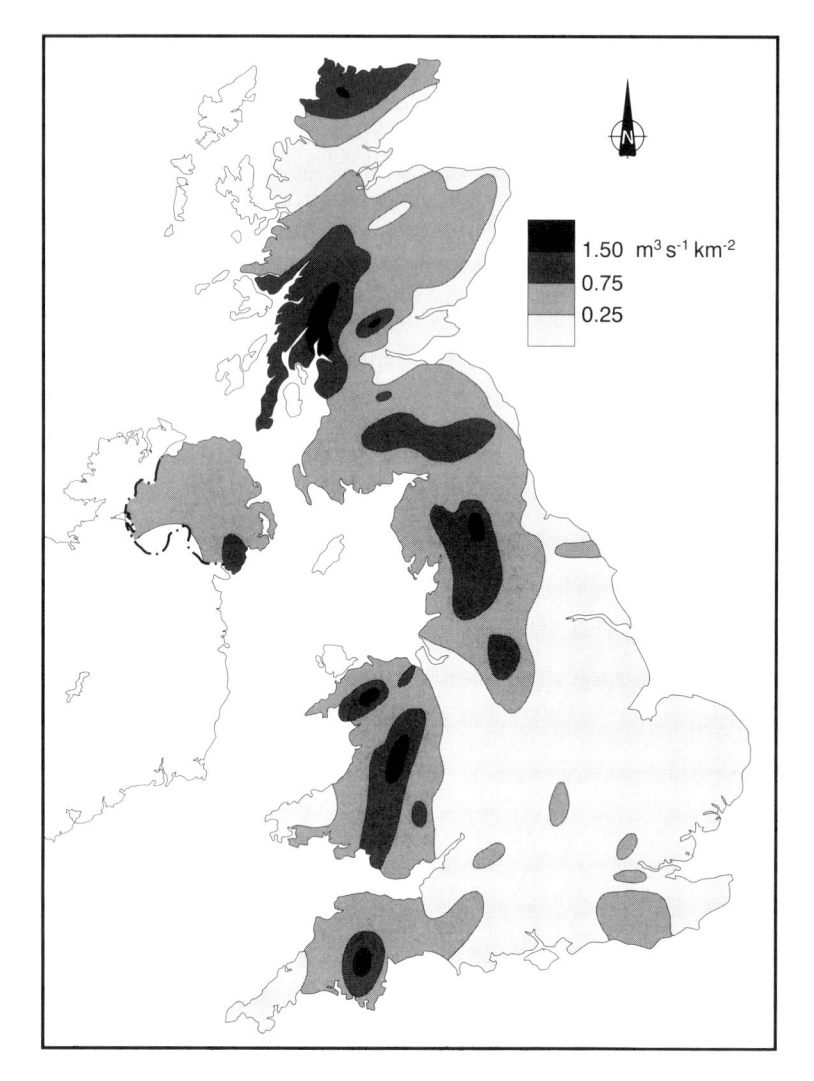

Figure 4.1 Spatial distribution of BESMAF (best estimate of mean annual flood) values in the UK
Source: NERC (1975). Reproduced by permission of Institute of Hydrology

annual flood (*BESMAF*) shows a marked gradient from values of specific discharge exceeding $1.50 \, m^3 \, s^{-1} \, km^{-2}$ in the north and west to values well below $0.25 \, m^3 \, s^{-1} \, km^{-2}$ in the south and east (Figure 4.1). And although, as a global average, quickflow accounts for almost two-thirds of river flow, with baseflow contributing the remainder, significant variations at the continental scale mean that quickflow makes a more dominant contribution to total river flow and flood potential in Asia and Australasia (70% and 75% respectively) than in Africa and Europe (55% and 57% respectively).

Such evidence suggests that there might be broad regional differences in either flood process or flood type and that it may be possible to generalise about the effects of catchment characteristics on floods within defined regions, i.e. to 'regionalise' flood experience in a way similar to that used to define and map climatic zones or mean annual runoff. With reference to Figure 4.1, for example, the basis for using geographical proximity as a surrogate for hydrological similarity might be that the north and west are characterised by higher rainfall, impermeable rocks and steep slopes and the south and east by lower rainfall, shallow slopes and permeable rocks. Successful regionalisation could lead to major improvements in both the understanding of flood processes and also in flood estimation for ungauged catchments.

Although many attempts at flood regionalisation have adopted geographically defined regions enclosed by political, administrative or physiographic boundaries, such regions are likely to be hydrologically heterogeneous (Hendriks, 1990). Thus flood experience has often been compared by country on the basis of national datasets, each of which may have been developed using different instrumentation and methods of analysis. In an intra-national example, the UK *Flood Studies Report* (NERC, 1975) identified the 11 geographical regions shown in Figure 4.2 although subsequently, using a test statistic based on the expected sampling variability of the annual maximum series, Wiltshire (1986) found that only two of the regions were really homogeneous. In view of the complexity of climatic, physiographic and hydrological factors controlling river floods, this outcome may be regarded as less surprising than the original expectation that such a geographical unit would be hydrologically homogeneous.

A fundamental problem is that although geographical location may be an important determinant of some hydroclimatological variables, such as mean annual rainfall, evaporation, or flow regime (e.g. Krasovskaia et al., 1994), or even mean annual flood, all of which vary comparatively smoothly in space (e.g. Figure 4.1), it is a much less important determinant of short-term or discrete hydrological outcomes such as individual flood peaks. As a result, alternative methods have been developed for defining hydrologically homogeneous regions for flood studies. These may be based on either the partial or full disaggregation of the geographical entity represented by the drainage basin or by a group of adjacent drainage basins.

One possible approach is to use areas of similar hydrological response, rather than individual sub-catchments, as the basis for flood estimation (e.g. Naden, 1992; MAFF, 1994). A histogram of the number of channels at incremental distances from the basin outlet defines a network width function, which is used to route water through the channel network. This network response component is then combined with the hillslope, or small catchment, response component which describes the delivery of water to the channel network to define hydrologically homogeneous sub-areas. Since these do not have to be contiguous, the method represents a *partial disaggregation* of the geographical unit. For example, Figure 4.3 shows a subdivision of the basin of the River Thames above Teddington (London), UK, which is based on geology and related soil type. The channel network is superimposed on the geology so that network width functions can be separately identified for the limestone of the Cotswolds, the Swindon and Oxford Clay vale, the Chalk, and the London Clay. Although geology dominates this particular subdivision, with each geological area showing marked differences in hillslope response, other factors such as rainfall input could easily be accommodated within the method (MAFF, 1994).

When attempts to define hydrological homogeneity involve the *full disaggregation* of the river basin, they are normally based on either relevant catchment characteristics or

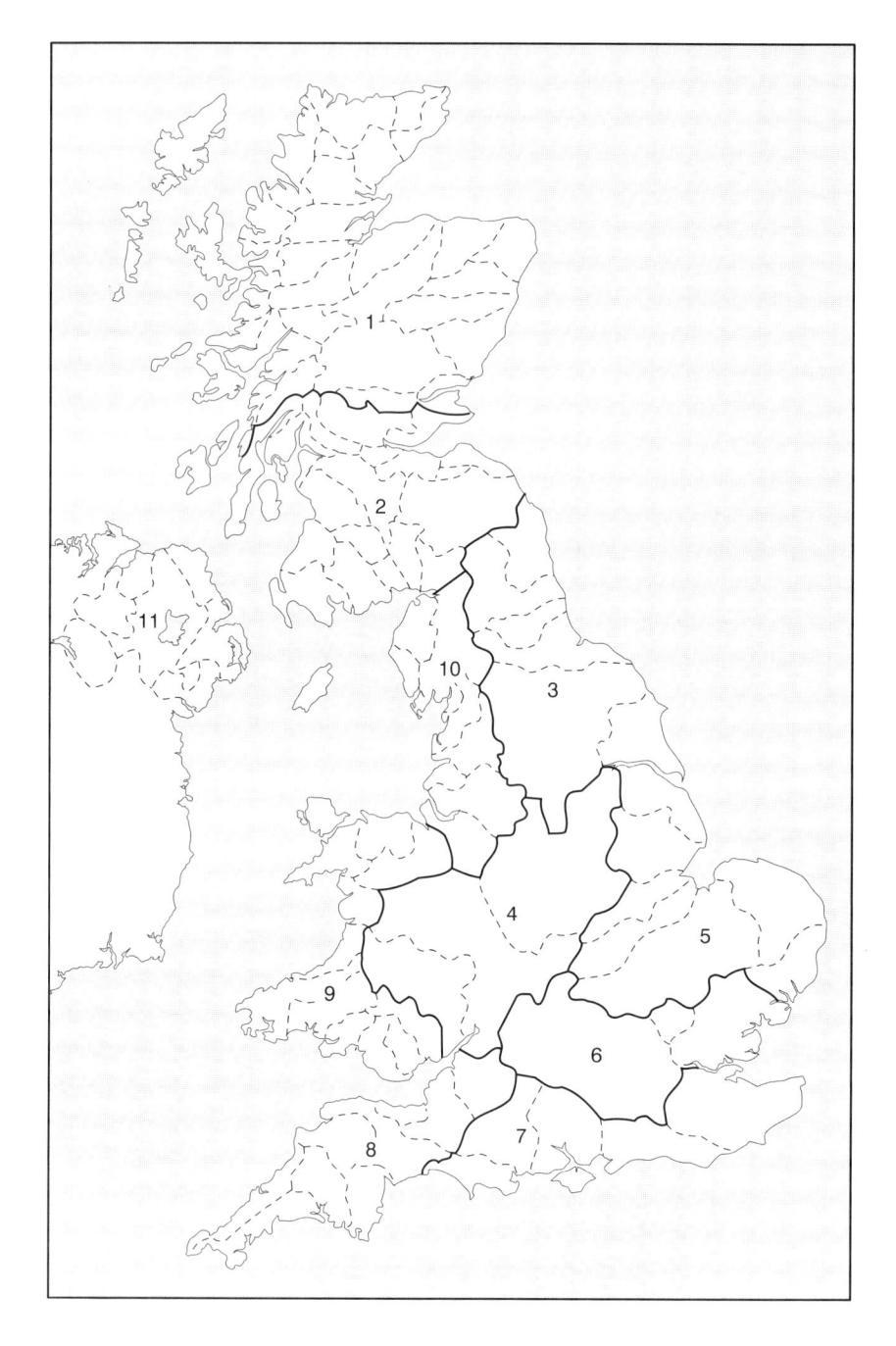

Figure 4.2 Hydrologically 'homogeneous' regions defined in the UK *Flood Studies Report*
Source: NERC (1975). Reproduced by permission of Institute of Hydrology

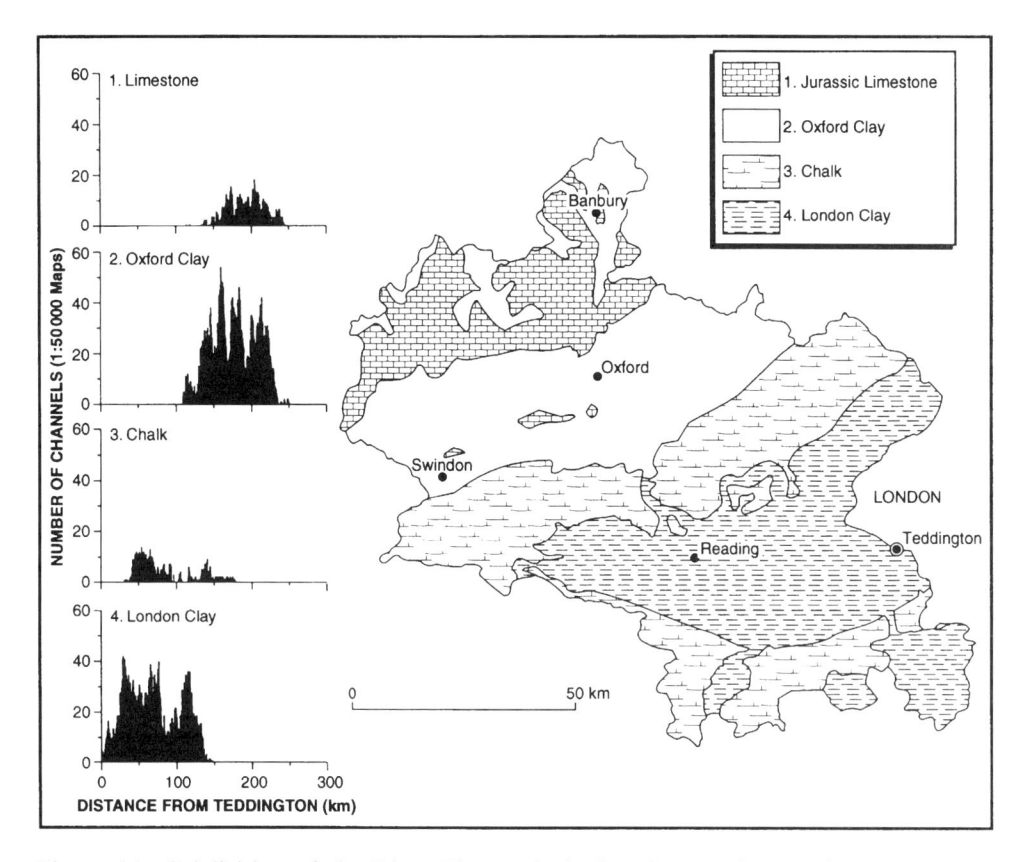

Figure 4.3 Subdivision of the River Thames basin based on geology and network width functions
Source: Based on a diagram in MAFF (1994)

relevant flood statistics. The use of relevant *catchment characteristics* involves allocating catchments to regions or groups on the basis of those basin characteristics expected to influence flooding (e.g. Acreman and Wiltshire, 1989; Hendriks, 1990; Solín and Polácik, 1994). At its simplest this might involve separating large and small, or wet and dry, catchments. A more detailed approach would be to use multivariate statistical techniques such as cluster analysis with a larger number of basin characteristics, e.g. area, slope, shape, network density, soil permeability, storage and average precipitation.

The use of relevant *flood statistics* to group catchments ensures that regionalisation is related directly to flood hydrology. Obvious basic statistics are mean annual flood ($Qbar$) in m^{3}s^{-1}, specific mean annual flood (Qsp) in m^{3}s^{-1}km^{-2}, and the coefficient of variation (CV) (e.g. Wiltshire, 1986; Acreman and Wiltshire, 1989; Roald, 1989). According to Wiltshire, Qsp is intuitively attractive and can be described as the spatial intensity of the mean annual maximum flood while the CV is a measure of flood variability from year to year so that, between them, they summarise the essential characteristics of a catchment flood series. For flood data from 376 UK catchments

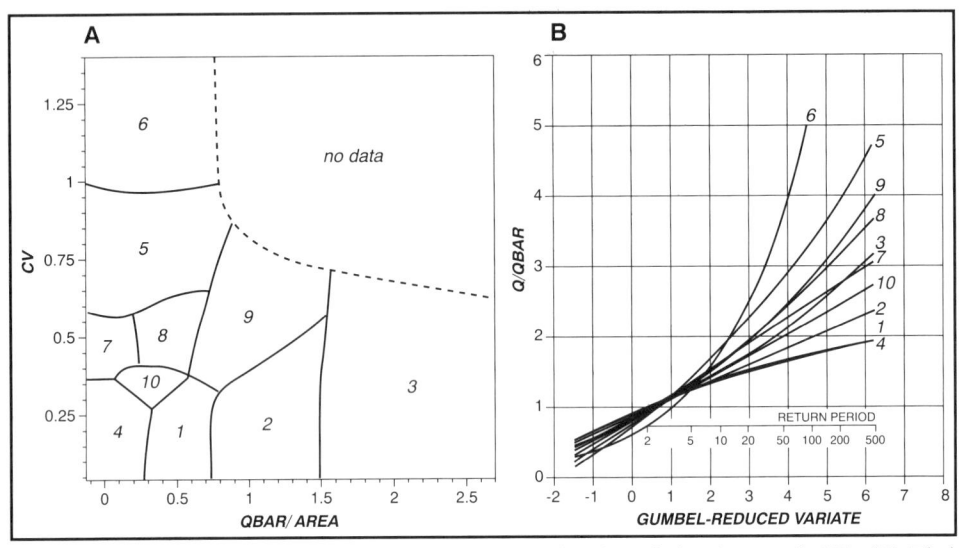

Figure 4.4 (A) Regions based on cluster analysis of flood statistics (Qsp and CV). (B) Their flood frequency curves
Source: Wiltshire (1986). Reproduced by permission of Blackwell Scientific Publications

cluster analysis produced 10 dataspace 'regions' (Figure 4.4A) and their associated frequency curves (Figure 4.4B).

However, the suitability of *Qbar* or *Qsp* in regionalisation has been questioned, unless causative relationships with geomorpho-climatic parameters can be established (Rossi and Villani, 1994). Also, it was suggested (IH, 1994c) that too much reliance would be placed on the annual maximum data if these were used not only to construct regions, but also to define growth curves, develop estimation methods for ungauged sites and judge success. Rather than using the flood values themselves, it might be better to consider covariate information such as land use or soil type; an intermediate approach would be to use flood dates (Reed, 1994a). These were adopted in research for the UK *Flood Estimation Handbook* (see also Chapter 6) in which regionalisation is based partly on the mean and standard deviation of flood date, which together summarise the seasonal distribution of flood events, and partly on the coefficient of variation of recurrence intervals (*CVRI*) which reflects the irregularity of flood occurrence in time (IH, 1994c). Interestingly, although use of these indices represents a basin-disaggregated approach to regionalisation, high values of *CVRI*, denoting irregular flood occurrence, show a strong geographical grouping in eastern England. This is an area where low effective rainfall and generally permeable soils lead to long periods of soil moisture deficit and consequently reduced quickflow production and flood formation.

Finally, the concept of *fractional membership*, i.e. allowing the catchment to belong partially to more than one 'region', may ultimately avoid the need to envisage *a priori* regions at all, certainly in respect of flood estimation at an ungauged site. This idea has been developed further, using a number of *statistical approaches*, e.g. by Cavadias (1990), Caissie and El-Jabi (1993) and Ferrari et al. (1993).

4.4 FLOOD CLIMATE ZONES

It is suggested in the preceding section that geographical location may be a reasonable determinant of many climatic variables. Consequently, classifications of climate and, by implication, definitions of hydroclimatic zones, have been based upon a variety of criteria such as net radiation, air temperature, precipitation and airmass activity. The usefulness of the resulting classification partly reflects the relevance of the chosen criteria to the user. Hydrologists have long recognised that the world's major climate types are reflected in the distinctively different hydrological characteristics of major river basins. Pardé (1955) classified the seasonal regimes of river flow caused by the interplay of precipitation, evaporation and melting, and the corresponding seasonal regime characteristics were later mapped by Guilcher (1965) and Beckinsale (1969).

However, although river regimes reflect the *mean* conditions which are central to the concepts of climate and climate classification, floods are essentially about *extreme* conditions and are therefore less likely to relate closely to conventionally defined climate zones. Certainly, until the work of Hayden (1988), there appeared to be neither a global classification of flooding nor a map of flood climate types on a global basis. Even so, there was a growing perception that some climatic environments are more prone than others to the sustained landforming effects of exceptional floods (e.g. Lewin, 1989) and that the occurrence of severe flooding may be related to both synoptic and mesoscale features of the atmospheric circulation (see Section 3.4.3).

Hayden (1988) attempted to incorporate contemporary understanding of such features into a classification of flood climate types. This recognised that the primary reservoir for floodwaters is the atmosphere, which may be in a state of either baroclinicity or barotropy, and that land-based snow and ice serves as a secondary reservoir. The nature of flooding reflects both the content of the reservoirs and also the meteorological mechanisms responsible for discharge from them. Hayden's approach generated 16 types of flood climate zone which, he argued, should be internally homogeneous in terms of the kind of flood-generating events, moisture availability and other relevant aspects.

The classification proposed represented an important development in flood hydrology and will undoubtedly serve as the starting point for many future studies. However, it is not used as the basis for organising the examples of flood form to which the remainder of this chapter is devoted. This is partly because of its complexity, partly because, like other approaches, it is based on mean conditions, and partly because it does not incorporate mountainous areas. Instead, a much simpler and more descriptive approach has been adopted in which examples of floods are discussed within a framework of cold, hot and temperate climates broadly reflecting that proposed by Supan in 1896 and subsequently modified by Köppen (Miller, 1950). Supan suggested the mean annual isotherm of 20°C as the boundary between hot and temperate climates and the isotherm of 10°C for the warmest month as the boundary between cold and temperate climates. Köppen used the same significant temperatures but differentiated sub-climates on the basis of the number of months that a station was above, between or below these critical values.

A temperature-based approach embraces a number of fundamentally important precipitation characteristics. First, the distinctiveness of flood production associated with the accumulation and melt of snow and ice is common to all cold climates irrespective of whether the temperature regime reflects a high-latitude or a high-altitude location. Second, not only are river floods in hot climates caused entirely by rainfall but over substantial areas the rainfall is the heaviest experienced anywhere and gives rise to most of the world's greatest rivers and to its greatest flood problems. Third, the temperate climates are characterised by intense interactions between tropical warm airmasses moving polewards and polar cold airmasses moving equator-wards. This means that major flooding may result from either rainfall or snow- and icemelt, or from a combination of rainfall and melt.

4.5 FLOODS IN COLD CLIMATES

In the context of flooding, cold climates are those in which the melting of snow or ice makes a significant contribution to the form and incidence of flooding on an annual basis. They therefore tend to be those areas, either in high latitudes or at high altitudes, in which the period of accumulation of snow and ice is sufficiently prolonged to dissociate the occurrence of flooding from the occurrence of precipitation for several weeks, or even months, at a time. Such areas may be defined in terms of the duration and frequency of continuous snow or ice cover, although the precision of such definitions is normally reduced by the sparsity of relevant data.

Church (1988) usefully mapped the nival region in the northern hemisphere on the basis of 15 years of satellite imagery. Figure 4.5 shows the region within which snow cover occurs in 50% of weeks in January. According to Church, it matches closely the same compilation for February, confirming the persistence of the mid-winter snow cover in the region shown and the relative stability of its climatological boundary. The nival area in the southern hemisphere is dominated by the continent of Antarctica and to complete the map it would be necessary to add the high mountain ranges in both hemispheres which also experience nival conditions.

The climate of high-latitude areas is largely controlled by polar and arctic airmasses and the annual precipitation is generally less than 300 mm. This is dominated by snowfall during the winter and melting with rainfall during the summer when flooding normally occurs. High mountains introduce significant complications within the nival area because they tend to experience the relatively high precipitation totals, characteristic of hot humid climates, and the mixed precipitation types characteristic of much of the area of temperate climate. Although their flood hydrology is discussed in this section, they also impact significantly on the downstream flood hydrology of river basins in other climatic areas.

Church (1988) classified cold-region floods into two main groups: (i) those caused by *hydrometeorological factors* such as the melting of snow and ice, rainfall, and rainfall on snow; and (ii) those caused as a result of *channel obstruction* by ice and snow, including winter or thaw ice blockages, the icing of channels from seeps and springs, and glacier outbursts or jökulhlaups. Examples of these main flood types are given in the remainder of this section. A third class of floods, caused by the failure of glacial moraine dams or by the blockage of watercourses by solifluction and landslides, is also characteristic of cold regions but such floods are also common to other climate zones and are discussed as a subset of flash floods in Section 4.6.2.3.

RIVER FLOODS: SPATIAL CHARACTERISTICS 105

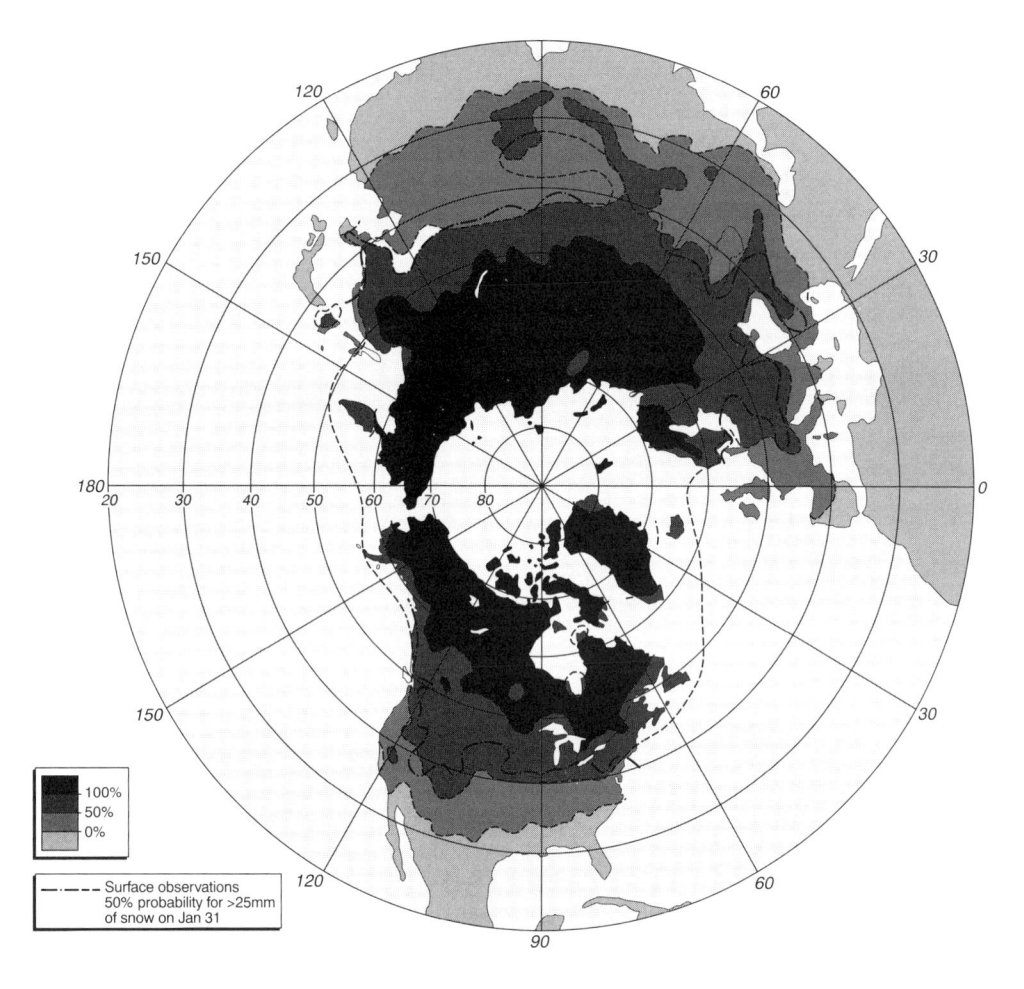

Figure 4.5 The nival region in the northern hemisphere
Source: Church (1988). Reprinted by permission of John Wiley & Sons, Inc

4.5.1 Floods Resulting from Hydrometeorological Factors

The *melting of snow and ice*, sometimes with an additional increment from rainfall, is a major cause of flooding in the high-latitude areas of Canada, the USA, Russia and other parts of Eurasia, and in high-altitude river basins in the major mountain areas. The production of meltwater is essentially seasonal, occurring each year at about the same time in corresponding latitudinal and altitudinal zones. Since the rate of melt depends on energy balance conditions (see Section 3.5.1), it may vary considerably from year to year within a given drainage basin. In addition, meltwater flooding tends to differ between rivers draining plains and mountains. The most severe flooding tends to occur in the areas of greatest snow accumulation when the temperature rises rapidly and remains high for several days or even weeks. The meltwater flood season on rivers draining extensive plains areas is comparatively early and severe (Figure 4.6A),

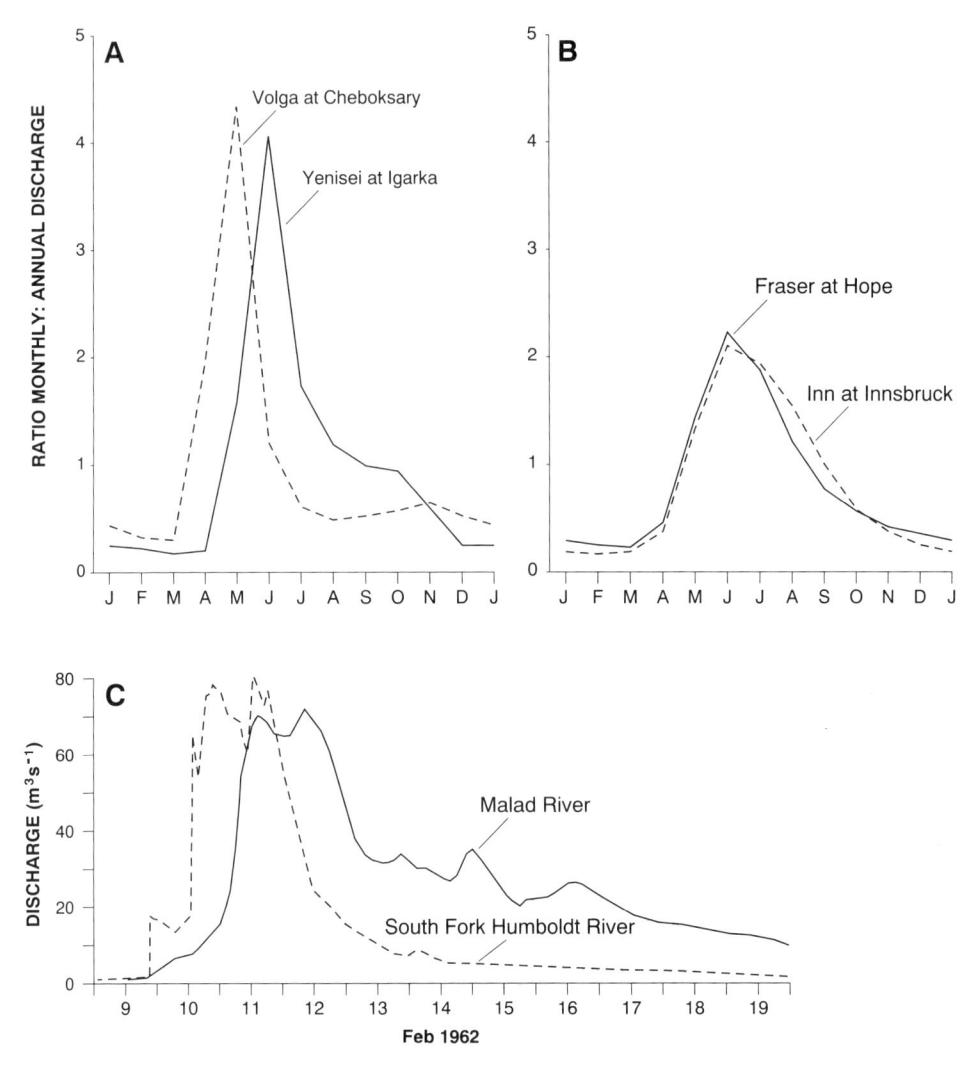

Figure 4.6 Snowmelt flood regimes: (A) rivers draining plains basins—the Volga at Cheboksary, the Yenisei at Igarka; (B) rivers draining mountain basins—the Fraser at Hope, the Inn at Innsbruck; (C) flood hydrographs for the Malad River at Woodruff, Idaho, and for the Humboldt River (South Fork) near Elko, Nevada, 9–19 February 1962
Source: (A) and (B) Pardé (1955); (C) Thomas and Lamke (1962), reproduced by permission of the U.S. Geological Survey

especially when föhn or chinook wind conditions occur, and maximum monthly flows are commonly at least four times greater than the mean annual flow. Some of the Russian rivers provide classic examples of plains snowmelt floods, with rapid thaws producing spectacular flows. For example, the mean June flow of the lower Yenisei at Igarka is $78\,000\,\text{m}^3\,\text{s}^{-1}$, which is exceeded only on the Amazon (Beckinsale, 1969).

Flooding is later and often more subdued on rivers draining mountain basins (Figure 4.6B), with maximum monthly flow typically being two to three times the mean annual flow. Severe flooding may, however, be caused by larger-than-normal snow accumulation, especially when this is accompanied by the late onset of melt. Kattelmann (1990b), for example, found that in all cases large snowmelt floods on rivers draining the Sierra Nevada, California, occurred when snow deposition was not only more than twice the average amount but also persisted into April or May, when the energy available for melt is much greater than in early spring. Again, one of the worst flood disasters in Canadian history took place on the Fraser River, whose normal flood regime is shown in Figure 4.6B. Heavy snowfall during the winter was followed by a late spring in 1948. Then suddenly, at the beginning of May, temperatures began to rise and remained high for several weeks leading to rapid snowmelt and prolonged severe flooding in the lower valley during which more than 20 000 ha of the best farmland in Canada were inundated and Vancouver was cut off from the rest of Canada, except by air (Sewell, 1969). Another cause of exceptional floods is snow avalanching, which rapidly transports very large quantities of snow from high to low altitudes in mountains like the Karakoram in Pakistan. This occurs mainly in the spring, with major implications for melt hydrology (Young and Hewitt, 1990).

Despite the expectedly important role of snowmelt in cold-climate flood genesis, the greatest flows on both high-latitude and high-altitude rivers are likely to be produced by *rainfall* and by *rainfall-on-snow*. The processes determining rainstorm magnitude, e.g. air humidity, convergence and uplift, are spatially variable and tend to diminish polewards. However, Church (1988) demonstrated that, even at arctic locations, extreme 24-hour rainfalls far exceed normal maximum daily snowmelt rates of about 40 mm (see Section 3.5.1). At subarctic locations they may be more than double the normal maximum melt rate.

Other things being equal, the relative contributions of rainfall and melt in the 'mixed' situation will reflect the scale of the rainfall event and the size of the river basin. Relatively modest synoptic-scale rainstorms may produce floods having much longer return periods. Thus the storm which occurred over about 80 000 km^2 of the Mackenzie Mountains in north-west Canada in July 1970 was estimated to have a return period of about 10 years, although the magnitude of the resulting flood had not been equalled in 100 years (Church, 1988). In very large river basins (Church suggested those exceeding 10^5 km^2) meltwater dominates the flood hydrograph, simply because the rainprint of individual storms covers only a small proportion of the basin area (see also Table 3.1).

Floods in mountain catchments have been investigated in the Sierra Nevada and the Rocky Mountains of the USA. Kattelmann (1990b) found that, on major rivers draining the Sierra Nevada in California, mid-winter rainfall on snow produced the highest flows and that, in general, summer thunderstorm rainfall makes a less significant contribution to flooding than in the Rocky Mountains. There is evidence that the relationship between annual snowmelt floods and annual rainfall floods depends partly on altitude, with snow accumulation being significantly greater above 2500 m (Kattelmann, 1991a). Jarrett (1990) also showed that, in the Rocky Mountains of Colorado, low-magnitude snowmelt floods predominate above about 2300 m and large-magnitude rainfall-produced floods predominate below that height, reflecting the greater frequency of large rainfall amounts at lower elevations (Figure 4.7).

Exceptional rain-on-snow floods may result from the presence of frozen ground, as in Idaho and Nevada, USA, in February 1962, when a 100-year flood resulted from prolonged low-intensity rain falling in warm weather onto moderate depths of snow at

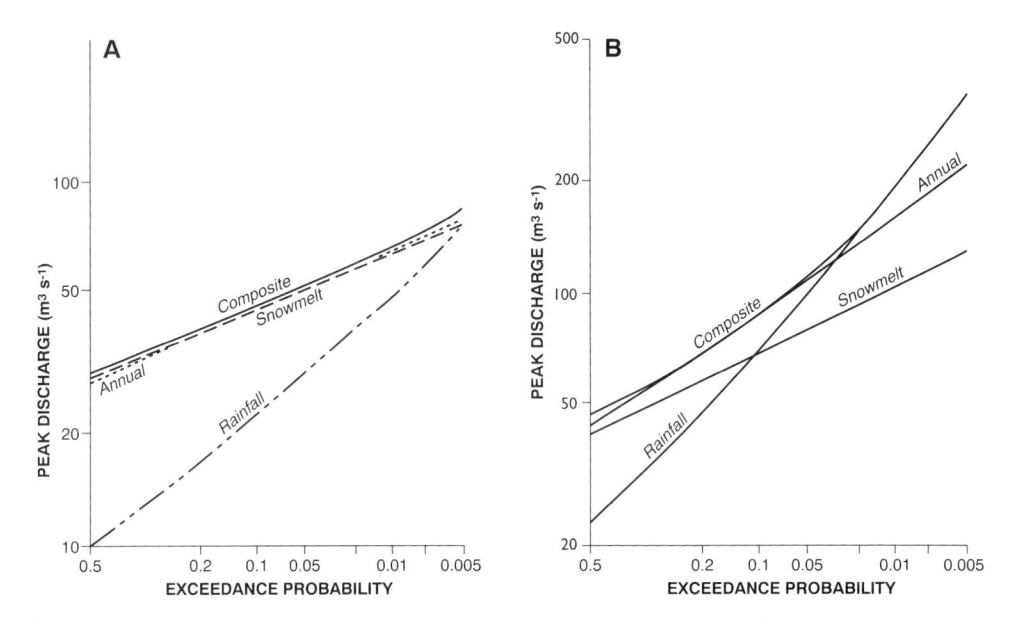

Figure 4.7 Flood frequency curves for snowmelt and rainfall floods in the Rocky Mountains, Colorado. (A) Clear Creek, near Lawson (gauge height 2463 m; catchment area 4.16 km^2); (B) Clear Creek, near Golden (gauge height 1748 m; catchment area 11.3 km^2)
Source: Jarrett (1990). Reproduced by permission of IAHS Press

low altitude when there was a surface glaze of ice over deeply frozen ground (Thomas and Lamke, 1962). Daytime temperatures were above freezing, rising to more than 10°C, and together with the rain were sufficient to melt most of the snow below 2000 m. The rain plus the melted snow ran off rapidly over the frozen ground causing a steep increase in river discharge (Figure 4.6C).

4.5.2 Floods Resulting from Channel Obstruction by Ice and Snow

Some cold-climate floods result not from hydrometeorological events, or even from high river discharges, but from the ponding back of water behind channel blockages of ice and snow during the winter or early spring (Church, 1988).

Snow is less massive than ice and is normally capable of ponding back water for only short periods of time so that its effects are significant mainly on small headwater streams. During years of above-average snow accumulation in the Colorado Front Range, for example, the outlets of small lakes, including beaver ponds, become blocked by snow dams (Caine, 1995). These collapse after a few days or weeks, because of the pressure of water or an increase in air temperature, giving rise to steep, short-lived small floods having instantaneous peak discharges of up to 2 m^3 s^{-1}. More substantial snow-dam floods have been reported from arctic environments, especially where rivers cease to flow after freeze-up, thereby allowing substantial depths of snow to accumulate as a plug in the channel, instead of the border or frazil ice which would be typical where flow continues after freeze-up (e.g. Clark, 1988). Again, in incised channels in the Canadian prairie and tundra, where snow drifting is frequent, the damburst floods which follow the rupturing of snow dams may increase downstream discharge by factors of up to five times in a few hours (Church, 1988).

Ice jams may cause disastrous floods on large rivers and were divided by Church (1988) into winter jams and breakup jams. The former are caused when slush and frazil ice accumulates at the downstream end of steep, turbulent reaches and may be drawn below the surface ice cover to form 'hanging dams' which constrict the channel. Breakup jams, formed by the temporary accumulation of ice floes as the channel ice cover breaks up during the spring melt period, may be massive and produce large increases of water level upstream and major flooding downstream of the jam.

Some of the most dramatic examples are found on major northern hemisphere rivers in North America and Russia which flow towards the Arctic Ocean and where southern tributaries either remain unfrozen or thaw early while the northern channel reaches remain frozen. In Siberian rivers, for example, the upper and middle reaches usually thaw between late April and mid-May but the lower reaches remain frozen until early June. Ice jams tend to recur at significant channel constrictions, including bridges, and as a result ice-jam floods are typical of many cold-climate cities where they may cause considerable damage and inconvenience.

Major ice-jam floods often result from the sudden onset of a warm spell, especially when this is accompanied by rain, since in these conditions the ice floes are of substantial thickness and strength and runoff volumes are large. From the limited data available it seems clear that ice-jam floods have a more variable frequency distribution than open-water floods in cold climates and are responsible for most of the severe floods, i.e. those having a return period greater than about 10 years on, for example, the Peace River in Alberta (Gerard and Karpuk, 1979). The spectacular ice-jam floods which occurred on the Rhine in January 1784 resulted in water levels at least 3 m higher than in any known iceless flood, and the March 1838 floods on the Danube were 2 m higher than any previous or subsequent recorded levels (Pardé, 1964), destroying more than half the buildings in Pest.

Ice-jam floods may also occur outside the areas normally associated with cold climates if localised thawing occurs in an upstream area of a river basin. Hoyt and Langbein (1955) described a number of examples, including the flooding of the Missouri River in the Dakotas in the spring of 1952. Warm weather upstream in Montana caused ice in the tributaries to break up and move downstream to jam against the firmly frozen Missouri. When the ice jam eventually broke, the discharge at Bismarck, North Dakota, rose from $2120 \, m^3 \, s^{-1}$ to $14\,200 \, m^3 \, s^{-1}$ within a few hours.

Cold-climate floods may also be caused by *channel icing* which occurs when seepage under pressure continues to feed water through the bed of an otherwise dry channel. The water freezes at the surface and during the winter the channel may become largely or entirely blocked by ice, forcing the first spring meltwater onto the floodplain. After a few days a channel is normally eroded through the ice but some rivers may remain partly ice-dammed for much of the summer (Church, 1988).

Although the ice jams caused by floating ice, and discussed earlier, are a form of *ice dam*, the latter term is normally reserved for an ice barrier, caused by a glacier or ice-sheet, which ponds back meltwater and streamflow to create ice-marginal lakes and which may also hold considerable quantities of water in a honeycomb of tunnels and cavities within the ice. Total or partial failure of such ice barriers commonly occurs during the summer, as cavities and cracks extend and enlarge through melting and ice deformation, or as the ice mass begins to float hydrostatically or is overtopped, thereby releasing the ice-marginal and englacial waters and causing *glacier outburst floods*.

Glacier outburst floods are common in Iceland, where they are known as *jökulhlaups*, a term which now has widespread geographical usage wherever continental or valley glaciers exist. One of the best-known and most dramatic examples occurs at Grimsvötn, where periodic releases of up to $7500 \times 10^6 \, m^3$ of water have produced flood peak discharges of almost $50\,000 \, m^3 \, s^{-1}$ (Thórarinsson, 1953). The flood hydrographs of the 1922 and 1934 floods at Grimsvötn (Figure 4.8A) show the typical 'reversed' image of such events whereby, in contrast to the normal flood hydrograph, flow starts at quite low rates and increases exponentially as the water-transmitting channels through the ice are enlarged by the flowing water until peak flow is attained. After the peak so little water remains in storage that discharge falls off very abruptly (Meier, 1964). Detailed studies, however, have indicated that the smoothly changing pattern of discharge illustrated in Figure 4.8A may represent a simple case. Sturm et al. (1987) reported jökulhlaups from Strandline Lake, Alaska, which showed several temporary reductions in discharge on the rising limb of the hydrograph, in which flow dropped to virtually zero before increasing rapidly again. They suggested that these reductions may have resulted from the collapse failure of major drainage channels or from their temporary blockage by ice blocks and that, as the obstructions were melted or swept away, high discharge was rapidly resumed.

In some cases the flood outburst appears to be triggered when meltwater behind the glacial dam reaches a critical elevation. Then, if meltwater generation is similar from season to season, jökulhlaup recurrence may take on a quasi-periodic annual or biennial pattern which may be so regular that its date can be predicted (e.g. Konovalov, 1990). It has been found that 95% of jökulhlaups in the European Alps occur from June through September (Tufnell, 1984). However, Sturm et al. (1987) suggested that major blockage of the outflow channels can terminate a jökulhlaup prematurely and initiate the refilling of the adjacent meltwater lake. Critical water elevation, and jökulhlaup initiation, would then be reached correspondingly earlier in the following melt season. Large variations in the recurrence and magnitude of glacier outburst floods could be explained in this way without the need to impute changes of climate, ice thickness, or lake volume. Occasionally, Icelandic jökulhlaups are triggered by volcanic action and in these circumstances the outflow of water may be very large indeed (e.g. Maizels, 1989). The October 1996 eruption of Bardhabunga, 600 metres under the Vatnajokull ice-cap, melted more than 2 billion cubic metres of ice. Initially this was stored in the icecap, raising sub-ice water levels to 1500 m, the highest elevations for a century, before breaking out on 5 November to give a major flood over a period of several days on the Skeidhara River in southern Iceland. This swept away roads, bridges and much else in its path although loss of life was averted because of the long delay between the eruption and the ensuing floodwater break-out.

Glacier outburst floods are also characteristic of most high mountain regions. In the Karakoram, for example, several major events this century were documented by Hewitt (1982). Exceptional events occurred on the Shyok/Indus river system in 1926 and again in 1929 when, on 15 August, an ice dam estimated to be 1200 m long and 150 m high burst (Gunn et al., 1930). Even 1200 km from the glacier dam, river levels at Attock rose 8.1 m from a base discharge of about $9500 \, m^3 \, s^{-1}$ to a peak discharge of $19\,500 \, m^3 \, s^{-1}$ in

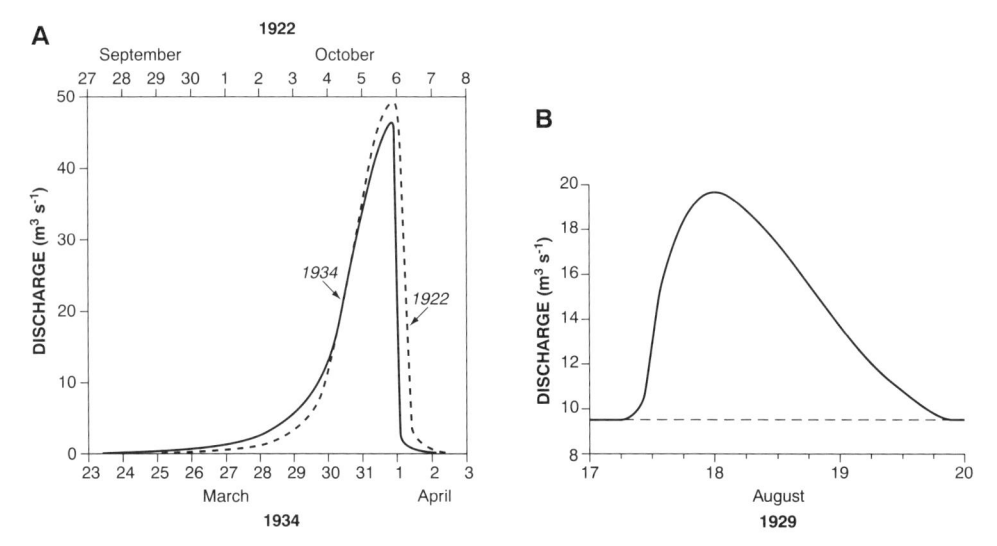

Figure 4.8 Glacier outburst floods: (A) The Grimsvotn, Iceland, flood hydrographs of 1922 and 1934; (B) flood hydrograph at Attock, 1200 km below breach of glacial dam on the Indus River
Source: (A) Thórarinsson (1953), reprinted from the *Journal of Glaciology* with permission of the International Glaciological Society; (B) Khan (1969), reproduced by permission of UNESCO/IAHS

18 hours (Figure 4.8B), the volume of the surge being about 1 230 000 m³ (Khan, 1969; Young and Hewitt, 1990). Similarly, the Mendoza River in Argentina, which drains some of the highest mountains in the Andes chain, is characterised by major glacier outburst floods, including the maximum recorded event of 10–11 January 1934 (Fernández et al., 1994). At Guido the normal monthly mean flow of the Mendoza for January is 103 m³ s⁻¹ but in 1934 the discharge rose within a few hours to a peak of 2317 m³ s⁻¹.

The failure of ice dams and moraine dams is a frequent cause of catastrophic flooding in China. Ding and Liu (1992) noted that 30 major glacier-dam failures had been recorded since 1956, mainly in the Karakoram, Pamir and Tien Shan mountains of western China. Floods caused by moraine-dam failures occur mainly in Tibet and 19 major events had been reported since 1935.

Possibly the largest ever floods were those associated with the outburst of glacial Lake Missoula which initiated the scablands of Washington, USA, about 15 000 years ago (Craig, 1987). These floods occurred as the depth of Lake Missoula reached more than 600 m before the failure of its impounding glacial dam released some 2500 km³ of water within an 11-day period. At least 35 such floods occurred and maximum discharges may have exceeded 13.7×10^6 m³ s⁻¹.

4.6 FLOODS IN HOT CLIMATES

Low-latitude hot climates are almost wholly dominated by equatorial and tropical airmasses. They are controlled by the subtropical high-pressure cells, regions of air subsidence which are basically dry, and by the great equatorial trough of convergence that lies between them (Strahler and Strahler, 1983). Precipitation values tend therefore to be either very high or very low. Mean annual precipitation is greater than 2000 mm in much of the wet equatorial belt and exceeds 5000 mm in limited areas of Central America, West Africa, and South and South-East Asia. Windward tropical coasts commonly experience average annual falls of more than 1500 mm. Tropical cyclones are important in some of the wet areas. In the tropical deserts annual precipitation is frequently less than 250 mm, although even some of these areas experience occasional heavy rainfall resulting from the incursion of tropical storms. This broad contrast between humid and arid environments is used to structure the ensuing discussion of floods in hot climates.

4.6.1 Humid Environments

The humid tropics are either continuously or seasonally very wet and are the source of most of the world's major rivers: the Amazon, La Plata, Orinoco and Sao Francisco in South America; the Congo and Zambesi in Africa; and the Mekong in South-East Asia. Other major rivers, such as the Nile and the Irrawaddy, originate outside the tropics but gain greatly in volume as they flow through humid tropical environments.

Rainfall in these areas is characterised by its intensity as much as by its duration. The intensity of a 15-minute tropical rainfall can be two and a half to four times higher than a rain of similar duration in Western Europe; a 60-minute storm may be three to six times more intense (Gladwell and Low, 1993). Such prodigious falls are a reflection of the large masses of near-saturated air and the strong convergence associated with tropical storms. Major cyclonic storms, alternatively known as cyclones, hurricanes or typhoons, occur with varying frequency (see Figure 4.9) but are especially important in the southern Atlantic and Pacific Oceans. Tropical cyclones combine high intensity and prolonged duration of rainfall with very high windspeeds, exceeding 120 km h^{-1} in coastal locations, and are a principal cause of river and coastal flooding in hot humid environments (see also Section 5.3.1). They are also a focus of much-needed research and study intended to achieve a better understanding of their mechanisms and the flooding to which they give rise. Non-cyclonic rainfall may also cause flooding in hot humid environments, especially in areas affected by monsoons, although in other areas it may be of lower intensity and shorter duration than cyclonic rainfall. Examples of cyclonic and non-cyclonic storm rainfall in Cuba are given in Table 4.1 and show that the equivalent of two or three months' average rainfall in southern England may fall in Cuba in one hour.

The flood peaks generated by such combinations of rainfall duration and intensity are often spectacular but vary greatly depending on the size of river basin and on whether the rainfall is an isolated event or part of a seasonal or quasi-continuous sequence of events. Volker (1983) compared flooding caused by monsoon rainfall on selected rivers in South-East Asia and Figure 4.10 shows hydrographs for the Mekong at Phnom Penh (basin area 663 000 km^2) and the Huai Bang Sai at B. Nong Aek (basin area 1340 km^2). The contrast between the flood season during the south-west monsoon and the dry season during the north-east monsoon is clear in both cases but while the resulting floods on the Mekong appear as a single, rather gradual event, with a peak discharge of about $45\,000 \text{ m}^3\text{s}^{-1}$ ($68 \text{ l s}^{-1}\text{km}^{-2}$) those on the Huai Bang Sai are flashy with separate responses to individual rainfalls giving peak discharges of more than $150 \text{ m}^3\text{s}^{-1}$ ($110 \text{ l s}^{-1}\text{km}^{-2}$) on several occasions. For comparison, the mean annual

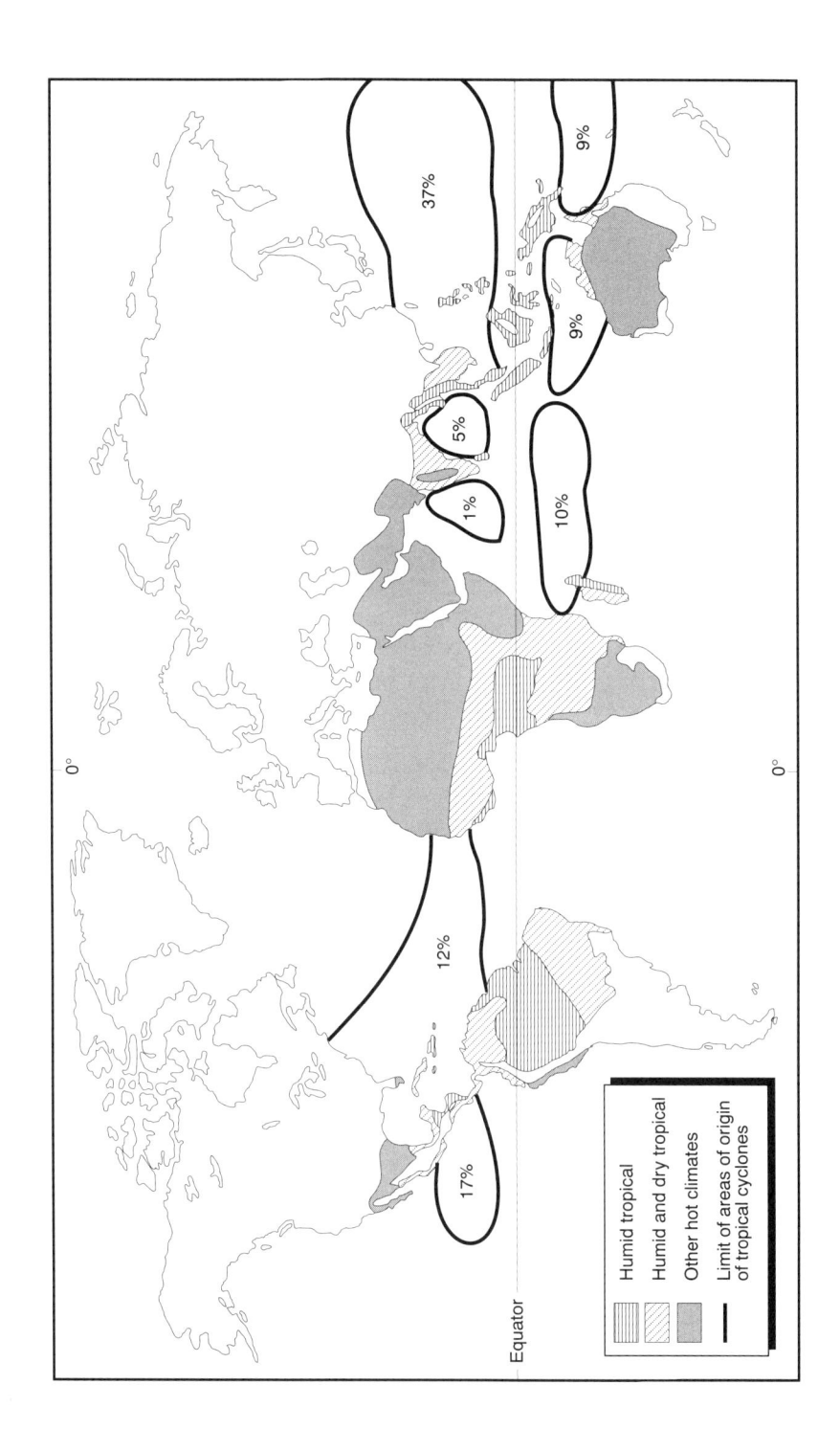

Figure 4.9 Generalised map of hot climates according to the Köppen classification showing the percentage frequency of tropical cyclones
Source: WMO (1983). Reproduced by permission of IAHS Press

Table 4.1 Precipitation intensities (mm h^{-1}) during major cyclonic and non-cyclonic storms in Cuba

	\multicolumn{10}{c}{Period of continuous rainfall (minutes)}
	5	10	40	60	90	150	300	720	1440	2800
Cyclonic storms (by date)										
Oct. 1926	342	282	144	115	87	65	43	23		
Oct. 1963	132	114	84	72	66	55	48	37	29	26
Nov. 1971	264	216	165	150	133	96	64	33	30	26
June 1982	168	150	135	127	115	102	80	52	31	31
Non-cyclonic storms (by zone)										
Western	180	165	138	109	82	50				
Central	122	122	101	75	57	44	23			
Eastern	163	162	91	82	65	45				

Source: Based on data in Díaz Arenas (1983)

Table 4.2 Selected flood data for Central Africa

River	Basin area (km^2)	10-year flood m^3 s^{-1}	10-year flood m^3 s^{-1} km^{-2}	Max observed flood m^3 s^{-1}	Max observed flood m^3 s^{-1} km^{-2}
Zaire, Kinshasa	3 747 000	63 000	16.8	81 000	21.6
Blue Nile, Roseires	210 000			11 300	53.8
Ogooué, Lambaréné	204 000	12 500	61.3	13 600	66.7
Lobe, Cameroon	1 940	504	259.8	540	278.4
Mberé, Tchad	7 430	1 700	228.8	1 930	259.8
Wouri, Cameroon	8 250	1 800	218.2	1 845	223.6

Source: Based on data in Rodier (1983)

flood of the River Thames at Kingston, UK (basin area 9948 km^2), is 326 m^3 s^{-1} (33 l s^{-1} km^{-2}).

Humid, hot-climate flood data were given for Central Africa by Rodier (1983) and selected examples for some large and some small basins are included in Table 4.2. These again illustrate the great magnitude of the direct discharge from the larger basins and the very high specific discharges from the smaller basins.

The natural vegetation of hot humid areas is dominated by forest, now rapidly disappearing in many tropical countries. The inevitable concern about the hydrological impact of forest clearance has not been helped by the fact that flood-producing mechanisms in humid tropical forests have been poorly understood until comparatively recently. Many workers have assumed, perhaps intuitively, that runoff mechanisms would differ from those prevailing in humid temperate conditions, especially in terms of the expected dominance of overland flow. This certainly appeared to be confirmed by field experiments in the tropical rainforests of north-east Queensland, Australia (Bonell et al., 1983). However, Newson (1994) reported other experiments in the Amazon

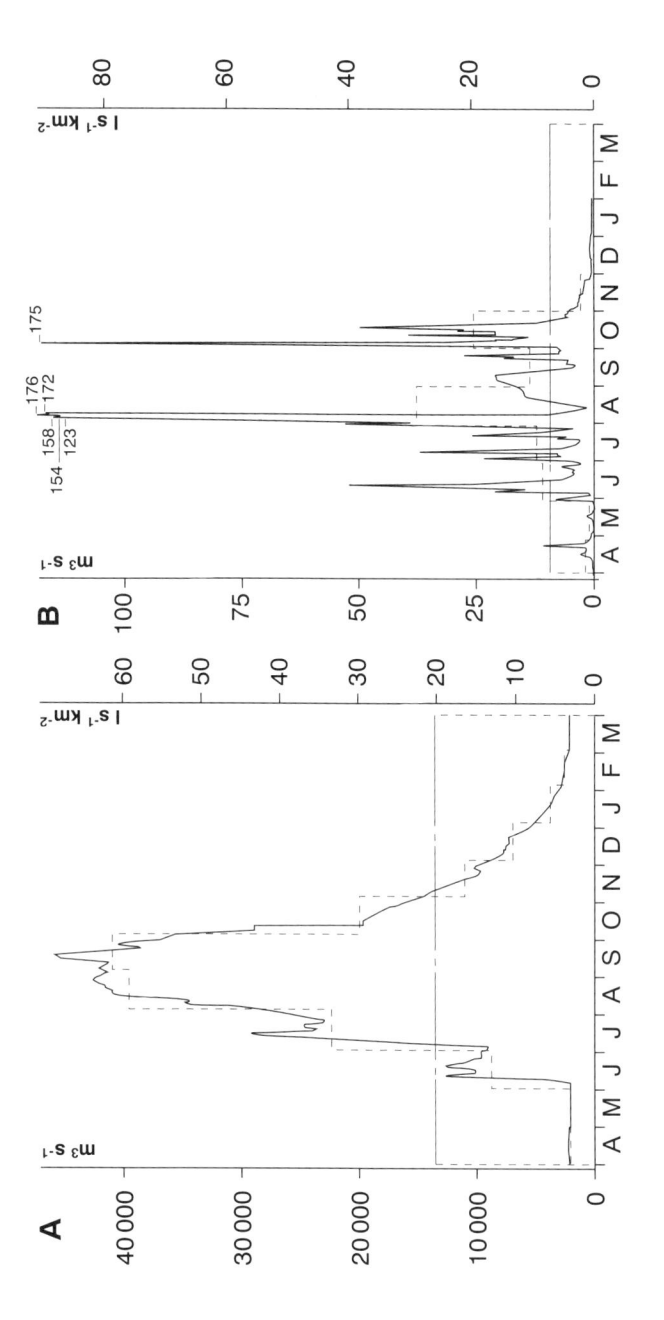

Figure 4.10 Hydrographs showing floods on: (A) the Mekong River at Phnom Penh (Kampuchea); and (B) the Huai Bang Sai at B. Nong Aek (Thailand) in 1972. Reproduced by permission of IAHS Press

Source: Volker (1983). Reproduced by permission of IAHS Press

rainforest of Brazil, by Ross et al. (1990), and in south-east Nigeria, where throughflow dominated the natural runoff processes. And studies in a forested basin in central Java, Indonesia, confirmed the applicability of a 'normal' variable source-area model, involving runoff contributions from the full range of channel precipitation, Horton overland flow, saturation overland flow and subsurface flow (Bruijnzeel, 1983). There is ample evidence that soil erosion in the humid tropics increases dramatically when the protective forest cover is removed (e.g. Lal, 1983) and in Burma the siltation of Meiktila Lake began only after ancient laws prohibiting forest clearance within two miles of the stream banks fell into abeyance under British administration (UNESCO, 1991). However, the rapid growth of scrub after deforestation means that the latter has much less impact on flood characteristics (Qian, 1983).

In a valuable review of tropical forest hydrology, Bruijnzeel (1990) suggested that the most dramatic changes in basin hydrological response result when land-use change is associated with a shift from the dominance of subsurface flow to the dominance of Horton overland flow in flood peak generation. However, he emphasised the need to avoid simplistic conclusions which ignore the complexity of factors influencing the generation of flood discharges.

4.6.1.1 The 1987 and 1988 Bangladesh Floods

Bangladesh straddles the Tropic of Cancer with 80% of its area occupied by the floodplains of the lower reaches of the Brahmaputra (Jamuna) River, the Ganges (known as the Padma River below its confluence with the Jamuna), and the Meghna River which joins the Padma south-east of Dhaka (see Figure 4.11). In total, these rivers drain a catchment area of about 1 500 000 km², on both the northern and southern slopes of the Himalayas and the southern slopes of the Assam plateau. The predominantly monsoon climate delivers a mean annual rainfall, most of which falls between May and September, ranging within Bangladesh from 1250 mm in the centre-west to more than 5000 mm in the extreme north-east, and exceeding 11 000 mm at Cherrapunji in the Meghna catchment north-east of Bangladesh (Brammer, 1990a).

Not surprisingly, in this geographical setting river flows are large (see Table 4.3), with the average peak flow of the combined rivers (110 000 m³ s⁻¹) being about 2.5 times that in the Mississippi River (Brammer, 1990a), and floods have become a regular feature, with 20–30% of Bangladesh being flooded in most years (Thorne et al., 1993), and more than 60% of the country being inundated during major events, of which there were 25 in the years between 1950 and 1988 (Khalil, 1990). The rivers begin to rise between March and May, as a result of snowmelt in the Himalayas and pre-monsoon rainfall in Assam and the north-east of Bangladesh, and then rise rapidly in June–July with the full onset of the monsoon rains. The Brahmaputra and Meghna normally peak in July–August and the Ganges about a month later, although sometimes a late Brahmaputra peak coincides with the peak on the Ganges and Meghna peak levels may extend into September because of backing up of water above the confluence with the Ganges (Brammer, 1990a).

In fact, flooding in Bangladesh is more complex than might be suggested by the foregoing description of the river regimes, and reflects four different mechanisms. *Flash flooding* occurs in response to exceptionally heavy rainfall over neighbouring hills and mountains, *river flooding* results from a combination of snowmelt in the high Himalayas

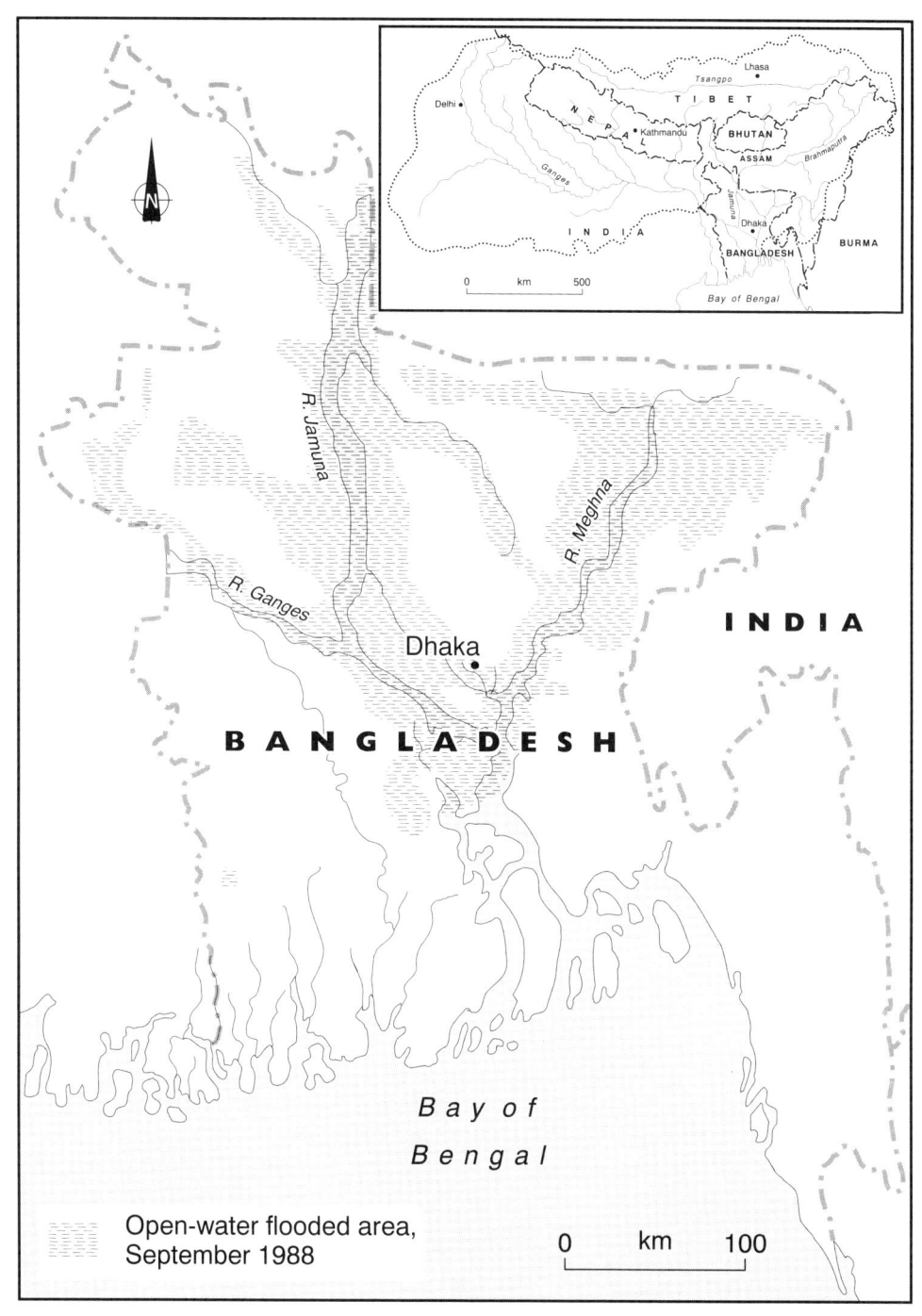

Figure 4.11 Map of Bangladesh, showing open-water flooded area in September 1988. Inset shows the relationship between Bangladesh and the catchment area of the Brahmaputra, Ganges and Meghna rivers
Source: Brammer (1990a). Reproduced from the *Geographical Journal* with permission

Table 4.3 Discharge data for the three major rivers of Bangladesh

River	Mean peak flow (m^3 s^{-1})	Highest peak flow (m^3 s^{-1})
Brahmaputra (Bahadurabad)	65 491	98 600
Ganges (Hardinge Bridge)	51 265	76 000
Meghna (Bhairab Bazar)	14 047	19 800

Source: Based on data in Brammer (1990a)

and monsoon rainfall, and *rainwater flooding* results from intense rainfall within Bangladesh falling on land which has low surface gradients and is traversed by embankments. In addition, *coastal flooding* results from storm surges caused by tropical cyclones in the northern Bay of Bengal (see Chapter 5). Furthermore, flooding in Bangladesh appears to have been aggravated by a number of slowly developing long-term factors. These were listed by Khalequzzaman (1994) and include:

- Local relative sea-level rise may result from either subsidence or inadequate aggradation in the delta area and on the major floodplains in conditions of generally rising sea level. This effect would become more pronounced if global warming develops and would severely intensify flood problems in Bangladesh.
- Aggradation of riverbeds in conditions of high sediment load and low bed gradient is a natural process in delta areas and in Bangladesh may have been accelerated by upstream deforestation and soil erosion. Aggradation reduces the water-carrying capacity of the channels and is especially pronounced for the Ganges and its distributaries which have aggraded by as much as 5–7 m in historically recent times (Alexander, 1989).
- Earthquake activity in northern Bangladesh and northern India has been associated with several recent floods, including those of 1988. Earthquakes may cause the breaching of embankments and changes of river course, both of which are likely to result in flooding. There is some evidence that floods may be a cause, as well as an effect, of earthquakes in tectonically unstable areas.

The 1987 Floods The floods in both 1987 and 1988 were reported to be the worst on record but according to Brammer (1990a) they had different causes, affected different areas and produced different effects. The 1987 floods were predominantly rainwater floods, although flash floods occurred in eastern hill-foot areas and on rivers in the north-west. Several periods of exceptionally heavy rainfall, with return periods of 100–150 years, occurred over northern Bangladesh and adjacent parts of India between June and September. For example, long-term averages were exceeded by 300–700 mm in all four monsoon months in parts of the north-west, and in July and September north-eastern border areas had excesses of 500–1000 mm. The magnitude of the 1987 rainfall is illustrated in Figure 4.12 which shows (A) isohyets for July-August and (B) departures from the average.

The resulting period of peak river flows was prolonged. For example, water levels on the Jamuna, Ganges and Meghna reached the danger mark on 27 July, 30 July and 2 August respectively, peak levels in these rivers were attained on 15, 19 and 7 August, and river levels remained above or close to the danger mark until near the end of September. With all the main rivers running at high levels for many weeks, runoff from the heavy local rainfall during this period was unable to drain off the land and rainwater flooding became very extensive, especially in embanked polder areas.

The 1988 Floods Brammer (1990a) described the 1988 floods as predominantly river floods, with the most serious flooding occurring in early September after 10 days of exceptionally heavy rainfall at the end of August. The pattern of rainfall in August is shown in Figures 4.12C and D and rainfall for the month exceeded previous records at several stations in the extreme north-east and north-west. The Brahmaputra flood peak on 30 August was the highest on record with an estimated recurrence interval of 100 years and the Ganges and Meghna reached record high levels at several locations in early September, with estimated recurrence intervals of about 50 years.

The sudden and virtually simultaneous rise of the major rivers meant that their flood peaks were synchronised within a two-week time period (Khalequzzaman, 1994) and this worsened the extensive overbank flooding (see Figure 4.11). At its peak the Brahmaputra flood was 50 km wide and Hoque (1994) estimated that about 67% of the country (97 000 km^2), the highest ever recorded, was inundated. The flooding, which included many rural towns and two-thirds of Dhaka city, would have been even more severe had not central Bangladesh received below-average rainfall in August (see Figure 4.12D), including the last 10 days.

Although the 1987 and 1988 floods were major examples of rainwater and river flooding, it is not reasonable to consider the 'freshwater' flood situation in Bangladesh in isolation from the 'coastal dimension'. The risk of coastal flooding (see also Section 5.3.2.2) in the delta that is Bangladesh is exacerbated by the triple threat of sea-level rise, delta subsidence and reduced delta growth caused by the effect of upstream dams restricting the downstream sediment flux. Milliman et al. (1989) estimated that the combined effects of these three influences might, by the year 2100, have led to a 3 m rise in sea level, a 2 km retreat of the shoreline, a 26% reduction in habitable land with 27% of the population displaced, and a two-thirds reduction in GDP.

4.6.2 Arid and Semi-Arid Environments

Although one-third of the world's land surface may be classified as arid or semi-arid, with nearly half the countries of the world facing problems of aridity (Pilgrim et al., 1988), the extent of such conditions within hot-climate areas is more restricted. There is also a lack of a satisfactory definition of what constitutes an arid or semi-arid environment. Indexes of aridity are based upon the balance between available water resources, of which precipitation is the most important, and water losses by evaporation. Many such indexes, both global and regional, have been proposed but none has achieved universal acceptance. Broadly speaking, however, the terms 'arid' and 'semi-arid' are applied to areas where the precipitation–evaporation balance means that rainfall cannot directly support sustained rain-fed farming (Rodier, 1985).

The tropical dry climates, which result from the subsiding stable airmasses of the tropical high-pressure cells, occur in the Sahara–Arabia–Iran–Thar desert belt of North Africa and southern Asia, much of Australia, and parts of Central America, South America and South Africa. In addition, Strahler and Strahler (1983) noted that

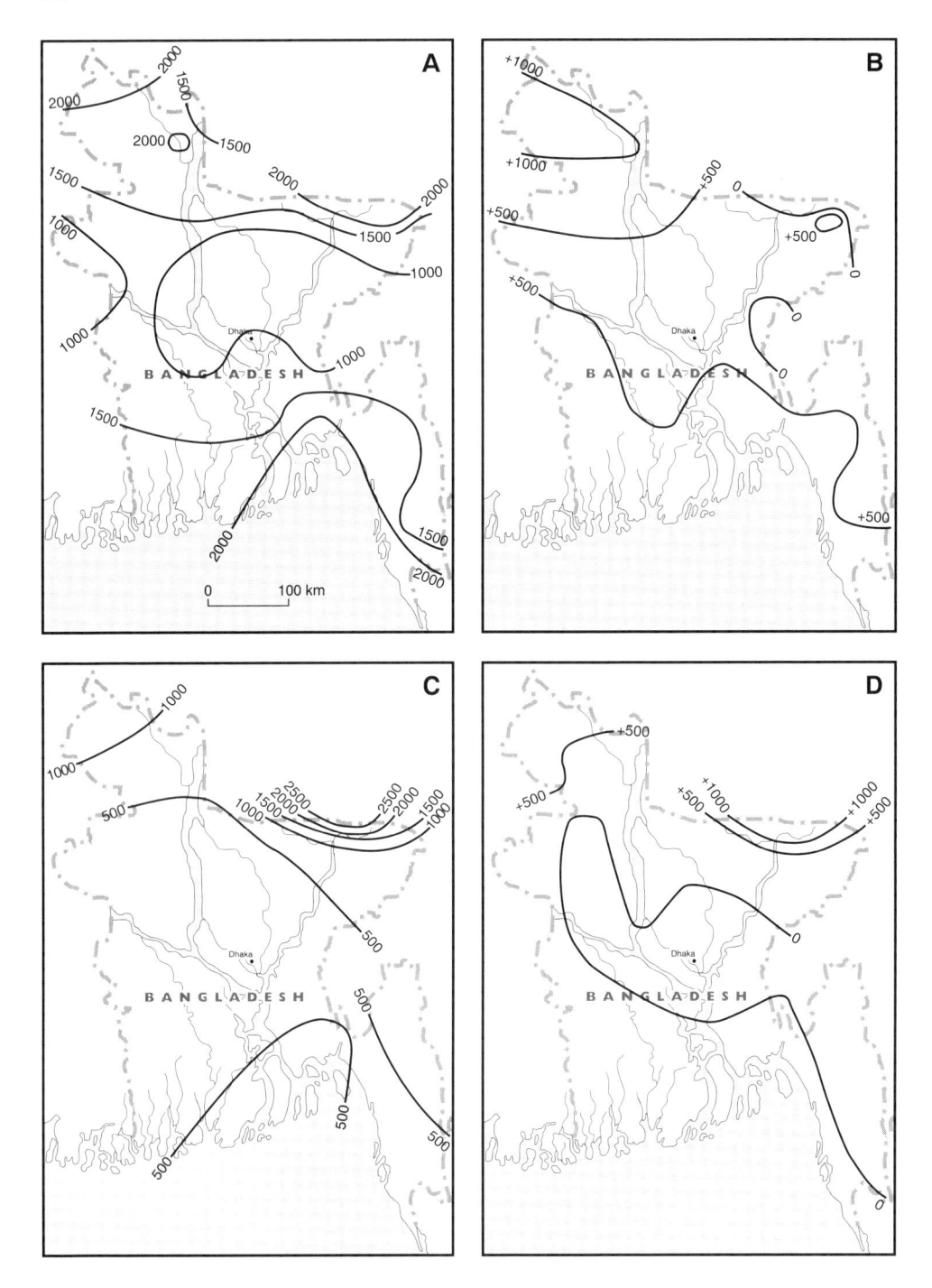

Figure 4.12 Rainfall distribution in Bangladesh (mm): (A) July–August 1987; (B) July–August 1987 departure from average; (C) August 1988; (D) August 1988 departure from average
Source: Brammer (1990a). Reproduced from the *Geographical Journal* with permission

important areas of a steppe sub-type of these climates are found in India and Thailand, with many small scattered dry areas on the lee side of uplands in the trade-wind belt. Furthermore, poleward extensions of the tropical dry climates, caused by the same airmass patterns but with a cold winter season, occur in North Africa, the Middle East (Jordan, Syria, Iraq), the south-western USA, the southern parts of Australia and South Africa, and in the Pampa and Patagonia of Argentina.

4.6.2.1 Hydrological Characteristics

Although there is a hydrological 'continuum' from very humid to very arid environments, some distinctive features of arid-zone hydrology have been recognised (e.g. Slatyer and Mabbutt, 1964; McMahon, 1979; Yair and Lavee, 1985; Pilgrim et al., 1988).

First, rainfall tends to be less predictable and more variable in both space and time than in humid regions and these characteristics intensify as the mean annual rainfall decreases. Spatial variability means that the rainprint of an arid-zone storm is normally small compared with the size of river basin on which it falls and that it will rarely be observed by an inevitably sparse raingauge network. Temporal variability is characterised mainly by prolonged periods without significant rainfall. Pilgrim et al. (1988) quoted examples from the north-western coast of Australia where, over four consecutive years at a location with a mean annual rainfall of 250 mm, the annual totals were 570, 70, 680 and 55 mm respectively. A nearby location, having an annual average rainfall of 330 mm, recorded 750 mm in one day during a tropical cyclone and a total of 4 mm in the whole of another year. Although the concept of 'mean annual rainfall' has little validity in arid environments, this latter example usefully counters the common misconception that rainfall in such areas falls largely in randomly scattered convective storms having high intensities over limited areas. Although such storms may be important in, for example, the south-western USA (e.g. Renard and Keppel, 1966), other mechanisms appear to dominate arid-zone rainfall elsewhere.

Examples include, convective storms which are spatially organised at the mesoscale (e.g. Sharon, 1981) perhaps, as in northern Kenya, in association with the inter-tropical convergence zone (e.g. Reid and Frostick, 1987), low-intensity frontal storms which move into arid areas such as the Negev, especially in winter (e.g. Sharon, 1980; Yair and Lavee, 1985), a variety of frontal and other synoptic-scale mechanisms in western New South Wales, Australia (e.g. Cordery et al., 1983), and tropical cyclones in Madagascar, the Indian sub-continent, north-western Australia, and some parts of the USA and Mexico (e.g. Dhar and Rakhecha, 1979). Alice Springs, on the Tropic of Capricorn, is more or less in the centre of the arid heart of Australia and has a mean annual rainfall of about 250 mm, but even here 80% of the daily falls are of less than 12.5 mm and such falls make up nearly half the annual rainfall (Slatyer and Mabbutt, 1964).

Second, plant cover, organic matter and litter at the ground surface are meagre so that the soil surface is largely the first point of contact by rainfall. This means that soil type and soil surface properties may play a very influential role in runoff production and explains the attention given in runoff studies in such areas to the way in which overland flow may be generated on hydrophobic and dispersive soils and where there has been crusting and pan formation.

Third, to some extent there is a different mix of hydrological processes in which baseflow is virtually absent and channel transmission losses assume major importance.

The latter is a well-documented phenomenon which probably affects all streams and rivers in hot arid and semi-arid environments. It involves influent seepage, i.e. the 'loss' of water into alluvial material in the streambeds, which leads to a downstream reduction in discharge and in some cases to the formation of playas or saltflats and of internal drainage basins.

Finally, especially in semi-arid areas, the hydrological balance is often a very delicate one so that the hydrological nature of an area may be significantly changed for a period by a prolonged sequence of wet or dry years.

4.6.2.2 Implications for Flooding

The distribution and form of floods in arid and semi-arid environments are determined by the way in which flood-forming processes are driven by the particular rainfall, surface and water-balance characteristics discussed above. This means that floods tend to be: (i) highly variable in distribution, (ii) generated by processes more obviously dominated than in humid areas by overland flow, (iii) of short duration, and (iv) generated on only a small fraction of a drainage basin, and progressively attenuated, often to extinction, by influent seepage so that the relationship between flood magnitude and drainage basin area is not only less consistent than for most humid areas but is also inverted.

Variability of Floods Although improved instrumentation and observation methods have continued to enhance the hydrological database in arid and semi-arid areas, earlier accounts of flood frequency and variability were often expressed in largely anecdotal terms. Describing the flooding of the southern Arava catchment in March 1966, Schick (1971) noted that there was no reliable means of estimating the period of recurrence of such a sporadic event and continued (p. 130):

> According to the report of the Government of Jordan, a flood similar in magnitude to the Ma'an catastrophe was neither within living memory nor within recorded history. Inhabitants elsewhere, such as an old Beduin from Wadi Misr and old-timers at Timna have never previously experienced such a flood.

The sparsity of the database and the typically large negative skew (long tail) of the annual flood distribution in such areas has encouraged the development of various techniques of palaeoflood analysis based, for example, on the interpretation of 'slack-water deposits' (e.g. Baker et al., 1979).

Although Yevjevich (1979) considered that the rainfall series of an arid area may be considered mathematically simply as the truncated series of a humid area and accordingly cast doubt on the validity of treating separately the hydrological time-series of arid and humid areas, there does appear to be a distinctiveness about the statistical characteristics of flooding in arid and semi-arid regions. Indeed, Yevjevich himself noted that the probability distribution functions for flood peaks of semi-arid and arid-zone rivers usually have large standard deviations, with a considerable increase of flood peak for a given increase of return period. Broadly similar conclusions were reached by McMahon (1979) on the basis of 63 flood series from six arid zones. McMahon found that the standard deviation of log peak discharge in Australia, the eastern

Mediterranean and southern Africa, is about double the humid-region value. For the same areas the skew of log peak annual discharge is generally negative. North American rivers appear to fit neither trend, having lower variability and a positive skew. Maps showing the high variability of streamflow over much of Australia were developed by the Australian Water Resources Council (AWRC, 1978).

A statistical model of flood frequency was developed for the Negev of Israel by Cohen and Ben-Zvi (1979) and was based on the complete series, using all the flood peaks, rather than on the annual series. Based on work in the Sinai desert, Schick and Lekach (1987) showed that flood frequency in arid environments decreases with catchment size. This presumably reflects mainly the effect of transmission losses and the resulting decline in specific discharge with catchment size which is discussed later in this section.

Role of Overland Flow Overland flow tends to play a more dominant role in the production of runoff in general, and of floods in particular, than it does in most humid areas. Overland flow commonly takes the form of sheetfloods, rather than of channel floods, especially on smoothly graded surfaces, and may occur either as hill wash and flow in small hill-slope gullies which floods out and coalesces at the slope foot, or as rainwater flooding on lowland areas (Joly, 1953; Slatyer and Mabbutt, 1964). Less spectacular forms of sheet flooding may occur on low-gradient, smooth plains, such as those which cover about 45% of the Australian continent, where there is a slow movement of very thin, discontinuous sheets of water downslope, which inundate extensive areas to a depth of generally less than 0.30 m (Slatyer and Mabbutt, 1964; AWRC, 1972).

Several factors, operating either singly or in combination, are responsible for the importance of overland flow, including intense rainfall, shallow patchy soils, low-density vegetal cover and low infiltration capacities.

The role of areas of low infiltration capacity in arid and semi-arid areas is particularly important in explaining both the non-uniform spatial generation of sheetflooding and also its generation by very small rainfall amounts. Yair and Lavee (1985) referred to a number of studies which confirmed that the strongly non-uniform spatial generation of runoff was related to great spatial variability in infiltration capacity. They observed, however, that whereas in humid areas this variability is attributable mainly to spatial differences in soil moisture, in arid and semi-arid areas it is most often controlled by the spatial heterogeneity of ground-surface properties. As a result, overland flow may be generated by rainfall amounts as low as 3–5 mm (Dubief, 1953; Yair, 1990).

Low infiltration capacities result from physical, chemical and biological factors. Physical factors include thin soils, exposed bedrock, and surface crusting by raindrop impact, especially on silty soils where final infiltration rates may be reduced by a factor of 10 (Reid and Frostick, 1987). Chemical factors are important in clay-rich soils where permeability depends partly on exchangeable sodium percentage (Yair and Lavee, 1985). For example, sodic soils in the Negev desert, Israel, crust easily during rainfall, so that often only the first few millimetres of rainfall are able to infiltrate before the crust begins to shed water with about the same efficiency as an asphalt road surface (van der Molen, pers. comm., 1983). Finally, Yair (1990) showed that a thin algal crust can so dramatically reduce infiltration into sandy desert areas that runoff generation may be expected for any storm exceeding 2–3 mm.

Low infiltration rates are also encouraged by the sparsity of vegetation cover in arid and semi-arid areas. Bonell and Williams (1986) emphasised the control of runoff processes by bare soil areas on smooth plainland in central–north Queensland and Pilgrim et al. (1988) described how the typically banded pattern of acacia trees over much of central Australia allows sheetflooding from the intervening bare areas to vitally augment water supplies to the groves of trees.

Duration of floods Floods in arid and semi-arid areas are often 'flashy', i.e. their hydrographs are of short duration and have steep rising and falling limbs. This partly reflects the way in which overland flow, generated on smooth or steeply sloping ground surfaces having sparse vegetation cover, may deliver sudden-onset flood flows and result in rapidly rising hydrographs in the wadis and other conveyances for surface water. Reid and Frostick (1987), for example, reported a time-to-peak duration of 4–16 minutes for a 7 km² catchment in Kenya which has a high wadi drainage density of 100 km km^{-2}. Often the hydrograph is so steep that the flood wave travels down a wadi as a shallow 'bore' (Hassan, 1990) or 'wall of water' (see the discussion of 'flash floods' in Section 4.6.2.3). However, because of the very high transmission losses into the alluvial bed materials, the recession limb of the flood hydrograph is substantially truncated and peak flows are of similar magnitude to those in humid climates, despite their very much shorter times-to-peak.

Some authorities have suggested that the 'wall of water' phenomenon in arid floods, although occurring from time to time, is not typical (e.g. Renard and Keppel, 1966). Others (e.g. Schick and Lekach, 1987) have argued that it is frequently caused by intra-event damming such as that caused by alluvial fans deposited in a main wadi channel over a period of time by flooding tributary streams. Subsequent flows in the main stream channel will be retained temporarily behind such fans before the latter are breached to release a sudden wall of water.

Not all arid-zone floods, however, are of short duration, especially in those arid and semi-arid areas subjected to monsoonal influences. For example, monsoon rains moving into northern Australia from the Timor Sea have caused major floods which have persisted for many weeks, as in 1974, and in some years, such as 1893, as many as five cyclonic systems have moved in (Bolt et al., 1975).

Specific Discharge To an even greater extent than in humid environments, floods in arid and semi-arid areas are generated on only a small fraction of a drainage basin. They then tend to be progressively attenuated by influent seepage, often to the point that the floodwaters dwindle to a mere trickle in an inland lake or playa. As a result, the relationship between flood magnitude and drainage basin area is less consistent than for most humid basins and may even be inverted, with larger flood peaks issuing from small upstream catchments than from larger catchments downstream in the same drainage system.

This is clear from Table 4.4 which illustrates the declining volume and peak discharge of a flood moving down the Walnut Gulch catchment in Arizona. An equally dramatic illustration from the southern Negev desert was described by Yair and Lavee (1985). Data collected over 18 years showed that a reduction in the runoff coefficient from 36% to 4.4%, was possible as flood peaks moved from a 0.05 km² catchment to one of

Table 4.4 Effect of influent seepage on a flood at Walnut Gulch, Arizona

Basin area (km²)	Flood volume (m³ × 1000)	Peak discharge (m³ s⁻¹)	Specific discharge (m³ s⁻¹ km⁻²)
95	92.3	41.9	0.44
114	79.9	27.2	0.24
149	40.1	15.6	0.10

Source: Based on a table in Pilgrim et al. (1988) by permission of Blackwell Scientific Publications and on data from Renard and Keppel (1966)

0.58 km² and up to 70% of the runoff events recorded at the upstream catchment failed to reach the larger downstream catchment.

Since such downstream reductions in actual peak discharge are common in arid and semi-arid basins it follows that the downstream reduction in *specific* peak discharge is likely to be even more marked than in humid areas, for an equivalent increase in catchment size. This seems to be confirmed by the specific discharges shown in Table 4.4, which decline by a factor of 4.4 for an increase in catchment size of 1.6. In comparison, a 4.2 times decrease in specific discharge between the Lobe and the Ogooué in Table 4.2 is associated with a 105 times increase in catchment size. However, because of the great spatial variability of flood flows in arid and semi-arid environments, such comparisons are essentially unsafe. Indeed, Wolman and Gerson (1978) found similar specific discharges for flood peaks in a wide range of climatic regions, from humid to arid, although with a large scatter within each type of region.

4.6.2.3 Flash Floods

Flash floods are often discussed in the context of arid and semi-arid environments. However, although criticised by Penning-Rowsell et al. (1986) for its imprecision, the term 'flash flooding' is even more widely used to signify sudden-onset flooding in a broad range of climatological and geographical conditions. These include potentially disastrous flooding resulting from the failure of dams or other control works but, perversely, normally exclude sudden-onset flooding resulting from glacier outbursts (jökulhlaups). Hall (1981) suggested that most flash floods are associated with intense localised thunderstorm activity, slow-moving or stationary cyclones, both tropical and extra-tropical, and intense monsoon rainfall. He also emphasised the dangers of flash floods in mountainous areas where orographic uplift may intensify rainfall and where the steep slopes may increase the potential for landslides and mudslides. In what follows the generic nature of flash flooding is emphasised by first considering flash floods in a wide range of environments, including flash floods in arid and semi-arid areas. The section is concluded by a brief consideration of catastrophic flooding resulting from structural failures.

The suddenness and unexpectedness implied by the term 'flash flood' may manifest either very intense storm rainfall (e.g. Woolley, 1946) or rapid melting, or it may indicate that a storm has occurred on steep, bare, impermeable surfaces such as a narrow mountain valley or a heavily built-up urban area, or in a small catchment through which the resulting flood peak passes too rapidly for adequate flood warnings to be given. These torrential floods are difficult to predict or forecast and often cause considerable damage which may be out of all proportion to the peak discharges actually experienced. Areas which are attractive to tourists and vacationers because of their dry sunny climates may nevertheless experience occasional intense storms and flash flooding of great severity. But because the river valleys, e.g. the 'ramblas' of Spain and the Canary Islands, are dry most of the time their floodplains have been used for buildings and campgrounds so that when floods occur damages are usually severe (e.g. Heras, 1974; Sempere et al., 1994). A flash flood, caused by intense thunderstorms associated with a southward-moving cold front, killed more than 80 people at a valley-bottom campsite near Biescas in the Spanish Pyrenees on 8 August 1996. The size and suddenness of the flood peak resulted partly from the collapse of a debris dam just upstream of the campsite. Late on Christmas night 1995 torrential rain, during which more than 100 mm was recorded in a period of half an hour, resulted in a flash flood

Table 4.5 Selected catastrophic floods in the USA and comparative data from the British Isles

No on Figure 4.13	Location	Area (km^2)	Peak Q (m^3 s^{-1})	Date	Type of flood
	Big Thomson Canyon, CO	155.00	884.0	31 July 1976	Flash
	Rapid City, SD	236.00	1433.0	09 June 1972	Flash
	Johnstown R, PA	1850.00	3260.0	20 July 1977	Flash
1	Monitor, WA	0.39	25.6	02 August 1956	Flash
2	Silver Springs, NV	0.57	47.6	20 July 1971	Flash
3	Old Irontown, UT	0.78	74.5	11 August 1964	Flash
4	Rockport, MO	1.97	144.0	18 July 1965	Flash
5	Rye Patch, NV	2.20	251.0	31 May 1973	Flash
6	Washta, IA	4.92	311.0	09 August 1961	Flash
7	Molin, OR	13.10	807.0	26 July 1965	Flash
8	Mitchell, OR	32.90	1540.0	13 July 1956	Flash
9	Wikieup, AZ	49.20	2080.0	18 August 1971	Flash
10	Nelson Landing, NV	59.30	2152.0	14 September 1974	Flash
11	Albany, TX	102.00	2920.0	04 August 1978	Flash
12	Fountain, CO	141.00	3510.0	17 June 1965	Flash
13	Loma Alta, TX	195.00	4810.0	24 June 1948	Flash
14	D'Hanis, TX	368.00	6510.0	31 May 1935	Flash
15	Syracuse, NE	549.00	6370.0	09 May 1950	Flash
16	Deer Trail, CO	782.00	7760.0	17 June 1965	Flash
17	Brackettville, TX	1813.00	15600.0	14 June 1935	Flash
18	Scotia, CA	8063.00	21300.0	23 December 1964	Regional
19	Comstock, TX	9300.00	27440.0	28 June 1954	Regional
20	Harrisburg, PA	62400.00	28900.0	24 June 1972	Regional
21	Vicksburg, MS	2964300.00	55600.0	12 May 1973	Regional
Comparative data from British Isles					
	Yellow R, Ireland	14.80	170.0	29 June 1986	Flash
	Allt Mor, Scotland	16.40	74.0	04 August 1978	Flash
	Lynmouth, England	101.45	252.0	15 August 1952	Flash

Source: Based on a table in Hirschboeck (1987) and on data in NERC (1975), McEwen and Werritty (1988) and Coxon et al. (1989)

which swept water and mud through a township outside Pietermaritzburg in the KwaZulu/Natal province of the RSA, killing at least 130 people.

A spate of disastrous flash floods occurred in the USA during the 1970s (see Table 4.5). At Rapid City, South Dakota, in June 1972, 237 people died when torrential rains of up to 400 mm, close to the annual average amount and having an estimated return period of several thousand years (Bolt et al., 1975), fell in less than six hours. The basic cause of this event was a stationary supercell thunderstorm and air with a high precipitable water content (Hirschboeck, 1987). Many lives were also lost at Big Thompson Canyon, Colorado, in July 1976, and at Johnstown, Pennsylvania, in July 1977.

The study of flash floods has attracted increasing attention throughout the world and has provided the focus for major programmes of international research such as IHP Project H-5-2 (Flash Floods). Analysing data from the USA, Hirschboeck (1987) recognised flash floods as that subset of devastating floods generated largely from 'small and medium size' basins, in comparison with the rarer large-scale regional floods, and noted the concentration of flash floods in the western states (see Table 4.5 and Figure 4.13). An earlier classification of the meteorological characteristics of flash flood events in the USA by Maddox et al. (1979; 1980), largely confirmed by Hirschboeck's work, had indicated basic contrasts between flash floods in western and eastern states. Those in the west appeared to be associated with relatively weak large-scale atmospheric

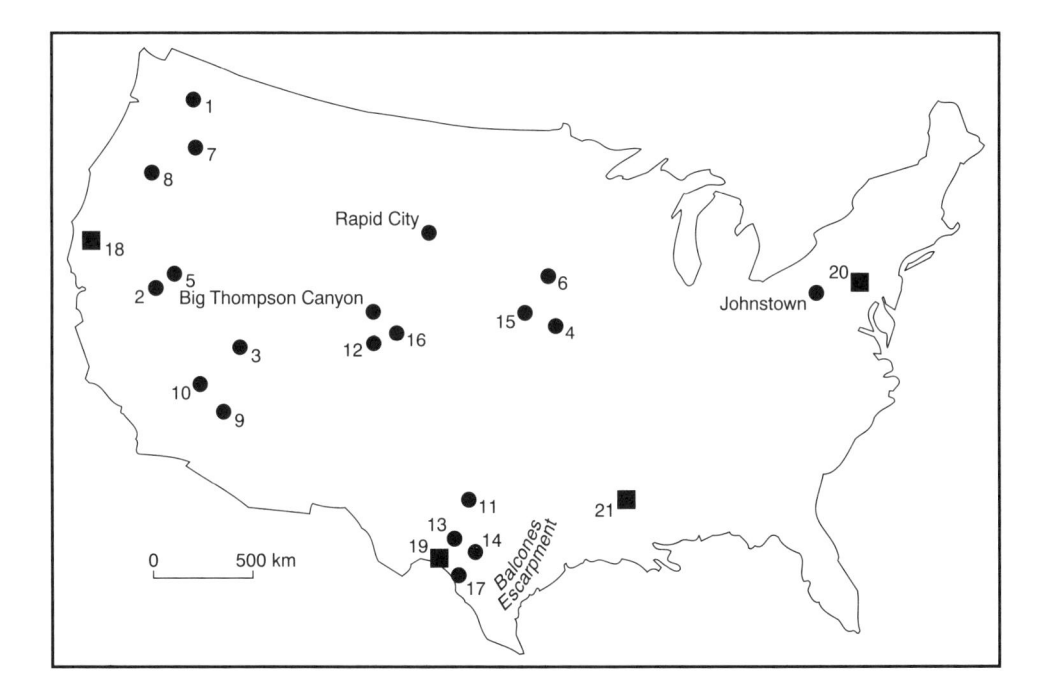

Figure 4.13 Location map of small-basin flash floods (filled circles) and large regional floods (filled squares) in the USA. See also Table 4.5
Source: Based on a diagram in Hirschboeck (1987)

patterns, and to show a seasonal concentration in July and August, perhaps indicating a link with the intrusion of moist air during the south-west monsoon season, and a diurnal concentration during the afternoon and evening. Eastern flash floods tended to occur at night and to involve larger amounts of rainfall.

In the British Isles flash floods and references to 'walls of water' (e.g. Archer, 1994) are quite common. However, these events are relatively modest in scale compared with flash floods in the USA (see Table 4.5). By far the best known is the Lynmouth flash flood of 15–16 August 1952 which resulted from an exceptional rainfall of up to 300 mm. This was one of the three heaviest 24-hour rainfalls ever recorded in the British Isles and may have been amplified by recently disclosed rainmaking experiments carried out by the Ministry of Defence. In any case, flood response was greatly intensified by antecedent and catchment conditions. Rain had fallen on all but two of the preceding 14 days and had totalled almost 90 mm. The ground was therefore unusually wet and this, together with thin soils underlain by bedrock over much of the catchment, would have encouraged rapid and massive expansion of quickflow source areas (see Section 3.2) immediately after the onset of the flood-producing storm. At the height of the storm the runoff rate probably equalled the rainfall rate over the high ground for a period of several hours (Bleasdale and Douglas, 1952). The large volumes of water moving over the moorland plateau of the Lyn catchment flattened the bog grass and effectively 'thatched' the ground surface, before pouring off the moor and into the deeply incised, steeply sloping valleys of the River Lyn system. Within the valleys it seems clear that temporary dams developed when trees and boulders piled up behind bridges and other obstructions and that the sudden breaching of these dams greatly intensified the devastating nature of the floods (Kidson, 1953).

Obviously no record of the flood hydrograph could have survived, but local residents described the flood's onset as resembling a 'wall of water' or a 'tidal wave' (Marshall, 1952). Immediate post-event reconstructions (e.g. Dobbie and Wolf, 1953) showed that peak discharge probably occurred about 3.5 hours after the beginning of rainfall and yielded estimated values which were considerably larger than the $252 \, m^3 \, s^{-1}$ which was subsequently used in the *Flood Studies Report* (NERC, 1975). Even this revised estimate, however, places the Lynmouth flash flood second amongst recorded UK floods in terms of the ratio (18.4:1) between peak discharge and median annual flood.

The Allt Mor, Scotland, and the Yellow River, Ireland, flash floods were described by McEwen and Werritty (1988) and by Coxon et al. (1989). Although, in both cases, heavy rainfall was the immediate trigger for the flooding, an equally important factor, as at Lynmouth, was the very wet antecedent conditions. This seems to be a recurring theme in the descriptions of flash flooding in temperate areas, where there is normally a widespread vegetation cover and where the *wetness* of the catchment, rather than the widespread occurrence of steep, bare, impermeable surfaces, or even the sheer intensity of the rainfall, is likely to be the main predisposing factor. Further specific examples were the disastrous floods in Virginia and West Virginia, USA, in November 1985 (Clark et al., 1987) and a succession of flash floods in forested valleys in the Cévennes, France, which seem to be triggered only when a threshold of about 270 mm groundwater and soil storage has been exceeded (Cosandey and Didon-Lescot, 1990).

Almost by definition, antecedent wetness is rarely a significant factor in the initiation of flash flooding in arid and semi-arid environments. Instead, as indicated in Section 4.6.2.1, low infiltration, the dominance of overland flow, and high drainage densities approaching 100 km km^{-2}, mean that the temporal parameters of the flood hydrograph (time of start, time of concentration) follow each other without delay so that the time of rise of the flood hydrograph is typically between 4 and 16 minutes (Reid and Frostick, 1987).

Because of the short-lived nature of desert flash floods, the 'wall of water' phenomenon, or flood bore, has not often been observed and the rare published eyewitness descriptions have been largely qualitative. Hassan (1990) observed the arrival of a bore, some 0.30 m in height, after heavy rainfall in the Judean desert. The bore was described as steep and noisy and peak stage was attained within 10 minutes. In order to increase the amount of quantitative data on desert flash floods, fully automatic monitoring stations were established on two wadis draining the south Hebron Hills in the northern Negev desert in Israel (Reid et al., 1994). The hydrograph of a typical flash flood bore in one of them is shown in Figure 4.14. The bore was 0.20–0.30 m high, moved at a speed of about 1 m s^{-1}, and reached a depth of 0.90 m in less than three minutes. After about an hour, the arrival of a second bore led to a further rapid increase in flow depth. This hydrograph, together with photographs of the arrival of the flash flood in the Nahal Eshtemoa (see Plate D), probably constitute the most detailed documentation yet available of a flood bore advancing down a completely dry channel.

Very substantial sediment loads may be transported and deposited by short-lived desert floods. As a result alluvial fans may be formed where tributary channels join a main wadi and may partially block the main wadi channel. Subsequent flood flows in

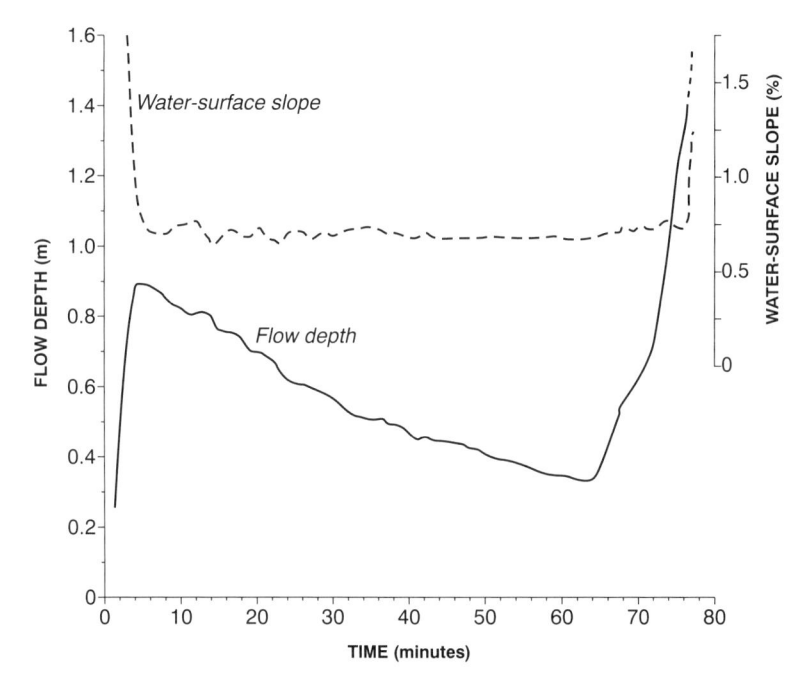

Figure 4.14 Hydrograph of flash flood of 11 November 1993 in the Nahal Eshtemoa wadi, Israel. The reach average water-surface slope is also depicted
Source: Reid et al. (1994), copyright by the American Geophysical Union

the main wadi may then be temporarily impeded until the alluvial fan is breached (Schick and Lekach, 1987). Alluvial fans in the lower reaches of large desert channels, however, may develop into larger and more persistent features, especially where the channel debouches at a major break of slope (e.g. Woolley, 1946). Such features create a special type of flash flood threat, especially when they form the site for urban development as in parts of the Middle East and the south-west of the USA. Drainage channels on the fans can shift unpredictably across the relatively steep slopes, bringing high-velocity flows which commonly reach speeds of 5–10 m s^{-1} and which are highly charged with sediment (Smith, 1996).

Floods resulting from dam failures are often much larger than those originating from rainfall or snowmelt (Costa, 1988). According to Jansen (1980) about 200 significant failures of constructed dams have occurred in the last 100 years, accounting for more than 11 000 deaths. Some of these failures were caused by deliberate military action, as in the 'Dambusters' raid on the Eder and Mohne dams in May 1943 which flooded an extensive area and caused more than 1300 deaths. Most, however, have resulted either from overtopping when floods exceed the spillway capacity, or from foundation defects, or from piping and seepage (Biswas and Chatterjee, 1971; International Commission on Large Dams, 1973). Many have occurred during initial filling because this is the time when design or construction flaws or latent site defects tend to appear (Costa, 1988).

Although information on deaths and damage is often readily available for such events, it is more difficult to reconstruct the flood peak, e.g. from geomorphological evidence, or to predict the hydrological characteristics (such as peak discharge, stage, volume and flood wave travel time) of a dam failure flood. Dam-break models are often simple and empirically based, and relate peak discharge to dam height and/or reservoir volume, e.g.

$$Q_{max} = a\, H^b \tag{4.1}$$

$$Q_{max} = a\, V^b \tag{4.2}$$

$$Q_{max} = a\, (HV)^b \tag{4.3}$$

where Q_{max} is peak discharge, H is dam height, and V is reservoir volume (e.g. US Soil Conservation Service, 1981; Hagen, 1982; Costa, 1988). However, more complex models have also been developed, e.g. the US Corps of Engineers HEC-1 program and the US National Weather Bureau DAMBRK model (Fread, 1980).

A spectacular dam failure occurred on 2 December 1959 when the left foundation of the 66.5-m-high Malpasset dam near Fréjus, in southern France, collapsed releasing 25×10^6 m^3 of water which swept down the Reyran valley, partly destroying the town of Fréjus and causing the loss of 421 lives. One of the most infamous examples of a dam-failure flood was the Johnstown, Pennsylvania, disaster of 31 May 1889 when the Conemaugh earth dam 25 km upstream burst after several weeks of very heavy rain. Water surged through a 100-m breach and moved downvalley picking up an enormous amount of debris which caused temporary damming and further surges as these dams broke. The flood wave destroyed villages and industrial plant before surging through

Conemaugh, Franklin and finally Johnstown where the highest estimated death toll was 2280.

Smaller, but in some ways similar, was the 1864 Dale Dyke flood in northern England. As at Johnstown, an earth dam holding back a long shallow lake was subjected to a period of above-average snow and rain. The resulting 100-m breach created a sudden outflow which demolished the hamlet of Lower Bradfield and caused nearly 250 deaths. Another British example was the Dolgarrog, Wales, disaster of 1925 when the foundations of a recently built hydropower dam gave way, resulting in a flood which destroyed the small village on the floor of the Conwy valley. This event was especially important to the history of flood hydrology in Britain since, as Newson (1989) observed, it was instrumental in prompting the seminal report of the Institution of Civil Engineers on *Floods in Relation to Reservoir Practice* (ICE, 1933).

In some cases dam flood disasters have occurred even when the dam did not fail. The 265-m high Vaiont Dam on the Piave River in north-eastern Italy, for example, was commissioned in 1960 as the third highest concrete dam in the world. Three years later seismic activity caused a massive landslip on the slopes of Mount Toc, which dumped $115 \times 10^6 \, m^3$ of rock into the reservoir and created a giant wave which overtopped the dam. A 70-m wave of water passed downvalley, destroying the village of Longarone, and much else in its path, and killing nearly 3000 people (Biswas and Chatterjee, 1971; Davidson and McCartney, 1975).

'Natural' dams may be formed in a variety of ways, e.g. by landslides across valleys or by ice blockages (see also Section 4.5.2). In areas of snow and ice, global warming is likely to increase the risk of flooding from the collapse of natural dams, especially where these are formed by essentially unstable terminal moraine. Fukushima et al. (1991) cited the flood disaster in 1985, ensuing from such a dam collapse in the Khumbu region of East Nepal during a summer-season typhoon, as an example of what might happen under the loading of additional meltwater runoff following global warming.

4.7 FLOODS IN TEMPERATE CLIMATES

As described briefly in Section 4.4, temperate climates are separated equator-wards from the hot climates by the mean annual isotherm of 20°C and polewards from the cold climates by the isotherm of 10°C for the warmest month. The so-called temperate zone is therefore broadly conterminous with the mid-latitudes and indeed temperate climates are now more usually referred to as *mid-latitude climates*. Especially in continental interiors such climates are far from 'temperate', being characterised by large extremes of temperature and by intense interactions between tropical warm airmasses moving polewards and polar cold airmasses moving equator-wards. This means that flooding in temperate climates tends to exhibit a mixture of the processes and forms already considered in Chapter 3 and in Sections 4.5 and 4.6 as being typical of cold-climate and hot-climate floods. Accordingly, in this section, selected examples of major floods in different temperate areas are used to illustrate some of the main threads of the discussions on the processes and forms of river flooding which have formed the subject matter of Chapters 3 and 4.

4.7.1 Susquehanna River Floods, 1972

Record-breaking flooding was caused by the exceptional rainfall associated with Hurricane Agnes that devastated the mid-Atlantic states of the USA in late June and early July 1972. Described in a contemporary assessment as 'the greatest natural disaster ever to befall the Nation' (US Department of Commerce, 1973), the event was remarkable for the great areal extent of the flooding which affected 12 states and was particularly severe in New York, Pennsylvania and Virginia, and which resulted in many very large drainage basins experiencing record-breaking floods (Hirschboeck,

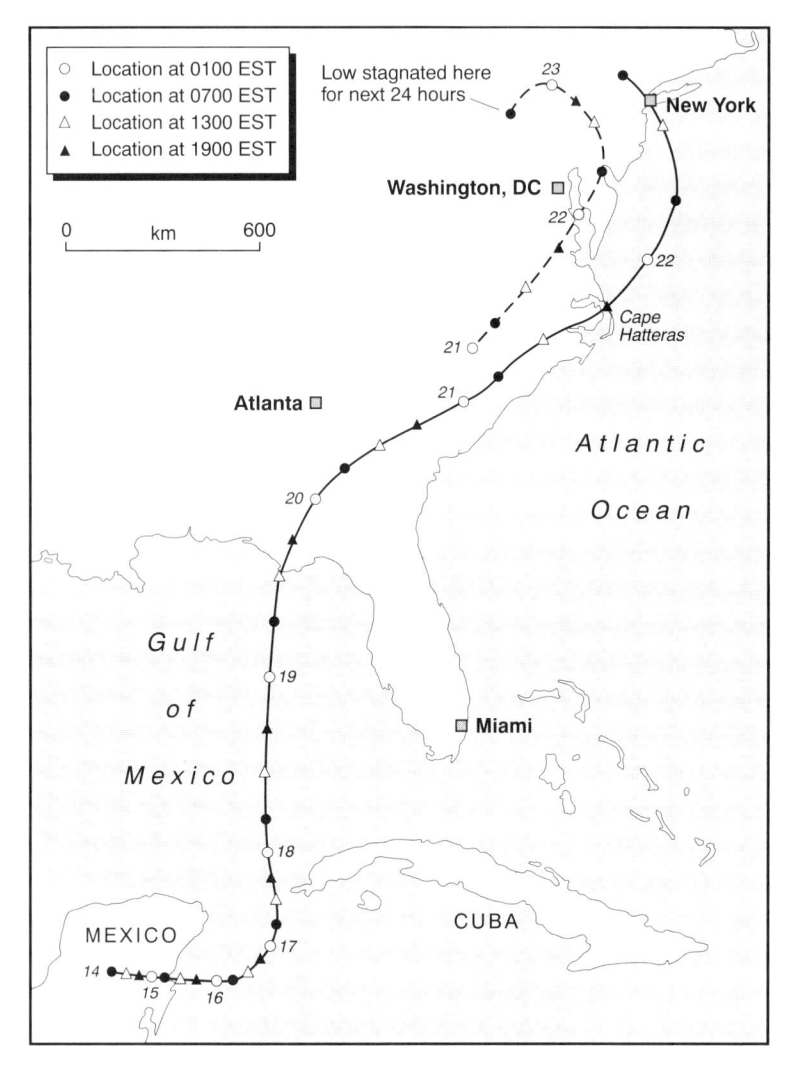

Figure 4.15 Tracks showing six-hourly locations over the eastern USA of the central low-pressure areas of Hurricane Agnes, 14–23 June 1972 and of an extra-tropical low, 21–23 June 1972

Source: Bailey and Patterson (1975). Reproduced by permission of the U.S. Geological Survey

1987). The estimated recurrence intervals of peak flows exceeded 100 years on many major rivers (Bailey and Patterson, 1975).

Ironically, although its effects on flooding were so severe, Agnes was a comparatively weak hurricane with windspeeds appropriate to a tropical storm, rather than a hurricane, for much of its life span. It was exceptional because of its relatively large diameter of 1600 km, its long overland track through populated coastal areas (Figure 4.15), and the fact that during its waning stages it merged with an extra-tropical cyclone and then stagnated over western Pennsylvania for about 24 hours as a result of an abnormal blocking configuration over the North Atlantic (Hirschboeck, 1987). This fed large amounts of moisture into the storm in its later stages. It is also believed that anomalously warm sea-surface temperatures in the western North Atlantic, possibly linked to the 1972–73 El Niño event (see Section 3.6.3), played a major role in sustaining the blocking configuration and in directing the hurricane's path to unusually high latitudes (Namias, 1973). As a result, Agnes rainfall amounts exceeded 375 mm in several locations from New York to Virginia and set a record for areas greater than 25 000 km^2 and durations longer than 24 hours (Bailey and Patterson, 1975).

Although, in the middle and lower reaches of the Susquehanna River, flood flows exceeded the recorded maximum at many locations, this was not the case in the headwater areas where flood recurrence intervals were small, e.g. two years at Vestal, NY (gauging station GS 102) (see Figure 4.16A), and 40 years at Waverly, NY (GS 107). However, according to Bailey and Patterson (1975), flood flows below Waverly increased rapidly downstream as all tributary streams experienced very high rates of runoff and the maximum peak discharge was determined as 31 640 m^3 s^{-1} at Conowingo, MD (GS 302). So remarkable was the downstream increase of flow that peak specific discharge was actually higher at Harrisburg, PA (GS 283) than at any of the upstream stations shown (Figure 4.16B). Peak flows had recurrence intervals greater than 100 years and greatly exceeded previous recorded peaks (see Figure 4.16C), e.g. by a factor of 1.67 at Towanda, PA (GS 169), 1.49 at Wilkes-Barre, PA (GS 183), 1.12 at Sunbury, PA (GS 247), and 1.38 at Harrisburg (GS 283).

Peak stages also exceeded previous known maxima by a considerable margin, e.g. by 2.56 m at Towanda, the highest level recorded since at least 1865, and by 2.38 m at Wilkes-Barre, the highest level recorded since at least 1784. It is estimated that maximum flood levels would have been up to 0.6 m higher had it not been for the effects of reservoir storage in the Susquehanna basin (Bailey and Patterson, 1975).

4.7.2 Mississippi River Basin: The Great Flood of 1993

From May to September 1993 catastrophic flooding occurred across much of midwestern USA. This was the greatest flood event in the Mississippi basin since records began. Approximately 600 river forecasting points were above flood stage simultaneously and nearly 150 major rivers and tributaries were affected by record high flows over a channel length of 2900 km, including 840 km of the Mississippi and 670 km of the Missouri (see Figure 4.17A). The floods inundated a larger area (more than 1×10^6 km^2) with a greater volume of floodwater and for a longer duration than any earlier event on record (Koellner, 1996).

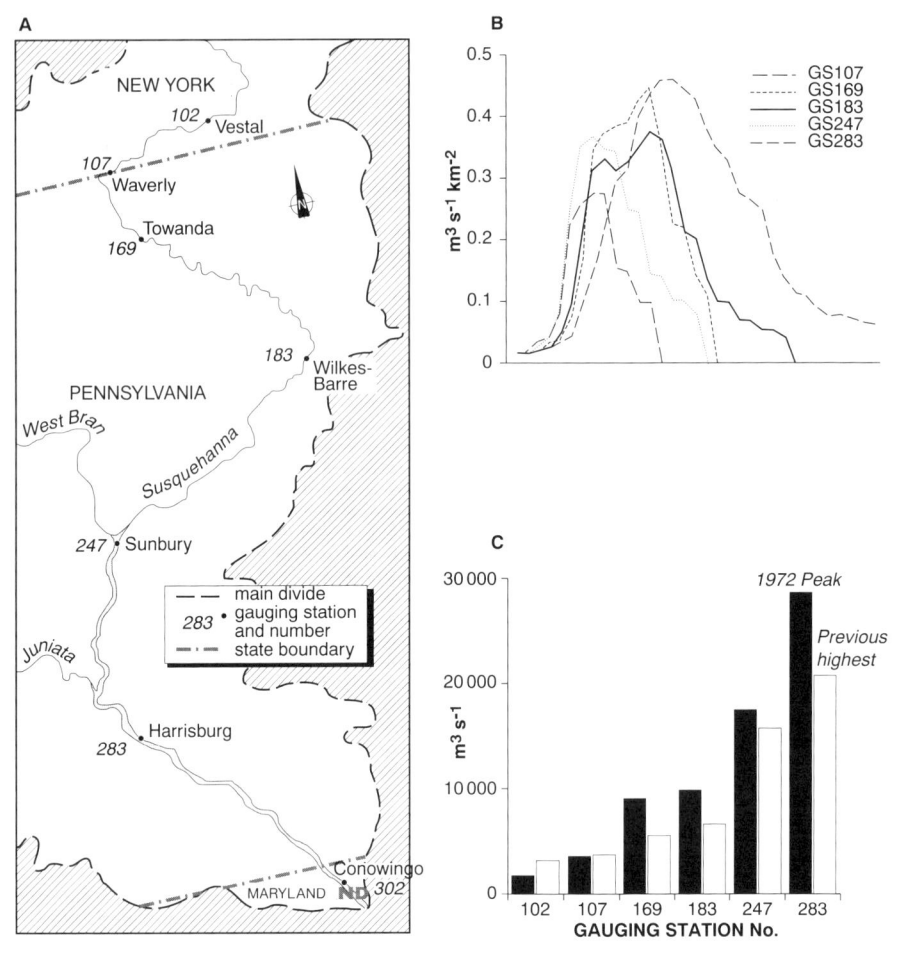

Figure 4.16 Flood discharges at selected gauging stations on the Susquehanna River, 21–28 June 1972: (A) location of gauging stations; (B) hydrographs showing downstream increase of specific discharge ($m^3 s^{-1} km^{-2}$); (C) bar-graphs showing relationship between 1972 flood peaks and previous highest flood peaks ($m^3 s^{-1}$)

Source: Bailey and Patterson (1975). Reproduced by permission of the U.S. Geological Survey

4.7.2.1 *The Climatological Context*

The 1993 flooding resulted from a combination of several factors, including wet antecedent conditions, an abnormal weather pattern, and increased runoff response. Although precipitation during the winter of 1992/93 was near normal, July, September and November rainfalls in 1992 were well above normal so that by late March 1993,

---

Figure 4.17 *(opposite)* The Mississippi River basin floods of 1993: (A) the extent and severity of flooding in July–August; (B) jet stream location and dominant atmospheric conditions in June–July
Source: NOAA (1994). Reproduced by permission of the U.S. National Weather Service, Office of Hydrology

RIVER FLOODS: SPATIAL CHARACTERISTICS

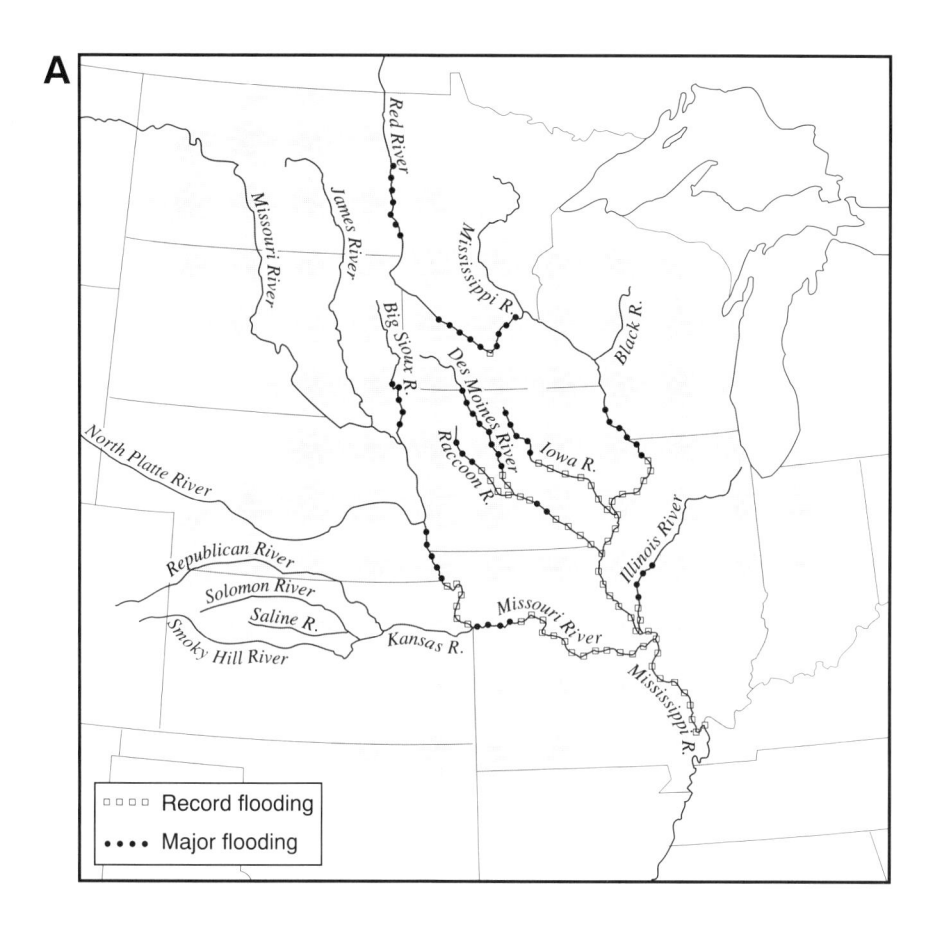

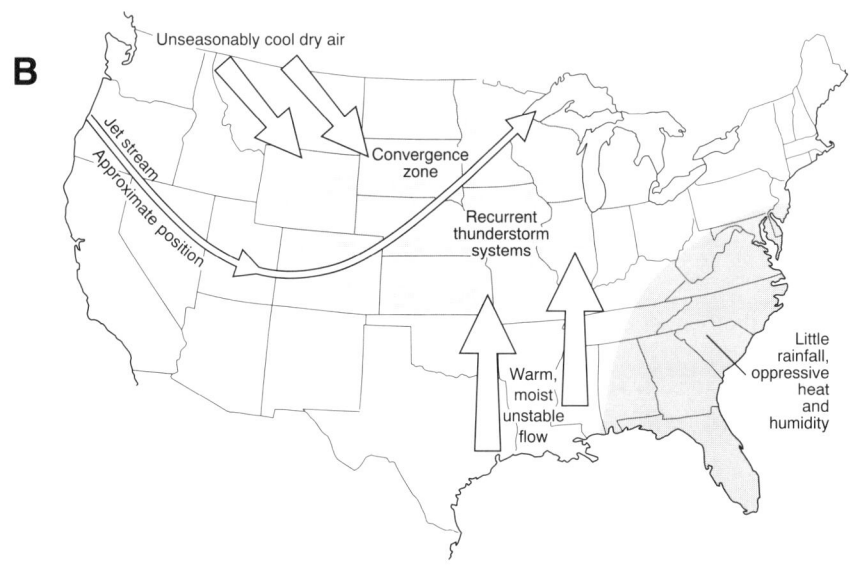

Table 4.6 Precipitation amounts and return periods for some states in the Mid-West of the USA during the summer of 1993

	April–July		June–July	
State	Amount (mm)	Frequency (years)	Amount (mm)	Frequency (years)
Iowa	688	300	460	260
Illinois	582	45	373	85
Wisconsin	559	200	312	75
Minnesota	480	70	310	100

Source: From a table in NOAA (1994). Reproduced by permission of the U.S. National Weather Service, Office of Hydrology

following spring snowmelt, extremely moist conditions (Palmer Drought Index >4) occurred over much of the Mid-West (NOAA, 1994). Above-normal precipitation over the upper Mississippi River basin during April and May meant, therefore, that even before the onset of heavy summer rains, most of the upper Mid-West had saturated soil and well-above-normal streamflows.

During the period from the beginning of June to the end of August 1993 rainfall totals exceeded 300 mm across much of the area. More than 600 mm of rain fell on central and north-eastern Kansas, northern and central Missouri, most of Iowa, southern Minnesota, and south-eastern Nebraska, with up to 975 mm in east-central Iowa. These amounts were approximately 200–350% of normal (Larson, 1995) and were associated with large return periods (see Table 4.6).

One of the principal reasons for this sequence of events was the anomalous pressure pattern and circulation system which established itself across the area. By the summer of 1993 the polar jet stream had moved south from its normal position close to the Canada–USA border and had become firmly established over the northern part of the Mississippi River basin with a south-west–north-east orientation (see Figure 4.17B). This quasi-stationary jet stream was associated at the surface with a stationary front that allowed nearly continuous overrunning of the cooler air to the north by the very moist air from the south. The front also acted as the preferred location for unusually strong and frequent cyclones which originated from the combination of the unseasonably vigorous jet stream overhead and the relatively strong frontal boundary at the surface (NOAA, 1994; Prestegaard et al., 1994; Bell and Janowiak, 1995).

Such was the magnitude and persistence of these atmospheric circulation anomalies that the authors of the subsequent Natural Disaster Survey Report (NOAA, 1994) concluded that they constituted 'a significant climate variation rather than simply a sequence of meteorological incidents'. The causes of this climate variation are unclear but may have been related to a long-lived ENSO episode (see Section 3.6.3.1) during 1992 and 1993 (Lott, 1994).

4.7.2.2 The Role of Human Interference

Although the prime causes of the great flood of 1993 were climatological, the magnitude and intensity of flooding were undoubtedly exacerbated by human action.

Two aspects in particular, i.e. agricultural drainage and levee construction, appear to have played a significant role.

In certain conditions agricultural drainage may intensify flooding (see also Section 3.6.1.3) and Prince (1995) referred to several studies which have shown this to be the case in parts of the Mid-West. Extensive drainage of wetlands such as sloughs, marshes and prairie potholes, has occurred since the Swamp Land Acts were passed by the US Congress in the mid-nineteenth century and has been accompanied by channel 'improvement' and levee construction in the drained areas. In Illinois, for example, almost all wet prairies have been drained for agriculture and in Iowa only 1% of the surface is now occupied by wetlands. Prince (1995) drew attention to a postscript added to a White House paper on environmental policy on 24 August 1993, in which the flood-ameliorating role of wetlands was acknowledged and which concluded:

> We must be cautious not to repeat the policies and practices which may have added to the destruction caused by these floods. One way to assist landowners while alleviating some flood risks is through funding wetlands restoration and acquisition programs . . .

The failure of flood levees causes disastrous inundation of adjacent land, although this will tend to reduce flood levels further downstream. Not surprisingly, during the 1993 flood 18% of Federal levees and 78% of non-Federal levees failed or were overtopped (NOAA, 1994) (see also Section 7.2.2). In urban areas the former are normally designed to withstand floods having return periods of about 200 years; the latter, mainly protecting agricultural land, are rarely designed against return periods of more than 50 years (Bhowmik et al., 1994).

4.7.2.3 Description of the Floods

Essentially the great flood of 1993 consisted of three flood waves which travelled down the Mississippi and Missouri rivers in mid-June, early July and late July. Such was the magnitude of these sequential events that, at some locations, flood stage was exceeded for a continuous period of five months. The most severe flooding occurred on an 800 km reach of the Mississippi River between Cairo, Illinois, and Minneapolis, Minnesota, and along a 640 km reach of the Missouri River between Omaha, Nebraska, and St Louis, Missouri. In places the Mississippi River expanded to a width of 12 km and the Missouri to 16 km although, near their confluence just north of St Louis, the Missouri attained a width of 32 km (Lott, 1994).

The 200 mm of precipitation which fell across the upper Mid-West in mid-June resulted in flooding on rivers in Minnesota and Wisconsin. This eventually contributed to the 12 July flood peak at St Louis which almost equalled the previous record stage of 13.2 m (see Figure 4.18A). Storm totals of 200 mm in early July caused severe flooding on rivers in Iowa. Their flow then combined with the near-record flows on the Mississippi River to push the stage at St Louis to a new record of 14.3 m on 20 July (Larson, 1995). In mid- to late July heavy rains farther west caused record flooding on rivers in Missouri, Nebraska, Kansas and the Dakotas. The Missouri River crested at 14.9 m on 27 July, breaking the 1951 previous record by more than 0.8 m (see Figure 4.18B). As this flood peak travelled downstream it set new records at many other

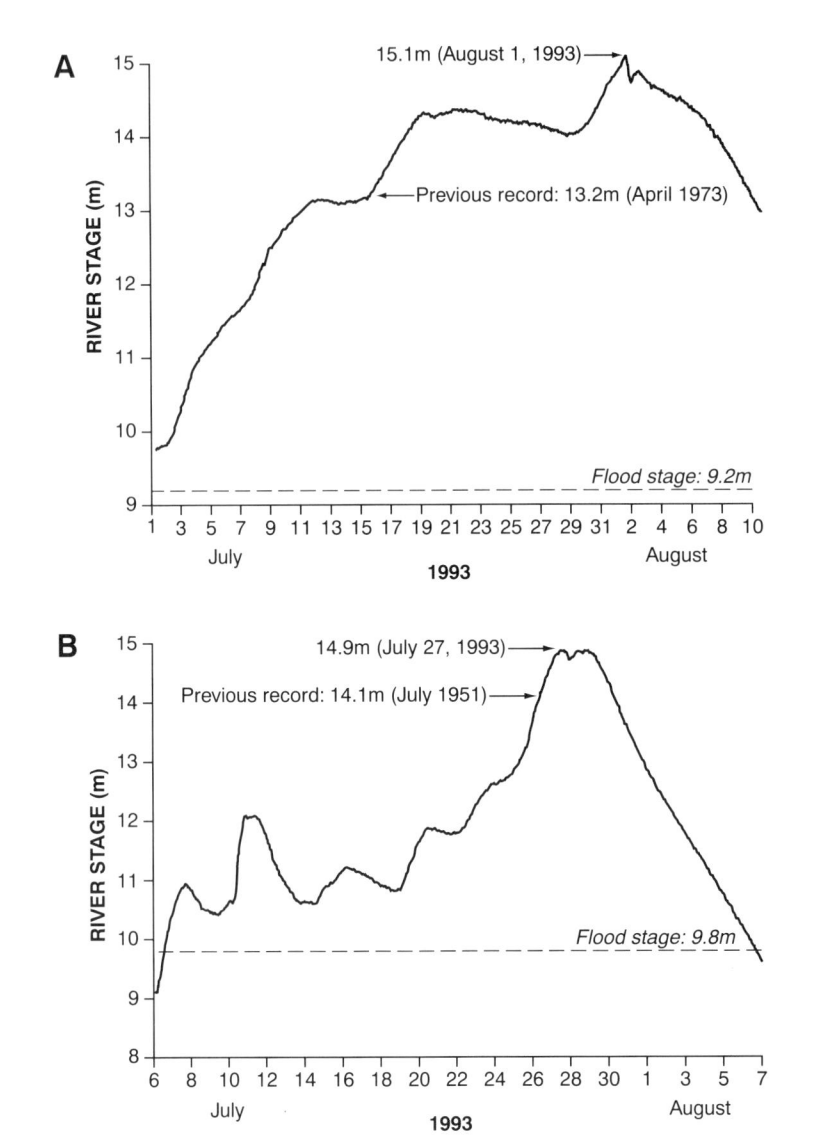

Figure 4.18 Flood hydrographs July–August 1993: (A) the Mississippi River at St Louis, Missouri; (B) the Missouri River at Kansas City, Missouri
Source: From diagrams in Larson (1995) with permission

locations. It then joined the already full Mississippi River just north of St Louis to push the Mississippi to another record stage at St Louis of 15.1 m on 1 August. At the peak level the floodwaters came within a metre of overtopping the floodwall protecting St Louis and designed to withstand the 500-year flood (Evans, 1994). Because of the high flood stages experienced and the large number of levee breaches that occurred, flood duration during 1993 was exceptionally long at many locations, especially on the Mississippi River (see Table 4.7).

Table 4.7 Flood stages, new stage records and flood durations at selected locations on the Mississippi and Missouri rivers during the summer of 1993

Location	Flood stage (m)	Record set in 1993 (m)	Increase on previous record (m)	Flood duration (days)
Mississippi River				
Quincy, IL	5.2	9.8	1.0	152
Hannibal, MO	4.9	9.7	1.0	174
Louisiana, MO	4.6	8.7	0.4	186
Clarksville, MO	7.6	11.5	0.4	187
Winfield, MO	7.9	12.1	0.9	183
Grafton, IL	5.5	11.7	1.6	195
Melvin Price, IL	6.4	13.0	1.8	179
St Louis, MO	9.2	15.1	2.0	146
Chester, IL	8.2	15.2	2.0	186
Missouri River				
Jefferson City, MO	7.0	11.8	1.3	62
Hermann, MO	6.4	11.1	0.2	77
St Charles, MO	7.6	12.0	0.6	94

Source: Based on data in Larson (1995)

Although a great deal of information is available on flood stages during the great USA floods of 1993, the sheer magnitude of these events and, again, the number of levee breaches that occurred, means that comparatively few hard data are available on flood discharges and still less on their estimated return periods. Observations indicated that there had been a significant increase in Mississippi discharge at St Louis during the three decades before the 1993 floods. Of the 10 maximum flood discharges since measurements began, the 1993 peak of 28 840 m^3 s^{-1} was the highest, ahead of that for June 1903 (28 364 m^3 s^{-1}). Other 'estimated' record flood peaks at St Louis occurred in 1875 (37 800 m^3 s^{-1}), 1844 (36 400 m^3 s^{-1}), 1850 (29 680 m^3 s^{-1}) and 1855 (29 400 m^3 s^{-1}) (Bhowmik et al., 1994). The important role of levee breaching in the redistribution of flood discharges is illustrated by the estimate that on the day of peak discharge some 20–25% of the Mississippi River flow passed through a single breach 665 m wide which was scoured by the floodwaters to a depth of 22 m (Bhowmik et al., 1994).

It is known that some of the areas flooded were outside the mapped 100-year floodplain boundaries (Lott, 1994) and a US Geological Survey analysis of flow data from 154 gauging stations in the upper Mississippi basin showed that the 1993 flood peaks at 46 of the gauging stations had a return period of more than 100 years and at 42 stations they exceeded all previous maxima (Parrett et al., 1993).

4.7.3 River Rhine: The Christmas Floods of 1993/94

The drainage basin of the Rhine (185 000 km^2) is only about one-tenth that of the upper Mississippi. However, the Rhine has one of the highest annual discharges of any river in Europe and its floods can constitute a major hazard, particularly in the lower reaches close to the borders with Belgium and the Netherlands. From December 1993 to

January 1994 a number of European countries experienced severe flooding caused by heavy precipitation over an area stretching from south-west Ireland to Poland. Severe damage was widespread and the events on the Rhine ranked first or second among the major floods on the river over more than a century (Engel, 1994).

4.7.3.1 The Climatological Context

The Christmas floods were caused primarily by two periods of precipitation which occurred from 7 to 18 December and from 19 to 21 December 1993. In the first of these the rapid passage of low-pressure systems from the west caused frequent daily falls of between 5 and 10 mm. In addition, snow which had fallen on ground above 600 m thawed in mid-December. The combined effect of the rainfall and snowmelt was to saturate the ground surface over wide areas of the middle and lower basin and to increase river levels to well above the long-term mean, e.g. the Rhine attained levels 1.6 m and 2.0 m above normal at Koblenz and Cologne respectively and the Moselle was 2.4 m above normal at Trier (Engel, 1994). The second period of precipitation began in mid-December with an advance of mild Atlantic air which was accompanied by heavy precipitation and by further thawing, even at high elevations. From 19 to 21 December precipitation intensities in some places reached 120 mm in 48 hours and again, the combination of rainfall and snowmelt resulted in major flood peaks on the Rhine and on some of its major tributaries.

The combined magnitude of these two periods of rain is reflected in the comparisons of the total monthly rainfall for December 1993 with mean December rainfall for the period 1951–80. In the Saar and Neckar basins this exceeded 400% of the December mean (Engel, 1994). Of the 39 stations in the middle and lower Rhine basin, 13 received rainfall of more than 200 mm in December 1993, including Oderen with 568 mm and Freudenstadt with 636 mm. At 18 stations the December 1993 precipitation ranked first in the long-term December series and at another 16 stations it ranked second or third (Engel et al., 1994).

4.7.3.2 The Role of Human Interference

Although record-breaking rainfall, combined with snowmelt, was undoubtedly the primary cause of the Rhine Christmas floods, much of the contemporary and subsequent media comment implied that the severity of the flooding was also partially the result of human interference and called for remedial measures to be taken to minimise the effects of future floods. It was suggested that these might include the restoration of 'natural river landscapes', the removal of 'paved areas' and an increase in 'water retention' in the catchment areas and along the rivers.

In a review of such possibilities, Ebel and Engel (1994) noted that 950 km^2 of floodplain have been lost to development from an original 1400 km^2. It would not be possible to restore all of these former floodplains but even so there is clearly scope for creating a significant volume of water storage in these areas. Although about 15% of the ground surface in Germany is impermeable, only a small part of this has been artificially sealed and even optimistic estimates suggest that only 5% of the surface could be made more accessible to infiltration. In any case this is unlikely to have a significant effect on major floods which often result from growth of surface-saturated

RIVER FLOODS: SPATIAL CHARACTERISTICS 141

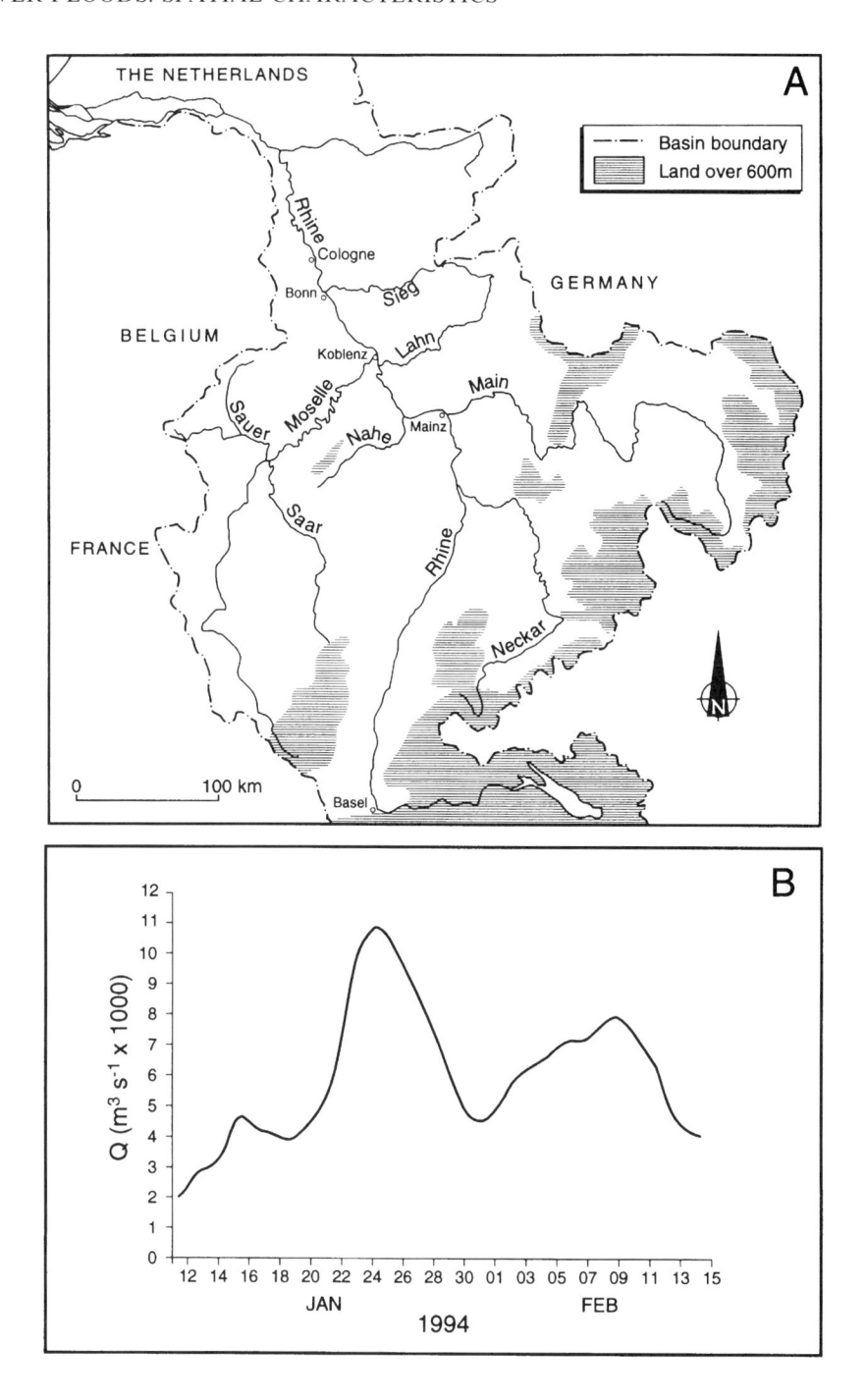

Figure 4.19 The Rhine Christmas floods 1993–94: (A) map of the Rhine basin showing the river and its major tributaries; (B) flood hydrograph of the Rhine at Cologne
Source: From diagrams in Engel et al. (1994) with permission

Table 4.8 Flood stages, discharges and return periods on the Rhine and its major tributaries, December 1993

Location	Flood stage (m)	Flood discharge ($m^3 s^{-1}$)	Return period (years)	Date of flood peak
River Rhine				
Maxau	7.5	3020.0	2.5	21 Dec. 1993
Worms	6.9	4759.0	10.0	22 Dec. 1993
Kaub	7.7	6500.0	40.0	23 Dec. 1993
Andernach	10.5	10600.0	65.0	23 Dec. 1993
Emmerich	9.5	11100.0	80.0	25 Dec. 1993
River Neckar				
Rockenau	9.9	2400.0	50.0	21 Dec. 1993
River Nahe				
Grolsheim	5.1	1370.0	>100.0	21 Dec. 1993
River Moselle				
Cochem	10.3	4170.0	80.0	22 Dec. 1993

Source: Based on a table in Engel et al. (1994)

contributing areas which are effectively impermeable, irrespective of their original infiltration characteristics (see Section 3.2). The considerable reservoir storage (some $106 \times 10^6 m^3$) that is available in the upper Rhine catchment was of little value in the 1993/94 flooding in the lower parts of the basin. This was partly because there was no flooding along the upper Rhine and partly because use of retention capacity on the upper Rhine to ameliorate flooding farther downstream would have required better forecasts of flood peak arrival times than were then achievable. Furthermore, the scale of the problem was immense, for example, the combined capacity of all major dams in the Ruhr catchment ($475 \times 10^6 m^3$) is only about one-half of the Moselle flood peak volume at Cochem (Ebel and Engel, 1994).

4.7.3.3 Description of the Floods

The rainfall pattern described in Section 4.7.3.1 caused flood waves in the Rivers Neckar, Main, Nahe and in the Moselle and its tributaries the Sauer and the Saar. The main flooding on the River Rhine, therefore, occurred downstream of the Neckar confluence (see Figure 4.19A) and only downstream of the Nahe confluence did the flood become extreme. Ultimately, it was the inflow of the River Moselle which brought the Rhine to record flood levels, while the Moselle itself had the highest discharge this century downstream of the Saar confluence (Engel et al., 1994).

On the Rhine the flood occurred as two main peaks (see Figure 4.19B) of which the first, the Christmas flood, was the highest, having a return period at Andernach of about 65 years, compared with five years for the second peak. The flood peak on the Rhine at Maxau on 21 December was $3020 m^3 s^{-1}$ (two-year event). As shown in Table 4.8, this had swollen to $6500 m^3 s^{-1}$ at Kaub on 23 December following the inflow of 50-year flood peaks from the rivers Neckar (50-year event), Main and Nahe (>100-year event) and to $10 800 m^3 s^{-1}$ and $11 100 m^3 s^{-1}$ respectively at Cologne and at Emmerich on the Dutch border.

CHAPTER 5

Coastal Floods

CONTENTS

5.1 Introduction 143
5.2 Flood-producing processes in coastal and estuarine areas 145
5.3 Storm surges 148
    5.3.1 Factors affecting storm-surge magnitude 151
    5.3.2 Surge-prone coastlines 153
        5.3.2.1 Storm surges in the southern North Sea 153
        5.3.2.2 Storm surges affecting Bangladesh 157
        5.3.2.3 Storm surges on the Atlantic and Gulf coasts of the USA 159
5.4 Tsunamis 164
    5.4.1 The main causes of tsunamis 165
    5.4.2 Tsunami-prone coastlines 168
        5.4.2.1 Tsunamis in the Pacific basin 168
5.5 Sea-level change and coastal flooding 170
    5.5.1 The likely magnitude of sea-level change 171
    5.5.2 The effects of sea-level change on coastal flooding 172
5.6 Floods and coastal geomorphology 175

5.1 INTRODUCTION

Globally, flooding along coasts and estuaries has tended to be less newsworthy than the fluvial flooding discussed in Chapters 3 and 4. This is partly because a large proportion of total coastline length is backed by high ground and partly because far more people are directly affected by the inundation of riparian land than by the inundation of coastal land. However, where floods occur in populated coastal and estuarine areas, their impact is often spectacular and disastrous. And in countries where the ratio of coastline to surface area is large, e.g. the UK, or where much of the coastline is backed by low-lying land, e.g. the Netherlands, coastal flooding may represent a greater flood risk than fluvial flooding (Parker, 1985). Even in the USA, where the ratio of coastline to area is relatively small, 50% of the population lives near the coast; and the fastest growing populations in the less developed countries show strong migratory tendencies towards coastal plain cities (Spencer and French, 1993).

Water levels along the coastline show short-term fluctuations, associated with wave period and tidal rhythm, which make even the flashiest river look quite sluggish. Longer-term fluctuations reflect climate change, tectonic movements, and isostatic movements caused, for example, by sedimentation or by the removal of ice-loading after the last glaciation. In natural conditions, therefore, the position of the coastline in low-lying areas would be in a state of continual change and in many parts of the world, especially where substantial sediment accretion has occurred during the last 10 000

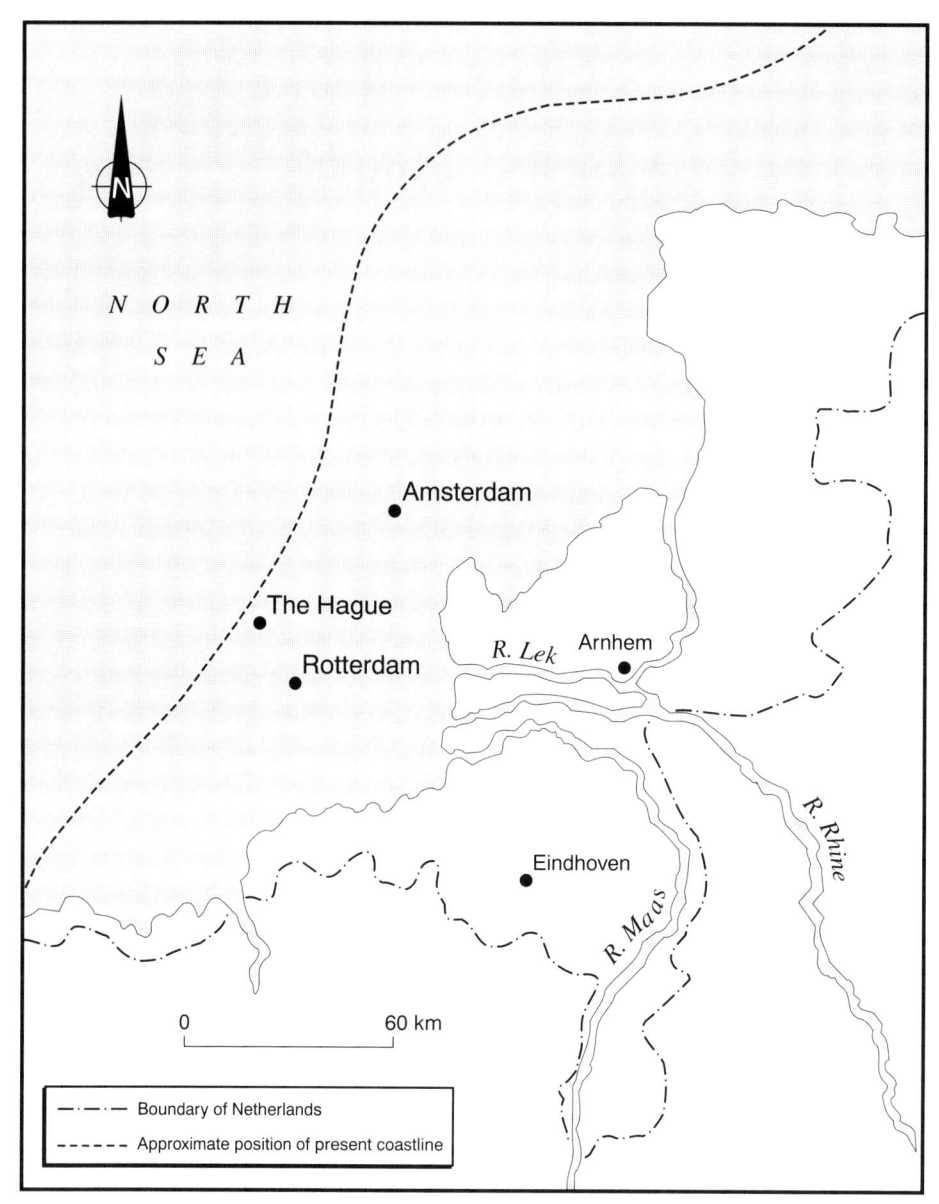

Figure 5.1 The position of the Netherlands' coastline in the absence of flood defence structures (the broken line indicates the approximate position of the present, protected, coastline)
Source: From a diagram by Graftdijk (1960) with permission

years, as in Britain, coastal lowlands are already under threat of inundation (Shennan, 1993). Often the present position can be held only by the construction of coastal defences (Spencer and French, 1993). For example, some 1300 km of the 3700 km of open coastline around England and Wales are protected against flooding by some form of sea defence (Burgess and Reeve, 1994) and Figure 5.1 illustrates that much of the

Netherlands would be regularly under water if the coastal flood defences were absent (Hillen and Jorissen, 1995).

Inevitably, in such cases, floods occur when defences fail as a result of either overflow, i.e. when sea level exceeds the height of the flood defence, or overtopping, i.e. when the combined effect of waves and water level results in waves running up and breaking over the defence, or structural failure which usually occurs when beach erosion undermines the toe of the defence structure (Burgess and Reeve, 1994) (see also Section 1.2.1.2). Using these three mechanisms of failure, the insurance industry in England and Wales has classified coastal flooding into three risk bands, the most vulnerable being where flooding would result from an event having a 50-year return period, and the least vulnerable being where flooding would result from an event having a return period of more than 200 years (Maddrell et al., 1995).

Increasingly, where the threat of failure is high and the value of protected land is low, 'managed retreat' or sea-defence setback, which allows the sea to inundate a previously protected coastal area, is becoming the preferred response to the flood risk (see also Sections 5.6 and 7.4). In the Netherlands, however, not only is managed retreat a non-viable option but the Dutch Government has declared any coastline retreat to be unacceptable and intends to maintain the coastline at its position of 1 January 1990 (Van Overeem and Peerbolte, 1994).

5.2 FLOOD-PRODUCING PROCESSES IN COASTAL AND ESTUARINE AREAS

Coastal and estuarine flooding, whether entirely natural or as a result of the failure of flood defences, can arise from the operation of several processes, including storm surges and tsunamis. As with river floods, the operation of these coastal flood-producing processes may vary with time as a result of climate change, especially through its effect on sea levels, and as a result of morphological changes along the coastline. However, flooding may also result from the occurrence of extreme events within the normal range of tides and waves, i.e. exceptionally high tides or exceptionally large waves. Forecasting such extremes is therefore important (see Section 8.11), especially in a country like the Netherlands.

More usually, flooding in coastal areas is caused by 'something extra', over and above the normal tidal and wave conditions. If this additional factor, e.g. a storm surge, happens to coincide with high tide or large wave conditions, the resulting flooding may be even more severe. Flooding in estuaries is further complicated by the interaction of the seaward flow of freshwater river discharge and the alternating seaward ebb and landward flow of saline water caused by the tidal oscillation. The estuary is the only part of a river channel in which freshwater discharge encounters an opposing landward water flow and, similarly, estuaries are the only portions of a coastline where the normal tidal currents meet a concentrated seaward flow of freshwater. Furthermore, because most estuaries become shallower and narrower with distance from the sea, they normally experience a greater tidal range from high to low tide than do adjacent sections of open coastline or the open sea. In the open ocean, for example, mean tidal range may be less than 2 m but at Hull, on the Humber estuary of eastern England, it is about 5 m and at Avonmouth, on the Severn estuary in south-west England, more than 10 m.

For most of the time, this estuarine interaction between fresh and saline water occurs uneventfully and without flooding, although this is often because of the flood defence structures present in many estuaries. For example, substantial areas of the city of Hull lie well below the mean height of high tides and this situation is repeated in virtually all estuaries where land has been reclaimed from former fringing saltmarshes and mudflats. Even in unprotected areas floods tend to occur only in conditions of large tidal range (i.e. spring tides) or peak freshwater discharge. Very high spring tide levels can so reduce the storage capacity of an estuary that even modest freshwater inflow on the rising tide can exceed the remaining storage capacity and produce flooding. Alternatively, flooding may result from very high river discharges into an estuary, even though high tides are below their maximum levels.

The greatest risk of flooding in estuaries is therefore at spring tides during the season of peak runoff, i.e. during winter in Britain, and exceptionally high tides due to storm surge effects (see Section 5.3), especially if these occur during the peak runoff season. For example, periodic flooding in the St Petersburg area of the Neva River estuary results from wind-induced surge effects which substantially increase high-tide levels in the Gulf of Finland (Labzovskii, 1966). An excellent example of the combined effects of a large freshwater discharge and high spring-tide levels intensified by a storm surge was the 1928 flood on the Thames estuary in south-east England (Brooks and Glasspoole, 1928). A heavy snow accumulation melted rapidly during a fast thaw on 2 and 3 January, causing widespread flooding in the Thames valley above London which was further exacerbated by exceptionally heavy rainfall. By 7 January the flow of the Thames at its tidal limit at Teddington was 499 m^3 s^{-1}, or more than twice the normal bankfull discharge. At the same time, a deep depression moving rapidly across Scotland on 6 January caused southerly gales over the North Sea which produced a surge effect on the Thames estuary spring tides. The combined effect of storm surge, spring tide and near-record river discharge was an increase in water height in the estuary to nearly 2 m above the predicted level. This caused widespread flooding, especially downstream of London.

Major river deltas are also areas of interaction between river and sea water. Because of their often high agricultural potential, they tend to attract large populations and therefore experience even more severe flood problems, as in the deltas of the Huang Ho (Yellow) River in China, the Hong (Red) River in Vietnam, the Mekong River in Cambodia, and the Ganges–Brahmaputra–Meghna rivers in Bangladesh. Gentle freshwater floods, with a slow rise to peak flow, normally prove beneficial but rapid river flooding is harmful, as also is the saltwater flooding induced by very high tides or storm surges.

On open coasts, as well as in estuaries and deltas, flood-producing processes normally operate in a situation where, in addition to wave, tidal and surge fluctuations, sea level is slowly but continually changing with time. Some of these changes are isostatic, reflecting regional crustal movements which involve either uplift, caused by the removal of ice-loading following deglaciation, or subsidence caused by the addition of water and sediment loads to areas of continental shelf, including estuaries and deltas. Uplift and subsidence may occur in close spatial proximity, as in Britain where estimated current rates of crustal movement were mapped by Shennan (1989). Figure 5.2 shows an effective tilting of the land area, with uplift (and diminishing flood risk) in the north-west and subsidence (and increasing flood risk) in the south-east. Palaeoenvironmental

COASTAL FLOODS 147

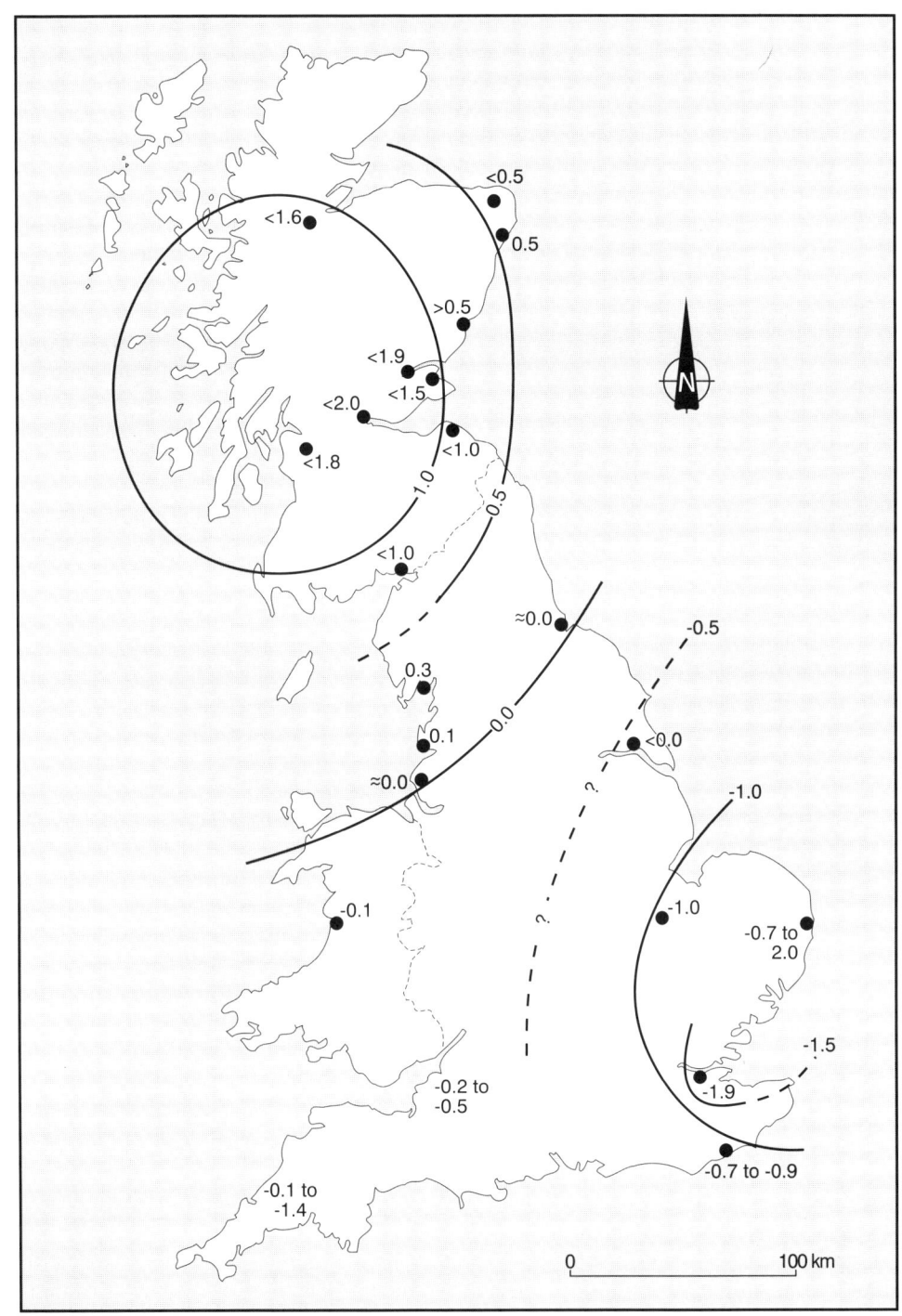

Figure 5.2 Estimated current rates of crustal movement in Great Britain (millimetres per year). Isolines cannot be drawn for much of southern England, although point estimates are shown
Source: Shennan (1989). Copyright John Wiley & Sons Limited

analysis indicates that these rates have remained fairly constant over the past 7000 years (Shennan, 1993).

Other causes of subsidence include massive deltaic deposition, as in the case of the Mississippi delta which is subsiding slowly in response to the submarine downwarping of the Gulf of Mexico caused by the additional loading of Mississippi sediment at the rate of about 0.278 km^3 per annum. Some coastal subsidence results from hydrological changes. For example, groundwater over-abstraction has led to significant coastal subsidence in several localities in the USA, including the Gulf of Mexico where it appears to have increased the area now prone to flooding during hurricanes.

Other long-term sea-level changes are eustatic, reflecting actual changes of sea level which occur globally (in contrast to the regional, isostatic, changes of land elevation discussed in the preceding paragraphs). Over long geological periods eustatic change has been largely controlled by plate tectonics. In recent geological time the main control has been the accumulation and melting of the Pleistocene ice-sheets, a process which continues to contribute to a slow increase in sea level. Of even more immediate concern, however, is the fact that sea-level rise is almost the only sure outcome of global warming, largely because of thermal expansion rather than further icemelt.

Accordingly, the operation of flood-producing processes in coastal areas will be conditioned by the interplay of isostatic and eustatic changes. In some locations these will operate in the same direction to accelerate the increase in sea level. At other locations the eustatic rise in sea level will be moderated by a local or regional isostatic increase in land elevation. Before these issues are discussed further, in Sections 5.5 and 5.6, detailed consideration is given to the ways in which storm surges and tsunamis cause coastal flooding.

5.3 STORM SURGES

It has already been noted that coastal and estuarine flooding results, not from the normal regime of waves and tides but from some extra factor which adds to the height of the sea surface, especially when that addition coincides with high-tide conditions. On open coastlines, a common cause of flooding is the severe meteorological conditions which produce abnormally high sea levels, known as storm surges. Flooding is likely to be most severe when a storm surge coincides with spring tides.

Most coastlines are subjected to a semi-diurnal tidal rhythm, with two high and two low tides each day, and with tidal amplitude or range reaching a maximum at spring tides and decreasing to a minimum at neap tides (Figure 5.3). A storm surge may be considered as the difference in elevation between the observed and predicted tide, e.g. the 1953 storm surge at Dover, England, which is shown in Figure 5.4. This approach is not entirely accurate since there is an interaction between the surge and the ordinary tide whereby, because water depth is increased by the storm surge, the ordinary tide will arrive earlier than expected and will be lower in amplitude, although the total tide will be higher (Bretschneider, 1967).

Two major types of storm surge may be recognised. On open coastlines, such as the Atlantic coast of the USA and the coasts of the Bay of Bengal, surges travel as running waves over very large areas of sea. Damaging surges of this type are mainly confined to those caused by intense storms such as tropical cyclones, hurricanes or typhoons and in such cases the storm surge may be considered as a dome of water, up to 100 km wide,

COASTAL FLOODS
149

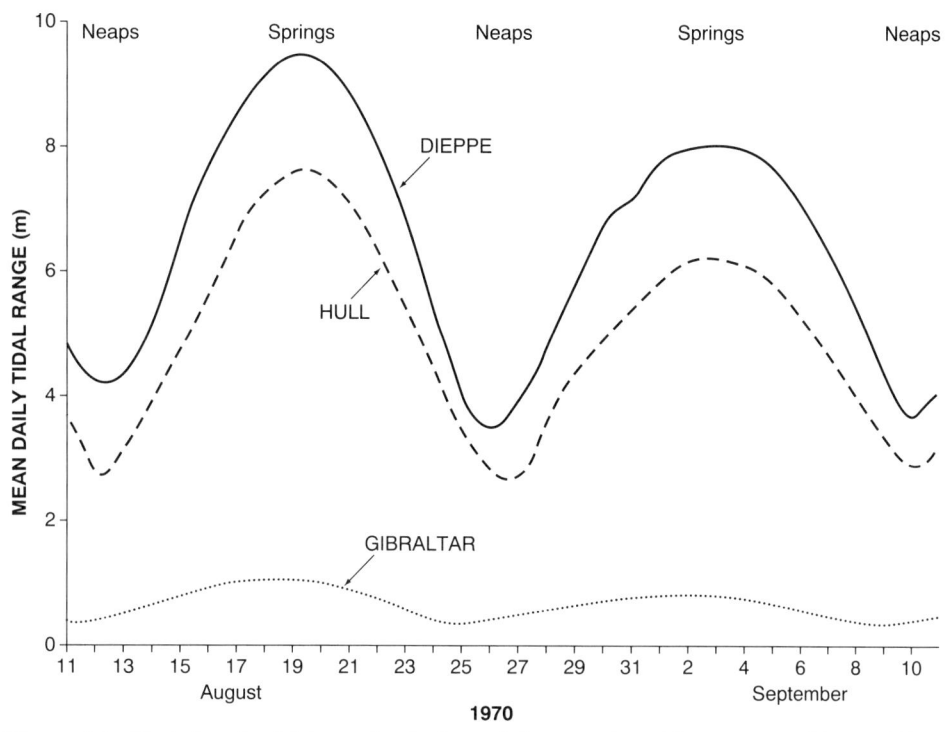

Figure 5.3 Variation in tidal amplitude (range) from spring tides to neap tides at Dieppe, France, Gibraltar, and Hull, England
Source: Based on data in *Admiralty Tide Tables* (1970), vol. 1

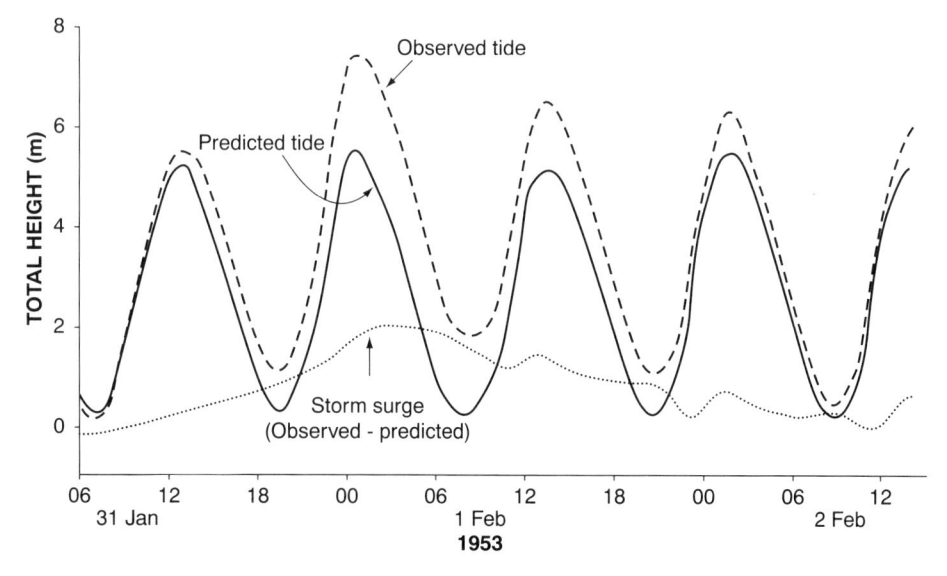

Figure 5.4 Graph of the 1953 storm surge at Dover, England, expressed as the difference between predicted and observed tides
Source: Based on a diagram in Steers (1953) with permission

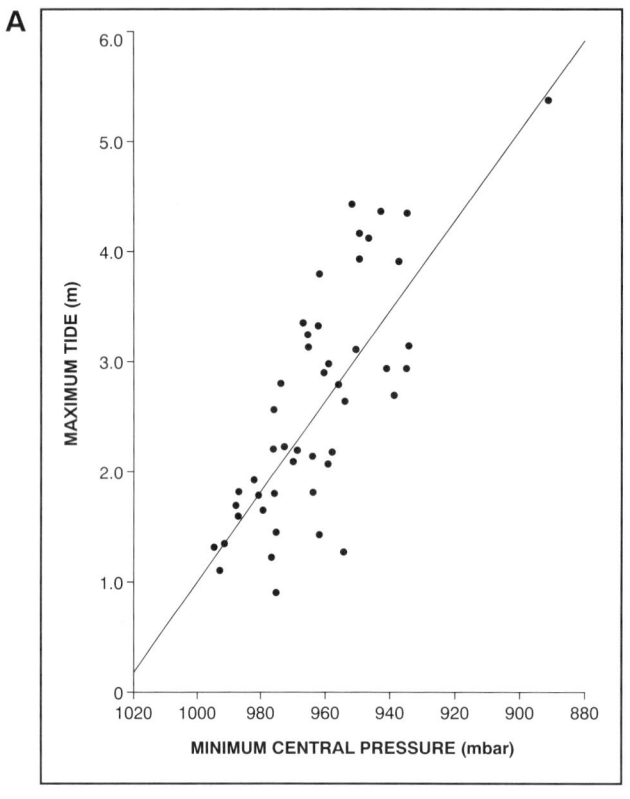

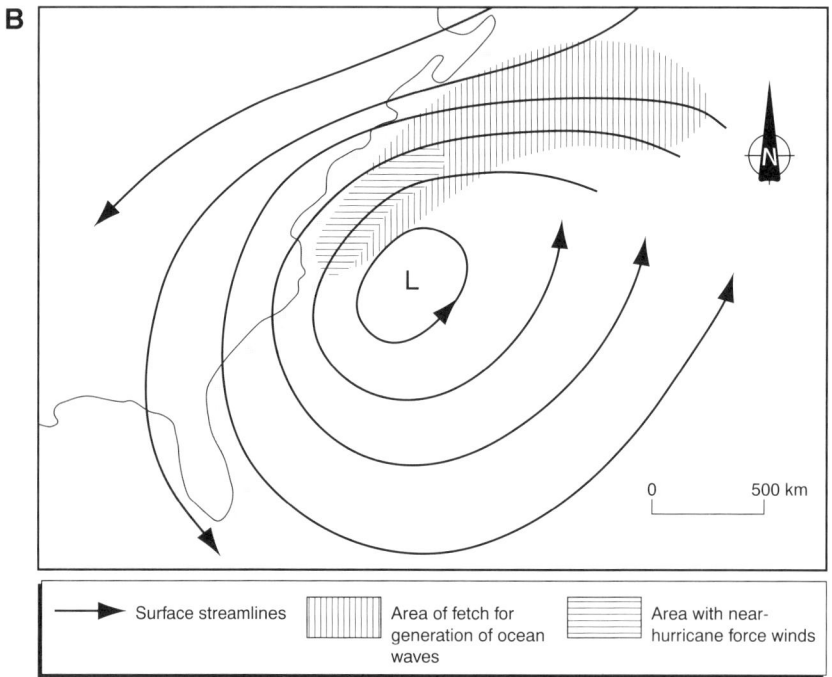

and 1–6 m high that sweeps onto the coastline as the cyclone makes landfall (WMO, 1990b). In November 1970, such a surge drowned at least 200 000 people in Bangladesh (WMO, 1990a). A second type of surge occurs in more or less enclosed seas, such as the Baltic Sea and the Adriatic Sea. Here, because the sea area is small in comparison with the spatial dimensions of the atmospheric disturbance, the surge will affect virtually the entire sea simultaneously. Such surges may be frequent and very damaging; Leningrad, for example, has been flooded periodically by storm surges and in 1924, 2000 people were drowned there in a surge 4 m in height (WMO, 1990a). Many surges share the characteristics of both the open and enclosed sea types, as in the seas around the UK, including the North Sea and the Firth of Clyde, western Scotland, where generally more than one-third of any surge is generated within the Firth (i.e. estuary) rather than propagating from the Irish Sea (Curran, 1995). Even very severe storm surges may be quite localised in area. This was illustrated by the January 1991 surge in the Firth of Clyde which had a probable recurrence interval of several hundred years, and caused severe damage at Rothesay on the Isle of Bute, but whose recurrence interval was only about 15 years some 30 km further up the estuary in Glasgow (Curran, 1995).

5.3.1 Factors Affecting Storm-Surge Magnitude

The magnitude of a storm surge is determined partly by storm intensity and track and partly by the configuration of the coastline and seabed. The most influential indications of intensity, for both tropical and extra-tropical storms, are the pressure gradient and its associated windspeed and the depth of low pressure at the storm centre (Harris, 1967; Jelesnianski, 1967). *Windspeed* serves to pile seawater against the coastline and to generate large-scale turbulence and waves which add further to the maximum sea surface elevation. Tropical storms are considered to have attained cyclone intensity at windspeeds of 120 km h^{-1} although windspeeds of more than 240 km h^{-1} are not uncommon. A rise in sea level normally accompanies a localised *reduction in atmospheric pressure* and is a marked feature of surges produced by intense storms, particularly of the tropical type (see Figure 5.5A). This 'inverted barometer' effect means that a decrease in atmospheric pressure of one millibar should produce an increase in sea level of about one centimetre. However, the effect is rarely experienced in full, for a number of reasons, including the dynamic response of the shallower waters of the continental shelf to the movement of the atmospheric pressure field (Pugh, 1987).

The threshold windspeeds of 120 km h^{-1} are generally associated with a central pressure drop of about 34 hPa, which is equivalent to about 0.36 m of hydrostatic head (NB in SI units 1 Pa (pascal) is a pressure of 1 kg m^{-1} s^{-2}, and 100 Pa = 1 hPa = 1 mbar). Pressure as low as 891.85 hPa has been recorded during a hurricane off the Florida coast. If normal pressure is taken to be 1013.20 hPa this gives an anomaly of 121.35 hPa and is equivalent to a sea-level rise of 1.25 m. An even lower pressure of 886.56 hPa was recorded during a typhoon off Luzon and represented an anomaly of 126.64 hPa and an

---

Figure 5.5 (*opposite*) Storm surges, eastern USA: (A) peak storm-surge elevation as a function of the central pressure of the associated hurricanes; (B) surface streamlines, area of fetch and area of near-hurricane force winds producing storm-surge effects from Canada to Florida, 7 March 1962
Source: (A) Based on a diagram in Conner et al. (1957) with permission from the American Meteorological Society; (B) based on a diagram in Riehl (1965) with permission of The McGraw-Hill Companies

Table 5.1 Hurricane characteristics (Saffir–Simpson scale) and surge magnitude

Scale no.	Hurricane characteristics		Surge magnitude (m)
	Central pressure (hPa)	Windspeed (km h^{-1})	
1	> 980	120–149	1.2–1.6
2	979–965	150–179	1.7–2.5
3	964–945	180–209	2.6–3.8
4	944–920	210–249	3.9–5.5
5	< 920	> 249	> 5.5

Source: Based on data in Oliver (1981) and Smith (1996)

equivalent sea-level rise of 1.31 m (Bretschneider, 1967). Table 5.1 summarises the relationships between hurricane characteristics, expressed in terms of the Saffir–Simpson scale, and surge magnitude.

The size and track of the storm and, in the case of open coastline surges, its proximity to the coastline, are further important factors which, together with wind direction and fetch, influence the magnitude of a storm surge. A period of constant *wind direction*, which may be only a few hours for shallow coastal water, may be sufficient to initiate large-scale drifting of the surface layers of the underlying ocean. In certain circumstances wind-drifted water is deflected up to 45° from the wind direction (to the right in the northern hemisphere). In this way maximum surge height may sometimes be achieved with lower windspeeds when the wind direction is parallel to the coast rather than when it is perpendicular to the coast.

Windspeed and *fetch* determine both the amount of water piling up at the downwind coastline of the ocean basin and also the height and energy of the waves driven on to the coastline. In the northern hemisphere intense storms often originate in the ocean areas around Iceland in the Atlantic and the Aleutians in the Pacific. As these storms move away south-eastwards, they may generate persistent gale-force and hurricane-force winds blowing without appreciable change of direction for 1500 km or more, thereby creating potentially severe storm surges along the eastern coasts of the USA and Japan (see Figure 5.5B).

Storm-surge magnitude, especially in relatively enclosed seas, is also influenced by the *shape of the ocean basin* and by its *bottom topography* and *coastal configuration*. Coastlines fronted by a wide, shallow continental shelf are more susceptible to damaging surges than those where the shelf slopes steeply. Bed roughness is also important and especially on low-lying coasts, vegetation may have a marked effect because, in these conditions, the rate at which water can be transported across coastal marshes and the effect of bed friction on water surface slope are decisive (Bretschneider, 1967). Coastal configuration is particularly important where converging coastlines create a funnel shape into which the storm surge moves and intensifies. This occurs in the Gulf of Mexico, where the coasts of Louisiana and Mississippi create a funnel, and in the North Sea, which is open at its northern end and almost closed in the Flemish Bight at its southern end. In addition, basin shape may affect tidal oscillation, particularly where the combination of basin shape and wind characteristics favours the formation of a standing wave oscillation which may intensify the effects of a wind-driven storm surge. The shape of the North Sea basin, for example, is such that the normal tidal oscillation, about three amphidromic points, results in a progressive southward movement of high or low water along the east coast of Britain. This is also the direction normally followed by storm surges and means that a coincidence of storm surge and high tide will sweep down virtually the whole length of the coast.

5.3.2 Surge-Prone Coastlines

Although storm surges can affect any coastline, there are some combinations of the meteorological and geographical factors discussed in the preceding section which make certain coastlines more susceptible than others to large and damaging surges. Clearly, one such combination is that of a low-lying, semi-enclosed coastline exposed to the effects of cyclones or other intense storms. Prime examples are the Bay of Bengal and the Queensland coast of Australia exposed to cyclones, the Gulf and Atlantic coasts of the USA exposed to hurricanes, and the east coasts of Japan and China affected by typhoons. Other vulnerable coastlines include those of the North Sea influenced by deep Atlantic depressions, and the northern Adriatic, where sudden storms pose a continuing threat to the city of Venice.

The most devastating loss of life occurs in the Bay of Bengal, where surge-induced flooding caused as many as 300 000 deaths in November 1970 and 140 000 in April 1991 (Askew, 1992; Flather, 1994). But other coasts experience a similar frequency of flooding, some of it of great severity. For example, in the period 1916–65, the Japanese coast was affected by five particularly severe typhoon surges having a maximum departure of more than 2 m above normal tide level (Miyazaki, 1967). Between 1900 and 1980 the coast of Florida, USA, experienced 50 major hurricanes, and even as far north as Maryland on the Atlantic coast there is an average of one hurricane per year which has direct or fringe effects upon the coastline.

5.3.2.1 Storm Surges in the Southern North Sea

The semi-enclosed funnel shape of the southern North Sea makes it particularly susceptible to storm-surge effects, either entering from the North Atlantic or generated by strong wind systems moving over the North Sea itself. Recurrent flooding from surges has occurred over many centuries. In 1099, a storm surge caused coastal flooding in England, the Netherlands and Belgium with a loss of nearly 2000 lives (Bolt et al., 1975) and about 50 000 lives are believed to have been lost in 1287 during the worst flood in Dutch history. At least seven floods since the thirteenth century have been classifiable as disasters (Jensen, 1953; Perry, 1981). Major surges have occurred much more frequently and since the southern North Sea is fringed by low-lying coastlands where the ground surface is often several metres below the level of spring high tides it is perhaps inevitable that substantial surges usually have disastrous results. Most of the severe surges of the recent past appear to have been associated with strong north-westerly winds accompanying the passage of deep depressions moving from the Atlantic towards the west coast of Norway. The surge of January–February 1953, the most disastrous in the twentieth century, was a notable example.

A depression formed south-west of Iceland on 29 January and deepened rapidly as it moved between Scotland and Norway (see Figure 5.6A), the central pressure falling to less than 968 hPa during the morning of 31 January. During this initial period the intensifying winds had a strong westerly and south-westerly component and this is believed to have caused a flow of water from south to north out of the Flemish Bight, thereby resulting in abnormally low tides at Southend (Robinson, 1953) (see Figure 5.6D). Although the depression filled slightly, to 976 hPa, as it continued to move south-eastwards, a strong ridge of high pressure developing over the Atlantic maintained a steep pressure gradient over the North Sea. As a result, the winds which had now veered to north-west and north continued to blow at gale force or storm force (see Figure 5.6B). Their greatly increased fetch reversed the flow of water previously expelled from the southern areas and encouraged a southward movement of water along the east coast of England as rotational force deflected the wind drift to the right (Robinson, 1953).

The northerly winds raised the level of the entire southern part of the North Sea, by more than 2 m south of the Humber estuary, and by more than 3 m off the Dutch coast (see Figure 5.6C). The peak of the surge had reached the Thames and Scheldt by the early hours of 1 February and then slowly, as the winds moderated, the 'external' water flowed back out of the North Sea, with rotational deflection ensuring that most of the movement took place along the coastlines of the Netherlands, Germany and Denmark.

The relative importance of external and internal components of the 1953 surge was not entirely clear. Corkan (1948; 1950) had shown that surges of external origin result from atmospheric disturbances well to the north of Scotland and that such surges have a rate of progression around the coasts of the North Sea which is identical to that of the tide. Surges of internal origin are caused by local effects within the North Sea basin and are not dependent on 'imported' water. Some authorities attributed the 1953 surge almost entirely to the influx of external water. Rossiter (1954), for example, estimated that $422\,000 \times 10^6$ m^3 of water was forced into the North Sea from the Atlantic, thereby raising the mean level of the North Sea by more than 0.6 m. However, the residual surge graphs for a number of east coast locations (Figure 5.6D) show clearly that the rates of progression of the surge and the tide were not identical and that the peak of the surge occurred almost simultaneously over some 500 km of coastline between the River Tees and Southend. Furthermore, the surge was maintained at a high level for five or six hours at some locations. These surge characteristics led Robinson (1953) to conclude that the greater part of the 1953 surge originated internally, within the North Sea basin itself, and was accompanied by abnormal sea surface gradients established directly by the high windspeeds.

Along the east coast of Britain the height of the storm surge at most locations varied between 1 m and 2 m (Figure 5.6D) but was much greater on the Dutch coast, averaging 3 m along the open coastline and reaching almost 4 m in some inlets (Robinson, 1953; Van Ufford, 1953). Minor oscillations, shown on most of the surge graphs, may reflect a standing wave oscillation which developed in the North Sea through the previously noted combination of wind strength and wind reversal. It is not clear, however, whether this same feature was responsible for the anomalous surge conditions in the Humber estuary in eastern England. Here, despite the classic funnel shape of the estuary and the size of the 1953 surge, there was no evidence of intensified surge effects at successive locations in the estuary upstream of Hull. Instead, the surge which was between 1.2 m and 1.5 m at Grimsby and Immingham was negligible at and above the confluence of the River Trent (Edwards, 1953a).

COASTAL FLOODS 155

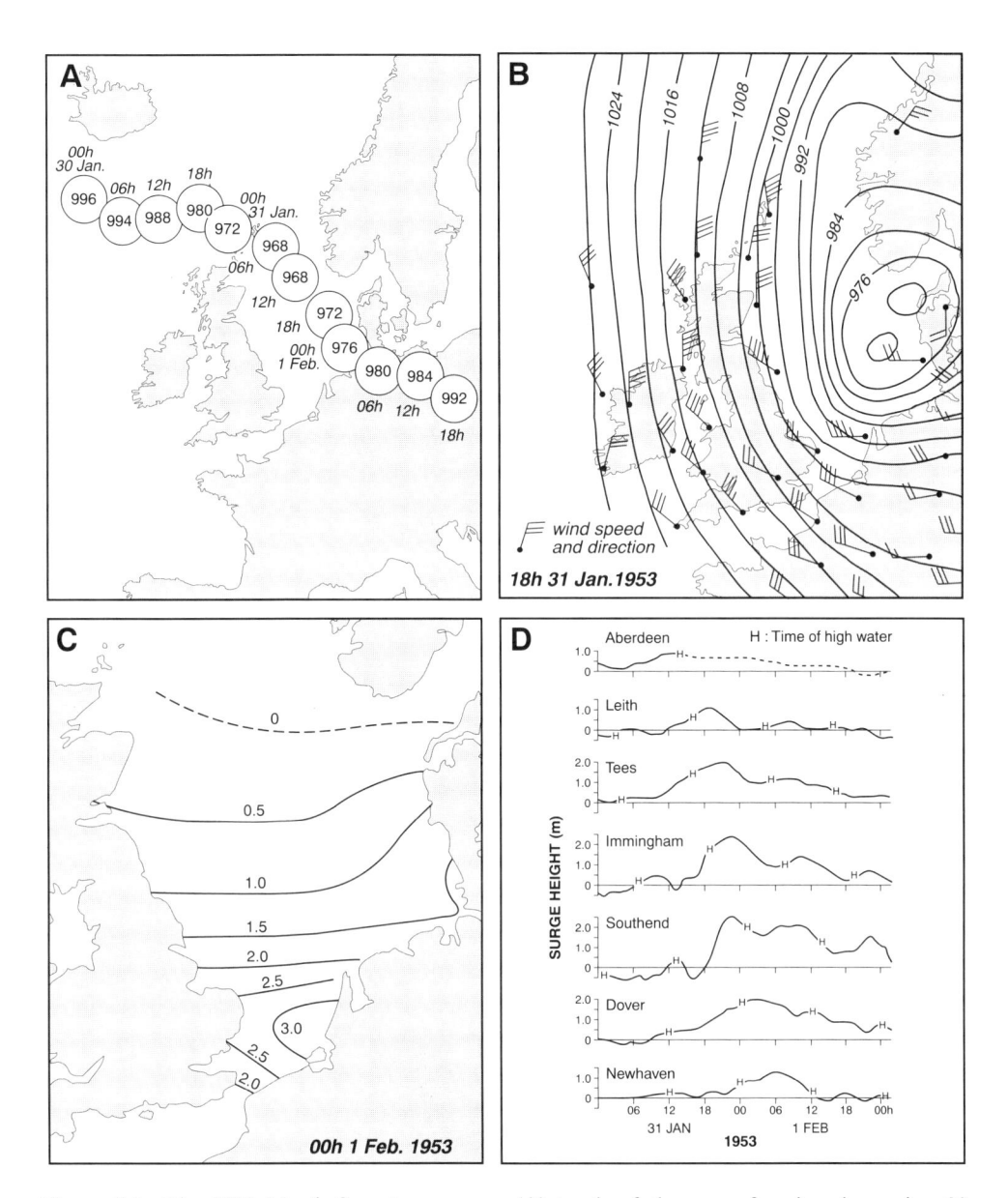

Figure 5.6 The 1953 North Sea storm surge: (A) track of the surge-forming depression 30 January–1 February (central low pressure in mbar); (B) synoptic chart for 1800 h on 31 January; (C) sea surface disturbance (m) at 0000 h on 1 February; (D) surge graphs for selected locations on the east coast of Britain
Source: From diagrams in: (A) and (D) Robinson (1953) with permission of the Geographical Association; (B) UK Meteorological Office Daily Weather Report. Crown copyright, reproduced with the permission of the Controller of HMSO; (C) Groen and Groves (1962). Reprinted by permission of John Wiley & Sons Inc

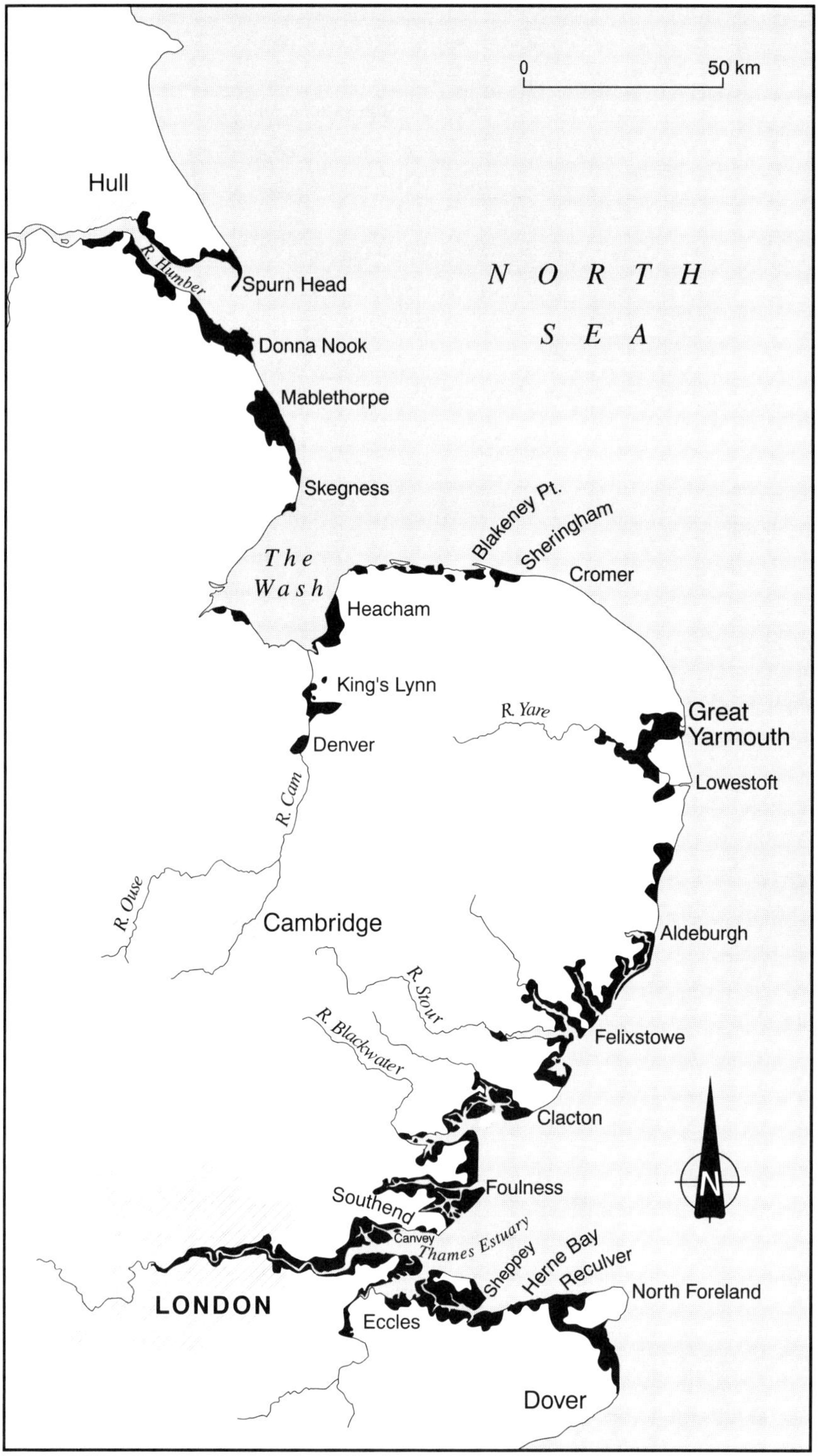
0
50 km
Hull
R. Humber
Spurn Head
N O R T H
S E A
Donna Nook
Mablethorpe
Skegness
The Wash
Blakeney Pt.
Sheringham
Cromer
Heacham
King's Lynn
R. Yare
Great Yarmouth
Denver
Lowestoft
R. Cam
R. Ouse
Cambridge
Aldeburgh
R. Stour
R. Blackwater
Felixstowe
Clacton
Foulness
N
Southend
Canvey
Thames Estuary
Sheppey
Herne Bay
Reculver
North Foreland
LONDON
Eccles
Dover

Despite being one of the largest North Sea surges on record, the 1953 event would have been worse if it had coincided with flooded rivers, particularly in the low-lying fenland areas of eastern England, or with the equinoctial spring tide, whose predicted levels for Dover, in south-east England, and Immingham, on the Humber estuary, were 0.61 m and 1.36 m higher than those predicted for the beginning of February (Robinson, 1953; Steers, 1953).

As it was, extensive flooding occurred on both sides of the North Sea (see Figures 5.7 and 5.8). In England flooding was concentrated between the estuaries of the Humber and the Thames (Figure 5.7). Some was natural flooding, as when the sea broke over dunes near Easington and flowed across a narrow neck of land into the Humber, turning Spurn Head temporarily into an island—a condition which has been repeated several times since and which threatened to become permanent in February 1996. In most cases flooding occurred because sea defences failed, allowing the inundation of more than 850 km^2 and the loss of 307 lives (Steers, 1953). Worse flooding occurred in the Netherlands where the higher surge levels, resulting partly from the funnelling effects of the Flemish Bight and from the directly onshore winds, led to the inundation of about 1600 km^2 and the loss of 1800 lives (Volker, 1953). The areas most affected were concentrated between Rotterdam and the Scheldt and substantial areas remained flooded five months later (Figure 5.8).

5.3.2.2 Storm Surges Affecting Bangladesh

The Indian and Bangladeshi coastlines of the Bay of Bengal are especially vulnerable to storm-surge flooding caused by the disproportionately frequent severe tropical cyclones (Murty and Neralla, 1992). Some 60 severe cyclones, mostly accompanied by storm surges, occurred between 1797 and 1991 (Khalil, 1992). These usually originate in the southern parts of the Bay or in the Andaman Sea, first moving westwards before curving to the north and north-east (Pugh, 1987). Surge effects are accentuated by a large astronomical tide, the seabed bathymetry, with shallow water extending to more than 300 km offshore in the northern part of the Bay, and the coastal configuration whereby the general funnel shape of the Bay is exaggerated by the right-angle change of coastal direction near Chittagong. This produces maximum storm-surge levels which are higher than would be produced by the same storm impacting on a straight coastline because the direct surge contribution is reinforced by a contribution reflected from the neighbouring coast (Pugh, 1987). The vulnerability of this area (and its inhabitants) is dramatically illustrated in Table 5.2.

The November 1970 cyclone, which heads this table, was described by Frank and Husain (1971) as originating in the remnants of a tropical storm that moved westward over Malaysia and spawned a depression over the south-central Bay of Bengal. This depression deepened and moved increasingly rapidly northwards and then north-eastwards before crossing the coast some 80 km north of Chittagong (see Figure 5.9), by which time it is estimated that the central low pressure had fallen to between 950 and 960 hPa. The cyclone was not, in this case, the most severe storm to affect

---

Figure 5.7 (*opposite*) Areas on the east coast of England flooded by the 1953 storm surge
Source: Based on diagrams in Steers (1953)

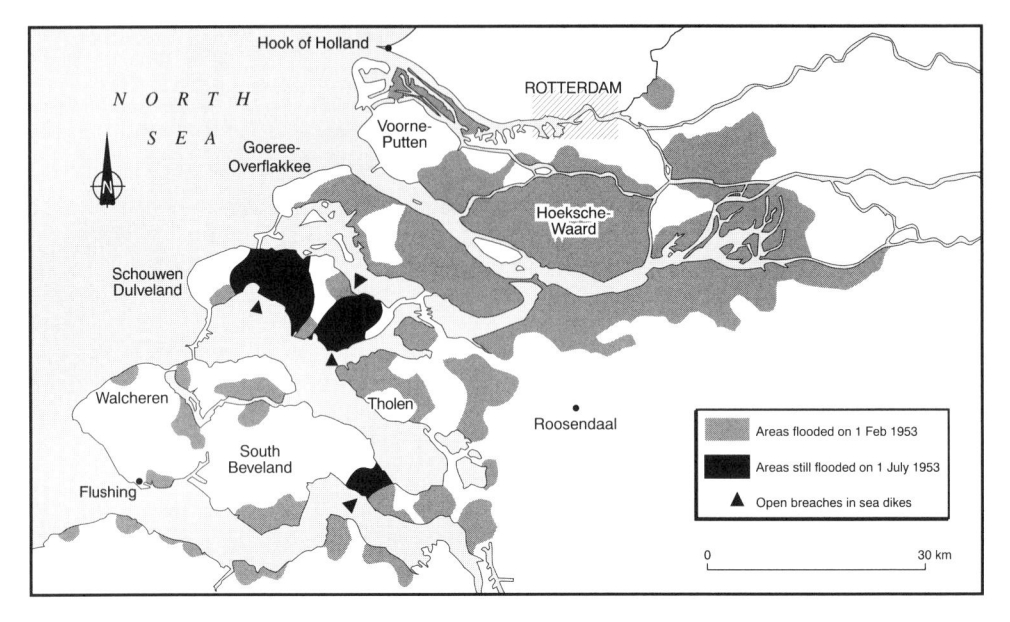

Figure 5.8 Areas in the south-west Netherlands flooded by the 1953 storm surge, 1 February and 1 July
Source: Based on diagrams in Edwards (1953b) by permission of The Geographical Association

Bangladesh. The 'Bakerganj' cyclone in 1876, which had a recorded central pressure of 930 hPa, generated a storm surge of more than 12 m and resulted in more than 100 000 deaths.

The storm surge caused by the 1970 cyclone is noteworthy not only for the disastrous loss of life involved but also because it was the first to be adequately recorded by the recently established network of tide gauges (Frank and Husain, 1971). A simplified reconstruction of surge water levels, based solely on coastal data, is shown in Figure 5.9. The circled values show the height by which the predicted astronomical high tide was exceeded and the isolines indicate the depth of water above the ground surface. Both sets of data confirm that, generally and as expected, the severity of the surge diminished away from the coastline. The maximum recorded exceedance of predicted high-tide level was 3.8 m on the north coast of Bohla Island where the normal diurnal tidal range is between 3.7 m and 4.3 m. In other words, at this point the maximum surge level was some 6 m above mean sea level (Frank and Husain, 1971). As in the North Sea where the normal diurnal tidal range is also very high in places, the severity of storm surges in the northern Bay of Bengal is substantially affected by the relative timing of storm surge and high tide. In November 1970 the two coincided. A decade earlier a surge of similar magnitude in October 1960 coincided with low tide and resulted in 5000 fatalities (Dunn, 1962).

Two decades later, a similar devastating storm surge occurred as an intensifying cyclone made landfall near Chittagong on 30 April 1991 with a sustained maximum surface windspeed of more than 200 km h^{-1} and an estimated central pressure of 940 hPa. As in 1970, landfall coincided with high tide and the resulting combined storm

Table 5.2 Major tropical cyclone disasters

Year	Location	Deaths
1970	Bangladesh	300 000
1737	India	300 000
1881	China	300 000
1923	Japan	250 000
1897	Bangladesh	175 000
1991	Bangladesh	140 000
1876	Bangladesh	100 000
1864	India	50 000
1833	India	50 000
1822	Bangladesh	40 000
1780	Antilles	22 000
1839	India	20 000
1789	India	20 000
1965	Bangladesh	19 279
1963	Bangladesh	11 468
1985	Bangladesh	11 000

Source: Based on data in Frank and Husain (1971), Askew (1992) and Smith (1996)

surge and astronomical tide reached heights of 6–8 m, exceeding previous records and impacting along almost the entire coastal belt of Bangladesh (WMO, 1992a).

5.3.2.3 Storm Surges on the Atlantic and Gulf Coasts of the USA

As in the North Sea and the Bay of Bengal, the susceptibility to storm surges of the Atlantic and Gulf coasts of the USA is partly the result of physiography, including a wide shallow continental shelf and a number of coastline embayments which give a funnelling effect, and partly of meteorological conditions, whereby these coasts are exposed to tropical storms moving from the south-east and mid-latitude storms moving south-westward from the North Atlantic. Studies of the effects of major storms, particularly hurricanes and other tropical storms, have indicated how susceptibility to storm surges varies with location along the coastline. For example, Figure 5.10A shows relative storm-surge potential in terms of a topographic variable (θ) which is based on the distance from the shore of the 50-fathom submarine contour. In conjunction with data on the central pressure of the hurricane, θ can be used to provide a quantitative estimate of storm-surge magnitude.

The Great Atlantic Hurricane of September 1944 was one of the most violent hurricanes recorded in the USA. As it moved northwards along the coast from Puerto Rico, via Cape Hatteras and Long Island, and into New England, enormous damage was done and the coastline of New Jersey was particularly badly hit (Truitt, 1967). Surge graphs for Atlantic City and for five other Atlantic coast locations are shown in Figure 5.10B and show surge values of about 2 m. These graphs are very typical of hurricane-produced surges along the Atlantic coast which, according to Redfield and Miller (1957), show three distinct successive phases: the forerunner, the hurricane surge and the resurgences. The forerunner is the gradual change in water level which takes place many hours before the storm itself arrives and which appears to result from the

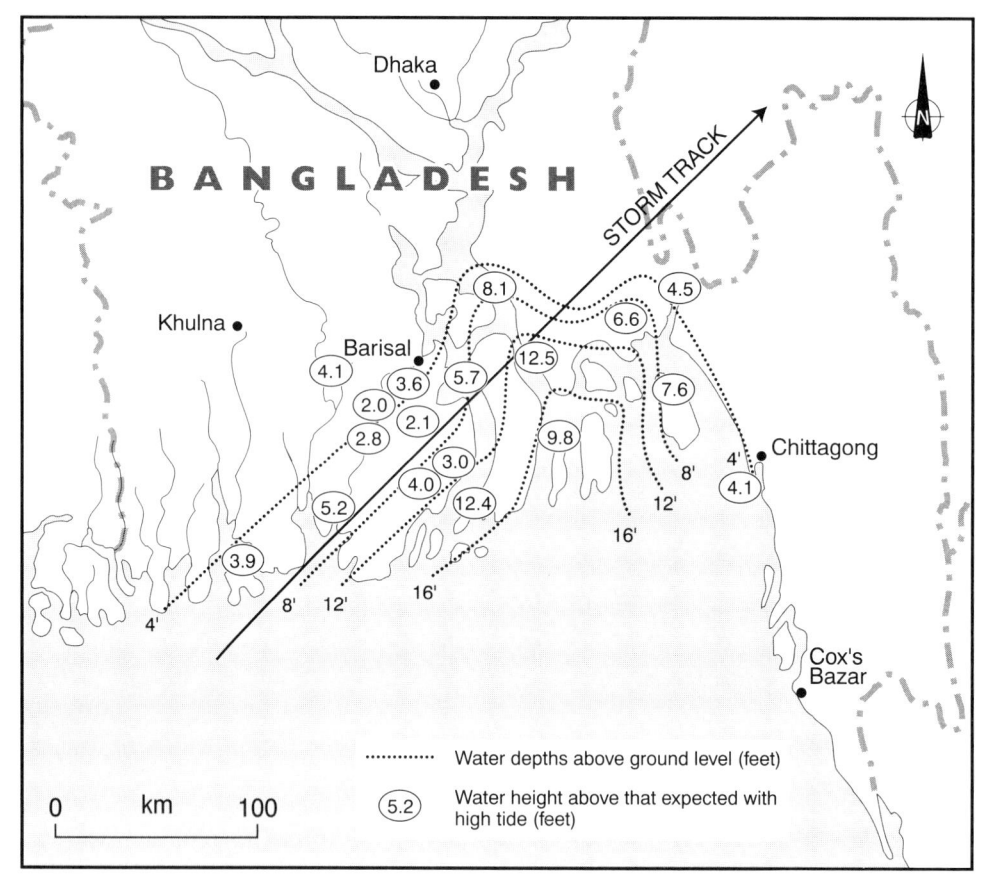

Figure 5.9 A simplified reconstruction of the November 1970 storm surge in Bangladesh. Circled data indicate the height (feet) by which predicted astronomical high tide was exceeded; isolines show the depth of water (feet) above the ground surface
Source: Based on a diagram in Frank and Husain (1971)

windfield over a more extended region than that of the hurricane proper. The forerunner may be either a rise or a fall in sea level, depending on whether the hurricane is moving northwards or southwards along the coast. The hurricane surge is the sharp rise in water level that occurs when the hurricane centre approaches the coast. This stage is usually brief, lasting up to five hours, during which peak surges of up to 4 m have been recorded. Finally, the resurgences are oscillations which occur after the passage of the hurricane and hurricane surge and are well-illustrated by the Atlantic City graph. The resurgences can be especially hazardous, partly because they may arrive unexpectedly as the storm appears to be subsiding and partly because coincidence with the astronomical tide may result in one or more of the resurgence peaks being higher than the original hurricane surge itself. Munk et al. (1956) regarded the resurgences as a 'wake' of waves in the trail of a hurricane (analogous to the wake of a ship). These 'wakes' progress along the coastline but are little affected by the hurricane once they have been generated.

161

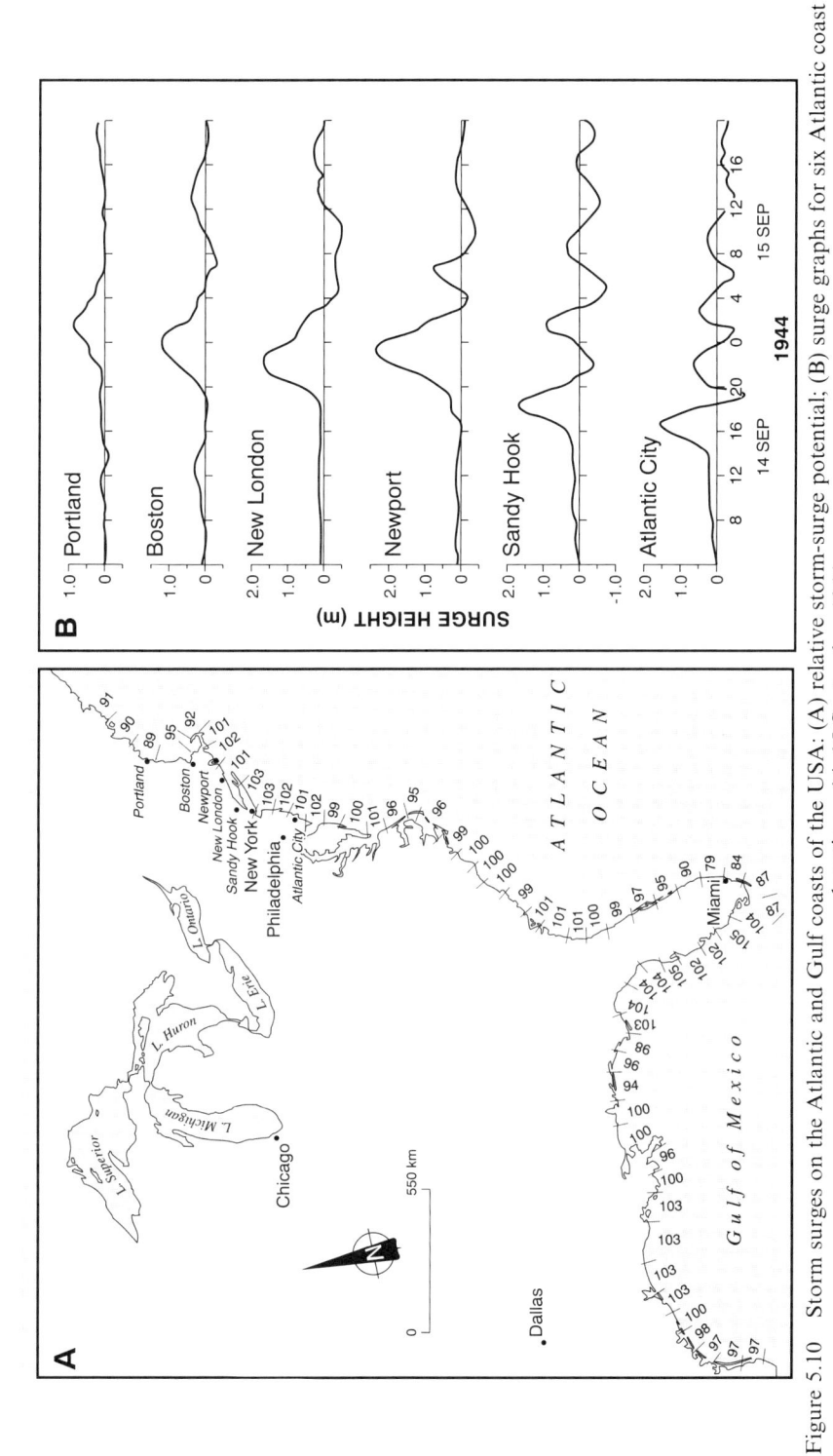

Figure 5.10 Storm surges on the Atlantic and Gulf coasts of the USA: (A) relative storm-surge potential; (B) surge graphs for six Atlantic coast locations, 14–15 September 1944

Source: From diagrams in (A) US Weather Bureau, reproduced by permission of the U.S. National Weather Service (1959) and (B) Redfield and Miller (1957), reproduced by permission of the American Meteorological Society

Table 5.3 Storm-surge heights on the Mississippi coast of the USA, 1700–1975

Year	Surge height (m)
1722	2.4
1740	1.5
1772	3.7
1812	4.6
1819	4.6
1821	3.1
1852	2.7
1855	4.6
1860	4.9
1870	2.4
1886	2.4
1893	3.4
1906	2.7
1909	3.7
1915	4.0
1916	3.4
1926	1.8
1947	4.6
1965	3.7
1969	7.5

Source: Based on information in WMO (1990b)

The Great March Storm of 6–7 March 1962, alternatively known as the Ash Wednesday Storm, was not a hurricane though its windspeeds attained hurricane-strength in some localities. However, its effects were more severe than those produced by many tropical storms and it was the most destructive extra-tropical storm recorded in the USA, affecting the coastline from Florida to New England (Truitt, 1967). Yet its origins were sudden and unexpected, as the prevailing weather pattern over the eastern USA on 4 March 1962 gave no cause for alarm (Burton et al., 1969). A small, cold-front storm between Florida and Bermuda was expected to move harmlessly out into the Atlantic and another storm centre in the Mississippi valley was filling and moving slowly northwards. However, both storms behaved unexpectedly. The Florida storm moved northwards along the coastline, the Mississippi storm moved eastwards and the two met near Cape Hatteras, increased in strength, and produced the Great March Storm. The storm intensified rapidly and then stagnated, producing the situation shown in Figure 5.5B, whereby windspeeds of 80–100 km h^{-1} developed along a fetch of about 2000 km. A storm surge of between 0.5 and 2 m was produced along the entire coast from Canada to Florida, accompanied by very high waves, and persisted during three successive high tides (Riehl, 1965; Truitt, 1967). It was this persistence which proved so serious because although offshore bars, barrier beaches, sand dunes and artificial coastal defences survived the first high tide in many places, the repeated attack of waves and tide during a second and third tidal cycle caused extensive breakthroughs, flooding and damage.

Another record-breaking storm surge was produced by Hurricane Camille which crossed the Mississippi coastline at Bay St Louis on 18 August 1969 (WMO, 1990b).

COASTAL FLOODS 163

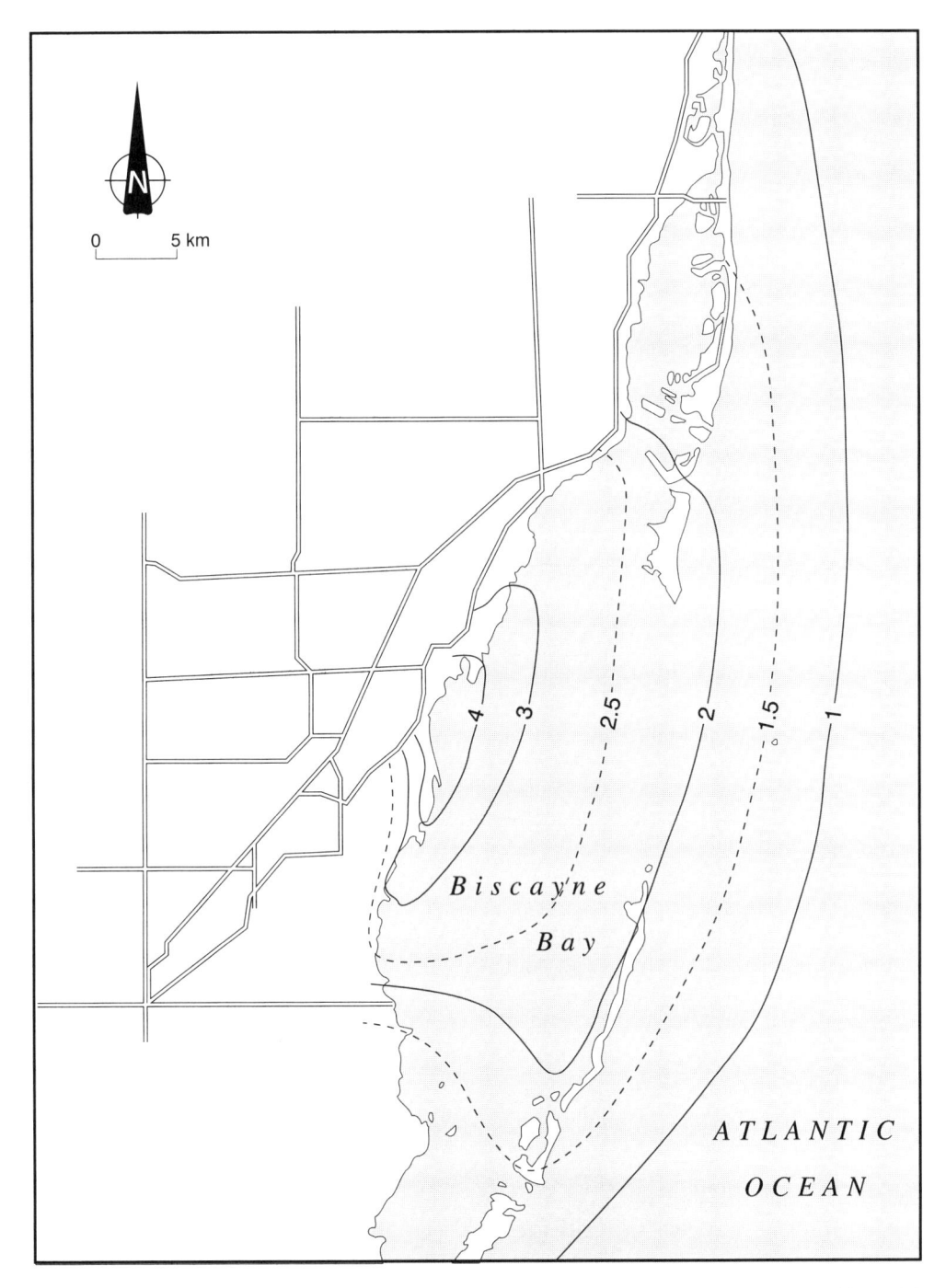

Figure 5.11 Maximum storm surge (metres above mean sea level) along the Florida, USA, coastline at the landfall of Hurricane Andrew, 24 August 1992
Source: Based on a diagram in Rappaport (1994) with permission from the Royal Meteorological Society

With maximum windspeeds estimated at 306 km h^{-1}, and central pressure of 905 hPa, Camille (like Gilbert in 1988) was a category 5 hurricane on the Saffir–Simpson scale (see Table 5.1) and produced a measured storm surge of 7.5 m, easily eclipsing any surge recorded in the previous 200 years (Table 5.3). In addition to the hydrodynamic loads imposed by water depths and velocities, and wave action, severe scouring undermined the foundations of structures leading to their collapse.

Although less severe in terms of both central pressure (922 hPa), maximum sustained windspeed (230 km h^{-1}) and surge height (5 m), Hurricane Andrew crossed the Florida coastline on 24 August 1992 to become the then most expensive natural disaster in the history of the USA (Rappaport, 1994). Storm-surge elevation above mean sea level was reconstructed from tide-gauge data, still-water marks and debris lines and is shown in Figure 5.11. The maximum surge occurred just north of the landfall of the hurricane's eye and values decayed rapidly along the coastline to both the north and south.

5.4 TSUNAMIS

Some of the most spectacular and disastrous coastal flooding is caused by tsunamis. These are waves (often wrongly called 'tidal' waves) which are set off by submarine earthquakes, landsliding or slumping, and by volcanic eruptions. Similar effects have also been caused by the explosion of nuclear bombs at the sea surface (Bascom, 1959). Tsunami is a Japanese word meaning 'harbour wave'. This emphasises that it is along the coastline, especially low-lying and densely populated coastlines, rather than in the open ocean, that the impact of tsunamis is felt most severely.

Most tsunamis comprise a train of waves generated by the rapid movement of the seabed. These waves are distinctively different from ordinary wind-generated waves, whose wavelength (i.e. the distance between successive crests) rarely exceeds 300 m, even in the Pacific, and whose speed rarely exceeds 90-100 km h^{-1}. By contrast, the wavelength of tsunamis is commonly between 150 km and 250 km and has been known to reach 1000 km. Furthermore, the water depth of even the deep oceans is less than one-half the wavelength, so given that the speed of a tsunami wave is proportional to the square root of the water depth, velocities may reach 700–800 km h^{-1}.

Because tsunamis are shallow in comparison with their wavelength, they are often barely detectable in the open ocean, where their amplitude may be as low as 0.5 m and rarely exceeds 3–5 m (Hindley, 1978), or even in deep water close to the coastline. Bascom (1959), for example, described the way in which the captain of a ship standing off the port of Hilo in Hawaii, watched astonished as the harbour and much of the city were demolished by waves he had not noticed passing beneath his ship. As the tsunami reaches shallow coastal water, its speed and wavelength diminish rapidly but the period (i.e. the time interval between successive waves in the train) remains unchanged, causing the marked increase in wave height.

Despite their very high speed, the enormously long wavelengths of tsunamis mean that wave periods are also very long, ranging from 15 minutes to an hour or more, so that wave successions may continue for many hours. The first waves are not usually the largest in the train but may raise sea level by a metre or two. Each wave is followed by a trough within 10 to 20 minutes which appears to take the tide out very rapidly. The heights of the crests and the depths of the troughs increase steadily until, immediately before the arrival of the main wave, the sea retreats far below the normal low-tide level, exposing reefs and other normally submerged offshore features. The main wave is usually a huge breaker, sometimes a 'bore' representing a near-vertical wave front 10–30 m in height which is believed to form when several of the later waves in a tsunami sequence catch up with each other (Hindley, 1978). Wave sequences for the first three waves of the Pacific tsunami of 22 May 1960 are shown in Figure 5.12A.

5.4.1 The Main Causes of Tsunamis

Earthquakes are a major cause of tsunamis, especially where these involve submarine thrust-faulting either at the subduction zones of island arcs, such as the Aleutians in the Pacific Ocean, or on mid-oceanic ridges, such as between the Azores and Gibraltar in the Atlantic. Tear-faulting, which has a horizontal rather than a vertical movement, rarely gives rise to tsunamis, even during major events like the 1906 San Francisco earthquake in which the San Andreas fault, that runs partly under the sea, was displaced by as much as 6 m (Bolt et al., 1975).

Tsunami magnitude may be expressed in terms of the logarithm of water-wave motion at a coastline close to the tsunami origin. One such formula is

$$m = 3.32 \log_{10} h \qquad (5.1)$$

where h is the maximum water height in metres at the coastline. A tsunami of magnitude 0 would have little impact and one of magnitude 3 would have a wave run-up as great as 12 m. However, such formulae give variable results and take no account of the important intensifying effects of coastal and seabed configuration (discussed later in this section). Various empirical relationships have been established between tsunami magnitude and the magnitude of the causal earthquake. One such set of relationships for shallow-focus earthquakes is shown in Table 5.4. If the earthquake focus is deeper, the tsunami magnitude decreases (Bolt et al., 1975). The relationships indicated in Table 5.4 must be treated with caution since some earthquakes release their energy slowly, over a minute or more, rather than near instantaneously and the Richter scale itself, based largely on the shorter seismic waves, may not fully reflect earthquake magnitude.

Submarine landslides, especially *long runout landslides*, where the debris extends well beyond the distance which would be possible under the influence of friction, are also important as an actual and potential cause of tsunamis. A major concentration of long runout landslides is found in the volcanic islands of Hawaii where there is evidence of 17 major slides, affecting most of the islands in the group, and each of which would have produced a tsunami. Almost half of some of the islands (e.g. Molokai, Oahu) appears to have slumped into the sea and debris outflows extend for many kilometres across the ocean floor. Although these Hawaii landslides probably occurred about 2 million years ago, there is evidence of similar events on parts of the Australian coastline about 800 years ago and again just before European settlement. The current and continuing movement of large volumes of rock at the 'great crack' in the main island of Hawaii appears to threaten a major tsunami event at some time in the future. One of the largest underwater slides may have occurred in the Norwegian Sea some 7000 years ago, producing tsunami waves of up to 10 m and extensive coastal flooding in the eastern North Atlantic (Dawson et al., 1994).

166 FLOODS

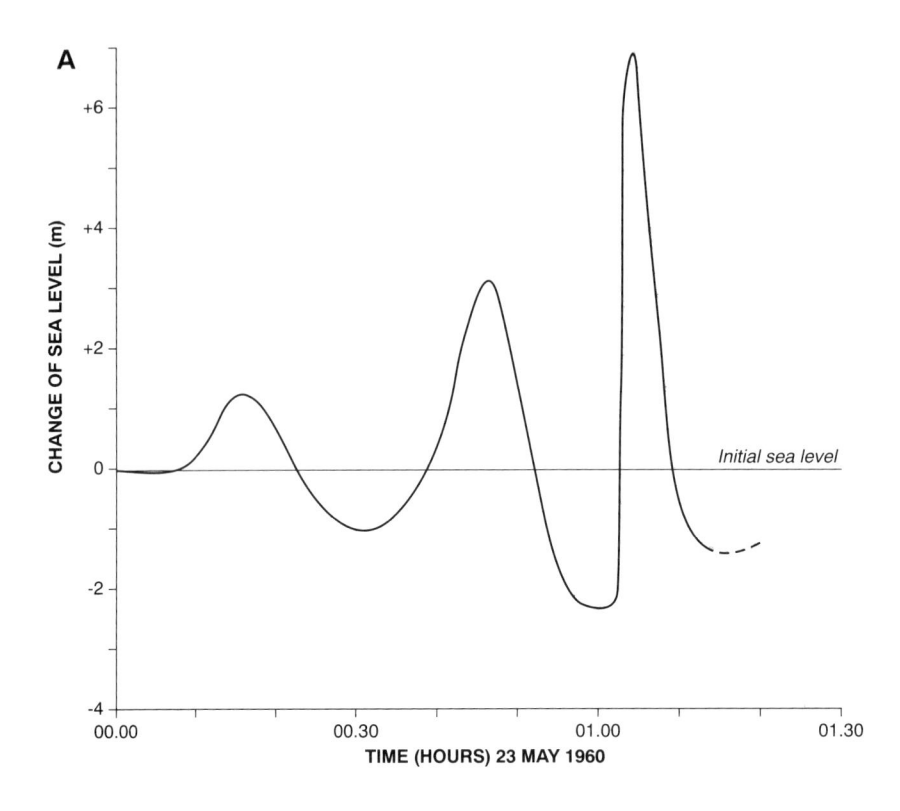

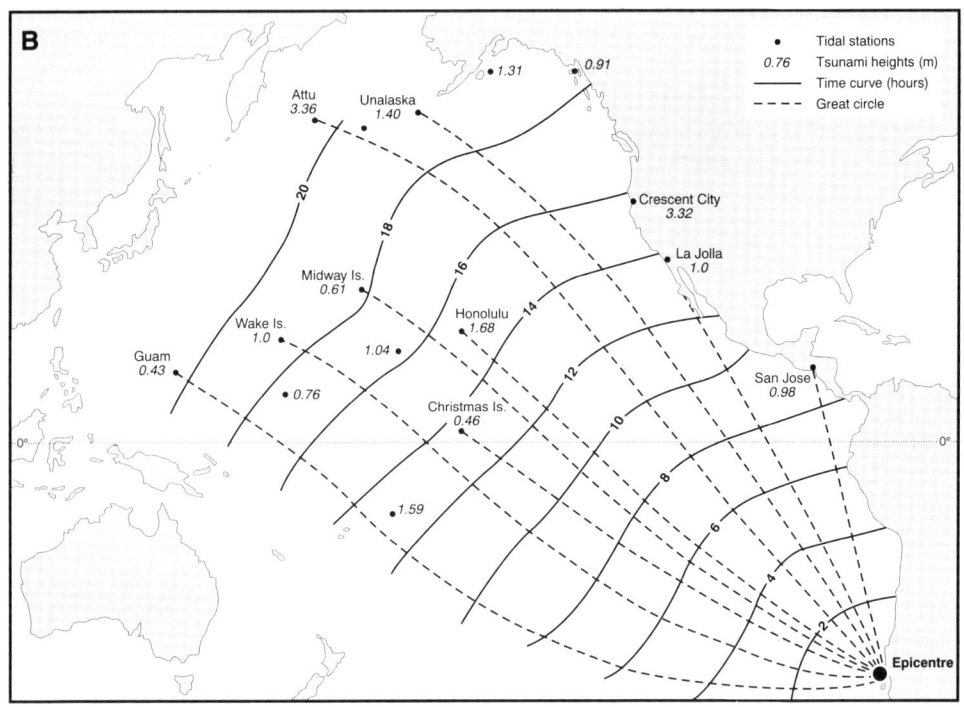

Table 5.4 Relationship between tsunami and earthquake magnitudes

Earthquake magnitude (Richter scale)	Tsunami magnitude	Maximum wave run-up (m)
6.00	Slight	
6.50	−1	0.50–0.75
7.00	0	1.00–1.50
7.50	1	2.00–3.00
8.00	2	4.00–6.00
8.25	3	8.00–12.00

Source: From a table in Bolt et al. (1975). Reproduced by permission of Springer-Verlag GmbH & Co. KG

The highest tsunami-type waves have been the result of landslides into more or less enclosed water bodies. By far the largest recorded event occurred when a rockslide, triggered by the Alaskan earthquake of July 1958, caused an estimated $30.5 \times 10^6 \, m^3$ of rock to fall from an altitude of almost 1000 m into Lituya Bay, Alaska. The resulting surge of water reached to a height of more than 500 m on the opposite coastline of the bay (Miller, 1960). In 1949 a volcanic eruption at Las Palmas, Canary Islands, caused a major slide which almost reached the sea. A further eruption at this site is likely and would almost certainly trigger a major slide into the sea and send a tsunami wave across the Atlantic.

In certain circumstances, explosive *volcanic eruptions* may cause tsunamis. Indeed, one of the best-known examples in historical times, well-documented by Baird (1884), was that formed by the explosion of the volcanic island of Krakatau (Pulau Rakata) in the Sunda Strait in August 1883. Waves 30–40 m high swept away the small town of Merak, more than 50 km away, and other settlements on the coasts of Java and Sumatra, killing 27 000 people. The man-of-war *Berow* was carried 3 km inland before being deposited some 10 m above sea level. The resulting tsunami crossed the Pacific and Indian Oceans as a train of waves at speeds between 500 and 700 km h^{-1} before traversing the Atlantic Ocean. Some 32 hours after the initial explosion the tsunami was recorded in the English Channel as a series of disturbances a few centimetres high (Robinson, 1961).

The severity with which tsunami waves impact upon a coastline is often determined at least as much by *intensifying factors*, such as bed topography and coastal configuration, as by the size of the trigger event. For example, because of the relationship between wave speed and water depth, to which reference has already been made, tsunami waves are much higher on flat and gently sloping coastlines than on steep ones with deep water offshore. Also, the effects of wave refraction may mean that certain coastal configurations concentrate tsunami wave energy at key locations, such as Hilo in Hawaii. Again, when a tsunami wave moves into a funnel-shaped inlet, estuary or harbour, it will increase considerably in height and in the case of an enclosed

Figure 5.12 (*opposite*)   Pacific tsunami, May 1960: (A) tide-gauge record of the passage of the first three waves of the tsunami at Hilo, Hawaii, 23 May; (B) isochrones of arrival times (hours) of the initial tsunami wave after the Chilean earthquake of 22 May

Source: (A) Based on a diagram in Hindley (1978) by permission of *New Scientist*; (B) based on a diagram in Robinson (1961) by permission of the Geographical Association

bay, resonance and interference effects may complicate the pattern of water-level fluctuations as successive waves arrive. Cumulative data from Japan and Hawaii indicate that coastlines facing the area in which the tsunami originates are subjected to the largest waves (Bolt et al., 1975). The dominant effect of bed topography and coastal configuration on tsunami waves which reach the coastline is confirmed by spectral analysis. This shows that the spectra of different tsunamis at any one coastal location generally look alike but that the spectral features of a given tsunami at different coastal locations vary widely (Munk, 1962). The great variation of tsunami magnitude along the North American Pacific coast was modelled by Whitmore (1993) for earthquakes in the Cascadia subduction zone.

5.4.2 Tsunami-Prone Coastlines

The preceding discussion of causes of tsunamis indicates clearly that some coastlines are likely to be more prone than others to tsunamis and that, along a given coastline, one location may experience more severe tsunami waves than another location nearby. Coastlines near zones of crustal instability, e.g. in the Mediterranean, Caribbean and the east coast of Asia, are particularly tsunami-prone (Bascom, 1959). The ancient Greeks recorded several catastrophic inundations by huge waves (Bernstein, 1954) and an Arabian Sea tsunami in 1945 prompted further thoughts about the biblical stories of Noah's flood and of the parting of the waters described in the book of Exodus. Also at risk are locations where tsunamis tend to converge. For example, the Pacific has many seismically active seabed areas, especially in the submarine trenches off South America, Japan and the Aleutian Islands, and these tend to face towards the central Pacific. As a result, many tsunamis originating in the Pacific move towards the Hawaiian Islands (Bolt et al., 1975).

The tsunami-proneness of the Hawaiian Islands is clearly illustrated in Table 5.5 which summarises some of the world's great tsunamis. This table also emphasises that, more generally, the Pacific Ocean appears to experience the majority of major tsunamis. Indeed, 270 damage-causing Pacific tsunamis were listed by Heck (1947) for the period 279 BC to 1946 and more than 40 events, mostly small, were reported for the decade 1980–90 (UNEP, 1991a). Accordingly, some further consideration is given to tsunamis in the Pacific basin.

5.4.2.1 Tsunamis in the Pacific Basin

The frequency of tsunamis in the Pacific basin is a reflection partly of its seismic instability and partly of the vast expanse of ocean, virtually uninterrupted by landmasses. This means that with any submarine earthquake having a magnitude greater than 7 on the Richter scale (see Table 5.4) and its epicentre within the Pacific, there is a real possibility of a tsunami spreading out and striking coastlines many thousands of kilometres away. As a result, at least 22 countries on the Pacific rim are estimated to be at risk from tsunamis (NAS, 1987). Although it could take up to 24 hours to travel the full length of the ocean (see Figure 5.12B), a tsunami may still arrive with enough energy to devastate many hundreds of kilometres on the farthest coastlines. However, travel times of three to five hours are more common and if the earthquake is nearby the travel time may be only a few minutes (Askew, 1992).

Table 5.5 Great tsunamis

Date	Source region	Wave run-up (m)	Report from
15 BC	Santorini	?	Crete
1707	West Pacific	Several	Japan
1 Nov. 1755	Eastern Atlantic	5–10	Lisbon, Portugal
21 Dec. 1812	California	Several	Santa Barbara, CA
7 Nov. 1837	Chile	5	Hilo, Hawaii
17 May 1841	Kamchatka	< 5	Hilo, Hawaii
2 Apr. 1868	Hawaiian Islands	< 3	Hilo, Hawaii
13 Aug. 1868	Peru–Chile	> 10	Arica, Peru
10 May 1877	Peru–Chile	2–6	Japan
27 Aug. 1883	Krakatau	30–40	Java
15 Jun. 1896	Honshu	24	Sanriku, Japan
3 Feb. 1923	Kamchatka	$c.5$	Waiakea, Hawaii
2 Mar. 1933	Honshu	24	Sanriku, Japan
1 Apr. 1946	Aleutian Islands	10	Wainaku, Hawaii
4 Nov. 1952	Kamchatka	< 5	Hilo, Hawaii
9 Mar. 1957	Aleutian Islands	< 5	Hilo, Hawaii
9 July 1958	Alaska	524	Lituya Bay, Alaska
23 May 1960	Chile	> 10	Waiakea, Hawaii
28 Mar. 1964	Alaska	6	Crescent City, CA
28 Feb. 1967	Eastern Atlantic	> 1	Casablanca
16 Aug. 1976	Moro Gulf	5	Philippines
1983	NW Pacific	14.5	Noshiro, Japan

Source: From a table in Bolt et al. (1975), reproduced by permission of Springer-Verlag GmbH & Co. KG, supplemented by data from various sources

One of the most thoroughly investigated Pacific tsunamis was that of 1 April 1946, mainly because of the large number of oceanographers who were in the area to observe the effects of an atomic bomb test at Bikini atoll (Berstein, 1954). The tsunami struck the Hawaiian Islands particularly severely, causing 159 deaths and damage amounting to US$ 25 million. The origin was in the Aleutian submarine trench, where a submarine landslide and seabed displacement were associated with an earthquake of magnitude 7.5. At Hilo, Hawaii, some 3500 km to the south, the first sea-level rise occurred about six hours later and was followed by a series of larger waves, some appearing to be 6 m or so in height. These varied greatly from place to place, some rising gently and some resembling a tidal bore, with a steep front and flat wave crest behind. In some localities the sixth, seventh and eighth waves were said to be the highest (Bolt et al., 1975).

The same tsunami wave-train caused considerable destruction throughout the islands of Oceania, 6400 km from the epicentre, and on the coast of South America. However, its effects were most spectacular at Scotch Cap in the Aleutian Islands of Alaska, where a lighthouse whose base was 10 m above sea level and a radio mast whose foundations were 30 m above sea level were both demolished (Bascom, 1959; Shepard, 1948).

Another major Pacific tsunami was that of 22 May 1960. This resulted from an exceptionally severe earthquake which affected southern Chile and caused a large area of the seabed to founder off the Chilean coast, with large-scale coastal deformation extending from latitude 38°S to 43°S. Locally the tsunami reached the coast in the form of three large waves which caused major flooding and more than 900 deaths, mostly in the coastal cities of Saavedra, Mehuin, Corral, Maullan and Ancud (Bolt et al., 1975). The resulting waves spread throughout the Pacific Ocean (see Figure 5.12B), reaching Hawaii at about 0100 h on 23 May after travelling for 15 hours at an average speed of some 710 km h^{-1}. Hindley (1978) described the sequence of sea-level variations at Hilo, Hawaii, which are shown in Figure 5.12A. After a minor initial crest and trough, a second wave about 3 m in height caused some flooding, and the succeeding trough drew the sea down to 2.5 m below normal level. This was followed by a roaring bore of seawater which reached the coastline as a wall of water 7 m in height, sweeping away large areas of the city. Although not shown in Figure 5.12A, further smaller waves followed, some crested and some bores, before further damage and flooding were caused by another large wave which arrived about 45 minutes after the main wave.

Some 22 hours after the Chilean earthquake the tsunami waves first reached Japan. The waves attained heights of 10 m along the coastlines of Hokkaido and Honshu and destroyed or flooded thousands of houses and hundreds of ships, as well as causing the loss of about 120 lives (Bolt et al., 1975).

5.5 SEA-LEVEL CHANGE AND COASTAL FLOODING

Of all the types of flooding discussed in this book, that in coastal areas is the most certain to be influenced by the assumed effects of climate changes induced by global warming. This is largely because, if global warming occurs or is indeed already occurring, the resulting thermal expansion of the oceans will cause a significant worldwide increase in sea level above and beyond the increase which would in any case result from the continuing retreat from the last Ice Age. And if global warming also melts substantial masses of land-based ice, the additional freshwater input to the ocean basins will lead to an additional increment of sea-level rise.

It is not appropriate here to discuss the merits of the predictions about global warming and its effects on sea level, except to observe that there appears to have developed a tendency towards alarmism in many areas of environmental reporting which, on this question as on many others, has caused unnecessary confusion among scientists and lay-persons alike. This is especially unfortunate in relation to sea-level change since the global database of sea-level measurements is not robust. For example, the derivation of mean values of sea level for 75% of the global surface has been hindered in the past by the comparative sparsity of gauges. In addition, it has always been difficult to measure sea level accurately at a given location, particularly in conditions of isostatic adjustment of the land surfaces. Only in recent years has the advent of the satellite-based Global Positioning System (GPS) made it possible to determine the vertical displacement of tide gauges, which can then be removed from the tide-gauge records, to determine absolute changes in mean sea level (Ashkenazi et al., 1995; Edge, 1996). It is estimated that the implementation of this new methodology and its use to connect tide-gauge benchmarks to a global geodetic reference framework will take several decades.

In this section, therefore, discussion of the relationship of sea-level change to coastal flooding is restricted to a brief consideration of two related issues. These are: the likely magnitude of sea-level change which may result from global warming, taking into account isostatic changes of land surface elevation; and the implications of that sea-level change for coastal flooding.

5.5.1 The Likely Magnitude of Sea-Level Change

Global sea level is in a state of perpetual change. On a geological time-scale it is clear that there have been variations of many hundreds of metres. But even on a time-scale of centuries, and during the period in which tide-gauge observations have been made, significant changes of sea level have occurred, mainly as a result of thermal expansion and icemelt. Actual tide-gauge records for more than 100 countries are archived and analysed at the Permanent Service for Mean Sea Level (PSMSL) at Bidston, England. These data need careful interpretation since plots of annual means, for example, are quite 'noisy', with variations of 0.10 m from year to year being common and variations of up to 0.20 m occurring at some locations. Graphs of some of the longest tide-gauge records were presented by Woodworth (1990) and Woodworth et al. (1990). Those for Cascais in Portugal (1880–1985), Brest in France (1810–1985), and Newlyn in England (1915–85), in common with the majority from other parts of the world, showed an average sea-level rise of approximately 0.15 m to 0.20 m per century. However, in parts of northern Europe, where isostatic adjustment is still continuing, the pattern is very different, with Stockholm in Sweden (1895–1985) showing an average fall in sea level of 0.4 m per century, and Aberdeen in Scotland (1860–1985) showing a more or less constant sea level.

Such evidence for long-term sea-level change, and the mechanisms which may have been responsible for that change, have been extensively reviewed (e.g. Warrick and Oerlemans, 1990; Woodworth, 1990; Woodworth et al., 1990). Although the picture that emerges lacks total clarity, there is now general agreement that it is 'highly likely that global mean sea level has been rising' (Warrick and Oerlemans, 1990). Estimates of the rate of rise range from 0.05 m to 0.30 m per century, with most lying in the range 0.10 m to 0.20 m per century. The main contributing mechanisms appear to have been thermal expansion, the melting of glaciers and small ice-caps, and the melting of the Greenland ice-sheet. Quantitative estimates of the contributions of each of these mechanisms are shown in Table 5.6 and emphasise that there is little evidence that the Antarctic ice-sheet has made a significant contribution. As the 'best estimate' column shows, if it is assumed that the contribution from Antarctica has been zero, then the combined contribution from thermal expansion and icemelt over the past century amounted to 0.105 m which is within the lower part of the range of observed sea-level rise referred to earlier (0.1–0.2 m). However, the range of uncertainty, illustrated by the 'high' and 'low' columns in Table 5.6, is large (−0.005 m to 0.22 m).

In light of the difficulties of estimating accurately what has already happened to sea level during a period in which tide-gauge data have been available, it is hardly surprising that there is even greater uncertainty about how mean sea level will change in the future. Assuming a 'business-as-usual' scenario, the IPCC forecast a continuing sea-level rise over the next century at rates higher than those experienced during the past century (Warrick and Oerlemans, 1990). The change for the period 1985–2030 is shown in Table 5.6 and indicates that by 2030 the best estimate is that global sea level would be 0.18 m higher than in 1985, although given the range of uncertainty about the contributing mechanisms, the rise could be as little as 0.08 m or as high as 0.29 m. By

Table 5.6 Estimated contributions (m) to sea-level rise over the last century and for the period 1985–2030

	Past 100 years			1985–2030		
	Low	Best estimate	High	Low	Best estimate	High
Thermal expansion	0.020	0.040	0.060	0.068	0.101	0.149
Glaciers/small ice-caps	0.015	0.040	0.070	0.023	0.070	0.103
Greenland ice-sheet	0.010	0.025	0.040	0.005	0.018	0.037
Antarctic ice-sheet	−0.050	0.000	0.050	−0.008	−0.006	0.000
Total	−0.005	0.105	0.220	0.087	0.183	0.289
Observed	0.100	0.150	0.200			

Source: From tables in Warrick and Oerlemans (1990). Reproduced by permission of the Intergovernmental Panel on Climate Change

the year 2070, the best estimate is a rise of 0.44 m (range 0.21–0.71 m) and by the year 2100, the best estimate is a rise of 0.66 m (range 0.31–1.10 m).

Although considerably more pessimistic earlier estimates were reviewed by Boorman et al. (1989), the IPCC estimates are regarded here as the most carefully reasoned and most widely accepted guide to what may happen and even they may be subject to some further downward revision, in due course, in the light of improved understanding of the effects of aerosols and sulphur particles on rates of global warming. Warrick and Oerlemans (1990) emphasised, however, that sea level will not rise uniformly around the world, partly because of the local and regional effects of isostatic adjustments and partly because dynamic processes in the ocean and atmosphere circulations may cause sea level to change regionally. In formerly glaciated areas, for example, isostatic uplift (glacial rebound) (e.g. Figure 5.2) may either significantly reduce the projected sea-level rise, as in Scotland, or even exceed it, as in parts of Scandinavia. In other locations, such as south-east England, downward tilting may further exacerbate the effects of eustatic sea-level rise. Again, because of the way in which air temperature changes drive the ocean circulation, model predictions show that there may be major departures from the projected mean sea-level rise of 0.18 m. Mikolajewicz et al. (1990), for example, noted that a rise of 0.40 m is projected in the North Atlantic, because of a reduction of deep-water formation, but that other changes in ocean circulation mean that the level of the Ross Sea, on the Antarctic margins of the South Pacific, may actually fall.

5.5.2 The Effects of Sea-Level Change on Coastal Flooding

The impact of sea-level rise on coastal flooding will amount to much more than the re-drawing of the coastline at a higher contour level. This is because the presence of deeper water offshore will modify wave, surge and tsunami magnitudes as well as the pattern of tidal water movement and these, in turn, will modify rates of erosion and deposition at the coastline. In addition, since sea-level rise is the predicted outcome of climate change, it must be expected that the climate conditions which generate waves and surges will themselves change, leading to more extreme sea-level events in some locations. Indeed, Warrick and Oerlemans (1990) suggested that, even with modest increases in global mean sea level, changes in the frequency of extreme sea-level events may be very significant in their impact on coastal flooding. At present, such events are difficult to quantify because of the uncertainties in regional predictions of climate change. Where flood-prone coastal areas are protected by flood defences, the effect of rising sea levels will be to increase the frequency with which those defences fail if left unmodified. In other areas, rates of sea-level rise which substantially reduce the protection now afforded naturally by beaches, dunes and marshes are likely to cause significant increases in coastal flooding. These issues are further complicated by wide discrepancies in the ability of communities, and even entire nations, to respond to such threats.

Quite small increases in mean sea level can have a significant impact on the frequency with which a particular threshold, such as the height of a flood defence structure, is exceeded. In Figure 5.13 hypothetical flood frequency curves are shown for two locations typical of the south coast of England (curve A) and the east coast of England (curve B). The return period of a 3.2 m sea level, which is currently 100 years in both cases, falls to 25 years with curve B and to just 5 years with curve A. This implies that a rise in sea level would have a greater effect on flood risk along the south coast of England than along the east coast (CCIRG, 1991).

These flood frequency curves would be altered even further if sea-level rise was also accompanied by an increase in the frequency of coastal storms. Relevant data are sparse and their interpretation far from straightforward. For example, results from the EU ENVIRONMENT research project showed that, although the storm climate in the near-coastal areas of north-west Europe has not systematically worsened during the past century, there has been a steady increase in significant wave height (the average height of the highest one-third of all waves present in a given time period) in the north-east Atlantic during the last three decades (Samuels and Brampton, 1996). In addition, higher sea levels, possible changes in storm tracks and more severe wave climates could magnify the height of waves when they break on the shoreline and against coastal defences (IH, 1994b). The modification of wave height will depend on beach morphology, such that the shallower the beach gradient, the greater the increase in wave height for a given increase in sea level. Sea-level rise over the past two centuries also appears to have been accompanied by an increase of almost 2 m in the maximum water levels of storm surges at London Bridge (Carter, 1988). Further studies of changing storm-surge magnitude around the coastline of the UK are currently in progress (Woodworth, 1990) but current global climate models do not provide an adequate basis for simulating storm-surge conditions. Modelling studies in the wider Caribbean region (Mercado et al., 1993) showed that, because of the subduing effect of offshore deep water, a sea-level rise of the order of 0.20 m would not directly produce a significant change in storm-surge magnitudes. However, storm surges could become more severe if hurricane intensity increased because of changes in air or sea temperatures. This situation was projected for the Bay of Bengal by Murty and El-Sabh (1992). They postulated an increased frequency and intensity of tropical cyclones and argued that the new breed of 'hypercanes' would generate storm surges on the coast of Bangladesh which could attain heights of 15 m compared with 6 m at the present time.

Tidal changes in shallow-water seas, such as the continental shelf of north-west Europe, are an expected outcome of changing water depth, although there is only limited evidence that significant changes have accompanied the rising sea levels of the

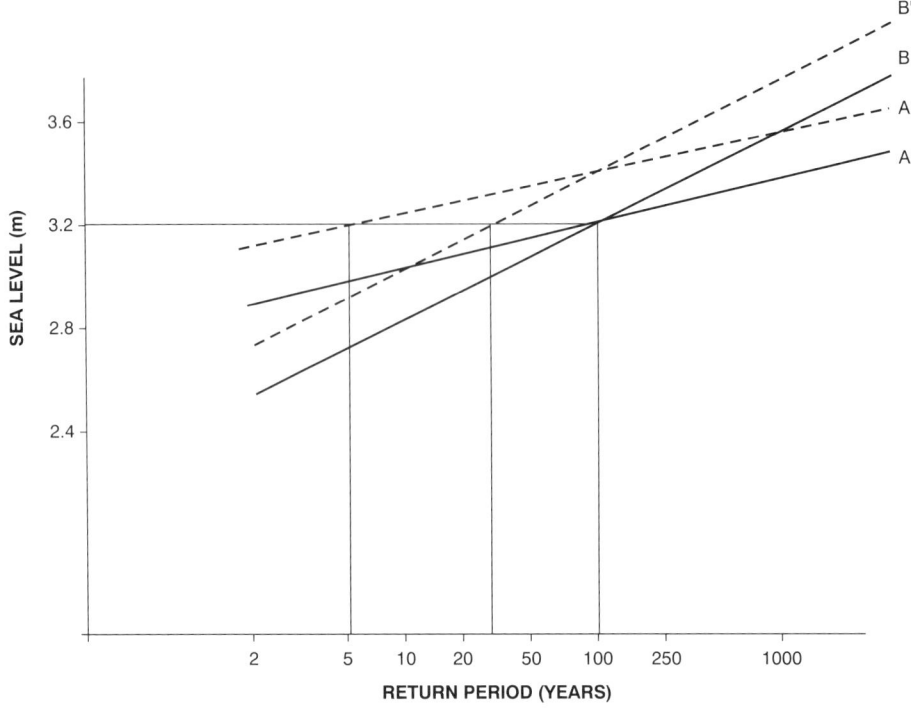

Figure 5.13 The effect of a 0.2 m change in sea level on flood frequency at two contrasting locations. See text for explanation
Source: Based on a diagram in CCIRG (1991). Crown copyright, reproduced with the permission of the Controller of HMSO

last century. At Newlyn, in south-west England, there was an increase in tidal amplitude of about 1% over the century, as mean sea level rose by about 0.18 m, and much larger increases occurred in Germany and the Netherlands (Woodworth, 1990). The effect of rising sea level on tidal regimes is not a simple one. Even though a first approximation may be made by simply adding the sea-level rise as a constant to the tidal curves, it is clear that the various components that make up the tide are affected differently by the change in water depth, with outcomes that it may be possible to interpret only with detailed model studies. Such studies on the Bay of Bengal (Flather and Khandker, 1993) showed that, following a rise in sea level, tidal amplitude and storm-surge height could increase or decrease at different points, depending on the relative timing of tidal high water and peak surge. The combined tide and surge elevations reached a maximum of about 4 m above mean sea level near Cox's Bazar, with reduced elevations in the north-east and north-west corners of the Bay and an increase in elevation in the intervening area. In an estuarine situation, changes in freshwater flows as a result of climate change could add a further complication.

Densely populated or highly developed, low-lying coastal areas, such as Bangladesh, the Nile delta and Florida, coral and similar-profile islands, including the Maldives, and major estuaries, including some in the UK, will be especially at risk from increased flooding following sea-level rise. Where flood defences already exist, they will need to be improved or set back; where none now exist, they may need to be developed, or some form of retreat, whether managed or unmanaged must be adopted. In most cases, the solutions will be costly and some may not be affordable. And yet the increase in risk may be considerable. Studies have shown that at a typical port location on the east coast of England, a 0.15 m rise in mean sea level doubles the probability of a sea wall or dike being overtopped; a 0.30 m rise quadruples the probability; and a 0.45 m rise increases the probability eight-fold (Woodworth, 1990). As already indicated, the risk will increase even more on the south coast of England and in many other coastal areas. Again, in an environment already susceptible to storm-surge flooding as a result of land reclamation, river diversion and groundwater abstraction, the city of Venice in Italy will suffer greatly increased flood risk such that a sea-level rise of 0.40 m will lead to Piazza San Marco being flooded by 87% of all tides (Spencer and French, 1993). Already, during the twentieth century, the estimated return period of the major flood event of 1966 has been reduced by 75% (Scotti, 1993).

5.6 FLOODS AND COASTAL GEOMORPHOLOGY

The importance of seabed morphology, coastal configuration and beach profiles in determining the impact and severity of storm surges and tsunamis is noted elsewhere in this chapter. Conversely, just as river floods interact dynamically with the sediment systems of river channels and floodplains (Section 3.6.2), so too do coastal flood events influence coastline sediment systems. Such impacts may modify the nature and extent of inundations resulting from subsequent events and may be detectable in the geological record after many thousands of years (e.g. Cullingford et al., 1989). Furthermore, coastal sediment systems may be significantly modified by human impacts, such as the construction of flood defences or the dredging of mineral aggregates, which may alter flood risk at other locations along the coastline, and also by changing sea levels.

The severity of flooding resulting from storm surges is partly dependent on coastal geomorphology (see also Section 5.3.1). The bathymetry of the seabed up to about 60 km from the coastline, for example, helps to determine surge and wave size, with shallow water being conducive to a larger storm surge. And bays and inlets can cause funnelling effects, raising the level of the open-coast surge by a factor of up to two (WMO, 1990b). In addition, the presence of offshore reefs and islands may modify the size of storm-surge or tsunami waves reaching the coastline. For example, as a result of the presence of an offshore reef which protects the northern coast of Oahu in the Hawaiian Islands, the size of the 1946 tsunami wave there was much smaller than on the comparatively unprotected northern coast of Molokai (Bolt et al., 1975). In low-lying coastal areas protected by a sea wall, the volume of water overtopping the wall during severe events depends, among other things, on wave height in front of the wall. This, in turn, is partly dependent on beach height so that changes in beach height will modify the severity of flooding resulting from a given storm event (Tinnion et al., 1995).

Beach profiles change almost continuously, with the accumulation of sand and other beach material during periods of quiet sea state, and the cutting back of that material during storm periods. Major storm events may severely erode or completely destroy protective features such as offshore reefs, dunes or high beaches. In this way, flooding from an ensuing event of similar magnitude may be more severe. Similarly, the tsunami waves which reached the Hawaiian Islands in 1946 tore loose coral fragments up to

1.3 m across and deposited them on the beach 5 m above sea level. The backwash of these waves exposed coastal mudflats for a distance of 150 m beyond the normal coastline (Bolt et al., 1975). Comparably, the tsunami of May 1960 deposited a layer of sand up to 0.3 m thick and up to 4 km inland in parts of south-central Chile (Wright and Mella, 1963). Tsunami wave run-up often causes widespread deposition of large boulders and the creation of boulder fields. Dawson (1994) cited several contemporary and Holocene examples from around the Pacific involving boulders and blocks up to 750 m^3 in size and occurring up to 30 m above present sea level.

The geomorphological consequences of a rise in sea level would depend on the nature of the existing coastline and were discussed by Boorman et al. (1989). Hard-rock, cliffed coastlines would be little affected, even by a substantial sea-level rise. Low-lying, soft-rock coastlines would experience increased erosion as sea level rose although the redistribution of the eroded material, mainly by longshore drift, would probably allow the coast to reform more or less unchanged. This process of 'sea-level transgression' is well-illustrated by the barrier island coasts of Norfolk in eastern England, and of the northern parts of the Netherlands and Germany, and would result in the character of coastal flooding remaining relatively unchanged, apart from its gradual migration inland. Only if sea-level rise occurred more rapidly than the processes of transgression could operate would the risk of coastal flooding increase significantly. Where soft-rock coastlines are also cliffed, as in parts of eastern England, already high rates of erosion would be greatly enhanced and this would provide additional supplies of sediment to saltmarsh and sand dune areas down the coast which would then be better protected against coastal flooding than they are at present.

Since many low-lying coastlines are protected against flooding by sea walls and dikes, there is a risk that accelerated erosion of material on the seaward side of such structures, which would result from sea-level rise, would rapidly exhaust the supply of sediment for longshore drift. Similarly, attempts to moderate increased erosion from soft-rock cliffs, or to protect coastal settlements exposed to increased flood risk, would also reduce sediment movement and result in beach starvation and increased erosion and flood risk at locations farther down the coast. In estuaries and deltas, the interactions between river and coastal flooding can add further complications. For example, attempts to control monsoonal river flooding in Bangladesh have restricted sediment inputs to inter-distributary areas. Since these sediment inputs have normally counterbalanced the sea-level rise resulting from sediment compaction and downwarping, their reduction has led to a dominance of subsidence which threatens increased coastal erosion, flooding and saline intrusion (Spencer and French, 1993).

The adverse, 'knock-on' effects of some attempts to reduce coastal flood risk at specific locations are already well-documented and have led to an increasing interest in managed morphological change and beach and shoreline management techniques which are intended, amongst other things, to moderate the severity of coastal flooding (see also the discussion of coastal flood defence strategies in Section 7.4). In many countries, coastal 'engineering' has relied heavily on procedures set out in the *Shore Protection Manual* of the US Army Corps of Engineers, to be succeeded in due course by the *Coastal Engineering Manual* (Pope, 1994). However, a basic problem in coastline management has been an inadequate understanding of physical processes and a lack of basic data, for example on beach and nearshore levels over several decades. Attempts to rectify these deficiencies have led to a number of major research programmes and policy developments intended to improve understanding of the relations between coastal erosion, deposition and flooding (e.g. Lee and Marker, 1995).

One such programme is CAMELOT (Coastal Area Modelling for Engineering in the Long Term), established in 1993 by the Ministry of Agriculture, Fisheries and Food in England and Wales (Soulsby et al., 1994). Part of this programme is concentrating on modelling waves, tides, surges and sediment transport and includes a coupled wave–tide–surge model for the eastern coastal area bounded by Whitby in the north and Cromer in the south.

The output from data collection and modelling programmes such as CAMELOT will provide a valuable input to shoreline and coastal management plans. Although these have been used for many years in some countries, such as Australia (Kay et al., 1995), they were initiated in England and Wales only in 1995 (e.g. Hutchison, 1994; Young and Pos, 1995) in order to strengthen further the development of coastal defence methods, such as raised beach levels, offshore breakwaters and set-back, which, in the words of the Minister of Agriculture, Fisheries and Food, 'work with the forces of nature rather than against them' and help to ensure that potential knock-on effects elsewhere on the coast are addressed. Often such options are feasible only in wealthier nations and are less easily implemented in some of the countries, such as Bangladesh, which suffer most from the adverse effects of large-scale coastal flooding.

CHAPTER 6

Flood Estimation

CONTENTS

6.1 Introduction 178
   RIVER FLOODS
6.2 Estimating river floods 179
   6.2.1 Deterministic methods 179
      6.2.1.1 Simple empirical methods 180
      6.2.1.2 Other empirical methods 180
      6.2.1.3 PMP and PMF 182
   6.2.2 Probabilistic methods 188
6.3 Enhancing the floods database 192
6.4 Contemporary approaches to river flood estimation 194
   6.4.1 The UK *Flood Estimation Handbook* 197
6.5 Meltwater floods 198
   COASTAL FLOODS
6.6 Estimating coastal floods 199
   6.6.1 Extending the database 201
   6.6.2 Flood estimation in estuaries 201

6.1 INTRODUCTION

Flood estimation, also known as flood prediction, involves using an often incomplete set of information on meteorological, hydrological, catchment and coastal conditions to estimate extreme flood conditions for a given riverside or coastal location. Sometimes the objective is to calculate the maximum flood which is likely to occur. More usually it is to estimate either the return periods of floods of specified magnitude, or the magnitudes of floods of specified return periods. What, for example, was the return period of the Great Mississippi flood of 1993 when the peak flow at St Louis reached a highest-ever magnitude of $28\,840\,m^3\,s^{-1}$? Or, what is the height of the 300-year storm surge on the coast of Bangladesh at Chittagong? Such estimates form an important basis for the design of structures such as flood embankments, tidal barriers, dams, bridges and culverts. They are also an essential input to the design of channel 'improvement' schemes, aimed at increasing the flood-carrying capacity of river channels, and to attempts to plan and manage land use in order to minimise flood risk and flood damage in both fluvial and coastal situations. Techniques have been developed which enable flood estimation to be carried out for catchments with no available hydrological data. Inevitably, however, the predicted values are likely to carry less conviction than those for catchments where a solid database exists.

Although there are important areas of common ground between flood estimation and flood forecasting, the latter is essentially carried out in real time, uses a wide range of available data, and provides detailed information, often of a very precise nature, on the magnitude, depth, timing and duration of the forthcoming flood event. Flood forecasts are normally associated with appropriate warning and evacuation procedures and are discussed in the context of responses to the flood hazard in Chapter 8.

Flood estimation techniques range widely from the crudely simple to the mathematically complex. There is no obviously correct approach. Nor is it surprising, since one is trying to estimate flood conditions towards the upper extremes of the magnitude–frequency distribution, that it is very difficult to validate the results of the methods which are currently in vogue. This is especially so in cases of design-flood estimation, i.e. the maximum flood against which a scheme of protection is designed. If, for example, a flood embankment designed against the 100-year flood fails because it is overwhelmed by a 500-year flood, it does not follow that the original design was wrong. Alternatively, when a flood embankment has not failed after a long passage of time this does not necessarily indicate that the design was correct—the design flood estimate may simply have been far too large and the resulting embankment far too substantial and unnecessarily costly.

The most commonly used flood estimation methods are set out in manuals or incorporated into computer programmes. The latter, as in many other areas of application, have become increasingly user-friendly and interactive and often run successfully in a desk-top computing environment. The objective of this chapter is not to recapitulate these approaches; nor would this be feasible. In the UK, for example, the *Flood Studies Report* (NERC, 1975) comprised five volumes, as will its successor—the *Flood Estimation Handbook*. Instead, for river and coastal floods, the main categories of approach are briefly described and their fundamental weaknesses are discussed. Selected contemporary methods of flood estimation are briefly reviewed, the balance of the discussion reflecting the fact that flood estimation has developed largely in the context of river floods. The special problems of estimating meltwater floods are briefly mentioned.

Traditional approaches to flood estimation fall into two main categories. Those which are deterministically based treat floods as the product of particular physical processes, e.g. a specified precipitation falling upon a designated catchment area, or a particular direction and speed of wind impinging upon a coastline during specified tidal conditions. Those which are probabilistically based treat floods as almost purely random events susceptible to statistical analysis. Neither approach is entirely satisfactory. Quite often, therefore, elements of both are incorporated in a given flood estimation method or, as in the UK *Flood Studies Report*, the user is recommended to use more than one method and then to compare the results.

RIVER FLOODS

6.2 ESTIMATING RIVER FLOODS

6.2.1 Deterministic Methods

The deterministic approach assumes that the magnitude of a river flood varies with the nature of both precipitation and catchment area and many methods have been developed to estimate maximum flood discharge from information on precipitation and catchment characteristics. These range from simple empirical equations relating catchment area and flood discharge, through more sophisticated relationships incorporating a wide range of catchment characteristics, to attempts to estimate the maximum possible flood-producing rainfall event.

6.2.1.1 Simple Empirical Methods

There are literally hundreds of simple empirical flood estimation equations which take the general form

$$Q_{max} = CA^n \tag{6.1}$$

where Q_{max} is the estimate of maximum flood discharge, A is the catchment area, n is an exponent and C a coefficient which depends on the geographical and climatological characteristics of the catchment. An example from the UK was that proposed by Bransby Williams for catchments larger than 10 square miles (26 km²)

$$Q_{max} = 4600 A^{0.52} \tag{6.2}$$

where the maximum flood discharge is in cusecs (cfs). One of the oldest and certainly one of the most widely used equations in this category is the 'rational formula'

$$Q_{max} = CIA \tag{6.3}$$

where C is a runoff coefficient indicating the percentage of rainfall which appears as quickflow, I is the mean rainfall intensity, and A is the area of the catchment. The rational formula was first described in the USA by Kuichling (1889) and in the UK by Lloyd-Davis (1906) and, although long since dismissed by most theoretical hydrologists on both sides of the Atlantic, the method maintains a surprising popularity with practising engineers, especially for flood estimation in small urban or relatively homogeneous rural catchments. It is, for example, included in several modern software packages, such as the Watershed Modeling System (Scientific Software Group, 1997) which incorporates Digital Terrain Model (DTM) data to enable a more accurate definition of catchment characteristics. DTM data were also used in the application by Mokadem et al. (1989) of a rural-area hydrological model developed by Williams et al. (1985) and which was based on a modification of the rational formula.

Simple empirical methods adopt an essentially 'black box' approach to flood estimation. Different equations tend to give widely different results and may not always be internally consistent. They do have a limited usefulness in flood estimation, however, particularly in small homogeneous catchments, in areas where hydrological data are sparse, and in situations where DTM data can be used to provide an accurate definition of catchment characteristics.

6.2.1.2 Other Empirical Methods

More sophisticated empirical formulae may use a sufficiently wide range of hydrometeorological and catchment data to enable adequate modelling of the physical processes contributing to major flood peaks. Much of the impetus for the deterministic, process-based approach to flood estimation was undoubtedly given by early work on the unit hydrograph (e.g. Sherman, 1932) and on the role of infiltration in quickflow production (e.g. Horton, 1933). Although the fundamental thesis of each has since been superseded their contribution to subsequent flood estimation methods was substantial. For example, in the context of unit hydrograph theory, some of the most valuable developments were those concerning the derivation of the instantaneous unit hydrograph and of synthetic hydrographs, particularly in relation to flood estimation for ungauged catchments. Horton's infiltration theory subsequently contributed to flood estimation methodology both directly, in the form of infiltration indices, and indirectly in the form of indices of antecedent precipitation, soil moisture and the proportion of the catchment under different types of land use—all of which are relevant to the generation of quickflow from source areas of variable size and location within the catchment (see Section 3.2).

A useful empirical method of estimating flood peaks from specified rainfall and catchment conditions was contained in the *Flood Studies Report* (NERC, 1975). This involves two main steps, in the first of which the proportion of precipitation appearing as quickflow is calculated from

$$P_q = 95.5 SOIL + 0.12 URBAN + 0.22(CWI - 125) + 0.1(P - 10) \quad (6.4)$$

where P_q is the percentage of rainfall appearing as quickflow, $SOIL$ is a soil index related to the winter rate of rain acceptance, $URBAN$ is the percentage of urban development, CWI is a catchment wetness index calculated from soil moisture deficit and antecedent precipitation, and P is storm rainfall. The second step involves distributing this volume of quickflow according to the ordinates of the synthetic unit hydrograph to determine the flood peak. In common with many other flood estimation methods a simple triangular flood hydrograph is assumed. It was found that the peak discharge was inversely proportional to the time-to-peak

$$Q_{max} = 220/T_p \quad (6.5)$$

where Q_{max} is the peak discharge in m³ s⁻¹ 100 km⁻² and T_p is the time to peak in hours and can be calculated from catchment characteristics using

$$T_p = 46.6 \, MSL^{0.14} \, S1085^{-0.38} (1 + URBAN)^{-1.99} \, RSMD^{-0.4} \quad (6.6)$$

where MSL is main stream length in km, $S1085$ is main stream slope between 10 and 85% of stream length upstream from gauge in m km⁻¹, $URBAN$ is as defined in equation 6.4 and $RSMD$ is the net daily rainfall in mm having a return period of five years.

The time base (T_b) of the hydrograph is controlled by the assumptions of a triangle and unit volume

$$T_b = 2.525 T_p \quad (6.7)$$

By making assumptions about the relationships between the return periods of rainfall and flood discharge (see also Section 6.3) this method allows the estimation of Q_t, i.e. the peak discharge for a selected return period.

In common with much of the methodology of the *Flood Studies Report*, this particular empirical approach to flood estimation relies heavily on the use of catchment characteristics derived from standard maps. Now that there is a much wider availability of DTM data it should be possible to derive other characteristics (e.g. stream length, stream density, average catchment slopes) which may be more helpful in estimating flood response to a specified input of rainfall. This is certainly the approach to the use of catchment characteristics which is expected to be adopted in the new *Flood Estimation Handbook* (MAFF, 1996).

6.2.1.3 PMP and PMF

There is a range of circumstances in which it would be helpful to be able to estimate the biggest flood which it is physically possible for a specified catchment to produce. Such an event would presumably result when, in a given period of time, the maximum possible amount of rain fell when the catchment was in its maximum quickflow-generating condition. This implies, for example, that extremely heavy rain falls onto a catchment whose surface is either saturated or frozen or for some other reason is virtually impermeable. In practice the concepts of 'maximum possible' precipitation or flood events involve many untestable assumptions. Accordingly, a more cautious approach has developed involving attempts to estimate the *probable* maximum precipitation (PMP) and flood (PMF) events.

In order to determine the PMP it would be necessary to know the upper limits on (a) the humidity concentration of the air involved in the rain-generating event (e.g. storm cell, frontal system); (b) the rate at which humid air can move into the rain-generating event; and (c) the proportion of the inflowing water vapour which can be precipitated, i.e. the storm efficiency. Traditionally the estimation of PMP involves three main steps: identification, transposition and maximisation. First, maximum rainfalls for the target catchment and for other comparable areas are identified from the rainfall records. Second, where these rainfall events are for other comparable areas they are transposed to the target catchment. Third, the transposed values are maximised for the target catchment on the basis of the differences in meteorological conditions between the target catchment and the area over which the rain originated.

Inevitably, very high-magnitude rainfall events are also very rare, so that transposition is used to broaden an unavoidably narrow database. Its use should be restricted to relatively homogeneous areas, e.g. those having no significant topographic barriers, in which major storms of the same type have an equal chance of occurring. It would be unreasonable, for example, to transpose a tropical cyclone to polar latitudes but a major thunderstorm could be transposed with much greater confidence anywhere within the Great Plains area of North America.

The maximum average depth of rainfall and its duration (depth–area–duration curve) for specified areas can be derived from analysis of recorded major rainfall events. Knowledge of the precipitation process enables these data to be maximised since the rate of precipitation depends essentially on the rate of uplift of moist air and on the amount of moisture present in that air. Although the lifting processes are complex and not easily measured, atmospheric moisture content (precipitable water) is a more accessible variable, especially with the present widespread availability of weather satellite data. Even before the advent of remote sensing it was known that precipitable water is highly correlated with surface dewpoint. Maximisation was normally carried out, therefore, by adjusting for moisture content or precipitable water with reference to the ratio of the maximum moisture content observed over the target catchment to the moisture content actually observed in each storm. Early definitive work was described by Hershfield (1961; 1965), Wiesner (1970) and Miller (1973) and has subsequently been supported by a wide variety of detailed studies (e.g. Maddox et al., 1979; 1980; Austin et al., 1995).

The transposition and maximisation of recorded rainfall events may be reasonable for extensive areas of low relief, where storm types can often be transposed over large distances and for which large amounts of representative data are available. These techniques can also be used in some mountainous areas where the distribution of maximum water content is spatially uniform and where the database is good, for example, the Harz Mountains in Germany (Schulze et al., 1994). In most mountainous areas, however, data are relatively sparse and orographic effects tend to dominate the occurrence of precipitation. Accordingly, methods have been devised to separate out the orographic component of rainfall from the meteorological (e.g. frontal) component so that non-orographic storms may be transposed to mountainous areas and then maximised. More usually the PMP for mountainous areas is estimated using storm models to simulate the movement of moisture into a catchment and its release as precipitation. The upglide model describes forced ascent at a topographic feature or frontal surface and the convergence model describes the dominantly convective processes in thunderstorms and tropical cyclones. Such is the distinctiveness of many mountainous areas that standard methods must be used with great care. For example, the very dramatic orographic barrier represented by the Southern Alps of New Zealand gives rise to extreme rainfall gradients across the range but negligible gradients parallel to it. In this situation Thompson (1993) suggested that a two-dimensional atmospheric model might generate more defensible PMP estimates than the standard transposition techniques. Recent work in the UK, employing radar-based estimates of rainfall (Austin et al., 1995), used a storm model which incorporated solar heating, orographic uplift and mesoscale convergence.

Statistical methods of estimating the PMP have sometimes been proposed as an alternative to process-based methods. For example, Hershfield (1961; 1965) used general extreme value theory to propose that the PMP may be expressed as

$$PMP = Pbar + K\sigma \qquad (6.8)$$

where *Pbar* is the mean precipitation for a specified duration, K is a frequency factor or reduced variate which depends on the statistical distribution used, the number of years of record and the return period, and σ is the standard deviation. From analysis of daily rainfalls from 2645 stations in the USA, Hershfield suggested that K could be assumed to have a maximum value of 15, so that the PMP would then be the mean plus 15 standard deviations for a given duration of rainfall. This is a simple way of estimating PMP for short-duration storms over small areas but its usefulness is conditioned by the amount of data available and by the assumption that K can have a maximum value.

Generalised isohyetal maps of PMP for different storm durations and areas in the USA have long been produced and updated by NOAA and its predecessors (see Figure 6.1). These provide an accessible source of information, particularly useful in the early

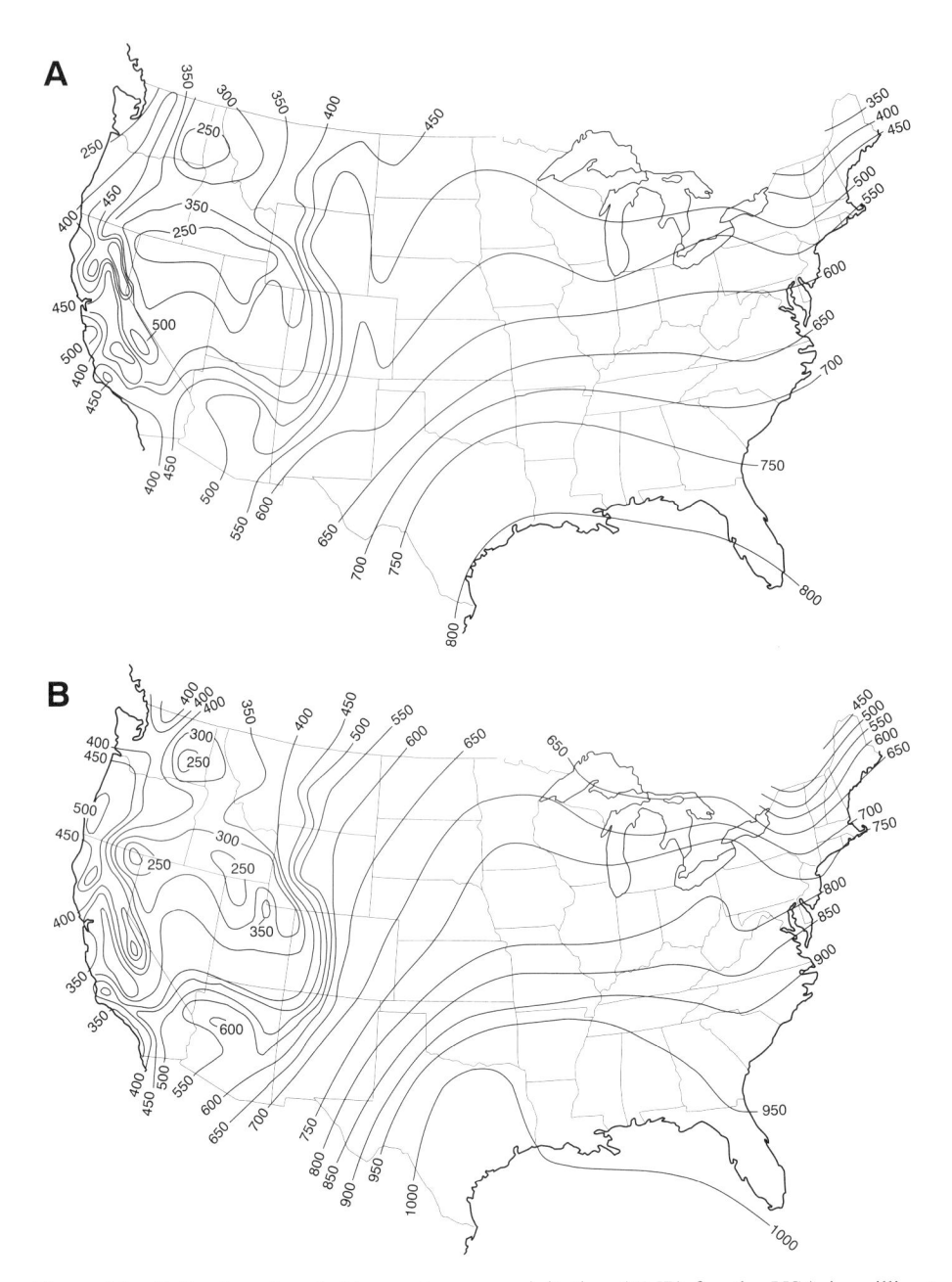

Figure 6.1 Estimates of probable maximum precipitation (PMP) for the USA in millimetres: (A) for 24 hours and 500 km^2; (B) for 24 hours and 25 km^2
Source: US National Weather Service

stages of project planning, ensure some consistency of PMP estimates between projects, and provide a basis for comparing individual estimates of PMP.

In the UK, maps of PMP for 2-hour and 24-hour rainfalls were included in the *Flood Studies Report* (NERC, 1975). These were developed with reference to the adopted

FLOOD ESTIMATION 185

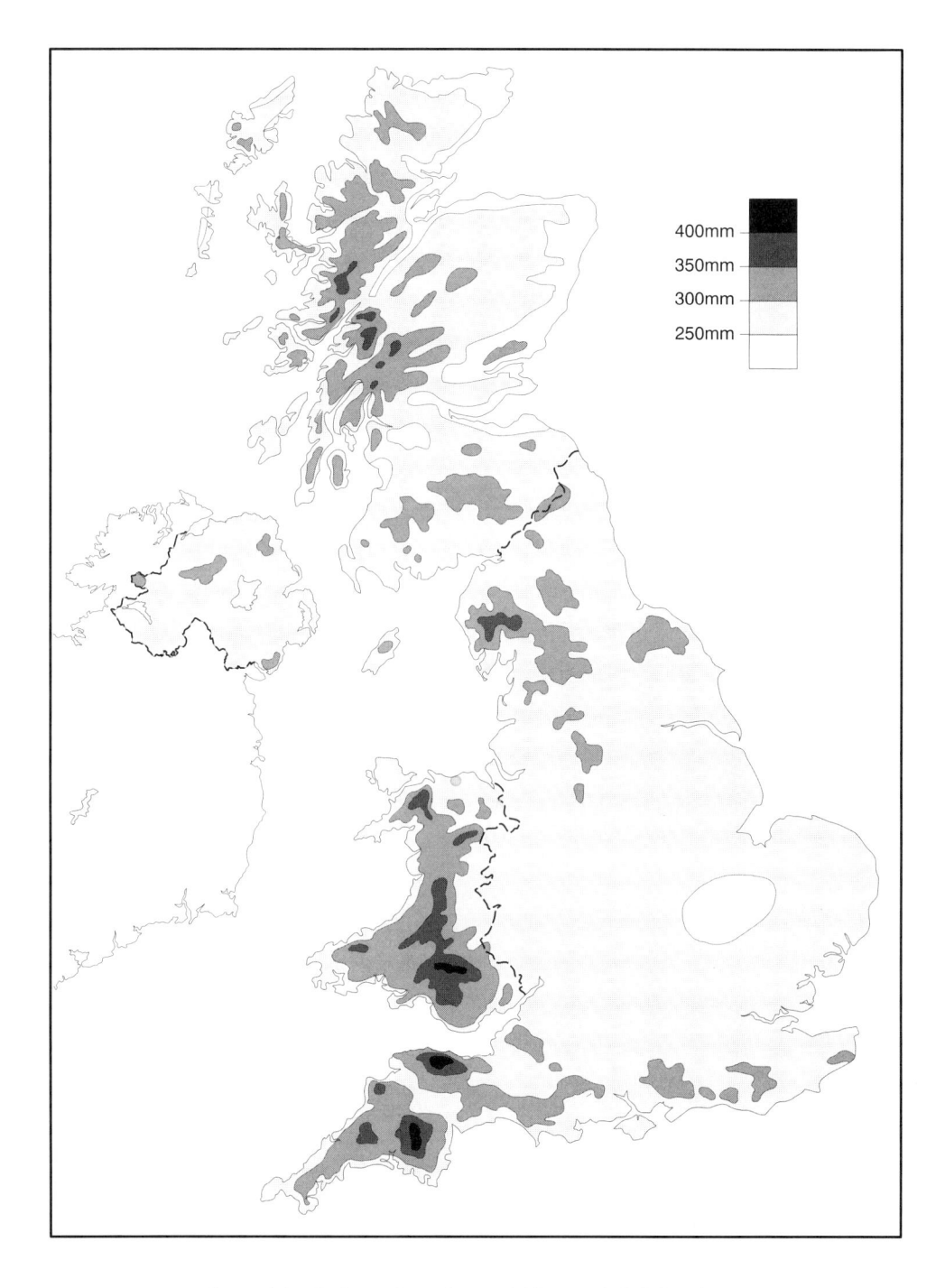

Figure 6.2 Estimated maximum 24-hour rainfall for the UK
Source: Based on a map in NERC (1975) by permission of the Institute of Hydrology

Table 6.1 A table for the rapid estimation of PMP in the UK from knowledge of the M5 rainfall

	\multicolumn{9}{c}{M5 (mm)}
	25	30	40	50	75	100	150	200	500	1000
England & Wales	189	215	253	280	297	326	388	460	812	1420
Scotland & N Ireland	158	178	212	241	280	326	388	460	812	1420

Source: Based on a table in NERC (1975) by permission of the Institute of Hydrology

standard rainfall frequency having a return period of five years (M5) and the estimated growth factor which relates the five-year rainfall for any given duration to rainfalls of any other return period for the same duration (see also Section 3.4.2). Values of the growth factor were found to vary slightly on a regional basis but to be independent of rainfall duration. The estimated maximum 24-hour rainfall shown in Figure 6.2 was developed by applying the highest recorded values of storm efficiency (i.e. 9.3 for the summer 24-hour storm at Cannington, Somerset, in August 1924 and 12.2 for the winter 24-hour storm at Loch Quoich, Scotland, in December 1954) to all major storms. This maximisation procedure showed that the highest estimated maximum 24-hour rainfalls in summer ranged from 250 mm in parts of eastern England and the Scottish lowlands to 410 mm in south-west England.

A rapid estimate of PMP for any duration, anywhere in the UK, can be obtained from Table 6.1 which uses M5 rainfall values and maximum values of the growth factors for England and Wales and for Scotland and Northern Ireland. The table shows, for example, that for a location in England or Wales which has a 75 mm M5 rainfall, of any duration, the estimated PMP is 297 mm.

PMP estimation has normally been difficult in tropical regions where the relevant database and comprehension of the processes producing extreme rainfalls were slower to develop than in the middle latitudes. A new chapter on PMP estimates for tropical regions was, however, included in the second edition of the WMO *Manual for the Estimation of Probable Maximum Precipitation* (WMO, 1986).

Direct estimates of the Probable Maximum Flood (PMF) may be derived by applying the empirical approach outlined in Section 6.2.1.1. This was the basis of the analysis by Francou and Rodier (1969) who plotted 1200 maximum recorded flood peaks from different parts of the world against catchment area on a log–log scale and found that for hydrologically homogeneous regions the upper envelope boundaries were straight lines which converged towards a single point. The equation for these converging envelope curves can be written as

$$PMF = 10^6 \, (A/10^8)^{1-0.1K} \qquad (6.9)$$

where A is catchment area in km² and K is a regional coefficient to which Francou and Rodier assigned generalised values for specific areas ranging from about 6.0 for the typhoon areas of the Pacific to values of less than 4.0 in the catchment areas of large tropical rivers such as the Amazon and the Niger (see also Section 1.1.2). Subsequent refinements to this empirical approach to the estimation of PMF have been made for various areas (e.g. Kovács, 1980).

Alternatively, PMF estimates may be derived either from statistical analysis of measured discharge data or indirectly by the conversion of PMP estimates using unit hydrograph procedures and taking into consideration the physical characteristics of the target catchment. Both approaches have serious defects. Frequency analysis, when applied to the estimation of the PMF, involves extrapolating to very long return periods (e.g. > 1000 years) from runs of data which rarely exceed 100 years (see also Section 6.2.2). The deterministic approach requires many assumptions about PMP and catchment conditions which can result in significant cumulative error. Furthermore, Kuchment et al. (1993) argued that with neither approach is it possible to predict the effect of unusual combinations of runoff factors, or of changes in runoff processes, including the effects of human activity. Instead they suggested using physically based runoff models to determine the PMF, although conceding that the data requirements are very high. A similar approach, recommended in 1990 guidelines for the estimation of extreme floods in Sweden, was described by Harlin et al. (1993). This uses maximum observed, rather than maximised, values of precipitation and other flood-generating factors. However, because it is based on the critical timing of those factors, the method generates design flood estimates that are typically twice as large as the highest recorded Swedish floods and have estimated return periods in excess of 10 000 years.

The assumptions which need to be made in order to estimate the PMF from the PMP, e.g. about the spatial variation of rainfall, the direction of storm movement, and the duration and profile of the flood-producing precipitation event, will differ in response to contrasting hydrometeorological and catchment conditions found in different areas. Accordingly, the conventional practice varies between, say, the USA and the UK, although the latter can be used to illustrate the underlying principles. The procedures were detailed in Section 6.8.3 of the *Flood Studies Report* (NERC, 1975) and summarised by Lowing (1995). They involve determination of

- the unit hydrograph, either from flow records or as a simple triangle (see equations 6.5 and 6.7), which is made peakier by reducing its time axis by one-third;
- the design storm duration;
- the 2-hour and 24-hour PMP (from maps, see Figure 6.2) which are used to produce estimates of maximum rainfall for different durations, adjusted by the appropriate areal reduction factor and increased by snowmelt;
- the storm percentage runoff separately for the urban and non-urban parts of the catchment.

The combined percentage runoff value is then applied to each design-storm increment and the result is convolved with the unit hydrograph to give a PMF hydrograph to which a token baseflow component is then added.

The PMP/PMF approach to flood estimation is at least partially based on current understanding of rainfall- and flood-producing processes in both atmosphere and catchment and provides a measure of the upper limit to the flood potential of a catchment. Such information has value for the hydrologist and provides an upper boundary to the design envelope within which the engineer must operate. The defects of the approach relate both to the inevitable sparsity of data on extreme rainfall/flood events and to the assumptions and simplifications inherent in the transposition, maximisation and modelling procedures used. PMP/PMF values are simply estimates

Table 6.2 Percentage probability of the N-year flood occurring in a particular period

Number of years in period	\multicolumn{7}{c}{N = Return period in years}
	5	10	20	50	100	200	500	1000
1	20	10	5	2	1	–	–	–
2	36	19	10	4	2	1	–	–
5	67	41	23	10	5	2	1	–
10	89	65	40	18	10	5	2	1
20	99	88	64	33	26	10	4	2
50	–	99	92	64	39	22	9	5
100	–	–	99	87	63	39	18	10
200	–	–	–	98	87	63	33	18
500	–	–	–	–	99	92	63	39
1000	–	–	–	–	–	99	86	63

and it is unsurprising therefore that they may greatly exceed design criteria in some areas and have been exceeded already by recorded rainfalls and floods in other areas. Nor is it surprising that gradually the PMP and PMF have come to be regarded not as impossible but as extremely rare events, to which a return period of 10 000 years or so would have to be assigned (e.g. ASCE, 1972). As with the estimates of PMP in the UK, this illustrates again the way in which river floods and the meteorological events which cause them are most usefully viewed as having both a deterministic and a probabilistic component.

6.2.2 Probabilistic Methods

The probabilistic approach assumes that floods are random quickflow events to which an underlying statistical distribution can be fitted and whose probability of occurrence during a given period of time can be calculated. As the time base of a flood dataset increases so too will the magnitude of the largest recorded flood become greater. In addition, a flood of specified magnitude is likely to occur more often in a long dataset than in a short one. The average interval between two floods which equal or exceed a particular discharge is the recurrence interval or return period. So the 100-year flood (Q_{100}) is expected to be equalled or exceeded once in every 100 years although it could occur within 50 years, five years or even one year. However, as Table 6.2 illustrates, Q_{100} has a greater probability of occurring during the next 100 years (63%) than during the next five years (5%) and a greater probability still of occurring during the next 500 years (99.3%). These statistical relationships between flood magnitude and frequency can be applied in two main ways to flood estimation. For a given return period, the flood peak which will be equalled or exceeded once can be estimated, or for a given flood magnitude the average frequency of exceedance (return period) can be estimated.

Unfortunately, a number of important assumptions which underpin the use of statistical methods may not be met by river flood data. For example, each discrete flood event should be independent of all other flood events. However, where multiple flood peaks are generated by successive storms falling on an increasingly wet catchment area, it is unlikely that the flood peaks will be independent of each other since the proportion of quickflow generated by successive storms will increase with catchment wetness. Use of the *annual flood series*, consisting of the highest flood in each year of the data period, may largely overcome this problem, except in the case of successive December and January flood peaks. But use of the *partial duration series*, consisting of all floods above a specified threshold (the 'peak-over-threshold' or POT approach), will probably involve some lack of independence, and use of the *complete duration series* (consisting of all recorded flood peaks) almost certainly will. Because of the greater number of flood events included, the peak-over-threshold approach is potentially attractive. Provided that the independence of individual peaks is rigorously checked or that adequate relationships between the partial duration and annual series have been developed (e.g. Langbein, 1949), POT datasets may be used with confidence in flood estimation. For example, the UK Institute of Hydrology POT database comprises more than 77 000 flood events, all of which have been checked for independence (Bayliss and Jones, 1993), and this database is likely to form the basis of methods to be developed in the *Flood Estimation Handbook* (Reed, 1994a).

Another underlying assumption is that each flood peak is drawn from the same (homogeneous) population. This will not be the case if, for example, some floods in the annual series result from rainfall and some from snowmelt or from a mix of the two (see Figure 3.8) and, as is shown in Section 3.4.3, flood-producing rainfalls may themselves come from different statistical populations. In a useful contribution, Schiller (1994) discussed tests for homogeneity of flood data.

One of the most important assumptions is that there is some underlying frequency distribution that describes the flood population. Both graphical and mathematical methods have been used to define this distribution from the data available and different distributions have been advocated by different individuals and agencies. An important criterion has been that, in the graphical case, the plotted values of flood discharge and associated return period lie on or near a straight line. This can be achieved by transforming the data and by using graph paper with distorted coordinates (e.g. probability, log–probability, Gumbel extreme value). In a mathematical analysis the preselected frequency distribution may be described by two or three statistics such as the mean, standard deviation and skew coefficient, which are computed from the data sample.

Ultimately any dataset can be plotted as a straight line since it is always possible either to fit a polynomial of degree n to an array of $n+1$ data or to scale a graph paper to fit a given data array. Accordingly, there are numerous possible distributions in current use. Cunnane (1989) listed 18, but gave no clear argument in favour of using one particular distribution, despite the fact that results will differ considerably between methods (Cunnane, 1985) (see Figure 6.3). Practice varies, often on a national or even agency basis so, for example, the general extreme value (GEV) distribution was advocated for use in the UK (NERC, 1975), the log–Pearson type III (LPIII) by the Water Resources Council in the USA (WRC, 1981) and the Gumbel extreme value distribution in Bangladesh (Hoque, 1994).

There has also been continuing interest, on the fringes of hydrology, in devising newer and better plotting formulae (e.g. Boughton, 1980; Ahmad et al., 1988; Karim and Chowdury, 1995). Klemes (1993), Baker (1994) and Bardsley (1994) have argued cogently against the further proliferation of such attempts. As Bardsley (p. 162)

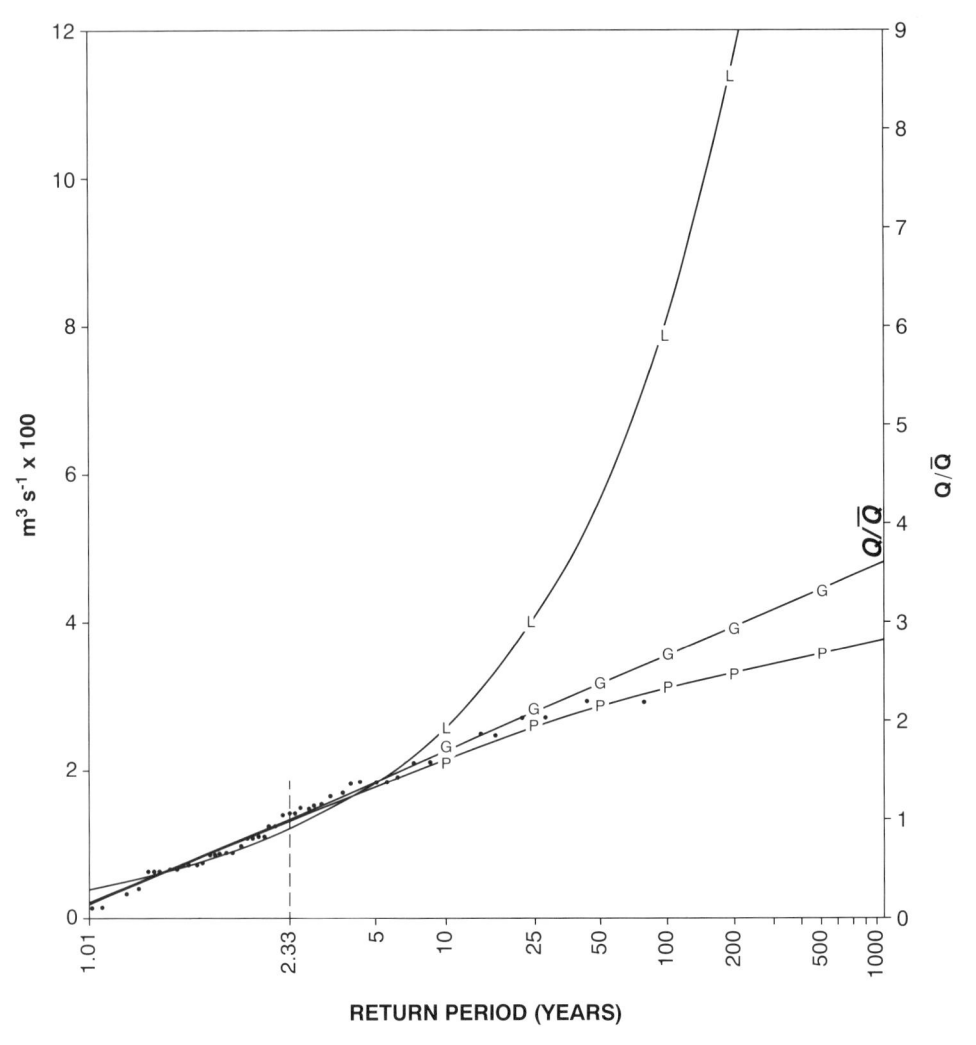

Figure 6.3 Gumbel (G), log–Gumbel (L) and log–Pearson type III (P) plots for the Raccoon Creek at Moffatts Mill, Pennsylvania, catchment area 461 km²
Source: From a diagram in Reich (1973)

emphasised, since the underlying probability distribution remains unknown, it follows that

> the only data relevant for extrapolating beyond the data range are those few large maxima...in the data record. Whether or not a fit can be achieved at the same time to the rest of the data is irrelevant.

The perceived advantage of the straight-line plot is the ease with which flood estimates can be either interpolated within the period of record or, more usually, extrapolated beyond the period of record. The risks of extrapolation, and the problems of attempting to estimate the tail of a probability distribution curve from a data sample which does not include data within this tail, have been stressed frequently (e.g. Melentijevich, 1969). Few investigators have adopted the cautious approach advocated by Wisler and Brater (1959) who suggested that the five-year flood was the rarest that could be estimated with any real certainty from 50 years of data. By contrast, there have been many instances of attempts to estimate floods having return periods of thousands or even tens of thousands of years from a few decades of data. As flood datasets increase in length and methods of regionalisation (see Sections 6.3 and 4.3) are refined, the risks of reasonable extrapolation are reduced. Even in 1975 it was implicit in the *Flood Studies Report* (NERC, 1975) that the 1000-year return period flood might be estimated in some circumstances.

Perhaps the most important assumption of all is that flood data recorded in the past and present may be used as a guide to the flood hydrology of the future. This implies, either that the flood data series exhibits *stationarity*, i.e. no discernible trend or periodicity, or that any trend in the data is explicable and capable of reliable extrapolation into the future. Analyses of the changing frequency of extreme hydrological events, for example in Europe by Arnell (1989), have thrown serious doubt on the ability of the past to serve as a model for the future. Such doubt must remain until a clearer understanding emerges of the nature of climate change and its effects on flood peak discharges (see also Section 3.6.3). However, some valuable progress has been made using rainfall–runoff models to generate flood peaks under different climatic and land-use scenarios. Using a fairly coarse (40 km grid) rainfall–runoff model and climate change scenarios from the UK Meteorological Office high resolution model scaled to the year 2050, Naden et al. (1996) showed likely increases in the 50-year flood for the Severn at Haw Bridge (catchment area 9895 km^2) of 7.9–11.5%, depending on how the projected rainfall increase was applied. Almost identical increases were projected for the Thames at Kingston (9984 km^2) and rather smaller increases of 4.4–9.6% for the Trent at North Muskham (8231 km^2).

Klemes (1993) found it unsurprising that flood peaks should fail to meet the rigorous criteria of statistics. He argued that high discharge values occur more often as a result of unusual *combinations* of influencing factors than of unusual magnitudes of the influencing factors themselves, and went on to advocate a multivariate, combinatorial approach to extreme-event probability in which 'the fine points of small sample theory...become spurious'. Similar arguments were put by Guillot (1993) in support of the gradex method which relies on the use of flood and rainfall data. The importance of combinations of factors is illustrated by floods in catchments having permeable rocks such as the chalk of England. Severe floods in such conditions are rare and tend to result only from the combination of heavy rainfall with either frozen ground (see Figure 3.2A) or exceptionally high groundwater levels, as in the case of the Chichester, UK, floods of 1994 (see Section 3.2). In these circumstances *joint probability analysis* is more appropriate, although far from straightforward. Joint probability analysis treats the floods as a joint distribution of rainfall and catchment conditions (e.g. extent of frozen ground or groundwater elevation) whose probabilities could both be calculated from observed data. Thus a 1000-year flood could be envisaged as the product of a 25-year rainfall and a 40-year groundwater level.

6.3 ENHANCING THE FLOODS DATABASE

Underlying the application of either deterministic or probabilistic methods of flood estimation is a persistent concern about the quality and quantity of the floods database. In many countries, including the UK, flow data were originally established to monitor water resource availability rather than flood risk so that many gauging stations are drowned or out-flanked by flood flows. In the UK, according to Samuels and Hollinrake (1995), this may result in errors of 30% or more for flows measured in out-of-bank flood conditions. Such a margin of uncertainty clearly exacerbates the already difficult problem of defining the relationship between flood magnitude and flood frequency (return period), although improved methods of rating-curve extrapolation and field assessment of flood flows have now been introduced (Samuels and Hollinrake, 1995).

The length of reliable flood records on individual rivers, even those on which gauging was established at an early date (see Section 1.1.2), is still far too short to support adequately either deterministic or probabilistic flood estimation methods. The problem is seen clearly in relation to estimation of return period. For example, two major floods occurred on the River Tay in Scotland in 1990 (maximum daily mean discharge 1647 $m^3 s^{-1}$) and 1993 (1965 $m^3 s^{-1}$) both of which easily surpassed the previous recorded maximum (1223 $m^3 s^{-1}$) in a record which began in 1952. The effect on estimated return periods was dramatic. For the period 1952–89 the estimated return period of a 2000 $m^3 s^{-1}$ flood was more than 1000 years and the 100-year flood was estimated to have a discharge of 1540 $m^3 s^{-1}$; for the slightly longer period 1952–93 the return period of the 2000 $m^3 s^{-1}$ flood had reduced to 105 years and the 100-year flood discharge had increased to 1990 $m^3 s^{-1}$ (IH, 1993b).

Such an example illustrates the fragility of return-period values and emphasises that estimation of even the 100-year flood event may benefit from the pooling of data from several sites in a region, i.e. *regionalisation*. This is even more necessary for longer return period events but raises important questions of which records should be pooled, how the extremes from different sites can be made comparable, and how a regional growth curve can be best derived and tested. Theoretical aspects of regionalisation and some examples of its application to flood studies are considered in Section 4.3. The approach is likely to play a growing role in the development of flood prediction methods and has undoubtedly been given a major boost by the increasing availability of GIS and DTM data (e.g. Aschwanden et al., 1993).

Regionalisation is normally used in the context of determining an 'index' flood (e.g. the mean annual flood ($m^3 s^{-1}$) or specific mean annual flood ($m^3 s^{-1} km^{-2}$)) and relating this by means of growth curves to floods having a wide range of return periods. The growth curves relating $Qbar$ to Qt which were used for the regions of the *Flood Studies Report* (NERC, 1975), for example, are shown in Figure 4.4B. However, there is a danger of over-reliance on flood data such as the mean annual maximum flow, if they are used to identify homogeneous regions, to develop growth curves and, as in the case of the *Flood Studies Report*, to develop prediction methods for ungauged sites. Accordingly, attempts have been made to incorporate other data, such as rainfall maxima, into the identification of homogeneous regions (e.g. Ferrari et al., 1993) or to replace the use of flood values by other covariate information such as land use, soil type, or flood date (Reed, 1994a). In the UK, for example, flood regions used in the new

Flood Estimation Handbook are constructed partly from flood date information (NERC, 1994).

Apart from regionalisation of existing flow records, there are other ways of extending the flood record at a particular site. *Flood marks* inscribed on walls and buildings to record exceptional flood levels have long been used as a means of extending the duration and range of flow record. In most countries flood marks cover a period of only a few hundred years (e.g. Benson, 1950; Archer, 1992; Georgiadi, 1993) but in China, for example, the record extends over 2000 years (Salas et al., 1994). Their evidence is often difficult to interpret, however, because of uncertainty about the way in which relations between water level and discharge have changed with time.

The flood dataset can be extended even further by using the evidence of geology and *palaeohydrology* (e.g. Baker, 1994). Baker et al. (1979) showed how slack-water deposits, up to the maximum water level of the largest floods, may be preserved in bedrock canyons in arid and semi-arid areas. Carbon dating of such deposits enabled a reconstruction of flood sequences for the Lower Pecos River in Texas over the past 9500 years. For a very different climatological environment Kale et al. (1994) reconstructed 5000 years of high-magnitude flooding on monsoon rivers in central India. Palaeohydrological evidence may be erosional, e.g. scour lines, palaeochannel size, as well as depositional, or may involve lichen and vegetation analysis.

Rather than attempt to extend the historical flood record, an alternative response to the inadequacy of the floods database is to generate *synthetic flow data*. This can be done either deterministically or probabilistically. The former implies an adequate understanding of flood-producing processes and the ability to reproduce these in an effective model of catchment hydrology. Improvements in hydrological understanding, model efficiency and computing capability have combined to make the deterministic generation of synthetic flow data an increasingly viable option. The models used may vary according to the objectives and even the nationality of the user. A sample selection is described briefly in Section 8.4.

Deterministic models of catchment hydrology are very data intensive (e.g. climatological and catchment data) and it is largely for this reason that the synthetic generation of flow data is more frequently achieved by probabilistic time-series models. These are based on analyses of existing river flow records but use the historical record as a sample of the total population rather than as the entire record. Trend and periodic components are isolated from the flow record, leaving a stationary, probabilistic component which contains a random element and a correlation structure, i.e. each flow value is affected by preceding flow values. Each part of the time-series can be simulated and then reassembled to give a stochastic model of possible rather than actual flow on which frequency analysis may be performed in the same way as for the (much shorter) historical record (see Section 6.2.2).

Assumptions incorporated in stochastic models (e.g. Monte Carlo and Markov chains, Fractional Gaussian Noise models, Broken Line models, etc.) usually include the following:

- the recorded historical sequence of flow values is unlikely to recur;
- it is unlikely that the maximum possible flood is included within the recorded flows of a given river;
- river flow exhibits persistence, whereby high flows tend to be followed by high flows and low flows tend to be followed by low flows.

In many countries flow records are briefer than rainfall records and in such cases attempts have been made, not to extend the flow record as such, but to base flood estimates on a statistical *analysis of rainfall* value. Serious error may be introduced, however, if the analysis assumes that the return periods of floods and their causal rainfall are similar. While the assumption may be reasonable for low-magnitude, high-frequency events, it is unlikely to be true for rarer floods since the response of a catchment to precipitation (see Section 3.2) depends on those antecedent moisture conditions which govern quickflow production and on the distribution of rainfall in time. This problem can be partially overcome by using joint probability analysis of rainfall and catchment conditions so that a 1000-year flood could result from a 40-year rainfall and a 25-year infiltration rate in a particular catchment.

6.4 CONTEMPORARY APPROACHES TO RIVER FLOOD ESTIMATION

Since floods are both deterministic and probabilistic in nature, the choice of an appropriate method for flood estimation is often not straightforward. Much will depend on 'local' conditions and as a result such guidance as has been published in, for example, the UK and the USA is often very general and rather arbitrary (Pilgrim and Doran, 1993).

For the UK, general guidance was given in the *Flood Studies Report* and is summarised in diagrammatic form in Figure 6.4. The initial choice between the deterministic and probabilistic approaches depends on whether or not it is necessary to evaluate the PMF or the shape of the flood hydrograph. If it is, then unit hydrograph methods are necessary; if it is not, then a choice of methods is possible. In both cases, the final choice of method would be determined by the length of flow records available.

If a flood peak of given return period (Qt) is required, and a detailed hydrograph is not needed, the probabilistic approach is the most straightforward. NERC (1975) recommended use of the general extreme value distribution. Regional curves based on this distribution (see Figure 4.4B) are then used to derive Qt from the mean annual flood ($Qbar$) for each selected site. Where no records exist a preliminary estimate of $Qbar$ and Qt, pending installation of a flow gauge, may be derived deterministically from catchment characteristics, using the equation

$$Qbar = C\ AREA^{0.94} STMFRQ^{0.27} S1085^{0.16}$$
$$SOIL^{1.23}\ RSMD^{1.03}\ (1 + LAKE)^{-0.85} \quad (6.10)$$

where C is a coefficient which varies from region to region, $AREA$ is catchment area in km^2, $STMFRQ$ is the number of stream junctions per km^2, $S1085$ is main stream slope between 10 and 85% of stream length upstream from gauge in m km^{-1}, $SOIL$ is an index based on the winter rain acceptance rate of the catchment, $RSMD$ is the net daily rainfall in mm having a return period of five years and $LAKE$ is the proportion of the catchment draining through lakes.

Where there are less than 10 years of data, the dataset may be extended by correlation with adjacent records and use of the POT series. More convincing estimates are likely to result with 10–25 years of data. Then $Qbar$ can be derived from the annual maximum series and the appropriate regional curve can be used to derive the flood

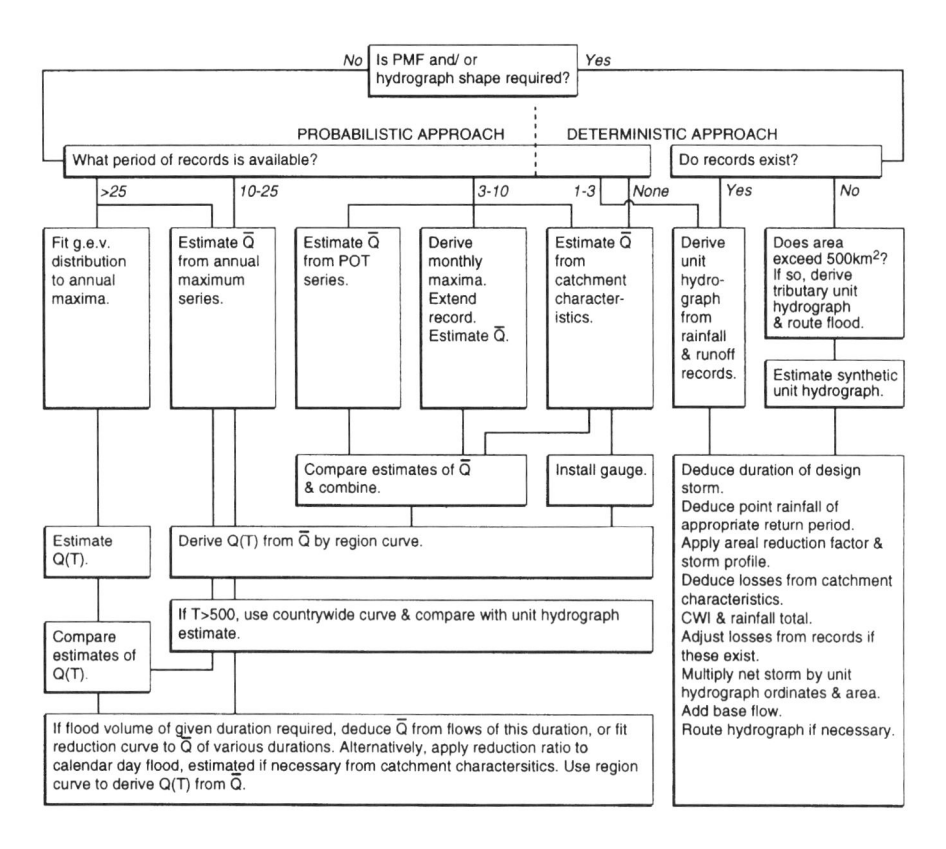

Figure 6.4 Decision diagram for selecting the method of flood estimation
Source: Based on a diagram in NERC (1975) by permission of the Institute of Hydrology

frequency for the site. With more than 25 years of record Q_T may be derived directly from fitting a frequency distribution to the annual maximum series though this result should be compared with that obtained by estimating *Qbar* and using a regional curve.

If it is necessary to specify the shape of the flood hydrograph or to calculate the PMF, then the unit hydrograph approach is used. Where the appropriate rainfall and flow data are available these are used to derive the unit hydrograph. Where no data are available the unit hydrograph may be estimated from catchment characteristics as outlined in Section 6.2.1.2. The proportion of storm rainfall contributing directly to quickflow is estimated from equation 6.4 although, if some records are available, this estimate may be adjusted suitably.

The *Flood Studies Report* emphasised the desirability of following a number of routes through Figure 6.4 and comparing and adjusting results accordingly in order to derive the best reasonable flood estimate.

A different approach to the selection of either deterministic or probabilistic methods for the estimation of flood peaks in Australia was described by Pilgrim and Doran (1993). The quantitative criteria outlined were recommended by the Australian Institution of Water Engineers (Pilgrim, 1987) but are not enshrined as a strict code of practice. The choice between flood estimation based on rainfall–runoff models or on flood frequency

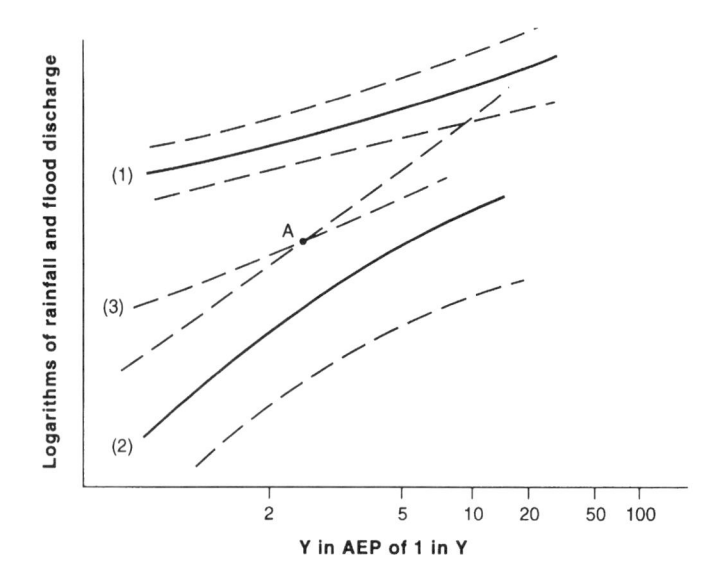

Figure 6.5 Frequency curves (solid lines) and confidence limits (broken lines) of design rainfalls (1) and observed floods (2). The upper confidence limit of floods estimated from the design rainfalls (3) is also shown
Source: Based on a diagram in Pilgrim and Doran (1993)

analysis is based on the confidence limits of the estimates obtained from the competing methods. This is illustrated schematically in Figure 6.5 which shows frequency curves for observed floods and for design rainfalls and their respective confidence limits. The upper confidence limit of floods estimated from the design rainfalls using rainfall–runoff models is also shown. Point A, where the two confidence limits intersect, is at the annual exceedance probability (AEP) of indifference. Smaller floods, having larger AEPs, plot to the left of A where flood frequency analysis has narrower confidence limits than rainfall–runoff modelling. For such floods statistical methods of flood estimation should be used. For larger floods, having smaller AEPs which plot to the right of point A, rainfall–runoff modelling has the narrower confidence interval and should therefore be used in preference to the probabilistic method. Clearly the actual criteria used in this example depend on the statistical characteristics of Australian rainfalls and floods. However, Pilgrim and Doran (1993) argued that, provided local values were substituted, the principles of the approach itself should be valid anywhere.

For many users in many parts of the world the choice of flood estimation method is conditioned largely by data availability and data-processing technology. Probabilistic methods of flood estimation are undoubtedly simpler than those based on rainfall–runoff modelling, their data requirements make fewer demands, and the methods are simpler to set up, calibrate and apply. Not surprisingly, therefore, they have come to be regarded as the mainstay of flood estimation throughout the world (Reed, 1994b). In the shorter term this situation is unlikely to change. Indeed, the accuracy of probabilistic methods should improve as the world databank of flood peak values continues to increase. In the longer term, however, progress in continuous simulation modelling should permit the development of powerful catchment models, incorporating inputs from GIS and DTM systems, which could be applied to a wide range of problems, such as predicting the hydrological effects of land-use change and real-time flow forecasting, as well as to flood estimation.

The impact of some of these technological and hydrological advances is already being seen in the development of a new flood estimation strategy for the UK which is intended to subsume and build upon the 1975 *Flood Studies Report*. The latter was itself produced as a direct response to the perceived inadequacies of flood estimation methods then available in Britain (e.g. ICE, 1967). It has been subsequently added to, in the form of supplementary reports published by the Institute of Hydrology, and regional refinement (e.g. Hanna and Wilcock, 1984; Acreman, 1985b) has resulted in its application in many parts of the world (e.g. Bree et al., 1989; Roald, 1989). Most importantly, in terms of ease of implementation, the methodology was made available by the Institute of Hydrology as a PC-based flood estimation package, Micro-FSR, which was updated to a Windows™ version by Water Resource Associates in 1997. Similar Windows™-based packages are also available for other flood estimation methods such as the National Flood Frequency Program developed in the USA by the US Geological Survey in cooperation with the Federal Highway Administration (FHA) and the Federal Emergency Management Agency (FEMA).

6.4.1 The UK *Flood Estimation Handbook*

The definitive successor to the *Flood Studies Report* (*FSR*) is the *Flood Estimation Handbook* (*FEH*), whose publication in five volumes is due for completion by 1999. The *FEH* benefits from the additional 20 years of runoff data which has been collected since publication of the original report and embraces new methods of flood estimation (MAFF, 1996). One of the main areas of improvement, for example, is in the use of catchment characteristics. This was deliberately limited in the *FSR* to those characteristics which could be derived easily from standard maps but in the *FEH* catchment characteristics can be derived from DTM and other digitised and gridded data. The urban and suburban fractions, for example, are derived from the refinement of data taken from the Institute of Terrestrial Ecology Land Cover Map of Great Britain and are more accurate than those derived for the *FSR* from the rather generalised depiction of urban areas in the 1:50 000 maps of the Ordnance Survey (Bayliss and Scarrott, 1996).

In many respects there are broad similarities in approach between the *FSR* and the *FEH* and indeed one volume of the latter comprises a clearer and more user-friendly restatement of the *FSR* rainfall–runoff method. In addition, a two-stage approach to the estimation of Qt is retained, involving an index flood and a growth factor, and it is still possible to estimate the index flood from catchment characteristics. However, there are also notable differences. For example, the median rather than the mean annual maximum flood is used as the index variable on the grounds that it is more robust, has a fixed return period and simplifies growth curve construction (Reed, 1997). Again, the *FEH* presents new generalised flood estimation methods based on statistical analysis which allow estimates to be made for any specified location. And it is anticipated that the generalisation of flood frequency in the *FEH* will be based not on geographical regions (see also Section 4.3) but on regions which are more hydrologically meaningful, involving groups of catchments that share similar POT flood and rainfall characteristics (Reed, 1994b; Jakob, 1997).

The *FEH* comprises five volumes as follows (MAFF, 1996):

1. Overview—summarises flood estimation methods, gives guidance on choice of method and links the *FEH* to areas beyond its remit.
2. Rainfall frequency estimation.
3. Statistical methods of flood estimation—reviews the statistical methods available and their application to different problems and takes account of new approaches to grouping for the determination of growth curves.
4. Restatement of the *FSR* rainfall–runoff method—narrower coverage than in the *FSR* is compensated by greater clarity and ease of use.
5. Catchment characterisation—with particular reference to the extensive datasets which have become available in recent years.

6.5 MELTWATER FLOODS

The techniques of flood estimation discussed in Section 6.2 are generally also applicable to the estimation of meltwater floods in the sense that both deterministic and probabilistic approaches are possible. However, there are additional problems associated with both approaches which result partly from the special circumstances in which melting occurs and partly from the limited database compared with that for rainfall-generated floods. Large-scale meltwater production tends to be associated with remote or inaccessible areas and the melt processes may occur coincidentally with quickflow production resulting from rainfall on the melting snow or ice (see Sections 3.5 and 4.5). In addition melting may be accompanied by outburst floods such as jökulhlaups (see Section 4.5.2). In such circumstances it is not surprising that a highly empirical emphasis dominated early flood estimation attempts in areas of winter snow-accumulation, such as the former USSR (e.g. Sokolov, 1969).

The gradual improvement of both databases and understanding of processes has encouraged the development of more sophisticated deterministic and probabilistic methods of meltwater flood estimation. Kuchment et al. (1993), for example, described the application of a detailed physically based model of snowmelt runoff formation which incorporated descriptions of the spatial dimensions of the snowpack, snowmelt, infiltration in frozen soil conditions, and overland and channel flow. The model was used to generate possible maximum snowmelt flood peaks using assumptions about extreme values of snowmelt, snowpack water equivalent, and rainfall during snowmelt. The estimated maximum possible discharge for the Sosna River, in the steppe zone of the CIS, which drains a catchment of $16\,300\,km^2$, was found to be $11\,600\,m^3\,s^{-1}$. This was 1.3 times higher than the 100-year flood derived from frequency analysis. A runoff model was also used by Harlin et al. (1993) to estimate design-flood peaks (having return periods in excess of 10 000 years) for the Swedish hydropower system. The model included snowmelt and rainfall floods, both separately and in combination (see also Section 6.2.1.3).

Flow data are available for the Mendoza River in Argentina from 1905 and provided a good basis for probabilistic flood estimates by Fernández et al. (1994) although it was not possible to desegregate flood peaks produced separately by rainfall and melting. The 100-year flood resulting from both factors was estimated for Guido as $580\,m^3\,s^{-1}$ compared with the historical maximum, recorded in 1984, of $450\,m^3\,s^{-1}$.

Table 6.3 Estimated 100-year snowmelt flood based on k values related to slope and area

Median snow depth (m)	0.05	0.1	0.2	0.3
Median water equivalent (mm)	6.5	13	26	39
k value related to slope				
Typical slope $S1085$ (m km^{-1})	1	2	5	10
Typical median k (mm h^{-1} °C^{-1})	0.066	0.09	0.15	0.22
100-year snowmelt runoff (m^3 s^{-1} 1000 km^{-2})	158	214	358	525
k value related to area				
Catchment area 100 km^2				
Estimated k (mm h^{-1} °C^{-1})	0.12	0.15	0.21	0.26
100-year snowmelt runoff (m^3 s^{-1} 1000 km^{-2})	286	358	503	622
Catchment area 1000 km^2				
Estimated k (mm h^{-1} °C^{-1})	0.051	0.066	0.11	0.14
100-year snowmelt runoff (m^3 s^{-1} 1000 km^{-2})	122	158	264	333

Source: Based on tables in NERC (1975) by permission of the Institute of Hydrology

Glacier outburst floods also occur on the Mendoza River but not at sufficiently frequent intervals to permit probabilistic flood estimation. Instead, Fernández et al. (1994) attempted to estimate the possible maximum flood from a model incorporating information on the ice dam and the size of flow tunnels through the ice.

In the UK *Flood Studies Report* (NERC, 1975) it was concluded that the separate analysis of snowmelt and rainfall floods is not only difficult, since many recorded winter floods contain some element of snowmelt, but also has statistical disadvantages and is largely unnecessary for design purposes. Median annual snow depth (the M2 snow depth) was mapped and median water equivalent was calculated, using typical density values, for snow depths of 0.05, 0.1, 0.2 and 0.3 m. A snowmelt coefficient (k) was calculated from catchment characteristics although its magnitude depended on whether slope or area was chosen as the independent variable. An example of typical 100-year snowmelt flood values is shown in Table 6.3.

COASTAL FLOODS

6.6 ESTIMATING COASTAL FLOODS

Deterministic and probabilistic methods are also used in flood estimation for coastal areas where flooding is largely the result of storm surges. The inadequacy of the database tends to constrain both approaches more severely than is the case for river flood estimation. In addition, special problems affect flood estimation in estuaries where both fluvial and marine processes interact. In the case of coastal floods caused by tsunamis the main issue, that of predicting the associated tectonic activity, is discussed briefly in the context of flood forecasting (see Section 8.11.4) and is not referred to again in this chapter.

Deterministic methods of storm-surge prediction may involve estimation of the PMF through maximisation of major storms, fetch and wave size and tidal level and, additionally for estuary flooding, the maximisation of river inflows. Maximisation

Table 6.4 Estimated magnitude–frequency relationships of annual maximum tidal elevation for the upper reaches of the Humber estuary

Return period (years)	Maximum tidal level (m)				
	1990	2000	2010	2030	2100
2	5.124	5.194	5.254	5.384	6.014
10	5.384	5.454	5.514	5.644	6.274
50	5.554	5.624	5.684	5.814	6.444
100	5.613	5.683	5.743	5.873	6.503
1000	5.765	5.835	5.895	6.025	6.655

Source: Based on a table in Pellymounter (1994) by permission of HR Wallingford Ltd

procedures, and indeed attempts to model marine flood processes, must also recognise the cyclic nature of astronomical influences on surge magnitude (see also Section 8.11.2) and the effects of climate change and of changing land surface elevations on past, present and future sea levels.

Changes of climate and land surface elevation may impose distinctive regional differences on the pattern of coastal flooding. This was demonstrated for the Humber estuary on the North Sea coast of England in a study reported by Pellymounter (1994). With almost 600 station-years of data available, annual maximum tidal levels and the five highest tides in each year were analysed using the GEV method. A uniform increase of high-tide elevation was demonstrated over the entire length of the estuary at a rate of 3.57 mm per year for the past 60 years. This was larger, by between 1.57 and 2.57 mm per year, than the general global sea-level rise for the same period. By using this historic trend to adjust past events, the magnitude–frequency curve was established for 1990. Magnitude–frequency relationships were then established for future dates either by assuming a continuation of the historic trend or by using the IPCC business-as-usual scenario of future sea-level rise (see Section 5.5.1) plus an additional local rise of 2.07 mm per year. Table 6.4 shows magnitude–frequency relationships, calculated by the latter method, for the upper reaches of the Humber estuary near Goole (see Figure 8.16 for location map).

Because of the generally smaller datasets for coastal flooding, the need to extend the database is at least as important as for river flood estimation and frequently involves the development of numerical and other models. Such models may be used to improve understanding of coastal flood processes as well as to generate synthetic flood data and in some cases models may be applied to both flood estimation and flood forecasting. Not surprisingly, in such circumstances, it is often difficult to make simple and clear-cut distinctions between deterministic and probabilistic approaches to storm-surge flood estimation. Even more than in the case of river flood estimation, both approaches may be used in tandem.

An illustration, provided by Können (1995), was associated with attempts by the Dutch government to improve protection against storm-surge flooding following the 1953 North Sea flood disaster (see Section 5.3.2.1). In this case a major objective was to upgrade the sea dikes to provide protection against the 10 000-year event but it proved difficult to estimate the magnitude of the event having this return period. One approach was to optimise the 1953 depression track and to maximise the astronomical tide. This yielded a 1 m addition to the 1953 sea-level elevation but it was difficult to know what return period to assign to this sea level. A direct statistical approach also failed, partly because of the implied statistical extrapolation of two orders of magnitude and partly because of the inadequate state of the art of statistics in the 1960s and 1970s. Although a partial solution was obtained in the 1980s with the development of a distribution-free model of extreme value statistics which was run with the observed time-series, Können felt that no further substantial improvement could be expected using the statistical approach. Instead, he suggested that future advances would require the use of physical–meteorological models and that when high-resolution atmospheric general circulation models become available, it may become possible to calculate the return periods of extreme sea levels by direct simulation.

6.6.1 Extending the Database

As with river floods, various ways have been proposed for extending the database on coastal floods, especially those resulting from storm surges. For example, Jelgersma et al. (1995) used sequences of shell deposits in coastal dunes to indicate the range of variation of storm-surge elevations on the Dutch coast during the Holocene. On the basis of their findings they concluded that statistical analysis of storm-surge data would fail to yield reliable return period estimates if such palaeo-data were excluded from consideration.

Storm-surge models, including hurricane models, have been used not only in a forecasting context (see Section 8.12) but also as a means of generating synthetic storm-surge data in order to improve probabilistic prediction of coastal flooding. For example, Siah and Lasch (1989) described a model having inputs of peak surge height, storm duration and shape coefficients, which was capable of simulating a variety of coastal storm events and associated flood hydrographs for the USA. The peak surge information is available for most coastal communities and can be obtained from the Flood Insurance Studies section of the Federal Emergency Management Agency (FEMA).

Despite the relative longevity of models such as SLOSH and SURGE (see Section 8.12.2), storm-surge models are still in an early stage of development and improvements continue to be proposed and described (e.g. Davies and Jones, 1993; Mastenbroek et al., 1993).

Henry and Murty (1990) drew attention to the obvious but often neglected need for improved data archiving and management of storm-surge data. Their work related specifically to storm surges in the Bay of Bengal but is more generally applicable to all countries where organised storm-surge monitoring programmes do not yet exist. A wide variety of data is required as input to numerical predictive models and this may take several decades to accumulate. It is therefore important that proper consideration is given to the documentation and preservation of the observations from each surge episode.

6.6.2 Flood Estimation in Estuaries

The particular problems of flood estimation in estuaries was outlined by Thompson and Frith (1995). Not only are floods within the tidal reaches of a river or estuary influenced by both freshwater and sea conditions but, for a given estuary, the relative importance of these two varies along the estuary and from time to time. For example, flood conditions near the mouth may be totally dependent on sea conditions, whilst the upper reaches of the estuary, which are tidal at normal flows, may be totally river dominated at high flows.

This complicates the task of defining a flood level for a given return period since, throughout much of an estuary, a given floodwater level can be obtained by a number of different combinations of events. For example, at a given location the same water level may be produced by the 100-year sea condition combined with a normal river flow, or by the 25-year sea condition and the five-year river flood, or by an average sea condition and a 50-year river flood. The situation is further complicated in smaller estuaries, where the responsive nature of the estuary and the short times to peak of flood flows mean that the timing of the river flood peak and high tide may also become significant.

Where river flood and coastal flood conditions are largely independent, as in the UK, the 100-year flood in an estuary is likely to be substantially smaller than that which would result from the combination of a 100-year river flood and a 100-year storm surge. Thompson and Frith (1995) noted that this is partly because the return period of that combination is significantly greater than 100 years, partly because the storage capacity of the estuary will attenuate the river flood peak as it moves towards the sea, and partly because of the improbability of the river flood and storm-surge peaks coinciding. In some major estuaries, however, the independence of river flood and coastal flood condition cannot be assumed, as for example in Bangladesh, where heavy rainfall and high storm surges may both result from the same tropical storm.

Thompson and Frith (1995) described a multi-level approach to estuary design-flood estimation which takes into account a wide range of interactions between river flood and coastal flood conditions.

SECTION THREE
Responses to the Flood Hazard

CHAPTER 7

Flood Defence

CONTENTS

7.1	Introduction to flood defence	205
7.2	River flood engineering	208
	7.2.1 Channelisation	209
	7.2.1.1 Channel resectioning	209
	7.2.1.2 Channel realignment	210
	7.2.1.3 Bank protection	210
	7.2.2 Levees	210
	7.2.3 Dams	213
	7.2.3.1 Retarding dams and detention basins	214
	7.2.3.2 Storage dams	214
	7.2.4 Environmental and ecological impacts	217
7.3	River flood abatement	220
	7.3.1 Topographic manipulation	221
	7.3.1.1 Terracing and contour ploughing	221
	7.3.1.2 Surface and underground water storage	222
	7.3.1.3 Gully control	222
	7.3.2 Vegetation cover management	224
	7.3.2.1 Grassland and crop cover	224
	7.3.2.2 Forest cover	224
	7.3.3 Flood abatement in urban areas	225
7.4	Coastal flood engineering	226
	7.4.1 Estuary and sea walls	226
	7.4.2 Barrages and barriers	227
	7.4.3 Environmental and ecological impacts	230
7.5	Coastal flood abatement	231
7.6	Flood proofing	233
	7.6.1 Elevation	234
	7.6.2 Dry flood proofing	236
	7.6.3 Wet flood proofing	236
	7.6.4 Adoption of retrofit flood proofing	236

7.1 INTRODUCTION TO FLOOD DEFENCE

The purpose of flood defence is to modify the hydrodynamic characteristics of river flows and coastal waters in order to reduce the flood risk. There is often a close relationship between flooding and erosion, so flood defence is frequently designed to influence both of these processes. Flood defence can be achieved in two broad ways. First, by traditional *flood engineering* methods using 'hard' defences which rely on river channel modifications or artificial materials, like concrete, shaped into dams, levees or other structures specifically designed to control flood flows. In coastal areas, sea walls have been widely constructed to resist the energy of incoming waves and tides. Flood engineering can be readily adapted to defend property at a site-specific, or even building-specific scale, when it is known as *flood proofing*. Second, by *flood abatement* methods using 'soft' defences which rely on essentially natural materials, whether of geological or biological origin, and existing environmental processes to control flooding and land erosion. For example, this includes the terracing of hillslopes or the afforestation of land within drainage basins. In the coastal zone, saltmarshes have been used to protect the banks of estuaries and beaches may be recharged with sand or shingle to retain the shoreline.

In practice, flood engineering is a relatively 'intensive' approach which is deployed either within the river channel or at the most vulnerable parts of the floodplain or the coast. It relies on capital works and control structures in a direct physical attempt to reduce the most damaging floods. In recent decades, large-scale flood engineering works have attracted growing criticism for causing adverse environmental and ecological impacts. Flood abatement is a more 'extensive' approach which works best over comparatively large areas of a drainage basin or sections of a low-lying coast. It usually serves goals additional to flood mitigation—such as timber production or recreation— and is implemented to the best advantage on a basis which works with, rather than against, the environmental processes that create a natural interdependence between land and water resources. For rivers, it implies techniques operating mainly outside the channel and it is used primarily to reduce the higher frequency/lower magnitude floods in smaller, upstream watersheds, perhaps less than 2500 km^2 in area. Both flood abatement and flood engineering offer a variety of individual techniques, many of which can be used singly or in combination, and a mixed strategy is often preferred today.

Flood engineering has a long history. The first dams and levees were constructed in the Middle East some 4000 years ago and the flood defence of individual cities in China goes back to the same period (Wu, 1989). According to Starosolszky (1994), levees are the oldest and most widespread type of flood defence worldwide and play a fundamental role in the life and work of many countries. For example, the use of sea dikes (polders) to reclaim and protect agricultural land dates back to almost AD 900 in the Netherlands, a country with about one-third of its area below sea level. In Vietnam, levees have been built for more than a thousand years. Today there are approximately 5000 km of river dikes and 3000 km of sea and estuary dikes throughout that country without which the cultivation of rice, the staple crop, would be impossible. So-called 'silt fields' were created in twelfth-century China where extensive river dikes, many constructed since 1949, protect densely populated alluvial plains against floods with a 10- to 20-year return period. In Hungary, where almost one-quarter of the national territory is floodplain, 97% of this area is protected by levees (Toth, 1994). Control dams have also been important engineering tools and Mirtskhoulava (1994) claimed that the former USSR is the region with the largest reservoir storage in the world. For example, the Dnieper River is impounded in a cascade of reservoirs with the capability of storing over 25 km^3 of flood waters. The system ensures a 20–50% decrease of flood peaks in high-water years and a 70–80% decrease in low-water years.

Given the relatively late adoption of flood defence in the USA, many schemes have been characterised by the use of combined techniques, as in the Sacramento Valley, California, where flood protection began about 1851. Major floods occurred in 1907 and 1909 and a comprehensive plan was adopted in 1911 to retain most floodwaters within levees with the excess flows discharged into by-pass channels (Etcheverry, 1931). Along much of the Mississippi valley, seasonal wetlands have been transformed to productive agricultural land by levee construction, a policy first adopted in 1861 after a report by the US Army Corps of Engineers. In 1879 Congress created the Mississippi River Commission, which further endorsed levee construction and effectively committed the federal government to ongoing flood control measures (Moore and Moore, 1989). The associated formation of levee and drainage districts occurred mainly between 1879 and 1916 (Thompson, 1989) and in 1917 the Flood Control Act authorised the Commission to spend $45 million on a cost-sharing basis with the local districts. Until the major 1927 flood, no levee built to Mississippi River Commission standards had failed. Many early levees have since been removed because of failures or high maintenance costs whilst new ones have been built, often to higher design criteria. Although design standards are important, no engineering structure can provide total protection and Sheaffer et al. (1976) claimed that over two-thirds of total US flood losses were due to catastrophic floods that exceeded design criteria.

In the coastal zone, and along the margins of large inland lakes, the threat of flooding results not only from wind-driven waves, storm-surge inundation and tsunamis, but also from shoreline retreat (Platt et al., 1991; Lawrence and Nelson, 1994). Often buildings that were initially set back from the shore are eventually placed at risk due to erosion. Along some coasts, flood defence is closely bound up with the multi-purpose management of estuaries and deltas. For example, land reclamation has long been an associated goal of coastal flood engineering schemes. Since 1930 about 165 000 ha of land have been reclaimed from the Ijsselmeer, Netherlands, whilst other large-scale poldering projects for agriculture have been implemented in Bangladesh, Indonesia and in the Danube delta. Low-lying, flood-prone land has also been reclaimed for industry, as in the Massvlakte area near Rotterdam, and for transport, as exemplified by the airports at Los Angeles, USA, and Wellington, New Zealand. As a result of the closure of estuarine river mouths and land reclamation, the surface area of the inter-tidal zone has been drastically reduced around the North Sea coast.

Flood abatement is associated with a wider approach to flood problems. The initial rise of river flood abatement stemmed from an awareness of the ecological devastation caused by unwise land management. Marsh (1864) was one of the first to draw attention to the damage from flash floods and accelerated soil erosion caused by deforestation in the European Alps. He specifically cited Lower Provence, France, as an area where deforestation in the fifteenth to seventeenth centuries allowed upland torrents to inundate valley floors and either erode good agricultural land or cover it with coarse sediment. Eventually, a tradition emerged in scientific hydrology of comparative basin studies of rainfall, runoff and erosion under forest and non-forest covers, starting with the Sperbelgraben and Rappengraben study in Switzerland in 1903 (Engler, 1919) and the Wagon Wheel Gap experiment begun on the headwaters of the Rio Grande in 1910 (Bates and Henry, 1928). Ward (1971) drew attention to the practical importance of Public Law 566 (Watershed Protection and Flood Prevention Act of 1954) in the USA. This legislation identified some 8300 watersheds, comprising about half the land area of the United States, that were in need of flood protection and related land and water resource development. Conservancy, or Watershed, Districts responsible for soil conservation and flood control have since been authorised in at least 24 of the states in the USA.

In recent decades, there has been a growing recognition that flood engineering for both rivers and coastal waters should adopt a more environmentally sensitive approach. Past defences typically confronted natural processes and resulted in rivers confined to fixed channels, the drainage of valley wetlands and the loss of coastal habitats such as mudflats and saltmarshes. Quite apart from the ecological consequences, and the capital and maintenance costs of structural schemes, there have often been important social impacts, such as the visual intrusion of flood embankments, the loss of wildlife habitats or reduced recreational opportunities arising from the construction of sea walls. The public within the MDCs has become more active in questioning such impacts and most new schemes are now subject to consultation procedures and environmental impact assessment before approval. A greater awareness of environmental considerations has inevitably created the need for a better scientific understanding of the natural processes leading to both flooding and erosion. This requirement has reinforced the trend towards a more comprehensive approach to flood problems involving a mix of both abatement and engineering techniques for optimum land and water management. Such policies are now enacted in some countries like the UK by the use of catchment management plans and by integrated shore and beach management plans which place flood problems in a more general environmental context.

The legislative arrangements made for flood defence vary widely throughout the world. Historically, the protection of land and property against flooding and erosion was first carried out by individual landowners and then by local authorities with links to central government. Where coastal issues assume national prominence, as in the Netherlands, there has been strong control by the central government whereas in Australia the responsibility is allocated to state or regional governments. In England and Wales the driving force in flood protection has always been the grant aid made available to local authorities for capital works by the Ministry of Agriculture, Fisheries and Food (MAFF). Such central funds are allocated on the basis of project appraisal methods relating to levels of protection and investment efficiency. The Environment Agency has responsibility under the Water Resources Act (1991) to supervise all flood defence matters with a specific remit for all *main rivers* (as defined on statutory maps) and for sea defences in areas which are not privately owned. In all, the Agency is responsible for a total of approximately 31 000 km of main river and 2850 km of sea and tidal defences. In the USA the Army Corps of Engineers has been allocated a similar operational responsibility by the federal government for engineered water projects and has invested over US$ 23 billion in flood control projects nationwide (Anderson et al., 1995). The Corps operates some 368 reservoirs for flood damage reduction and has constructed nearly 17 000 km of levees and floodwalls.

7.2 RIVER FLOOD ENGINEERING

River engineering is used mainly to protect urban areas or high quality farmland by containing and diverting flood flows within or around artificial structures. Such works may also be used to achieve other goals, such as navigation. Many major rivers in the MDCs have now been modified by such river training works.

7.2.1 Channelisation

Channelisation has been widely undertaken to increase the carrying capacity of the natural river channel in order to contain flood peaks and facilitate land drainage. For example, Brookes (1988) claimed that over one-quarter of main rivers in England and Wales have been subject to such works. Channel capacity and conveyance efficiency can be improved by several methods including increasing the river slope, or widening and deepening the channel to increase the cross-sectional area. In addition, relief channels can be provided for flood flows and hydraulic conditions may be further improved by decreasing the roughness coefficient through clearing boulders and other obstructions from the natural riverbed. Channelisation can provide less environmental disturbance than other flood engineering methods, such as dam construction, provided that the geomorphological equilibrium is not radically changed. Levee construction, sometimes regarded as part of channel improvement, is treated separately because of its importance worldwide.

7.2.1.1 Channel Resectioning

Resectioning is the term given to the widening or deepening of river channels in order to increase the flood-carrying capacity. According to Brookes (1985), a channel should have a cross-sectional area which provides the maximum flow efficiency with a minimum of excavation, although factors such as bed porosity and bank instability can make this difficult. Each modification of the channel has certain advantages and disadvantages as far as hydraulic and environmental conditions are concerned (Hamid and Amaning, 1991).

Channel widening tends to avoid the problems of bank instability associated with channel deepening, especially in erodible materials, but it creates shallower depths of flow with an attendant risk of increased siltation. In urban areas, it may be difficult to acquire the additional bank-side land which is required and the shallower flows can lead to unsightly dry riverbeds in some cases.

Channel deepening can be achieved either by narrowing the natural channel or by dredging. This approach avoids the problems of lowering the flow depths and reducing hydraulic efficiency in the channel. In addition, no extra land is required and bank-side vegetation is preserved. On the other hand, there is usually more disturbance to the river regime and channel habitat, with the possibility of some headward recession of the bed. In addition, the deepened channel will need to be matched in with existing bed levels downstream, for example with respect to urban drainage outfalls.

Dredging may involve either the *in situ* disturbance of gravels or other bed materials, which are then transported downstream by the current, or the mechanical removal of sediment from the channel for dumping ashore. Sometimes partial dredging is undertaken, e.g. to create a narrow channel through gravel shoals. After the disastrous floods at Florence, Italy, in November 1966, systematic dredging of the River Arno lowered the river near two of the bridges within the city by about one metre and increased the channel discharge capacity from $2900\,\text{m}^3\,\text{s}^{-1}$ to $3200\,\text{m}^3\,\text{s}^{-1}$. Although this is still well below the maximum discharge of $4300\,\text{m}^3\,\text{s}^{-1}$ of the 1966 flood, it is estimated that such bed lowering could increase the flood return period from 1:120 to 1:150 years.

Channel clearing is necessary to remove obstacles, and the associated accumulation of silt, from a watercourse. For example, Haslam (1978) showed that the annual growth of aquatic vegetation can be an important factor in reducing channel capacity. The regular removal of large weeds and trees, especially willows, is an important means of retaining channel capacity and reducing roughness. This also includes the removal of household rubbish in urban areas which might block stormwater culverts.

7.2.1.2 Channel Realignment

Highly sinuous and meandering river courses are prone to flooding because of their low hydraulic efficiency and the backwater effect of bends. Straightening or realigning the channel by artificial cut-offs shortens and steepens the section, thereby increasing the flow velocity which, in turn, reduces the flood stage. Diversion, or relief channels, are used to by-pass existing urban areas where resectioning by channel widening is not possible. Such channels act as flood overflows only, with the natural channel carrying the normal discharge.

One of the most direct methods is to make relief channels from meanders or abandoned channels. On the lower Mississippi, 16 cut-offs made between 1929 and 1942 reduced the river length by 240 km and lowered flood stages by 2.0 to 2.5 m. Bank protection is subsequently required to ensure that the extra energy is not directed into accelerated channel migration. An often neglected aspect of diversion works is the provision of better stormwater management with higher capacity drains and sewers. In populated areas, any extension of the existing stormwater system may be difficult and expensive because of its impact on traffic flows and other urban functions.

7.2.1.3 Bank Protection

Bank protection is necessary where there may be abrasion and landslipping which would create an obstruction to river flow. On some large US rivers, like the Rio Grande, Missouri and Mississippi, channel improvement has involved stabilising the bank-line and also contracting the width to provide a deeper channel. Wooden pile dikes, or groynes, projecting into the flow and quick-growing plants, like willows, have been used to slow the near-bank velocity of the river and encourage sedimentation. But the rates of bank stabilisation on the Mississippi between the Missouri and Ohio tributaries were less than expected because land conservation and reservoir construction upstream reduced the source of sediment supply to the Mississippi (Tiefenbrun, 1965).

In some major cities, such as London and Los Angeles, it has been necessary to place certain rivers in wholly artificial channels lined with concrete. Such works reduce channel roughness, avoid the need for more land and are relatively maintenance-free. But they are expensive to construct, severely damage the river ecology and present a harsh visual appearance, especially at low flows.

7.2.2 Levees

Levees, stopbanks or dikes, are linear structures built parallel to the main stream in order to contain overbank flows. Often levees are constructed on one side of the river only but the entire floodplain may be protected by double-embanking as near as possible to the natural channel. This is a more controversial strategy because the embankments will be high and the increased flow velocity creates a risk of levee breaches unless the embankments are placed beyond the meander belt of a migrating river. Double-embanking also tends to lead to more flooding downstream and long-term sedimentation in the channel after the flood peaks have passed. This process eventually raises the river above the local floodplain and re-establishes a flood risk. In Vietnam the construction of dikes has gradually reduced overbank areas previously available for excess flood flows with the result that river levels have risen up to 5–6 m higher than the land to be protected.

Levees increase the local carrying capacity of the channel and are intended to prevent all flood damage to the adjacent river corridor until the water level exceeds the top of the structure. If overtopped, damage is incurred as if the structure did not exist and the resulting damage may exceed what would have occurred without the levee. This is because of an increased flood duration and the so-called 'levee effect' (see also Section 2.6.2 and 9.5). Main levees are, therefore, used to defend lives and high-value property from all but the most rare events and require careful design and construction. In any event, levees can never provide total protection and many schemes also depend on expensive pumping measures to drain the protected low-lying land from the ponding of surface rainfall. This is also important if the structure is overtopped when the duration of flooding may well increase.

Levee design needs to ensure the most favourable hydraulic conditions for the safe passage of floods of a given magnitude. In some cases quite exceptional protection is provided. After the 1953 flood disaster in the delta area, the Netherlands government approved the strengthening of the river dikes to meet an estimated return interval of 1:3000 years, although the level of recommended protection was subsequently reduced to the 1:1250 year event (Breusers and Vis, 1994). On the Mississippi River there is a basic difference between urban and agricultural levees: urban levees tend to be designed against 200-year floods, or greater, whilst agricultural levees seek to protect against the 50-year flood, or less. Such expectations require decisions to be taken on factors such as optimum height, distance apart and cross-sectional dimensions for stability. In practice, the height of levees very rarely exceeds 10 m and is often less than 6 m. Freeboard is normally added to the initial design height and should be sufficient to allow for some settlement and to contain wind-driven waves. This additional height is unlikely to be less than 0.5 m and is often 1.0–1.5 m. The average levee slope (vertical:horizontal) is normally 1:2 or 1:3 on the river side and 1:3 or 1:4 on the land side, depending on local soil permeability (Figure 7.1).

For major levees, the top should be at least 3 m wide and support a crushed stone road to facilitate maintenance and repair. The design should minimise seepage which, under a high hydraulic gradient, can create washouts and sand boils which undermine the structure. Levee position is important because, although the nearer the structure is placed to the river channel the greater the area of protected land, the structure is more likely to be breached by river action. The question of the optimum set-back distance is a major problem in the alluvial channels of aggressive, braided rivers, like the Brahmaputra, where the lateral movement of the channel can be 50–100 m per year (Russell and von Lany, 1994). Apart from maintaining the natural stability of the river, set-back flood banks help to maintain valley wetlands and have less ecological impact.

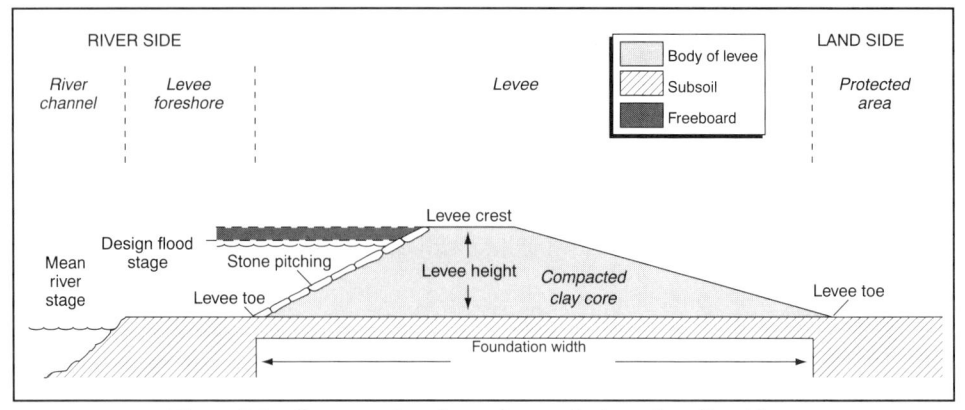

Figure 7.1 Cross-section through a typical earthen flood levee
Source: Modified after Starosolszky (1994)

As shown in Figure 7.2, levees can fail for a variety of reasons. The weakest point on most levees is the junction between the ground surface and the levee. Foundation seepage control is of great importance during construction and careful consideration needs to be given to the properties of the materials used in construction and to the substrate. All materials should be well-compacted to achieve a good bond. Floodplain deposits often consist of sand layers between clay or silt deposits and this allows seepage. In the worst cases, sand boils can undermine the structure. In Hungary the major cause of damage to levees is liquefaction of sand layers which occur at the intersections with ancient riverbeds (Toth, 1994). A gravel filter toe on the landward side of the levee, and a drainage ditch, can help to prevent piping and liquefaction in subsoil which might otherwise undermine the stability of the structure. Gilvear et al. (1994) in studies on the River Tay, Scotland, found that levees on the outside of bends and overlying old river courses are most vulnerable to erosion. In addition, microscale processes, such as scour around fence posts located along the top of the levee, can also be a significant cause of destruction.

The problem of calculating the design flood level is enhanced because levees tend to settle through time and flood stages may increase as other levees are constructed upstream and downstream. This means that many levees require additional height in

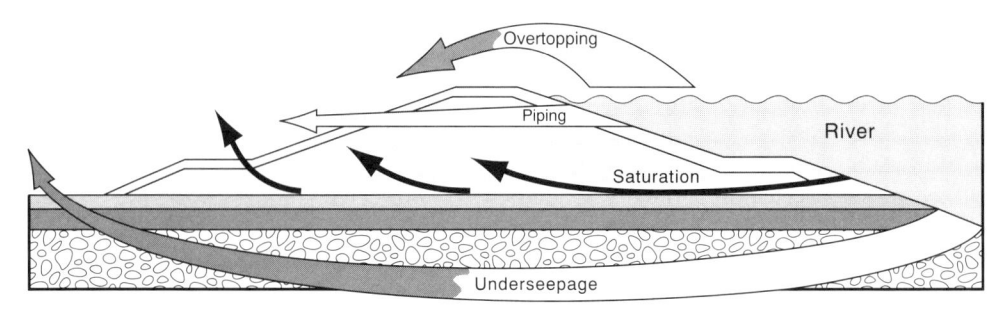

Figure 7.2 Some potential causes of levee failure
Source: Anderson et al. (1995)

A. Monsoon flooding affecting Nagaon City, Bangladesh. Despite the depth of the floodwaters, the event is sufficiently routine simply to impede—rather than totally disrupt—the more traditional forms of road transport. (©Gil Moti/Still Pictures)

B. Coastal flooding and flood damage at Jaywick Sands, Essex, UK. The surge of 31 January–1 February 1953 was the most disastrous to affect the east coast of England in the 20th century, raising sea levels by 2 m and resulting in major overtopping of sea defences on both sides of the North Sea. (©Aerofilms Ltd)

C. The 1993 'Christmas flood' on the River Rhine at Köln, Germany. The overtopping of both permanent and additional, temporary flood defences resulted in the flooding of more than 4500 households in the city. Another 9000 households, in parts of the city that lie permanently below the river level, were damaged by rising groundwater. (Courtesy of H. Engel, Bundesanstalt für Gewässerkunde)

D. Flash flood advancing rapidly along the dry bed of the Nahal Eshtemoa, Israel. The flood front, or 'bore', some 0.3 m in height and moving at a speed of about $1\,\mathrm{m\,s^{-1}}$, is carrying a substantial debris load. Within 3 minutes the water depth had reached 0.9 m. The hydrograph of this flash flood is shown in figure 4.14. (Courtesy of Professor Ian Reid)

E. A house under construction in 1997 at the coastal resort of Waihi Beach, North Island, New Zealand. The site is already under threat from storm waves and active beach erosion. The existing shoreline protection by boulders and wooden structures is unlikely to resist the increasing threat associated with further sea level rise. (Photograph by Keith Smith)

F. Storm surges pose an increasing flood risk to central London because of continuing land subsidence and rising sea level. Important, though gradually diminishing, protection is provided by the 520 metre Thames flood barrier whose four main gates are flanked by smaller subsidiaries. These rising segment gates are normally housed in recessed sills on the river bed and can be raised hydraulically in 15 minutes. The photograph shows one of the main gates in the raised position. (©David Hoffman/Still Pictures)

G. Concrete blocks deployed to channelise the flow and protect the banks of the River Po at Casale Monferrato Alessandria, Italy. The Magistrato per il Po (the regional river authority) spends about £100 m per year on such materials which are proving progressively less popular for ecological reasons and also because of the harsh visual appearance. (©Mark Edwards/Still Pictures)

H. Flooding in the Cité Soleil shanty town, Port-au-Prince, Haiti which is exposed to both storm surge and tropical rainfall floods. In the absence of any efficient system for drainage and waste disposal, raw sewage and other human debris contaminates the flood waters creating additional health hazards from waterborne diseases. (©Mark Edwards/Still Pictures)

order to maintain the original design standard. For example, flood stages on the Mississippi River have changed through time with the construction of levees and locks for navigation. Between the levees the river continues to erode its banks, a process contributing some 900 000 m^3 of material each year to the Mississippi sediment load before bank protection (Winkley, 1971). Belt (1975) quoted evidence that successive navigation works had reduced the capacity of the natural channel by about one-third since 1837 with the result that the 1973 flood stage reached the level of a 1 in 200 year event although the flow had only a 30-year average recurrence interval.

Levees are reliable only if they have been properly designed and well-maintained. Regular maintenance is much less likely in the LDCs, especially during periods of rapid social change. Wickramanayake (1994) highlighted the situation in Vietnam where local people were formerly required to devote 30 days of free labour each year to flood mitigation, mainly repairing damaged dikes. This commitment is now reduced to 10 days per year. In Bangladesh the failure of embankments is a cause for concern. Hoque and Siddique (1995) found that almost all the embankments they studied had been badly planned, and a quarter were poorly designed, such factors accounting for most cases of failure. In addition, about one-third of the sampled structures failed because of inadequate maintenance. In the MDCs, levee maintenance is often the responsibility of local statutory bodies. In the USA, the original Mississippi drainage districts had four main functions: (a) to construct a levee against flooding; (b) to construct and maintain drainage ditches to remove stormwater and lower the water-table for agricultural purposes in the protected area; (c) to provide and operate a pumping plant to remove any other excess water from the area; and (d) to maintain these facilities in good condition by charging the landowners and farmers in the drainage and levee district an annual fee for maintenance and improvement costs.

Levees can perform well, even when stressed beyond their design potential. For example, during the Mid-West floods of 1993 in the Missouri and upper Mississippi River basins, floods broke through some two-thirds of the levees but only one federal levee failed to perform as designed and no federal urban levee failed. According to Army Corps of Engineers records, 157 of the 193 Corps levees in areas affected by the 1993 floods prevented the flooding of almost 500 000 ha and some US$ 7.4 billion in damage (Anderson et al., 1995). When they do occur, levee failures can help to mitigate flood stages downstream by providing local floodplain storage and Bhowmik et al. (1994) demonstrated this for the 1993 Mississippi floods. Although many levees were damaged sufficiently in 1993 to need repair, the levee system did work overall, leading to an estimated saving of a further US$19 billion of potential loss.

7.2.3 Dams

Dams are engineered structures built across river channels for water control purposes. For flood mitigation the aim is to impound water in a reservoir during periods of high flow in order to maintain, as far as possible, downstream discharges within the safe carrying capacity of the channel. Although often called flood control reservoirs, these structures can only mitigate flood flows. Thus, in all cases, it is necessary to determine the degree of downstream flood protection required. Most schemes have an emergency spillway to protect the dam from being overtopped which can cause failure of the structure. Once the water level rises to the top of the spillway, the water stored above this elevation is in the surcharge pool (see Figure 7.3). The outflow of the surcharge is determined solely by the depth of water and the geometry of the spillway opening.

7.2.3.1 Retarding Dams and Detention Basins

These simple structures have no permanent impoundment and the reservoir space is held empty by a low-level, ungated outlet, the size of which is geared to a specified maximum outflow rate. The most basic type has an automatic discharge controlled either by a spillway or by a series of outlets at different elevations in the dam. Effectively it is a self-regulating system and the reservoir automatically empties after each flood and is then ready to receive the inflow from succeeding storms. The reduction of downstream peak flows is proportionately better for small floods because, during large floods, much of the storage will be taken up before the peak arrives.

Detention basins were first used in the USA on a large scale for the protection of the Miami and Franklin County Conservancy Districts from the Miami and Scioto Rivers in Ohio. The earliest schemes were built after a major flood in 1913 which claimed 360 lives. In such schemes the land within the basin may be used for agriculture, or other low-intensity purposes, but is not suitable for buildings or similar investments. Apart from flood mitigation, other benefits include improving infiltration, trapping river sediment and raising downstream water quality.

7.2.3.2 Storage Dams

These structures have gates or valves which enable the reservoir outflow to be regulated according to a set of operating rules. Such schemes may be operated as either single-purpose (flood control only) or multi-purpose (flood control plus other uses, such as water supply or hydropower generation) reservoirs. In practice, flood alleviation depends on the availability of any storage created by draw-down, which can be accidental in the case of water supply reservoirs.

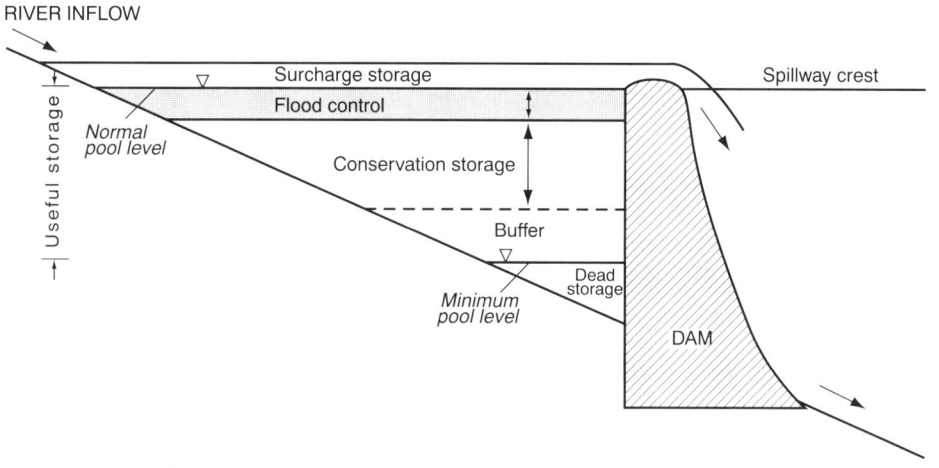

Figure 7.3 Water storage zones in a multi-purpose reservoir
Source: After Harmancioglu (1994)

Single-purpose, controlled outlet reservoirs have the advantage that they permit the immediate evacuation of the early, non-damaging part of the rising flood hydrograph, thereby maximising the storage available for any subsequent storm rainfall. This controlled outlet type can only function efficiently if accurate flow forecasts are available. If the reservoir operation leads to outflows in excess of natural discharges, and if these coincide with tributary peaks downstream, flood conditions will be worse.

Multi-purpose flood control reservoirs retain a permanent pool of water for a variety of uses, although not all this water is readily available. As shown in Figure 7.3, the flood control pool is stored above the multi-purpose pool level and the dam always stores some dead water. The requirements for optimum operation are often in conflict. For maximum benefits, flood control requires an empty reservoir in advance of floods, whilst power generation and navigation require water levels kept as high as possible for maximum production. Conversely, irrigation needs a seasonal release of water during the growing season and recreation requires a full pool during the recreation season.

The performance of a flood control reservoir depends on a combination of design and operation. Design includes a variety of factors including location, design flood, estimated flood hydrograph, storage capacity and spillway capacity. In turn, the design is affected by many economic factors which include the capital costs of construction, land purchase, annual costs of amortisation, interest, operation and maintenance, lump sum benefits for increased values and annual benefits due to storage of floodwater and the release of conservation storage.

There are several types of design flood. The *spillway design flood* is the flood which a dam will pass without failure when the reservoir is full and overflowing. This criterion is important in ensuring the safety of the structure when it is overtopped. Much more attention has been given to dam safety than to bridge safety in floods where often the vertical clearance is the only criterion and little allowance is made for structural loads imposed by wave action or floating debris. The *reservoir design flood* is the flood against which the structure will provide downstream protection. This flood is usually smaller than the spillway design flood. An essential feature is the assumption that adequate precipitation forecasts will be available on which to base the reservoir operating decisions. In particular, multiple storage schemes must be operated to ensure that the crest delay does not result in the simultaneous arrival of the flow from several tributaries at a vulnerable downstream location.

The location of the dam in the catchment is important because the degree of flood protection afforded by any reservoir diminishes with distance downstream, i.e. as the percentage of the total drainage basin under reservoir influence declines. This is especially so if the main river is augmented by flow from a number of unregulated tributaries. It has been suggested that at least one-third of the total basin should be under reservoir regulation for effective flood protection (Linsley and Franzini, 1964). In a large basin, a series of small dams on tributaries often offers better protection than the same volume of storage provided in a single mainstream reservoir. Thus, a reservoir needs to be located near to the area to be protected because it will have little downstream effect on large rivers, like the lower Mississippi, which experience floods from well-distributed rains.

The 1993 Mid-West floods provided a good example of the use of flood control storage, much of which is contained in 66 reservoirs in the upper Mississippi and Missouri basins. Many of these storages are in the Kansas River basin, which accounts for about 10% of the area of the Missouri River basin, and where reservoirs control streamflow from 85% of the drainage area (Perry, 1994). During 1993, flood discharges were reduced by 30–70%, although the inflow to some of the reservoirs was several times their total storage capacity. The greatest benefit was on the Big Blue River where Tuttle Creek Lake retained a daily mean flow of 3029 m^3 s^{-1} on 5 July 1993 (Figure 7.4). This greatly reduced the peak flow which would have caused more damage than the maximum controlled release of 1700 m^3 s^{-1} later in the month.

The maximum storage capacity behind a dam is relevant because the best flood reduction would be achieved by schemes capable of storing all the inflow water in excess of the safe capacity of the downstream channel until the inflow falls below this level and can be released. In practice, this is rarely achieved and reservoir operation during the flood is critical. According to Harmancioglu (1994), there are three release control strategies. *Automatic release*, as shown in Figure 7.5A, operates when Q_A defines the maximum safe channel flow and S_I is the total storage capacity of the scheme. In cases where Q_A is used to define the size of the outlet structures, this operation may not utilise the total storage capacity. *Constant release* (Figure 7.5B) produces a constant flow between t_1 and t_3. In this case, the maximum amount of storage occurs as S_{II}. *Adaptive release* is a step-by-step regulation at variable time intervals depending on the hydrological information available. Real-time monitoring through the use of radar and other weather forecasting tools is important. Figure 7.5C shows that when the reservoir capacity is small, and incoming flows are not known in advance, this may lead to Q_A being exceeded for a short period. Incoming flows are retained in storage or released at different times according to decisions made at variable time intervals.

Some of the disadvantages of dams include the potential for catastrophic failure and the flooding of upstream bottom land. New sites are often difficult to find, for example in eastern England where neither the terrain nor the geology is suitable. In the longer term, the efficacy of much dam construction for flood control decays as a result of the progressive loss of storage through sediment accumulation. This problem can be reduced either by deploying methods to stop suspended material entering the reservoir,

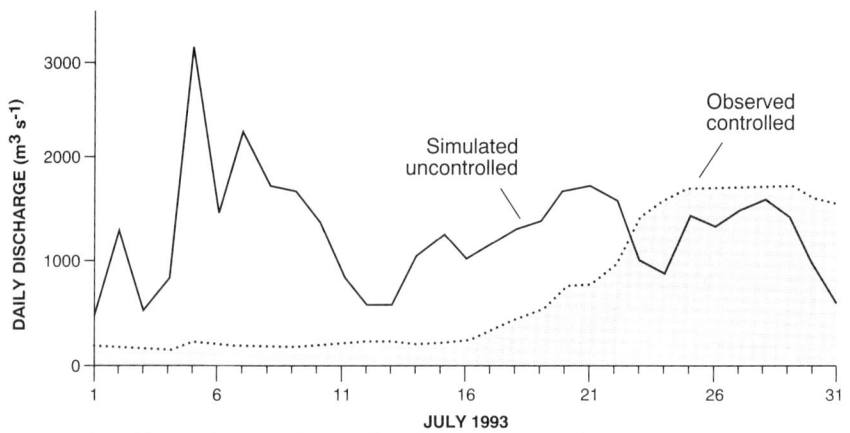

Figure 7.4 The effects of reservoirs on flood discharges during the upper Mississippi River floods of July 1993
Source: After Perry (1994)

FLOOD DEFENCE 217

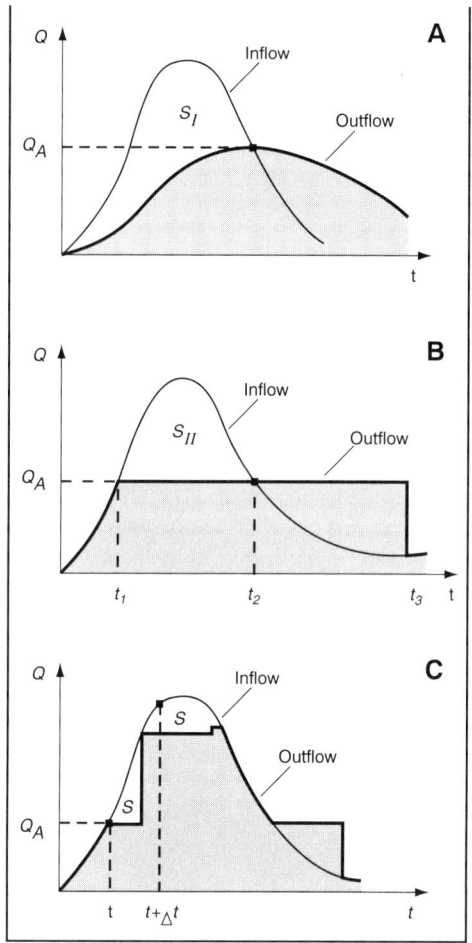

Figure 7.5 Three types of reservoir operating rules with Q_A representing the safe downstream channel capacity and S the storage in the reservoir: (A) automatic operating rule (B) constant release rule and (C) adaptive rule
Source: After Harmancioglu (1994)

for example by controlled sedimentation in the floodplain above the reservoir, or by flushing out the bottom water, for example by surplus irrigation water at the end of the season. Assuming a sedimentation rate of 0.50% of the capacity per year, one-quarter of the storage is lost in 50 years. This rate is exceeded for many storages on rivers with high sediment loads. For example, the Nizamsager reservoir in central India lost over 60% of its capacity ($840 \times 10^6 \, m^3$) in 40 years and was expected to lack any effective storage within a further 30-year period (Chettri and Bowonder, 1983).

7.2.4 Environmental and Ecological Impacts

The natural river environment is characterised by a highly variable flow regime, including seasonal flood pulses, which maintains key channel habitats, such as pools, riffles, meanders and steep banks. A further variety of habitats, such as depressions, oxbows, ridges and plains, exists in the floodplain beyond the channel and depends heavily on the frequency and duration of inundations from the river (Junk et al., 1989). River engineering works of all types, including flood control structures, have led to more uniform flow and channel conditions with serious consequences for the in-stream, freshwater ecology and the biological diversity of the floodplain. Reductions in overbank flows have often destroyed wetland flora and fauna on the floodplain. Levees and channelisation have isolated the river from its alluvial plain to the detriment of the riparian ecology and the environmental functioning of the entire 'river corridor'.

Major reductions in both the variability of the flood regime, and the geomorphological diversity of rivers and their floodplains, have occurred. For example, during the last 50 years, river training works along the River Murray, Australia, have reduced the magnitude of the average annual flood by over 50% at all stations, although large floods with an average annual recurrence interval of 20 years or more have been little affected (Maheshwari et al., 1995). A decreased sediment input to wetlands and delta areas has occurred through the construction of dams and the reduction of soil erosion. According to Sestini (1992), the sediment load of the Po River, Italy, decreased from 16.9 to 10.5×10^6 tonnes per year between 1965 and 1973 and dams have caused a reduction in sediment delivery to the Ebro delta, Spain, from 4.0 to 0.4×10^6 tonnes per year over the last 30 years (Marino, 1992). In Vietnam, engineered structures, mainly dams, have interrupted the natural supply of sediment to the coast to the extent that parts of the coastline have receded by as much as 1 km in 20 years. Watercourses have also changed. Extensive river training works on the upper Hunter River, Australia, over the last 40 years have transformed an actively migrating stream into a laterally stable channel (Erskine, 1992). Decreased resistance to flow and less sediment have caused the preferential erosion of sand and fine gravel fractions from the bed by over 1 m since the 1950s. A wider channel may create relatively little disturbance to existing riverbed habitat but it may affect the river corridor ecologically and aesthetically if bankside trees, or other vegetation, have to be removed.

Yevjevich (1994) identified a broad historical pattern whereby, because levees were built first on many of the large European and Asian river systems, the initial effect was to increase downstream floods and reduce low flows. In the later phase of dam construction, a shift to decreased floods and an increase of low flows downstream can be observed. In turn this can lead to a cycle of channel aggradation and degradation, such as has occurred on the Po River, Italy, in recent times. At first, channel aggradation was apparent due to increased erosion in the basin, and the riverbed, trapped between levees, rose above the level of the surrounding floodplain. Following reservoir construction upstream, with greater sediment retention, a degradation phase ensued leading to the collapse of some bridge supports undercut by channel erosion.

The cumulative ecological and environmental effects of all river training works have been well-documented, especially for heavily managed hydrosystems in the MDCs. Indeed, there is an increasing recognition that the ecosystems along river floodplains, one of the world's most precious natural assets, have been extensively damaged. For example, Fruget (1992) demonstrated the reduced biological diversity, expressed in terms of benthic invertebrates, fish communities and aquatic birds, which has resulted from over 200 years of human interference with the lower Rhone, France. Similar concerns are now being voiced for some of the LDCs, such as parts of southern China (e.g. Dudgeon, 1995), and the wisdom of embarking on large, new river control projects, such as the so-called Three Gorges project in central China, has come under considerable scrutiny (Edmonds, 1991). It is less usual to find studies confined to the specific impacts of flood engineering works. In an investigation of 18 flood alleviation schemes in the UK, Hey et al. (1994) found that conventional channel dredging, widening and straightening reduced the number of plant species and they recommended that new schemes should incorporate set-back flood banks and two-stage channels, where the floodplain adjacent to the river is excavated to create a flood berm while preserving the natural low-flow channel. Typical adverse effects include increased flow velocities, which may move spawning gravels downstream during floods, and the drainage of floodplain grasslands which is thought to create a lower diversity of plants and animals than in wetland habitats. However, detailed impacts can be complex, as those associated with the lowering of Lough Neagh and channelisation in Northern Ireland to protect agricultural land against 1:3 year flows (Johnston et al., 1994). A subsequent survey found some loss of wetland species, with the main effects on species favouring dry conditions, but there was an overall increase in the number of species.

Over the past two decades, there has been a growing emphasis on more environmentally sound methods of river training in order to achieve nature conservation and river rehabilitation. Disruption during the construction phase is not the full story and geomorphologists have concentrated on achieving a better overall union between river engineering and fluvial geomorphology which involves re-uniting river channels with the floodplain environment for the benefit of the river corridor as a whole. Restoration strategies for temperate rivers are required to enhance conservation, recreation and educational interests while being economically justifiable. Patt (1994) described the federal nature preservation law in Germany which is now the legal basis for water-related projects and instanced the River Roter Main, at Bayreuth, which was put into a concrete channel in the 1960s and is now rehabilitated. Similarly, Hill (1994) described a scheme on the River Thames, England, the largest to be undertaken in the UK, where the new channel is designed to appear within natural banks, with small islands and low wetland areas. Extensive bankside planting, woodland and hedgerows are planned to produce new wildlife habitats.

It is now accepted that all river engineering schemes have to be more environmentally sensitive. For example, the US Army Corps of Engineers has placed environmental values on an equal footing with economic and engineering concerns in support of more sustainable development in the future. The visual intrusion of a high flood wall running across the view from a riverside home or walkway is just one illustration of such issues. The severity of visual intrusion always has to be assessed on the basis of the type and quality of the landscape in which it is situated. Thus, very high levels of flood protection in the Netherlands have led to some public opposition to river-dikes based on landscape, ecological and cultural values. River-dikes have distinctive vegetation, cultural–historical significance and are an important element in the river landscape (Breusers and Vis, 1994).

Proposals for new flood engineering works in the MDCs now often include statutory public participation in the decision-making process. In many cases, the water authorities simply see participation as a means of gaining acceptance of the scheme as presented and public participation does not, in itself, lead to consensus. Some critics hold that participation leads to the emergence of elite groups and conflict which is against the general public good (Bruton, 1980). For the process to be effective, floodplain residents need to be consulted as early as possible in the planning process to ensure genuine dialogue, which may, for example, involve determining the level of trade-off between environmental values and level of flood risk. This dilemma was highlighted by Fordham et al. (1991) on the Thames floodplain, near London, where residents see their location as a positive asset, providing a river view and open spaces for walking and other recreation. Many would rather live with some flood risk than accept structures, such as flood embankments which block a river view, which they perceive will have an adverse effect on their local environment and amenity. This leads to conflict between floodplain residents and flood engineers who typically complain about spending more time on consultation and public relations than on design.

7.3 RIVER FLOOD ABATEMENT

It is sometimes asserted that the cost of flooding in non-metropolitan areas is underestimated. According to Goddard (1976), 47% of flood losses in the USA occurred in agricultural areas, mainly in the cropland of the Corn Belt and the northern plains. Of these costs, 62% were incurred by rural headwater streams, which are the main target for flood abatement. As Schwab et al. (1993) indicated, headwater flood abatement measures on farmland are different from those adopted on major rivers downstream. This is because the average rainfall intensity tends to be higher in the headwaters, often due to localised, convective storms, and the effects of crops, soils and tillage practices are much greater. For example, due to cropping patterns, the surface conditions of some tributary basins can change completely from one season, or from one year, to another.

Generally speaking, factors associated with river flooding, such as high rainfall intensity or rapid snowmelt, sheet and channel erosion, and high sediment transport, also lead to soil erosion. The water erosion of soils is an important agricultural problem in many parts of the world and an extensive literature exists on soil and water conservation practices in rural watersheds. Although these practices are geared primarily to agricultural production goals, any land treatment that reduces the total flow of water, or decreases its speed over and through the soil, will also help to control floods. In some cases, for example in forested areas, such delay is complemented by an increased opportunity for a net loss of water by evapotranspiration. Rural land practices are important because, due to the small area of most drainage basins which is occupied by stream channels or other surface water, the vast majority of precipitation has to be converted into flood flows through the intermediate media of vegetation and soils.

The link between over-intensive land-use practices, soil erosion and flooding is well-established for many environments, including the tropics (Aneke, 1985). Most of the evidence, however, comes from temperate areas. For example, the Obion, Forked Deer and Hatchie tributaries of the Mississippi River flow through unconsolidated formations and became choked with sediment as a result of deforestation and upland erosion associated with land clearance in West Tennessee during the late 1800s (Simon, 1990). This led to the streams frequently flooding for 3 to 10 days at a time and to a remedial programme of channel dredging and straightening which began around the turn of the century. Channel lengths have since been shortened by as much as 44%, gradients increased by 600% and beds lowered by up to 5.2 m. Boardman (1995) demonstrated how a change over the last two decades to a largely monocultural pattern of autumn-sown cereals on the South Downs of southern England, sometimes cultivated on slopes in excess of 20°, has led to erosion of the thin soils. This is because the soil surface remains open to winter rains for many weeks after sowing. The chief result has been flood damage to buildings and rural infrastructure, the costs of which have not been borne by the agricultural community. Agricultural incentives which encourage farmers to use all the available land for arable crops, even when the soil is unsuitable and low yielding, are a problem in parts of mainland western Europe too (Boardman et al., 1994). On the other hand, ecologically sound land use for soil and water conservation is implicitly practised over many rural basins.

Flood abatement attacks the flood problem as near to its source as possible with the aim of preventing damaging flows from developing downstream but the strategy has some important disadvantages. First, in order to have any chance of success, flood abatement must be able to detain flood-producing precipitation over significantly large areas of land, either by enhanced surface storage or infiltration, so that the water is released from the watershed more slowly over a longer period of time. As a result, the effectiveness of headwater measures decays rapidly with the magnitude of the storm and land treatments which appear successful in first-order watersheds may well be undetectable further downstream. Second, because most of the literature is focused on soil and water conservation, much of the evidence for the success of this strategy is of a rather indirect, even negative, kind. For example, the effects of deforestation in increasing streamflow in small watersheds are well-documented, but comparatively few investigations have been conducted into the effects of afforestation in decreasing flood peaks in large drainage basins. For all these reasons it is generally agreed that flood abatement is only effective in reducing floods with return intervals of 10 or 15 years rather than the larger events of 50 years or longer. For the largest storms, devices such as contouring and terracing agricultural land have less effect on flood flows than in reducing soil losses.

Flood abatement embraces a variety of specific techniques, many of which may be implemented in combination. For example, Hurni (1983) described attempts to control soil loss in the Ethiopian Highlands involving some 590 000 km of contour bunds, 600 000 km of diversion ditches, 1100 check dams and 470 000 km of afforestation terraces.

7.3.1 Topographic Manipulation

The degree of slope affects both the amount of runoff and the extent of soil erosion, other factors such as precipitation intensity and soil type being equal. A long slope also produces more runoff and erosion than a shorter one at the same angle because the water gathers momentum as it flows. Topographic manipulation seeks to minimise the flood-producing abilities of rural terrain and to reduce the runoff coefficient by changing, or optimising, the surface geometry in a variety of ways.

7.3.1.1 *Terracing and Contour Ploughing*

Terraces of various kinds have been used to control runoff and soil erosion in the steeper agricultural areas for thousands of years. In general they are built along the contour and are used to break up the length of hillslopes prone to water erosion. They also facilitate water infiltration and sediment deposition, thereby improving soil quality. The classic *bench terrace* has been adopted in both irrigated and dryland farming areas where the agricultural land is very steep (20–30%). In wet rice growing areas, the interception of water can aid surface irrigation. In drier areas these terraces are useful on gentle slopes of 5–8% with relatively coarse-textured soils where they help water to seep into the soil, thus increasing the water available for crops.

Conservation or *broadbase terraces* are more commonly used to control erosion. They are built to divert excess water from flowing directly down a slope and to conduct it off agricultural land at a non-erosive rate. The key feature is a flat channel section with a low embankment upslope of the cultivated area to provide the maximum opportunity for infiltration. Such terraces are not practical on slopes over 10–12% or where the topography is uneven. According to Schwab et al. (1993) a flood on a river basin extending over 2250 km^2 in Iowa, USA, could have been reduced by over half by constructing level or absorptive-type terraces covering little more than one-third of the area. If only 17% of the area had been terraced, the flow could have been reduced by 32%.

Strip cropping is a common practice on the gentlest slopes between 4 and 12%. It involves the growing of alternate strips of different crops along the contour to control water erosion. Rotations of close-growing perennial grasses and legumes alternating with grain and row crops is a common pattern. Narrower strips are required on the steeper and longer slopes, especially on clay soils with poor permeability and high erodibility or high rainfall impact.

7.3.1.2 Surface and Underground Water Storage

The surface detention of floodwater by farmers wishing to retain water on the land for stock supply, irrigation or other purposes has a beneficial effect on downstream flood behaviour. Most detention basins on farmland are a relatively 'soft' form of engineering, whilst small lakes and swamps also store floodwaters temporarily before release. This natural storage can be augmented by small dams, and Schwab et al. (1993) described the use of three types of farm ponds—dugout ponds, on-stream ponds and off-stream storage ponds—many of which operate in a manner similar to reservoirs.

Flood reduction by underground water storage, sometimes called water spreading, is applicable only in semi-arid and arid areas. On relatively dry rangelands, with 250–500 mm of annual rainfall, surface flows from flash floods can be diverted from drainage lines and spread by contour bunds to increase infiltration, as described by Branson (1956) and Miller et al. (1969) for areas in the western USA. As shown in Figure 7.6, floodwaters are diverted by concrete weirs placed in small stream beds and then spread by earthen bunds or dikes up to one metre high. The best results have been achieved with flood-spreading over flat valley bottoms which are used for hay production.

7.3.1.3 Gully Control

Gully control is directed to high-risk areas of erosion and flooding, such as where furrows made by ploughs or the wheels of vehicles run up and down the slope. Check

FLOOD DEFENCE

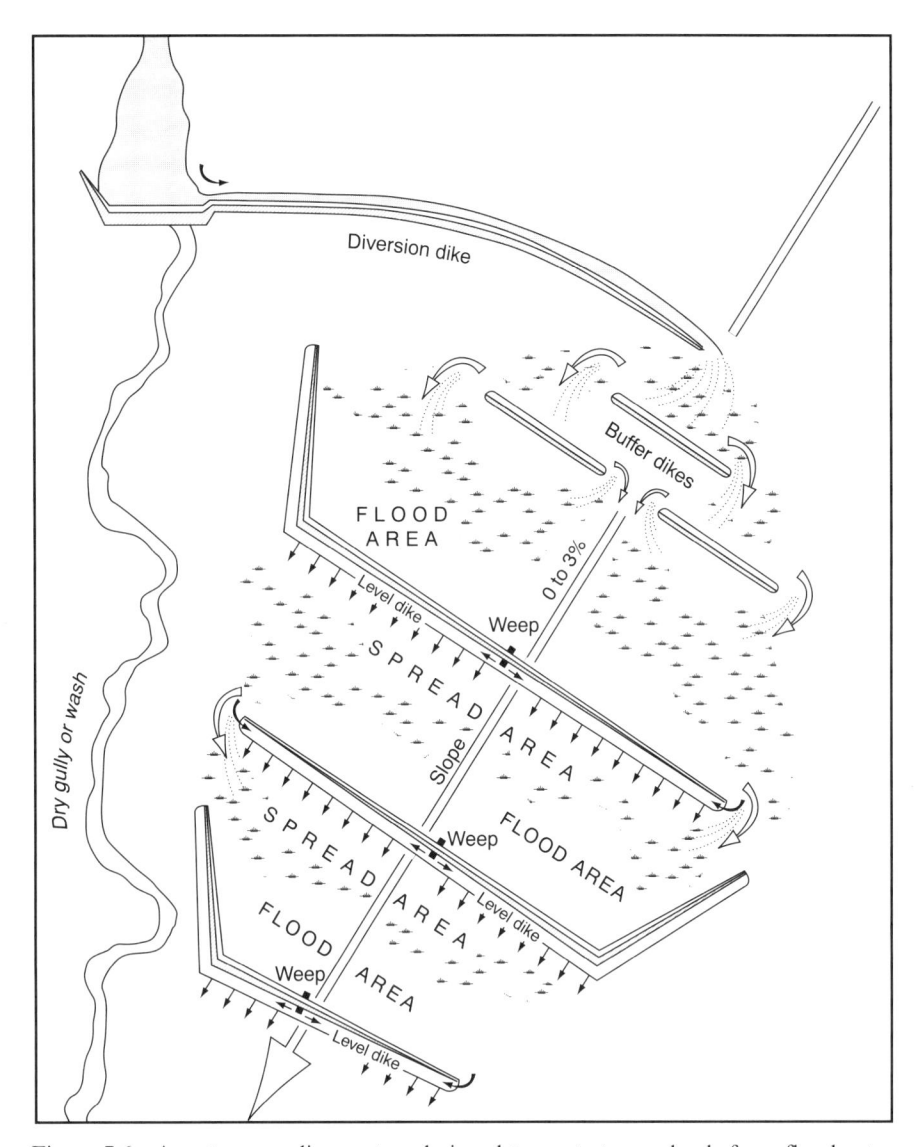

Figure 7.6 A water-spreading system designed to protect rangelands from floodwaters
Source: After Branson (1956)

dams have been widely constructed in mountainous regions with steep headstreams and high debris loads. The purpose is to retard the flow of water down the ravines in the upper watershed, thus reducing erosion, and to encourage the greatest absorption of water either in the ravines or in the debris cones at the mouth of the canyons. In addition, this may encourage vegetation growth and increase the duration of low flows. Although check dams are inexpensive and often built of loose rock, they control relatively small volumes of water and are subject to high rates of sedimentation. In flatter areas, check dams also help to spread water, reduce surface velocity and retain soil and moisture for vegetation. Such dams can be made of woven wire and fence posts, brush, post and planks and stake and straw (Agriculture Canada, 1975). A number of small dams, not over 0.5 m high, are preferable to a smaller number of larger dams because they are less expensive to build and are less likely to fail.

7.3.2 Vegetation Cover Management

Vegetation cover management, particularly when it involves a major change in land-use strategy, is one of the most effective methods of flood abatement (see also Section 3.6.1). Generally speaking, a complete vegetation cover helps to reduce soil erosion and flooding through the detention of rainfall by interception, increased infiltration, and reduced runoff through enhanced evaporation and evapotranspiration. The management of vegetation includes actions as diverse as the re-seeding of sparsely vegetated areas and the reduction of wildfires, which can have a devastating effect on sediment yields and flood flows.

7.3.2.1 Grassland and Crop Cover

Cropping patterns can be important in avoiding bare soil during the main seasons of precipitation stress on the land. Tillage practices also have a role to play as minimum tillage helps to retain the soil structure and an organic residue on the surface. For example, Kramer and Hjelmfelt (1989) reported that the ridge tillage of corn reduced runoff by two-thirds compared with conventional tillage. A grass or legume cover, such as timothy or alfalfa, is an effective control on soil erosion. In addition to providing above-ground protection from raindrop impact, such crops add organic matter to the soil and bind the soil particles together. This is because these crops produce a great deal of root growth, the weight of which is as great as that of the above-ground crop.

It has been estimated that more than 4000 km^2 of marginal land in the Great Plains of the USA have been converted to permanent grass with runoff similar to that from native meadow as a result of federal programmes (Dragoun, 1969). Grassed waterways are broad shallow channels designed to carry water away from farmland with a minimum of erosion and maximum potential for infiltration. The ideal profile is nearly flat-bottomed, from 3 to 10 m wide, with only a slight slope of about 1%. Such waterways are used mainly for hay production.

7.3.2.2 Forest Cover

Trees are one of the most effective land covers for reducing soil erosion but clear experimental evidence on the direct role of afforestation, or reforestation, in flood abatement is sparse. As with the issue of soil conservation and flooding, there is a large literature on forest hydrology but this is mainly concerned with the effects of forest cover on water yield for supply purposes rather than on peak discharges (see also Section 3.6.1.2). In addition, much of the evidence is that wholesale removal of vegetation has adverse consequences for storm hydrology, whereas forest replacement is not always beneficial in reducing floods.

Numerous experiments have demonstrated changes in runoff regime resulting from more selective forest cutting and removal practices, such as clear cutting, block or strip cutting. More specifically, Pereira (1973) described the effects on flood hydrology of forest destruction by fire on the steep Yarrangobilly basin in the Australian Alps. After the fire, storms which would have been expected to produce flows of 60–80 m^3 s^{-1} gave a peak of 370 m^3 s^{-1}. Similarly, land clearance for tea cultivation in Kenya led to storm flows of 27 m^3 s^{-1} km^{-2} compared with maximum rates of 0.6 m^3 s^{-1} km^{-2} for areas of undisturbed forest. Other evidence suggests that peak discharges from snowmelt also increase after forest removal, although careful interpretation is necessary. Although open and forested sites generate snowmelt at different rates and times, the influence of forest cover on snow accumulation and melt during upland rain-on-snow events can be complex (Berris and Harr, 1987). In Minnesota, Verry et al. (1983) found that when a mature aspen cover was cleared from one-half of the drainage basin, the snowmelt runoff peak actually declined but, when more than 70% of the stand was removed in the following year, the peak rate of snowmelt nearly doubled. Jakeman et al. (1993), working in central Scotland, found that forest clear felling and drainage work produced little hydrological change on a daily basis and attributed this to the fact that the quickflow component of runoff is less affected by land-use change.

Far fewer experiments test the reverse process of afforestation in reducing runoff peaks. Black (1968) described changes in streamflow as more than 137 000 ha of abandoned farmland in New York State reverted to mature forest. In one small watershed, reforestation of 58% of the area reduced winter and spring flood peaks by 16 to 66%, although no significant change was observed after canopy coverage had reached 90%.

Melting snow in the spring can cause erosion on sloping ground that is bare of vegetation, especially if rapid melting coincides with spring rains. In the USA the bulk of the rural flood damage is concentrated in the agricultural interior within, and downstream of, the annual snow-covered areas and some incentive exists to reduce the rate of snowmelt wherever possible. Universally valid management guidelines are difficult to establish but the general principle is to maintain a diversity of forest cover on watersheds. If a watershed is cleared by more than 60%, there is a risk of increased snowmelt peak discharge. Forest cover management will have less effect in mountainous areas. Higher rainfall and steepness of slopes will over-ride water retention by forests and snowmelt runoff will be more de-synchronised than in lowland basins because of differences in elevation, slope and aspect. In some high-altitude areas, such as the upper Indus basin, Himalayas, peak flows from glaciers and snowmelt can coincide with early season monsoon rains and little can be done by management.

7.3.3 Flood Abatement in Urban Areas

Limited, local flood abatement can be achieved in urban areas by encouraging the maximum infiltration rates in parks and water detention basins (see Section 3.6.1.1). For example, Ganoulis (1994) described the use of stormwater retention basins which can be used as gardens, picnic areas and stadiums with sports facilities. A series of such basins has been designed for the city of Rethymnon, Crete, with catchment areas up to 3.5 km^2. They reduce the peak flows and help to maintain the efficiency of the sewer system by retaining sediment. Groundwater storage can also be enhanced artificially by the use of water spreading in shallow basins and by means of injection wells.

7.4 COASTAL FLOOD ENGINEERING

According to Thorn and Roberts (1981), the main purpose of coastal flood engineering is to prevent further coastal encroachment and erosion by sea action. This includes the stabilisation and protection of land to the rear of any structural defences. The maintenance and strengthening of natural defences in order to safeguard lives and property are the first priority of the coastal engineer and, along many low-lying coasts and estuaries, it has been necessary to construct sea walls and more elaborate structures.

7.4.1 Estuary and Sea Walls

This type of flood embankment is used extensively in England and Wales where almost 5000 km^2 of land are below sea level and protected by defences. About one-third of these sea defences, mainly in the south and east of the country, are engineered. The construction of flood banks and walls started with earth banks built by the Romans over 2000 years ago. The current level of central government grant allows for the protection of an average 20 000 houses, 2500 commercial premises and 45 000 ha of agricultural land each year.

Coastal embankments, like river levees, are normally constructed from local surface deposits, such as clay or sand, excavated from borrow pits. Earth embankments are subject to erosion by wave action, together with scour on the landward face due to overtopping. In addition, the upper parts of clay embankments can dry out and experience fissuring which can lead to seepage and slumping unless the embankment is raised at least 1 m above the highest tide. Thorn and Roberts (1981) stressed that the design of sea walls depends on whether the incoming waves are reflected by the structure or break in front of it.

Estuary walls are expected to reflect limited wave action, so these structures are built rather like river levees with an emphasis on adequate height, thickness and stability. They often benefit from some natural protection associated with a restricted wave fetch and the presence of extensive saltmarshes on the seaward side of the structure. Most overtopping will involve small quantities of seawater, which can either be absorbed in the saltings or discharged back to sea via gravity outfalls. Where active wave attrition occurs, the seaward side of the structure can be faced with more resistant materials. In situations where the soil mechanics limit the maximum height of the earthworks, as in Figure 7.7A, any further height increase (in this case up to 6.10 m) has to be achieved by a reinforced concrete wall on top of the existing earth embankment.

Sea walls fronted by high foreshores have to face a relatively shallow depth of water at high tide and waves break on a shingle foreshore, rather than the wall, effectively dissipating excess energy in turbulence and the run-up. A slope of about 2:1 is the steepest on which waves will break on the foreshore before reaching the wall. A typical example is the Seasalter sea wall on the north coast of Kent (Figure 7.7B).

Sea walls fronted by low foreshores are placed in deeper water and storm waves break on the wall itself. This requires more robust construction, as shown in Figure 7.7C. As waves break on the structure, variations in pressure may lift off the facing material whether it is stone pitching, granite blockwork or concrete blocks. Figure 7.8 shows improvements made to the Sheerness sea wall to reduce overtopping. The wall is formed of a heavy stone facing laid on a clay foundation. In this case, the required reduction in

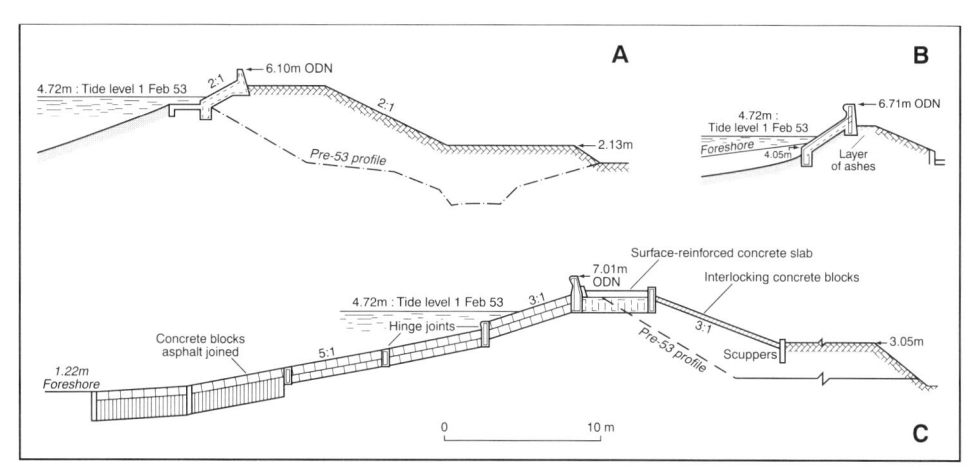

Figure 7.7 Examples of coastal floodwalls from southern England: (A) a concrete crestwall on an earth embankment, Isle of Sheppey, (B) cross-section of the Seasalter wall, north coast of Kent, (C) cross-section of Northern sea wall which runs from Reculver to Birchington, Kent
Source: After Thorn and Roberts (1981)

overtopping was achieved by raising the level of the main wall and by topping this with a wave wall on the landward side of the structure. The wave wall was restricted to a height of 1.4 m for amenity reasons.

As with river levees, the degree of flood protection offered by coastal walls depends primarily on the height. After the 1953 flood disaster around the North Sea coast (see Section 6.6), recommended practice in the UK was that structures should provide protection against any recurrence of such an event, although subsequent revisions of the standards have led to a degree of local interpretation. For example, in the former Southern Water Authority area, a 1 in 1000 year standard was adopted where major risks to human life were combined with a serious damage potential, compared with a 1 in 250 year standard for residential or industrial areas or large areas of valuable farmland and a 1 in 50 year standard for all other areas (Thorn and Roberts, 1981). The design is meant to take account of any secular changes in sea level within the design life of the structure and, if the recommended standard cannot be justified economically, most defences have to be built to a maximum height consistent with an acceptable cost–benefit ratio. As a result of subsequent raising, many embankments have reached the practical limit in terms of the stability of the constructional materials or the strength of subsoil foundations.

7.4.2 Barrages and Barriers

Sea walls tend to cut off the supply of sediment to coastal marshes and also lead to a reduction in the moisture content of the previously deposited silts and clays. Over several hundred years, the subsequent shrinkage of this material has created a difference in level of up to 2.5 m between the saltings on the seaward side of the defences and the marshes on the landward side of some UK sea defences. Another characteristic of sea walls has been the damming-off of many small tidal estuaries in order to reduce the length of the flood defences required. The elimination of the tidal currents which previously scoured the estuary has sometimes led to considerable siltation on the

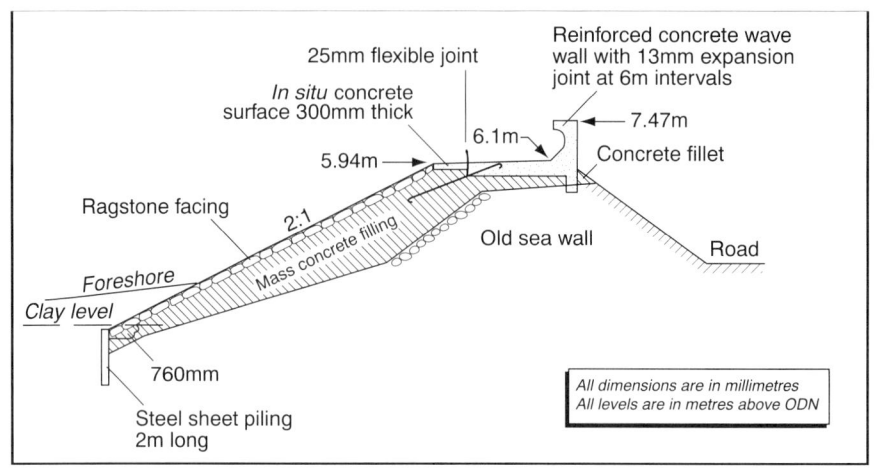

Figure 7.8 An example of a sea wall fronted by a low foreshore, as at Sheerness, Kent
Source: After Thorn and Roberts (1981)

seaward side. Where the estuary is used for navigation, dredging is necessary to maintain the channel and this also has disturbed the ecology which has changed from a marine to a brackish, or freshwater, state.

Barrages and barriers, as used extensively in the Dutch Delta Plan, have been deployed to reduce these problems. At the start of the present century, the tidal area of the Zuyder Zee covered 350 000 ha and flooded at high water to a depth of 3.5–4.5 m. In 1932 a dam was built across the estuary mouth to form a freshwater lake and to provide for some land reclamation. The Dutch Delta Plan was initiated in 1958 after the 1953 North Sea disaster but was not completed until 1986 (Spencer and French, 1993). Both barrages and barriers have been used and several of the inner distributaries have been cut off by barrages and are now freshwater lakes. In other cases, the estuaries have remained open with barriers closing only in the event of a storm. Some of these methods are known to disrupt natural estuarine processes (Carter, 1988).

As a result of new technology and environmental concerns, permanent means of estuary closure are now less favoured. The *Thames tidal barrier* is an example of a structure which is normally open to tidal flows but which can be closed quickly when exceptionally high water threatens. This barrier is designed to protect more than one million people, living in some 130 km² of Greater London, against high water with an annual probability of 1:1000 with an allowance for the adverse trend in sea level to the year 2030. Coastal protection to the same design standard for urban areas on the seaward side of the barrier was achieved by bank raising. The design of the structure was determined by the size of the main openings which are used for navigation. The Port Authority requires 1.5 to 2 hours warning for shipping, issued in association with the Storm Tide Warning Service (see Section 8.12.1.1), before the barrier is closed. The structure comprises four main sector openings of 61 m in length with six smaller openings of 31.5 m. Rising segment gates normally lie in a recessed sill in the bed of the river but can be lifted to close the main segments and two of the smaller openings (Horner, 1981).

A similar scheme has been proposed for Venice which is vulnerable to continuing sea-level rise. According to Bandarin (1994) a total rise in sea level of only 30 cm would flood

FLOOD DEFENCE 229

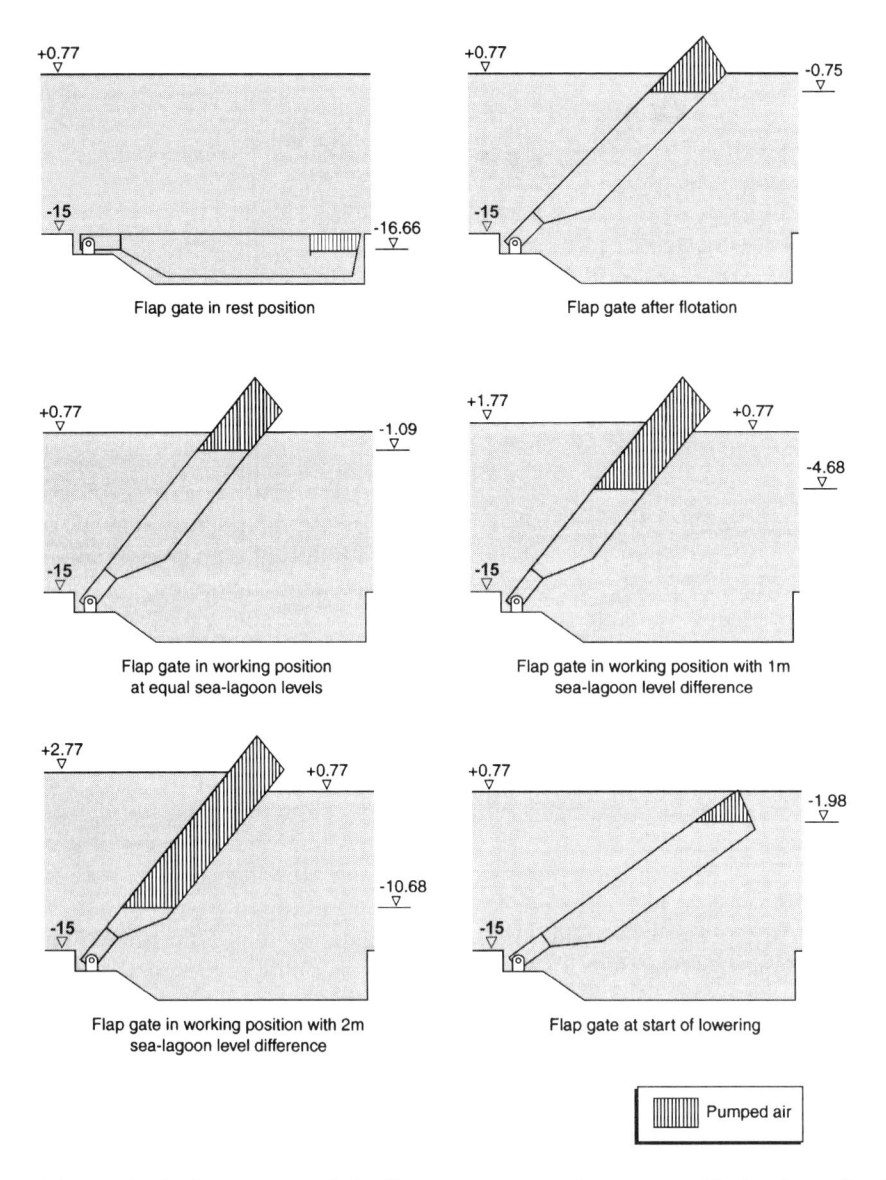

Figure 7.9 Method of operation of the flap gates proposed to protect Venice from flooding during periods of exceptionally high tides
Source: After Bandarin (1994)

St Mark's Square more than 360 times per year, a situation which could well arise in the second half of the twenty-first century. Following severe flooding in 1966, a system of rising gates has been proposed which could be used to block the three entrances to the Lagoon from the Adriatic Sea whenever the city was threatened by high tides or storm surges (Figure 7.9). As in London, these mobile gates would allow navigation for ocean-going ships to reach the port of Marghera (Lewin and Scotti, 1990).

Engineering works can also offer some protection against tsunami flooding. Several cities along the Sanriku coast of Japan are defended by a combination of offshore

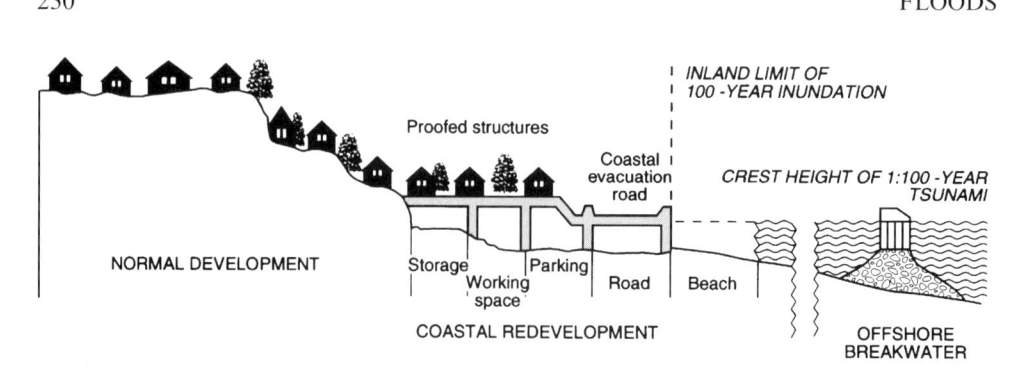

Figure 7.10 A representation of combined tsunami engineering defence, including an offshore breakwater and coastal redevelopment, as adopted along parts of the coast of Japan

breakwaters and onshore tsunami walls designed to protect coastal property and communication routes (Figure 7.10). These schemes are expensive and create a variety of problems. For example, although breakwaters do not take up coastal land and can provide shelter for shipping, they interfere with the tidal circulation whereas onshore schemes can occupy valuable land and lead to considerable visual intrusion to the detriment of a tourist industry.

7.4.3 Environmental and Ecological Impacts

Many human actions along coastlines, including those for flood engineering, have upset sedimentation processes and reduced the capacity of the natural flood defences. The steady process of land reclamation in estuaries has removed an essential part of the estuarine cross-section, thus preventing any natural morphological response to changes such as sea-level rise. Delta closure has often been followed by de-salination with a resulting change in the ecosystem. Some flood engineering has deliberately reduced the diversity of coastal habitats, for example by reducing cliff erosion and the supply of marine sediment. In other cases, impacts have been inadvertent. In recent decades there has been a large loss of saltmarsh in south-east and southern Britain due to a widening and extension of the saltmarsh drainage systems. Although the causes are not well-understood, they may well include sediment starvation from coastal protection works as well as sea-level rise, increased storminess and wave energy, dredging and pollution. Whatever the cause, accelerated wave erosion at the marsh edge can lead to average rates of recession as high as 6 m per annum.

In many tropical areas, pressures associated with aquaculture, tourism or other coastal developments have created greater vulnerability to flooding along low-lying coasts. Gordon (1988) showed how increases in soil salinity due to salt production and a restricted tidal range caused degradation of the mangrove forest which previously offered some flood protection in north-west Australia. In some instances, mangrove timber has been exploited directly for a variety of purposes, e.g. firewood in Thailand (Walsh, 1977). Along the Atlantic and Gulf coasts of the USA, barrier islands made up of concentrations of dunes have been extensively settled and managed despite tension between a recognised need to minimise interference with natural processes whilst encouraging these processes to contribute more to coastal protection (Viles and Spencer,

1995). The fixing of dunes with vegetation to create a dike, as at Cape Hatteras, North Carolina, means that islands erode at the front rather than retreat as a whole, as in the case of natural barrier systems. There is also some evidence that managed systems are eroded more by storms and take longer to recover (Dolan and Godfrey, 1973).

The total defence of the existing coastline is no longer seen as either possible or desirable. A large proportion of the coast of England and Wales is composed of soft cliffs which are receding and are creating a threat to clifftop properties. For example, the Holderness coastline in eastern England is retreating at an average rate of 1.7 m per year (MAFF, 1995a), resulting in a yearly supply of more than $1 \times 10^6 \, m^3$ of sediment to the North Sea. Erosion at the cliff base has previously been temporarily halted, rather than controlled, by 'hard' engineering structures which have often been installed without a complete understanding of the interrelationship between all the natural processes involved. The present view is that engineering should replicate and work with—rather than against—natural coastal processes. Any economic justification for engineered schemes depends heavily on accurate prediction of recession rates with and without structures.

The safety of all existing coastal and many lowland river defence systems will be challenged by projected sea-level rise as a consequence of climate change (see Section 5.5.2). The increase in mean water depth will lead to increases in mean wave height and wave-related effects and a greater risk of overtopping of both hard and soft defences. According to Wind and Peerbolte (1993), a massive 1 m rise in sea level implies a typical shoreline retreat of dunes by up to several hundred metres. Sea-level rise will also affect rivers through the back-water effect, perhaps extending tens of kilometres upstream, with more overtopping, seepage through embankments and salt intrusion of polder areas. The greatest problems exist in the densely settled delta areas, already protected by defence structures, where a combination of subsidence and eustatic sea-level rise will increase the flood threat. Broadus (1993) indicated that a rise in sea level of 1 m by 2050 could result in a cumulative loss of up to 2% of current GDP in a worst-case scenario which assumes no mitigating action. In practice, there is probably little option but to raise escape mounds and embankments, and use more costly pump drainage in countries like Bangladesh (Brammer, 1993). This scenario poses severe economic problems in the MDCs, such as the Netherlands where, with a 1 m rise over a century, the annual maintenance costs for coastal and river defences are likely to double from the present-day sum of US$ 30 million. The total cost of adapting to such a sea-level rise has been estimated at US$ 10 billion, which compares with about US$ 7 billion spent on the Delta Plan 1958–88 (de Ronde, 1993).

7.5 COASTAL FLOOD ABATEMENT

Today, 'soft' engineering strategies, with the aim of creating a more acceptable and sustainable coastline based on natural environmental processes, are more common. By helping to absorb wave energy, dune systems, sand and shingle beaches, mudflats and saltmarshes can all act as natural defences against erosion and flooding. For example, barrier islands provide some protection for the main shoreline, as in the Netherlands where they protect over three-quarters of the coast, although they are highly dynamic features and are migrating onshore due to relative sea-level rise. There will always be situations when a sea wall is needed, such as where densely populated urban areas occupy low-lying coastal plains, but there has been a move away from the routine use of concrete sea walls and timber groynes deployed across beaches towards a mixed approach exemplified by beach recharge schemes and other types of 'soft' defences.

The importance of the inter-tidal zone is now widely recognised in terms of flood protection and nature conservation. Saltmarshes and mudflats can play an important role in dissipating wave and tidal energy in a way which promotes further marsh growth and provides greater stability for existing flood embankments. For example, a zone of healthy saltmarsh 80 m wide in front of a sea wall can reduce wave height by half and reduce the overtopping discharge by a factor of five (MAFF, 1992). Specifically, saltmarsh vegetation provides a buffer against tidal currents which, when reduced in velocity, lead to accretion of the suspended sediment load. Recent research suggests that saltmarshes have roughly similar rates of instantaneous sediment deposition to mudflats but that the marsh offers crucial protection for the fresh deposits from wave and tidal scour, thus allowing more rapid long-term sediment accretion (MAFF, 1993).

Effective beach management is another important component of coastal flood defence. This might include the use of rock-armoured groynes and offshore breakwaters to control sand beaches with relatively flat slopes. Beach recharge schemes have been a popular, and more aesthetically pleasing, method of protecting the shoreline from erosion for some time. For example, Thorn and Roberts (1981) described the use of permeable barriers to encourage sand dune accumulation which can then be stabilised by planting marram grass. Shingle beaches respond rapidly to changing wave conditions and can absorb around 90% of all incident wave energy, thus protecting coastal land. But, during severe storms, shingle moves along the coastline causing erosion of some shingle beaches with the possibility of damage to coastal defence structures and flooding of low-lying land. The prevention of shingle beach failure depends on the maintenance of the beach profile above the failure threshold with extra amounts of beach material. In some areas, the high acceptability of artificial beach recharge is causing a problem. Marine dredged sand and shingle deposits are generally favoured for such schemes because they replicate the characteristics of natural beach materials and can be delivered in large quantities by sea to the point of use. Such supplies of beach material are beginning to run out locally in the UK, especially where the construction industry is in competition for the limited supply of sand and gravel.

Another example of a mixed approach is the set-back of sea defence structures to achieve a stable coastal protection system with inter-tidal habitat improvement. This again implies the need to understand the processes responsible for the establishment of salt-tolerant vegetation and the accretion of silt deposits. At an experimental site at Tollesbury on the Blackwater Estuary in Essex, England, managed jointly by MAFF and English Nature, the sea wall was deliberately breached (MAFF, 1995d). The main purpose was to monitor the hydrodynamics and sediment movement, especially to determine the rates of surface erosion or accretion, and changes in the chemical composition and stability of the soil, as well as those affecting plant species composition as the area is flooded with seawater.

Effective coastal flood abatement is dependent on better shoreline management which, in turn, requires a strategic approach that looks beyond the performance of individual flood defence schemes, and the geographical area of responsibility of particular local authorities, in order to understand long-term morphodynamic behaviour along coasts over decades of time and lengths of shoreline tens of kilometres in extent. Unfortunately, at the present time, the integration of coastal geomorphology into flood protection engineering is rather weak due to a relative lack of knowledge of natural shoreline processes. For example, there is a need for a better understanding of how hydraulic flood control structures interfere with the ecology of delta areas to ensure that the changed tidal range will not damage either the environment or the flood defences by the loss of buffer zones in front of artificial barriers. Ideally, modified deltas and estuaries should be managed by a separate agency to ensure that excessive demands are not made on the system and that long-term monitoring of the changed ecosystem takes place.

It is clear that both new and reconditioned sea defences have to be better engineered, even though costs will be incurred. As indicated by Brooke (1995), there are many environmental considerations which could result in extra expenditure for flood engineering projects. These range from a preliminary environmental review, through environmental assessment to mitigation works during implementation and post-project monitoring. For example, when a decision was taken to upgrade the level of protection offered to 500 ha of land near Christchurch, southern England, by the Pennington sea wall to the 1-in-50 year standard, the attendant engineering works raised questions about the local SSSI of marshes and brackish lagoons. English Nature proposed that part of the new sea wall should be moved 9–10 m seaward, thus leaving the lagoons largely undisturbed at an additional cost of £250 000 (Martin, 1994). This is one of the first examples in the UK of the use of public money for a conservation benefit. In Canada, Andrews (1993) reported on the success of the Tsawwassen dike project which provides flood protection to much of the municipality of Delta, British Columbia, while preserving the largest saltmarsh in the Fraser River estuary. This outcome has been achieved by allowing the free daily flow of saltwater in and out of the marsh at high tide. By facilitating tidal exchange between the saltmarsh and adjacent tidal flats, the biological productivity of the estuary has been maintained and the total saltmarsh–tidal mudflat ecosystem continues to provide a habitat for a wide range of fish species and other wildlife.

Any long-term strategic policy for open coasts is dependent on the maintenance of natural processes wherever possible. In some areas this may mean abandoning existing coastal infrastructures. In England and Wales, MAFF has become more concerned with the protection of the overall coastal environment, including its recreational sites, rather than the loss of individual properties or land from erosion by the sea. The Ministry has increasingly taken a more integrated view of coastal management, rather than a project-specific approach, as the problems arising from the separate management of adjacent coastal areas have become more explicit. As a result, there is now a move towards grouping local authorities and regionally based strategy studies as part of comprehensive Shoreline Management Plans. These trends reflect the underlying importance of the large-scale movement and transport of sediment for shoreline evolution which can be best understood in the context of coastal process cells which are unrelated to administrative boundaries. Shoreline Management Plans based on these requirements are currently in preparation for almost the entire coastline of England and Wales.

7.6 FLOOD PROOFING

The vulnerability of buildings to flooding depends on their resistance to the force of moving water and the response of building materials to immersion. Where existing structures are exposed to recurrent floods of comparatively shallow depth and short duration, householders have little option but to live with the hazard. In the past, floodplain dwellers tended to accept the inconvenience associated with seasonal inundation. For example, the occupiers of some low-lying houses with stone floors in the UK would adopt a simple avoidance strategy by minimally furnishing the ground floor and routinely moving upstairs for part of the winter. Today, a more formal response is appropriate. Flood proofing is achieved by any combination of structural or non-structural adjustments to the design, construction or use of individual buildings or properties that will reduce flood damages either to the structure or the contents. These adjustments may be incorporated into the original design or introduced subsequently through the process of retrofitting and compliance with new building codes. Some adjustments may be permanent or temporary.

Flood proofing is a relatively neglected flood response strategy, often due to a mix of householder apathy and a lack of technical knowledge about what can be done. In most cases flood proofing involves extra expenditure but this is likely to be less on a new structure than that incurred in retrofitting. For successful retrofitting, a lot depends on the original design and construction materials. For example, Marco and Cayuela (1994) outlined the concept of the flood-planned city based on sound hydraulic design and a knowledge of flood behaviour within the urban area. On the other hand, existing properties with basements tend to pose a problem and, because of their weight and tendency to crack under stress, brick or masonry structures are usually impossible to move bodily compared with some wood-framed houses that can be transferred to a new location. In general, relocation of properties is probably less appropriate in the LDCs where many traditional building materials, such as straw and adobe, lose strength and load-bearing capacity after immersion in water. Wooden buildings too may rot in the longer term.

7.6.1 Elevation

Raising properties above flood level is a common response, especially where there are small groups of detached buildings. Khairulmaini (1994) described how traditional Malay houses have been raised on stilts at least 1.5–2.0 m above ground, although it was noted that more than half the residents in the area surveyed would still relocate if given sufficient compensation. In the MDCs, raising properties is only likely to be cost-effective on a larger scale where land values are very high. Some buildings, such as timber-framed structures, can be elevated so that the lowest floor level is relocated above the design flood stage, typically the 100-year flood. This can be achieved by placing jacks and lifting beams below the structural members of the property, detaching the building and its utilities from the foundations, raising it to the desired level, building up the foundations to the new height and then re-attaching the property to the new foundations (Dozier and Yancey, 1993). Individual buildings can be elevated either on piles or columns, or on extended foundation walls or on earthen fill (Figure 7.11A).

In Lismore, New South Wales, over 90% of the 2000 flood-prone detached houses of weatherboard construction have been raised on columns by as much as 3–4 m, mostly without any financial assistance from government, in what has been seen as a cost-effective flood-proofing response (Penning-Rowsell and Smith, 1987). The raising of homes on engineered fill ensures the best protection because the height of the fill keeps floodwater away from the property. But introduced fill may increase flood stages

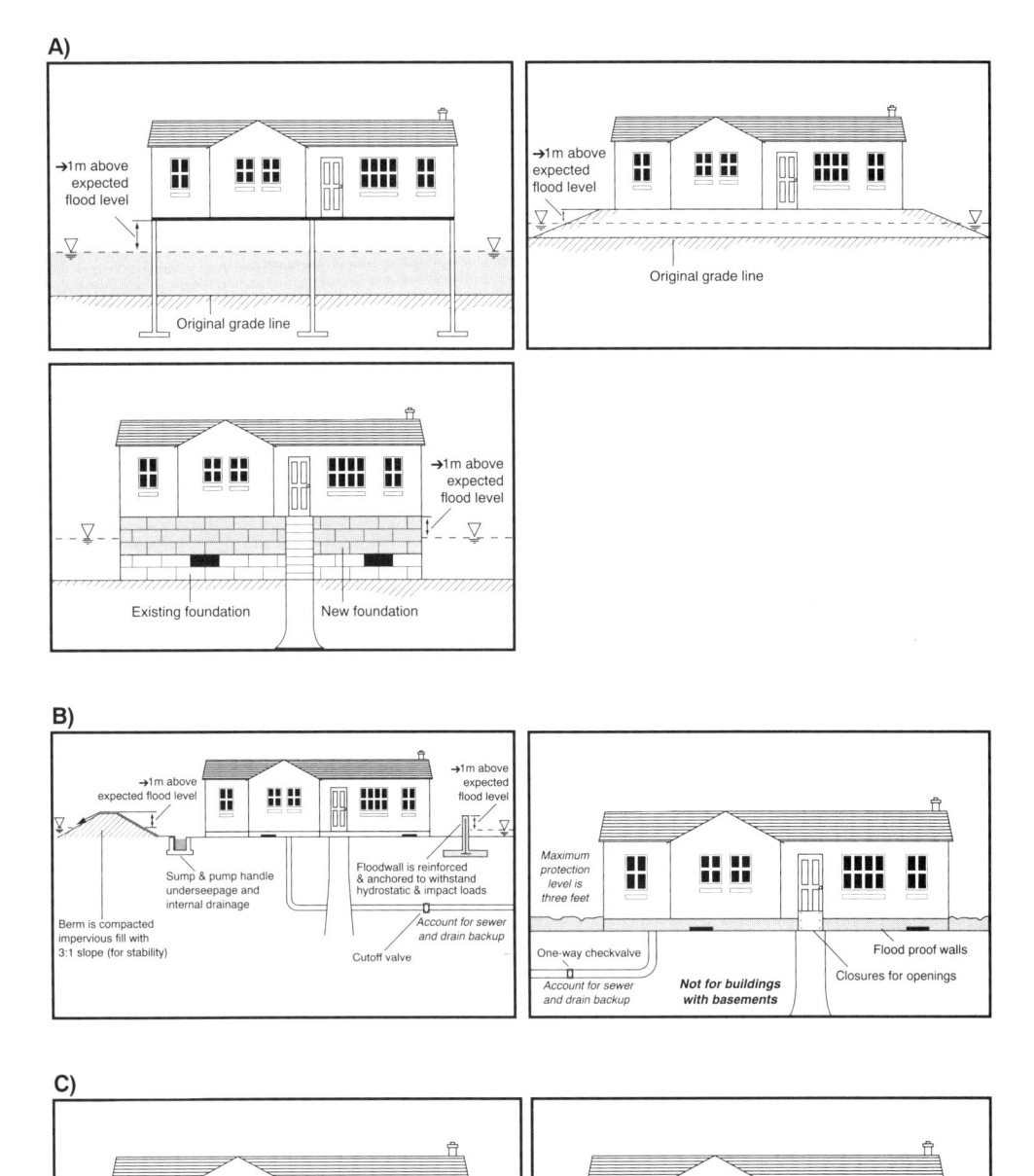

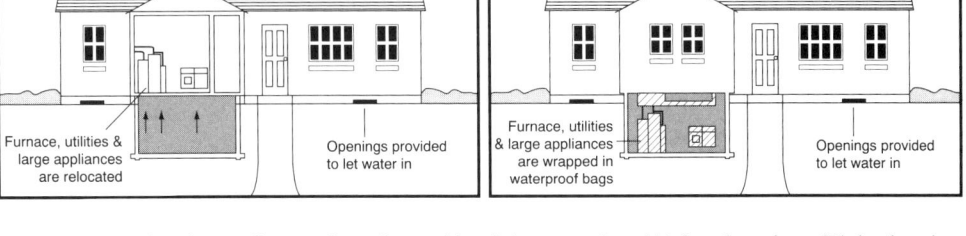

Figure 7.11 Flood proofing options for residential properties: (A) by elevation; (B) by barriers; and (C) by dry and wet flood-proofing measures
Source: After Dozier and Yancey (1993)

for adjacent properties and may not be permissible under local planning regulations. For example, in the USA homes constructed on earthen fill platforms are not allowed within the 100-year floodplain in the special hazard areas along coasts which are subject to storm waves.

7.6.2 Dry Flood Proofing

Dry flood proofing is the most common method and involves sealing a property so that floodwaters cannot enter (Figure 7.11B). It is really only suitable for frequent, low-level flooding because waters more than about 1 m high are likely to cause a collapse of the walls. In addition it is not suitable for properties with basements which are difficult to protect from under-seepage. Small-scale polders can be created around properties by dikes or walls and the ground can be further protected by a drainage and pumping facility to ensure that the building is not flooded by seepage from external waters.

7.6.3 Wet Flood Proofing

Wet flood proofing includes a range of modifications to the structure, utilities and contents to allow floodwaters to enter the building while ensuring that minimal damage results (Figure 7.11C). For example, the ground floors of certain expensive river-front properties in Koblenz, Germany, consist mainly of utility rooms or simple living rooms which are tiled throughout. Hooks are set into the ceilings to hoist small items above the periodic floods. When the water has receded, the ground floors are hosed clean ready for re-use. However, because of the inevitability of some damage to material possessions in most properties, wet flood proofing is relatively rare today in the MDCs. In high-risk coastal areas, bungalows should be built to withstand immersion without collapse and with floored lofts, loft ladders for access and dormer windows for potential escape. Wet flood proofing also includes relevant landscaping and the use of garden plants native to wetlands.

7.6.4 Adoption of Retrofit Flood Proofing

A survey by Laska (1991) of over 1500 homeowners flooded at three suburban locations in Louisiana, Illinois and Wisconsin in the mid to late 1980s provided important information on the adoption and effectiveness of the retrofit flood proofing of domestic properties. No more than 15% of the residents sampled had implemented retrofitting before ever being flooded, which indicates that flood-prone householders are not receptive to the idea of advance flood proofing. Table 7.1 shows that, of those that took action, the most widely adopted group of measures chosen by half the people was dry flood proofing of the basement, followed by the wet proofing of basements by raising certain items. The focus on basements reflects the fact that work there is less disruptive to the household than alterations to the ground floor. The amount of money spent varied greatly, from US$ 300 to US$ 1200, but most owners spent under US$ 500. The most costly item was a sewer back-up valve followed by wall construction around the house. Expert evaluation, together with homeowners' experience in subsequent floods, indicated that about half of the measures were effective in offering some flood

Table 7.1 Types of flood proofing retrofitted in the USA

Type	Percentage
Wet flood proofing	
Raised furnace/heater/appliances	10
Raised wiring/fuse box	4
Stopped using basement	7
Sub-total	21
Dry flood proofing of a basement	
Glass-bricked windows	4
Protected basement openings	6
Installed sump pump	18
Water-proofed basement walls	11
Added dirt fill next to house	11
Sub-total	50
Prevention of sewer back-up	
Installed back-up valve	8
Installed standpipe or plug	8
Sub-total	16
Protection from surface water	
Built wall around house	2
Built levee or berm	2
Improved drainage next to house	9
Sub-total	13
Total percentage	100
Number of responses	1538

Source: Modified after Laska (1991)

protection, but that installing a standpipe or a plug for basement drains resulted in the best return on expenditure. Conversely, sump-pump installation was ineffectual often because water entered basements too quickly for the capacity of the pump.

Most measures were undertaken immediately after the latest flood, which suggests that this is when official assistance should be available. Those measures that were installed in homes that were subsequently flooded provided 55–70% of the homes with protection. The best results were achieved by employing specialist contractors, which— although more expensive than self-help—suggested that the greater technical expertise would lead to improved effectiveness. Most homeowners obtained their information from persons who had already retrofitted and from specialist contractors. The interest in future retrofitting was especially high among those who had already undertaken some work and found it effective, but the vast majority (90%) of those who wanted to retrofit also wanted the government to protect them. This suggests that, if structural measures are not implemented, this may deter people from implementing their own actions. Certainly, there was a belief that government agencies should play a more active role in developing and publicising the technology of flood proofing and in providing financial incentives in order to motivate householders to help themselves.

CHAPTER 8

Flood Forecasting and Warning

CONTENTS

8.1 Introduction 239
                        RIVER FLOOD FORECASTING
8.2 Hydrological components 241
    8.2.1 Calculation of gross input 241
        8.2.1.1 Monitoring rainfall 242
        8.2.1.2 Monitoring snow cover. 244
    8.2.2 From gross input to time-distributed hydrograph. 245
        8.2.2.1 Unit hydrographs. 246
        8.2.2.2 Hydrological models. 246
8.3 Hydrograph routing 247
8.4 Real-time flood forecasting systems 249
    8.4.1 Channel routing models. 251
    8.4.2 Catchment models 251
        8.4.2.1 Flood forecasting models in the USA 252
        8.4.2.2 Flood forecasting models in the UK 253
        8.4.2.3 Snowmelt and ice forecasts 254
    8.4.3 Integrated systems incorporating meteorological models
        or forecasts 257
        8.4.3.1 River Flow Forecasting System 258
        8.4.3.2 Flood forecasting on large rivers 260
    8.4.4 Forecasting flash floods. 263
                        FLOOD WARNING
8.5 The nature of flood warning 264
8.6 Prior assessment for flood warnings 268
    8.6.1 Forecast accuracy 268
    8.6.2 Forecast lead time. 268
    8.6.3 Threat evaluation 269
8.7 Dissemination of flood warnings 270
    8.7.1 Institutional arrangements 270
    8.7.2 Message content 271
    8.7.3 Message delivery. 272
8.8 Response to flood warnings 274
    8.8.1 Preparedness 274
    8.8.2 Event characteristics 274
    8.8.3 Message recipients 275
8.9 Effectiveness of flood warning schemes 275
    8.9.1 Economic assessment 276
    8.9.2 Public survey assessment 278
    8.9.3 Performance-based assessment 280
                        COASTAL FLOODS
8.10 Introduction 280
8.11 The components of coastal flood forecasts 280
    8.11.1 Waves 281

        8.11.2   Tides 282
        8.11.3   Storms 282
        8.11.4   Earthquakes 282
8.12 Storm-surge forecasting and warning systems 283
        8.12.1   Extra-tropical storms 284
                 8.12.1.1   Storm Tide Warning Service 285
        8.12.2   Tropical storms 287
8.13 Tsunami warning systems 290
        8.13.1   The Pacific tsunami warning system 290
        8.13.2   Other tsunami warning systems 291

8.1 INTRODUCTION

The accurate forecasting of flood conditions in both fluvial and coastal environments is an essential prerequisite for the provision of reliable flood warning schemes. River forecasts are now commonly required over a wide range of discharge and stage conditions, for the purposes of navigation, power generation, pollution control, fisheries and water supply, and are not therefore usually concerned exclusively with flood conditions. However, river flood forecasting has benefited enormously from the attention which has been given to forecasting the complete river hydrograph and its accuracy has increased greatly with improvements in telecommunications and computerised data handling and processing.

Indeed, in the context of both river and coastal floods, the processes of flood forecasting and flood warning are increasingly being brought together in integrated forecasting and warning systems which operate in real time on desk-top computers. The main emphasis in this chapter is on the use of such models, although it is recognised that in many parts of the world the older and less sophisticated techniques of flood forecasting and warning are still in use.

The need for reliability in flood forecasting has been stressed frequently. Errors in the forecast of flood stage or of the time of arrival of flood conditions may lead to under-preparation and avoidable damage (if the forecast stage is too low and/or the forecast timing of inundation is late), or to over-preparation, unnecessary expense and anxiety, and a subsequent loss of credibility (if the forecast stage is too high and/or the forecast timing of inundation is premature). In the past, many forecasts were unreliable because of inadequate understanding of hydrological processes, sparse data collection networks, poor communication systems and slow data-processing.

Regular river and river flood forecasting began, with the advent of the telegraph, in 1854 in France, 1866 in Italy and 1871 in the USA (Hoyt and Langbein, 1955). Here it was initiated by the Weather Service of the Army Signal Service but, even when it was taken over by the US Weather Bureau in 1890, the stream gauging network was still quite sparse. In Britain, even rudimentary river flood forecasting was made difficult by the late development of stream gauging, reflected in the national total of only 27 regular gauging stations as late as 1936. Improvements in river flood forecasting have resulted partly from the global increase in stream gauging stations, especially during and since the International Hydrological Decade, and partly from major and accelerating advances in both the technology of data collection (e.g. weather radar, satellite imagery, electronic instrumentation) and in computer-based data-handling and telecommunication systems. As a result, it is now possible to make quite accurate forecasts of the downstream timing and magnitude of flood peaks in major river basins on the basis of data from rainfall and melt events which have either just occurred or which may still be taking place. Small river basins pose additional problems for flood forecasters because the achievement of an adequate lead time for flood forecasts is ultimately dependent on the accuracy of weather forecasts for periods of more than 24 hours. Unfortunately, in terms of their ability to forecast specific flood-producing rainfall or melt events, weather forecasts are generally accurate only for comparatively short periods of, at the most, one to three days.

Storm surges and tsunamis may be triggered by events occurring many hundreds or even thousands of kilometres from the location of the resulting coastal flooding. Their accurate and timely forecasting therefore relies heavily on sophisticated telecommunication systems, involving GIS, radar and weather satellites. Until comparatively recently the combination of sparse data and inadequate technological development meant that the forecasting of coastal floods lagged behind that of river floods. Long after integrated river flood forecasting and warning systems had been established in countries like the USA, there was still little provision for forecasting coastal floods and few properly established coastal flood warning systems. Flood forecasting deficiencies in the UK were dramatically exposed in 1953 during the severe flooding along the east coast (see Section 5.3.2.1). Despite the fact that floods in the Thames estuary had been forecast for a quarter of a century, there were no forecasts or warnings elsewhere on the east coast. Indeed, according to Grieve (1959), at the inquest on one group of flood victims, the jury expressed their strong feeling that '... the consequences of this disaster might have been avoided if warning had been sent down the east coast'. Belatedly, the Government set up the Waverley Committee which, in turn, recommended the immediate establishment of the Storm Tide Warning Service (see Section 8.12.1.1).

The forecasting of both river and coastal flooding would benefit enormously from improvements in long-range weather forecasting. However, the prospect of accurate long-term weather forecasts for months, or even years, ahead is still remote, although it is one of the aims of the World Meteorological Organisation's developing Global Oceanic Observing System (GOOS). It is hoped that, after 2007, when the system should be fully in place, its detailed monitoring of sea surface temperatures and ocean energy fluxes will permit weather forecasts for many months ahead. Even with present short-term forecasts, the improved processing of within-basin hydrological data and near-coastal data and the development of integrated, computer-based forecasting systems has resulted in greatly improved coastal and river flood forecasts and warnings, even for small river basins.

Further improvements in flood forecast accuracy are also being brought about by the use of satellite imagery to determine flood outlines. At first detailed delineation of flood outlines for given hydrological or coastal conditions, using Landsat imagery, could be used to improve detailed flood forecasting only for subsequent events (e.g. Brown et al., 1987). However, modern imagery such as the synthetic aperture radar (SAR) from the ERS-1 satellite is now processed so rapidly as to be available in near real-time and can contribute directly to the current flood forecasting operation (e.g. NERC, 1993; Wagner, 1994).

In this chapter, consideration of the main forecast components and of the principles underlying the down-channel routing of river flood hydrographs is followed by discussion of selected real-time flood forecasting systems in which technological and computer advances have been combined in powerful, integrated simulation and forecasting models. Examples are chosen to illustrate the problems of river flood forecasting at various scales. Next, consideration is given to some of the particular problems associated with the production and dissemination of flood warnings, whether in fluvial or coastal situations, together with the factors which create an effective loss-reducing response from the community at risk. The final sections of the chapter deal with coastal floods, for which the distinction between forecasting and warning is much less meaningful than it is for river floods.

RIVER FLOOD FORECASTING

8.2 HYDROLOGICAL COMPONENTS

Although the processes which generate river floods are well understood (Chapter 3) their spatial and temporal complexity (Chapter 4), even within small river basins, has meant that it is normally possible to incorporate them into flood forecasting procedures only in a generalised and largely empirical manner. The essential components of the basic flood forecasting methodology are:

1. Calculation of the gross input to the flood-producing system, in the form of either rainfall or meltwater generation.
2. Calculation of the net input to the channel system, i.e. the amount of direct runoff or quickflow resulting from that precipitation or melt.
3. Conversion of the direct runoff volume into a time-distributed hydrograph.
4. Routing procedures to estimate the change in shape of the hydrograph (and therefore changes in peak discharge, peak stage and period of inundation) as it moves to specified downstream locations.

Items (1) to (3) may be considered as the hydrological components of flood forecasting and are considered in Sections 8.2.1 and 8.2.2; item (4) is a matter of hydraulics and is considered in Section 8.3. However, all these items are now commonly incorporated into comprehensive flood forecasting models such as those discussed in Section 8.4.

8.2.1 Calculation of Gross Input

Calculating the gross input of water to the flood-producing system involves the measurement of catchment inputs of rainfall and snowmelt and the accurate estimation of the water equivalent of catchment snow and ice cover. Traditional methods, involving the manual analysis of raingauge and snow-course data, were slow and relatively inaccurate, so that only in large catchments could reliable flood forecasts be developed *before* the flood peak passed a selected location. A typical winter flood peak on the River Rhine, for example, takes only 1.5 days to move the 300 km downstream from Basle in Switzerland to Worms in Germany (Martinec, 1985). However, enormous improvements have resulted from the development and application of modern instrumentation and telecommunications methods. For example, the monitoring of rainfall still relies heavily on data being transmitted to the forecast centre from recording raingauges by means of telephone, radio, or computer links, but it has been significantly strengthened since the mid-1970s by continued improvements in the accuracy and reliability of weather radar. Similarly, in the case of estimates of snow-cover water equivalent, major advances have resulted from the development of space-borne and other remote sensors.

8.2.1.1 Monitoring Rainfall

The measurement of rainfall by radar (Radio Detecting and Ranging) has been well-established since the mid-1970s (e.g. Browning et al., 1977) and radar-generated images of the spatial extent and propagation of storms have become an accepted part of daily weather reports and forecasts on television. However, the increasingly successful incorporation of radar into flood forecasting has been possible only as a result of gradual improvements in the spatial and temporal resolution of radar imagery, even in mountainous terrain (e.g. Andrieu et al., 1989; Fatorelli et al., 1995), and in the accuracy with which a corresponding rainfall total can be determined. System improvements result from continuing research in these areas (e.g. Cluckie and Collier, 1991; Giuli et al., 1994; Chandler, 1995; Tilford, 1995; Roberts, 1996), including the development of dual wavelength, doppler and dual polarisation radar systems. The weather radar network for Europe in the early 1990s is shown in Figure 8.1 and planned improvements were set out in Newsome (1992). Similar networks have been developed elsewhere.

FRONTIERS (Forecasting Rain Optimised using New Techniques of Interactively Enhanced Radar and Satellite) is one of two radar rainfall forecasting systems in operational use in the UK. It was developed as a semi-automated forecasting tool, operating at national level, which combined the use of digital radar data from the UK network with Meteosat weather satellite data to produce forecasts at 30-minute intervals for up to six hours ahead (Moore, 1995). The advantages in flood forecasting of mesoscale monitoring of weather conditions, using radar and satellite measurements, is also being explored via **SHARP** in Canada, **PROFS** in the USA and **PROMIS 600** in Sweden (Takács, 1989). A logical development would then be to use the output from such systems as the input to rainfall–runoff models to generate flood hydrographs which can then be routed to downstream locations. Initial studies using FRONTIERS yielded disappointing flood forecasts when radar data were used as an input to traditional models, such as the synthetic unit hydrograph or a conceptual catchment model, but more encouraging results were obtained when inputs were made to models which did not use a mass balance approach and which had a real-time updating facility (Tilford and Cluckie, 1990).

A complementary development in the UK was HYRAD (Hydrological Radar) which was designed to use data from a single radar for use in flood forecasting and warning at regional level. HYRAD corrects automatically for anomalies in the radar image, calibrates the radar using raingauge data, constructs rainfall forecasts and calculates catchment average rainfall for input to the River Flow Forecasting System described in Section 8.4.3.1. Advantages of HYRAD for real-time flood forecasting are that the calibrated and forecast images are of high resolution, are available within one minute of receipt of the incoming radar data, compared with 40 minutes for FRONTIERS, and

FLOOD FORECASTING AND WARNING 243

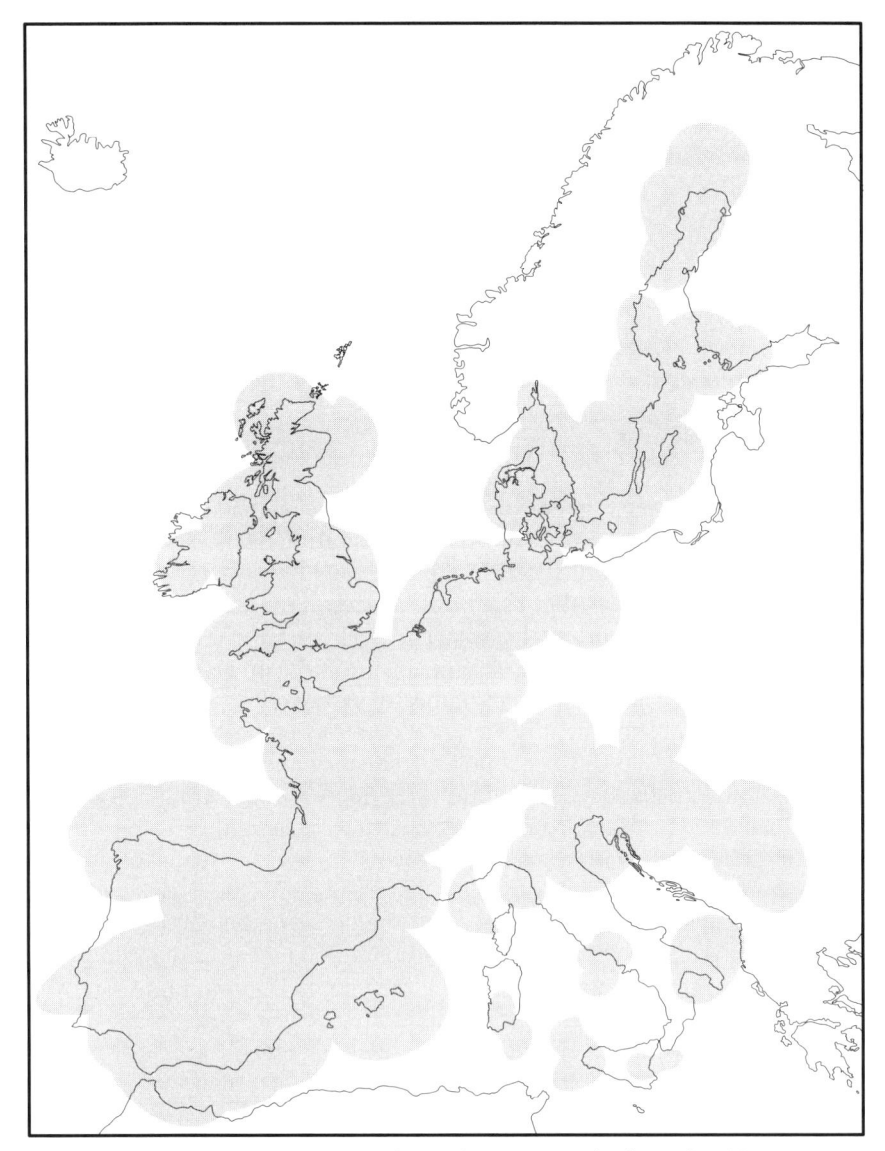

Figure 8.1 European weather radar coverage in the early 1990s
Source: Based on a diagram in Collier (1992). Crown copyright, reproduced with the permission of the Controller of HMSO

are generally more accurate. However, the dependence of HYRAD on a single radar restricts forecast lead times to about two hours, compared with as much as six hours for FRONTIERS forecasts based on a composite radar field (Austin and Moore, 1996; Moore et al., 1994).

A fully automated successor to FRONTIERS became operational in 1995. NIMROD, named for the biblical hunter figure, produces quality controlled rainfall reports and short-period forecasts. Combining information from radar and weather satellites with data from the UK Meteorological Office numerical weather prediction models, NIMROD generates detailed maps of rainfall distribution in the UK for up to six hours ahead and is able to distinguish rain from snow, even in conditions of freezing rain (Anon., 1996a).

Still further improvements will undoubtedly result from the opportunities afforded by radar to improve the robustness of weather prediction models and rainfall–runoff models. In the UK, for example, numerical weather prediction models should benefit from improved radar data on, say, moisture content fields and wind fields, especially in mesoscale events (Hardaker, 1996). Also, the inherent compatibility between the grid square characteristics of radar data and grid-based distributed models of river catchments, such as the IH Grid Model (see Section 8.4.3.1), has led to improved river flow forecasts in trial investigations (e.g. Hatton, 1994; Moore and Bell, 1996). Becchi et al. (1994) described similar developmental work in Italy.

8.2.1.2 Monitoring Snow Cover

In some respects snowmelt floods are easier to forecast than rainfall floods. For example, melting is heavily dependent on air temperature whose short-term variations are more predictable than are those of rainfall, and the often long delay between snowfall and snowmelt means that there can be several months in which to assess the volume and water equivalent of the snowpack. On the other hand, environmental conditions in areas of snow accumulation, particularly in mountains, make access difficult and this has meant that programmes of data collection which depend upon manual observations of snow depth and quality have often been very restricted in their scope and coverage, so that the distribution, frequency and accuracy of measurements has been far from satisfactory. In addition, the high spatial variability of snowpack thickness, quality and water content means that it is difficult to sample these characteristics adequately with a necessarily sparse 'manual' instrument network.

As with rainfall floods, technological improvements in instrumentation and communications have been responsible for major advances in the accuracy and timing of flood forecasts in recent years. Traditional weighing precipitation gauges have long been supplemented or superseded in the calculation of gross input. Snow pillows, for example, register the accumulated weight of snow as a pressure change, and radioactive probes give a direct measure of water content in the snowpack profile. Neither, however, addresses effectively the problems of replication and network density. In this respect the greatest contribution has been made by remotely sensed data based on NOAA/AVHRR or later, higher resolution imagery.

Remote sensing of snow area has been found to improve the performance of snowmelt runoff models such as that for the Rio Grande basin, Colorado (Rango et al., 1990), and has long been carried out operationally in regions of substantial seasonal snow cover, especially mountain regions in which there is a well-defined relationship between snow area and altitude (e.g. Rocky Mountains, European Alps). In major mountain areas information on snow area is much more useful when used in conjunction with GIS and digital terrain modelling (DTM), as in the Himalayan Snowcover Information System (HIMSIS) proposed by Ramamoorthi and Haefner (1991). But areal extent, although a useful indicator, is a less valuable input to snowmelt flood forecasting and warning systems than the snowpack water equivalent which is a function of snowpack quality and depth.

Kuittinen (1989) described a method to estimate water equivalents during snowmelt in lowland areas of Finland using NOAA/AVHRR images, airborne gamma-ray spectrometry, and field measurements (see Figure 8.2). A relationship was established between snow-free area and the water equivalent of the remaining snow-covered areas and satellite imagery was then used on wholly or largely cloudless days (about 50%) to measure snow-free area. The snow water equivalent was also inferred from the ratio between gamma emission from bare ground and emission from snow-covered ground and was checked periodically by field measurement.

Even in areas of temporary and discontinuous snow cover, such as the UK, NOAA/AVHRR imagery may facilitate an improved characterisation of the snowpack during the melting phase (e.g. Lucas et al., 1989).

8.2.2 From Gross Input to Time-Distributed Hydrograph

The derivation of a flow hydrograph from information on the input of rainfall and snowmelt to a river catchment continues to pose one of the greatest challenges in hydrology and to occupy a major part of hydrological textbooks (e.g. Bras, 1990; Wanielista, 1990; Ward and Robinson, 1990). Even assuming that there is adequate information about the physical properties of the catchment (e.g. soils, geology,

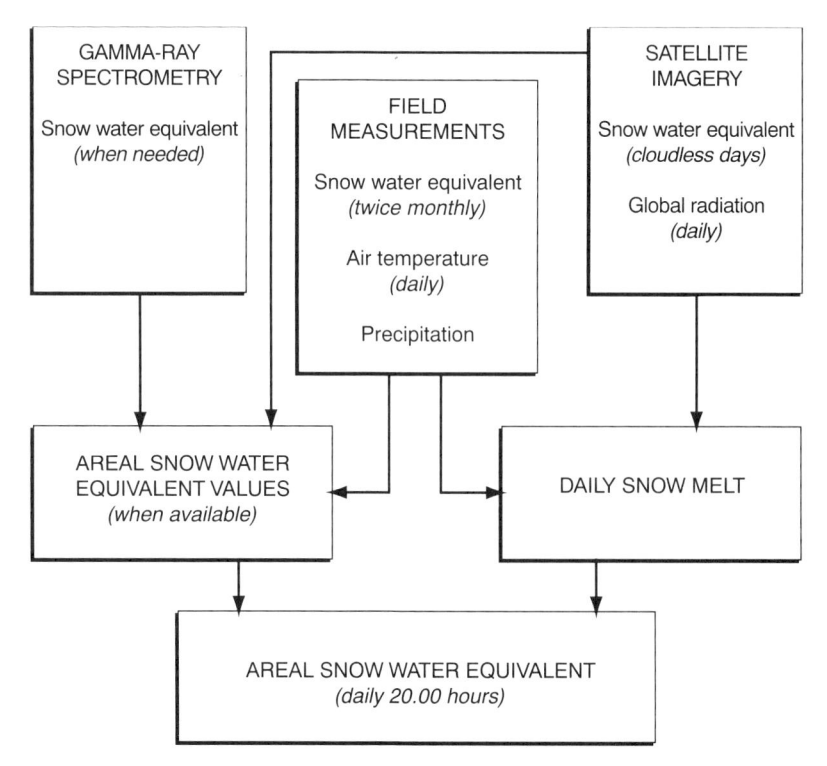

Figure 8.2 Diagram illustrating the daily determination of snowpack water equivalent values during the period of snowmelt
Source: Based on a diagram in Kuittinen (1989) by permission of IAHS Press

vegetation, topography) and about initial hydrological conditions at the onset of rainfall or melt (e.g. soil water content, groundwater levels), there remains the major hurdle of calculating 'losses' (infiltration, evaporation, interception, surface detention) in order to derive a value of 'effective' or 'excess' or 'net' rainfall. Although evaporation estimates now form part of the national weather recording and forecasting systems in some countries, for example MORECS in the UK (Meteorological Office Rainfall and Evaporation Calculation System), they are not readily available in many parts of the world.

8.2.2.1 Unit Hydrographs

Where adequate flow data are available a characteristic hydrograph response to a known input of rainfall may be determined. This approach formed the basis of the unit hydrograph method which developed from the work of Sherman (1932) and which identifies the characteristic hydrograph resulting from a given unit of precipitation, originally one inch, now usually one centimetre, falling during a specified time interval. Frequently, however, precipitation data are good but flow data are either inadequate or are not yet available, as for example where there is a need to predict the hydrograph likely to result from a proposed future change of land use. As a result there have developed a number of techniques for generating synthetic unit hydrographs, in which the shape of the hydrograph resulting from a specified unit of precipitation is related to catchment and precipitation conditions. In such cases the hydrograph is normally reduced to a simple triangle in which the key dimensions are its height, i.e. the peak discharge (Q_{max}), its time base (Tb), i.e. the duration of the flood, and its time to peak (T_p). In the UK *Flood Studies Report* (NERC, 1975), for example, the time to peak of a one-hour synthetic unit hydrograph was given as

$$T_p = 46.6(MSL)^{0.14}(S1085)^{-0.38}(1 + URBAN)^{-1.99}(RSMD)^{-0.4} \tag{8.1}$$

where MSL is the main stream length (km), $S1085$ is the slope (m km^{-1}) between two points at distance of $0.1MSL$ and $0.85MSL$ respectively from the catchment outfall, $URBAN$ is the fraction of the catchment in urban development, and $RSMD$ is the one-day M5 rainfall, less effective mean soil moisture deficit. The peak of the unit hydrograph in m^{3}s^{-1} per 100 km^2 is estimated from equation 6.5 and its time base is estimated from equation 6.6 (see Section 6.2.1.2).

The method was described in detail in NERC (1975) and simplified step-by-step calculations were given by Wilson (1983). A number of synthetic unit hydrograph techniques developed in the USA (e.g. SCS method, Contributing Area method) were described by Wanielista (1990).

8.2.2.2 Hydrological Models

Essentially, the unit hydrograph method presupposes a lumped catchment response, i.e. the whole catchment is assumed to be responding as a point, with a characteristic linear response function. However, catchments behave in neither a lumped nor a linear manner and so, increasingly, attention has been devoted to developing more realistic hydrological models of catchment flood response. The more sophisticated of these, which will probably incorporate radar rainfall data, remotely sensed data for soil moisture and evaporation flux, and digital terrain modelling, will be compatible with standard GIS systems and will be supportable by the computing power of desk-top machines. Ideally, they will also incorporate hydrograph routing processes and flood warning procedures. Some examples of recent developments of this sort are discussed in Section 8.4. Examples of simpler models in which flood hydrographs are generated from rainfall data or from catchment characteristics were described respectively by Ponte Ramirez and Shaw (1983) and Bergmann et al. (1990).

8.3 HYDROGRAPH ROUTING

Whether derived from rainfall or snowmelt and whether measured directly or estimated using the unit hydrograph or some other technique, flood hydrographs are usually available only for a limited number of specific channel locations within the drainage basin. As a hydrograph moves downstream it behaves like a wave and will attenuate, i.e. the time base of the hydrograph increases and the peak flow decreases, and the quantitative process which describes this attenuation is known as hydrograph routing or, in the case of a flood hydrograph, *flood routing*. Flood routing thus provides the means of forecasting the magnitude and arrival time of a peak flood discharge based on known measured or estimated hydrograph conditions upstream. It is particularly important in large basins where the channel flow times and distances may be large compared with those of hillslope flow moving towards the channels. The procedure is not confined to river channels but can be applied as well to reservoirs, lakes and swamps.

The movement of a flood wave in a stream channel is a complex process of non-steady and usually non-uniform flow (Prasuhn, 1987). Not only do flow velocity and depth vary with time as the wave progresses downstream, but channel properties and lateral inflow and outflow may also vary. In the simple case of a channel reach with no tributaries, no gain of water from direct runoff or groundwater flow, and no loss by evaporation, the effects of storage both within the channel and in and on adjacent floodplain areas are sufficient to dampen the flood wave as it moves down the reach, thereby attenuating the peak and extending the time base. A simple system of this type is illustrated in Figure 8.3 (inset) and a typical example, the Savannah River in Georgia, is shown in the main diagram. At first water enters the channel reach faster than it exits at the downstream end, the excess water being retained in the channel and floodplain storages thereby raising water levels within the reach. Later, the inflow into the reach drops below the rate of flow at the downstream end and the excess water comes out of storage.

There are theoretical, empirical and statistical approaches to flood routing. The theoretical approach involves either convection/diffusion equations or, more usually, the St Venant equations to model the one-dimensional bulk flow of water in a river channel. The St Venant equations are two partial differential equations of motion and continuity which can be solved by simplification and numerical integration with small steps of distance and time, although there are problems in determining the coefficients of the differential equations and in establishing the initial and boundary conditions (Miljukov, 1972). The application of first- and higher-order approximations of the St Venant equations was described by Ambrus et al. (1989) and a finite-difference solution was presented by Chau (1990).

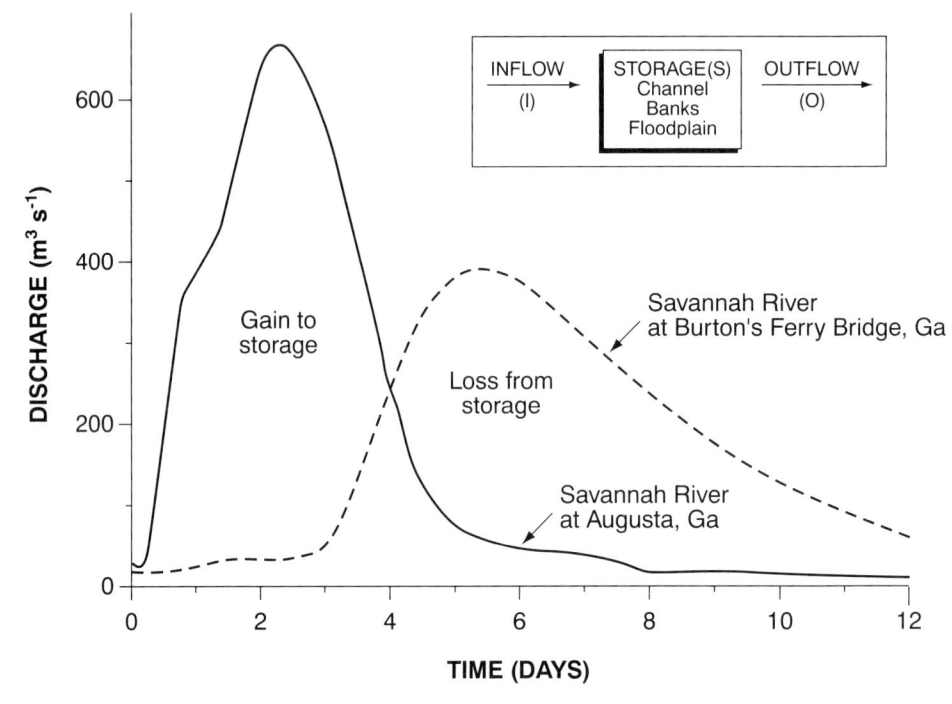

Figure 8.3 Attenuation of a flood wave on a reach of the Savannah River, Georgia, USA. Inset: Simplified systems diagram of a channel reach
Source: Fox (1965), reproduced with permission

Flood routing has been more commonly carried out using simplified empirical procedures which recognise that for any channel reach the difference between inflow and outflow is equal to the stored or depleted water in a given time interval. As a flood wave passes through a channel reach (Figure 8.4) outflow reflects only the storage beneath a line parallel to the streambed, the prism storage. Between this parallel line and the water profile is the wedge storage, which represents the excess of inflow over outflow during the advance of the flood wave, and the negative wedge storage which represents the excess of outflow over inflow during the recession of the flood wave. The Muskingum method, a widely used example of such a procedure, was developed by G.T. McCarthy and others in connection with a US Army Corps of Engineers flood control project in the Muskingum Conservancy District. McCarthy (1938) proposed that storage, S, should be expressed as a function of both inflow and outflow

$$S = K[xI + (1 - x)O] \qquad (8.2)$$

where I and O are inflow to and outflow from a given channel reach, x is a dimensionless constant for the reach and K is a storage constant with dimension of time. Outflow from the reach can be obtained by solving simultaneously equation 8.2 and the water balance for the reach (equation 8.3)

$$\Delta S = I - O \qquad (8.3)$$

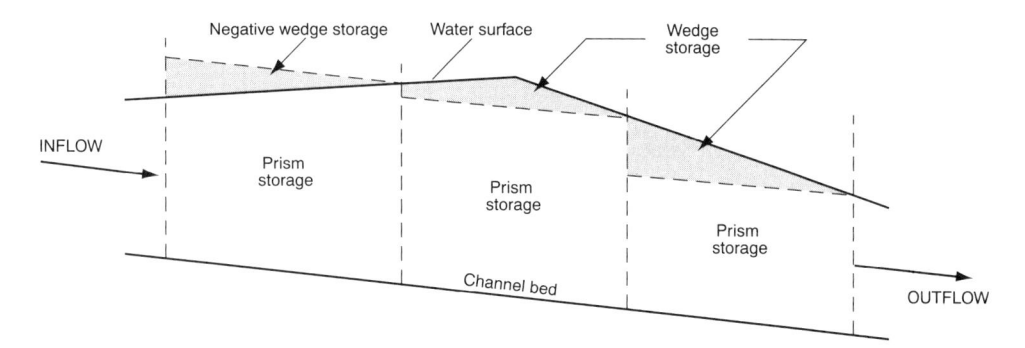

Figure 8.4 Flood wave profile illustrating wedge and prism storage

for a discrete time interval so that

$$O_2 = c_0 I_1 + c_1 I_2 + c_2 O_1 \tag{8.4}$$

in which O_1 and O_2 and I_1 and I_2 are respectively the outflow and inflow at the beginning and end of the routing period, and where the values of the coefficients c_0, c_1 and c_2 are found from the empirically determined values of x and K, and the value of the calculated time interval Δt. The necessary empirical determinations and examples of their application are described in the appropriate literature (e.g. Wilson, 1983; Prasuhn, 1987; Perumal, 1993) and some of the weaknesses of the Muskingum method were discussed by Koussis (1983; 1986) and Gill (1986). The problems of applying the method during periods when the flood wave is moving overbank down the floodplain were illuminated by empirical analysis of flood peak discharges in the reach of the River Tees between Broken Scar and Low Moor in north-east England. Archer (1989) showed that attenuation is highly variable over the range of discharge conditions experienced. Floodplain storage, and associated delays due to frictional resistance, suppress downstream flood growth at discharges above bankfull but channel storage effects at discharges below bankfull mean that as flood peak and volume increase, attenuation decreases and downstream flood growth becomes steeper. As a result, bankfull discharge marks a break of slope on the flood frequency curve.

Continued enhancement of computing power has facilitated widespread development in the conceptual modelling of flood routing processes (see Section 8.4). In addition, traditional routing methods in streamflow forecasting have been supplemented in recent years by an increased use of statistical estimation techniques, such as Kalman filtering (e.g. Ngan and Russell, 1986; Amirthanathan, 1989; Wilke and Barth, 1991).

8.4 REAL-TIME FLOOD FORECASTING SYSTEMS

The successful development of computer models of some of the principal hydrological and hydraulic components of river flood production is noted in Sections 8.2.2.2 and 8.3. Flood forecasting models, even in large catchments where the lead time is likely to be substantial, are implicitly of most value when they operate in real time. Ideally, therefore, such models should be capable not only of using the current output of continuous monitoring systems (e.g. for precipitation, temperature, streamflow, etc.)

but also of continuously updating model inputs as the hydrological and hydraulic situation changes during and following a flood-producing event. Even where such systems exist, they are still far from perfect and in many countries it is in this area that significant research and development activity is being concentrated. It is likely that, eventually, real-time flood forecasting systems will also incorporate, as a matter of course, systems for effective flood warning and control. Some already do; most do not. Accordingly, this section concentrates on systems for flood forecasting. Flood warning is discussed separately in Sections 8.5 to 8.9.

Types of real-time flood forecasting systems were identified by Feldman (1994) and their general characteristics are shown in Table 8.1. From this it is clear that each type of system has its advantages and disadvantages. For example, although the simplest are often the cheapest to set up and operate and tend to generate the most reliable forecasts, their lead time is usually the shortest of all. Since the length of lead time has the greatest influence on the potential success of a flood warning, in terms of lives saved or damages prevented, it is likely to be the more sophisticated systems with complex data inputs to which most attention will be paid in the future. However, lead time is not only a function of the amount and complexity of data used, since even simple systems can have a substantial lead time if the forecast target location is sufficiently far downstream from the flood-producing event. The categories shown in Table 8.1, which provide the framework for the ensuing discussion of flood forecasting procedures, should not be regarded as mutually exclusive. Simple channel routing models (categories 1 and 2), for example, may either stand alone or form an input to a forecasting system driven by a

Table 8.1 Types of real-time flood forecasting systems

Category	System description	Forecast lead time	Sophistication of forecast procedures	Expense	Potential uncertainty
1	Real-time measurement of STREAMFLOW with routing/correlation to key locations	Shortest	Simplest	Least	Least
2	Same as 1 but including PRECIPITATION in correlation to key locations	Short	Simple	Little	Little
3	Real-time measurement of Precipitation, Streamflow, Temperature, etc. plus a CATCHMENT RUNOFF MODEL	Long	More complex to very complex	More to very	More
4	Same as 3 but including a METEOROLOGICAL MODEL	Longer	Very complex	Very	More
5	Same as 4 but including WEATHER FORECASTS (precipitation, temperature, etc.)	Longest	Most complex	Most	Most

Source: Modified from a table in Feldman (1994)

catchment model (category 3). Similarly, the flexibility of integrated systems (categories 4 and 5) for flood forecasting, flood warning and river control is largely dependent on advances in software design. This enables such systems to accommodate the inputs not only from radar and satellite imagery, and therefore much better data on the spatial variability of the processes controlling flood runoff, but also from a wide variety of digital terrain models, and catchment and routing models through a generic model algorithm interface (Moore, 1993). It is in this area that radar and satellite data have enabled the greatest advances.

8.4.1 Channel Routing Models

Channel routing models (see categories 1 and 2 in Table 8.1) either rely entirely on streamflow data or use a combination of streamflow and precipitation data. Sometimes the flows measured at upstream locations are simply correlated to downstream forecast locations; more usually they are input to a channel routing model based on kinematic wave theory.

Feldman (1994) described the simple methodology employed in the HEC FORCST model (HEC, 1975) which essentially keeps track of water already in the river system and does not make use of precipitation data. Moore (1993) described a four-parameter non-linear kinematic wave model developed for the Yorkshire Ouse river in Northern England. This model assumes that the speed of the flood wave is dependent on discharge and it allows for local overflow from the channel onto the floodplain to be either routed back into the channel farther downstream, or lost to the channel system by evaporation or flow into old mine workings. An alternative channel flow routing model, used in the Severn–Trent river system in central England, allows for variable flood-wave speed and out-of-bank flows (Douglas and Dobson, 1987).

Flood routing in tidal channels is a more complex exercise. According to Moore (1993), routing in the tidal reaches of the Yorkshire Ouse is accomplished using the HYDRO algorithm which is based on the US National Weather Service DWOPER/NETWORK program (Fread, 1985). This uses a four-point implicit solution of the St Venant equations (Section 8.3) and has been adapted for real-time use. It has also been applied to the tidal Shenzhen River in Hong Kong.

Again, in rivers where flood routing is controlled more by downstream levels than by upstream flows, as in the Yamuna River in India, river-specific routing models have been developed (e.g. Rangachari et al., 1989).

8.4.2 Catchment Models

A large proportion of flood forecasting relies upon the use of conceptual catchment models (category 3 in Table 8.1) which have either been developed initially or adapted subsequently to operate in a real-time mode. Much of this activity is still at a developmental stage and even small catchments which are hydrologically complex, e.g. 'hybrid' catchments having a mix of urban and rural areas, may severely challenge models which work well in hydrologically homogeneous catchments (e.g. IH, 1993a). Of the many models that have been developed for river catchments throughout the world, only a comparatively limited number have been found to have a general or quasi-general applicability in flood forecasting. A small selection of these, including some early examples that have made a continuing contribution to flood forecasting over a long period of time, are described briefly in this section.

8.4.2.1 Flood Forecasting Models in the USA

HEC-1 (see Figure 8.5) is a flood hydrograph package developed by the US Army Corps of Engineers which enables the flood forecaster to develop unit hydrographs using various methods, calculate loss rates and route hydrographs, again using a variety of methods including Muskingum–Cunge and kinematic wave approaches. Best-fit unit hydrograph, loss rate, snowmelt, base freezing temperatures, and routing coefficients are derived automatically from data on precipitation, snowpack characteristics, temperature and runoff (Beard, 1971).

HEC-1F is an adaptation of HEC-1 for real-time operation and employs unit hydrographs and routing techniques to simulate runoff from a subdivided basin having both gauged and ungauged sub-catchments. Generally, the model is used in two stages, first to estimate parameters and calculate discharge hydrographs for gauged sub-catchments and, second, to calculate discharge hydrographs for remaining sub-catchments and to route and combine hydrographs throughout the basin. In the second step calculated and observed hydrographs can be blended, enabling use of observed data wherever it is available (Peters and Ely, 1985). The model does not approach the flood forecast with a knowledge of catchment conditions from previous storms, nor

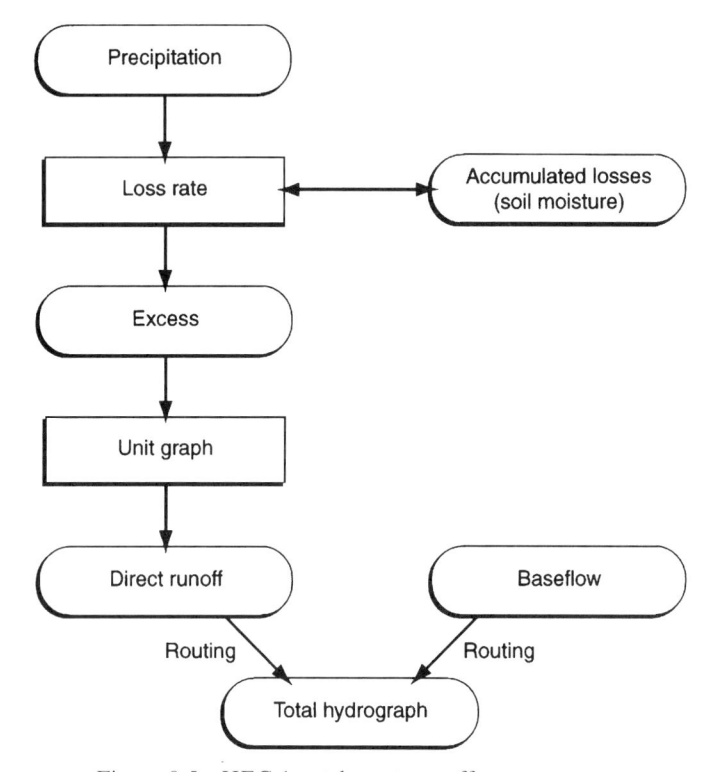

Figure 8.5 HEC-1 catchment runoff processes
Source: Based on a diagram in Feldman (1994) with kind permission from Kluwer Academic Publishers

FLOOD FORECASTING AND WARNING 253

does it have a continuous soil moisture accounting capability. Instead it relies on precipitation and streamflow data collected during the storm to adjust the infiltration parameters as the storm and flood progress (Feldman, 1994).

SSARR (Streamflow Synthesis and Reservoir Regulation) was originated in the mid-1950s by the US Army Corps of Engineers and has since been extensively developed and improved (e.g. Schermerhorn and Kuehl, 1969; US Army Corps of Engineers, 1972). Unlike HEC-1F, SSARR maintains a continuous accounting of catchment moisture by means of a simple soil moisture index, SMI (see Figure 8.6). The SSARR model was designed originally for use in planning and daily operational forecasting during the snowmelt flood period on the Columbia River in the USA Pacific North-West but was also used successfully on the Mekong River (e.g. Rockwood, 1969).

8.4.2.2 *Flood Forecasting Models in the UK*

Several effective catchment models have been devised for flood forecasting in the UK. For example, FLOUT, developed at the Hydraulics Research Station (Price, 1978) computes the flood hydrograph and routes it downstream using either a diffusion process approach or the Muskingum method as modified by Cunge (1969).

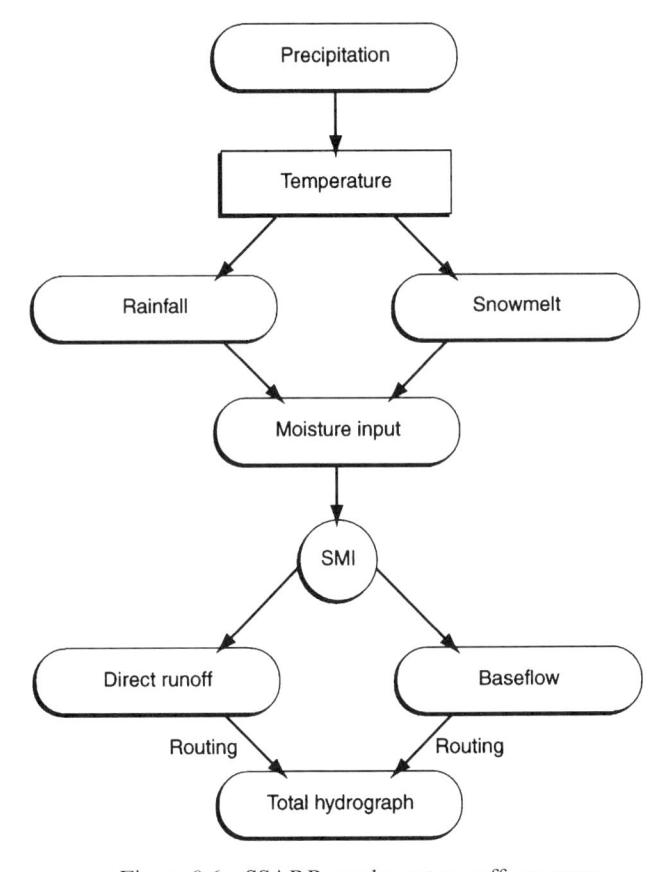

Figure 8.6 SSARR catchment runoff processes

Operationally, however, there are three main rainfall–runoff models currently in use for real-time forecasting in the UK (Moore, 1993): the Institute of Hydrology Conceptual Model, the Thames Conceptual Model, and the Probability Distributed Moisture model.

The *Institute of Hydrology Conceptual Model* (IHCM) was originated by Dickinson and Douglas (1972) and was adapted for the Severn and Trent River basins by Bailey and Dobson (1981). Subsequent modifications were described by Dobson and Cross (1994). Two linked models are used, one for generating flood hydrographs for each of the tributary catchments and the other for routing these flows down the main river network. In the catchment model, a number of storages control the simulation of runoff from rainfall (see Figure 8.7). These storages are modified dynamically, using an hourly time-step, reflecting inputs and outflows. Water is lost to evaporation from the surface storages, and the groundwater store is replenished by excess water from a saturated soil store and is depleted by baseflow. Soil moisture is represented by a soil moisture index (SMI), based on Meteorological Office MORECS data (see Section 8.2.2), and controls the proportion of rainfall which enters the stream channels rapidly as runoff. Runoff is then lagged and cascaded to produce a smoothed hydrograph for each tributary area which is then routed down the main river to the forecast location using a variable parameter Muskingum–Cunge procedure.

The *Thames Conceptual Model* (TCM) operates in a different way in order to accommodate the varied hydrology of a mixed urban and rural river basin having significant aquifers and artificial influences. The model was described by Moore (1993) as being based on a subdivision of the basin, not into discrete tributary catchments, but into different response zones representing, for example, contrasting geological units, aquifers, riparian and paved areas, and sewage effluent sources. Within each response zone soil water movement is controlled by the assumptions of the Penman root constant model. Runoff outflows from each zone are combined and if necessary routed through a kinematic wave channel flow routing model.

The third UK flood forecasting model, the *Probability Distributed Moisture* (PDM) model (Moore, 1985; 1986) has been widely used in northern and eastern England and in Scotland, as well as in Hong Kong. The model, which forms an integral part of the River Flow Forecasting System discussed in Section 8.4.3.1, was designed for flood forecasting at the basin scale, uses a short time-step (e.g. 15 minutes) and allows real-time measurements of flow to be incorporated in order to improve flood forecasts. A probability distributed soil moisture storage is used to partition rainfall into direct runoff and groundwater recharge (Figure 8.8). Direct runoff is routed through a surface store before contributing to basin runoff and the remaining rainfall enters soil water storage where it is depleted by evaporation. Soil water storage is also depleted by recharge which contributes water to basin runoff via the groundwater storage.

8.4.2.3 Snowmelt and Ice Forecasts

Improvements to the way in which snow cover is monitored, including the use of satellite imagery for determining snow-cover areal extent (see Section 8.2.1.2), have greatly improved the potential for successful real-time forecasting of floods resulting from snowmelt. Rango et al. (1990) showed how satellite imagery had been used to effect the transition from snowmelt runoff simulation to operational flood forecasts

FLOOD FORECASTING AND WARNING

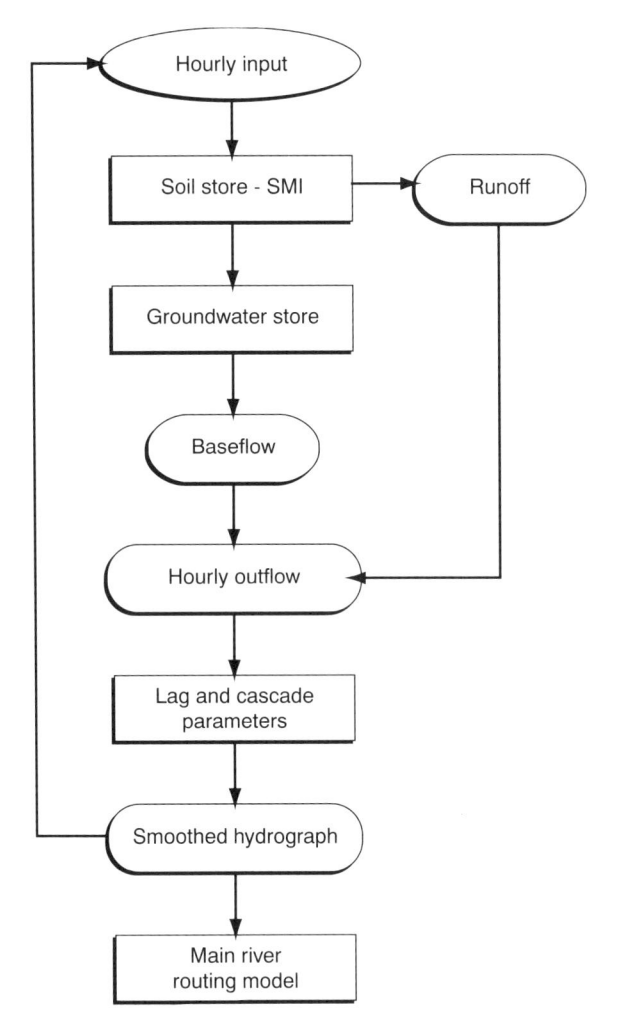

Figure 8.7 Principal storages and time lags in the Severn–Trent tributary catchment flood forecasting model
Source: Based on a diagram in Dobson and Cross (1994) by permission of HR Wallingford Ltd

in real time for the Rio Grande basin in the Rocky Mountains of Colorado, USA. A key element of the forecasting model was a 'family' of snow-cover depletion curves (Rango and van Katwijk, 1990) covering a range of areal snow water-equivalent values at the beginning of the melt season in each of three elevation zones (e.g. Figure 8.9). Satellite measurements of snow cover were used to select the appropriate snow-cover depletion curve for the melt season in question. As melting proceeded it was found that forecasts were improved when runoff estimates were updated with actual observations every seven days and when forecast, rather than observed, temperatures were used.

In many situations snowmelt contributes to floods in snow-covered areas, rather than being their sole cause, and accordingly snowmelt models are now increasingly

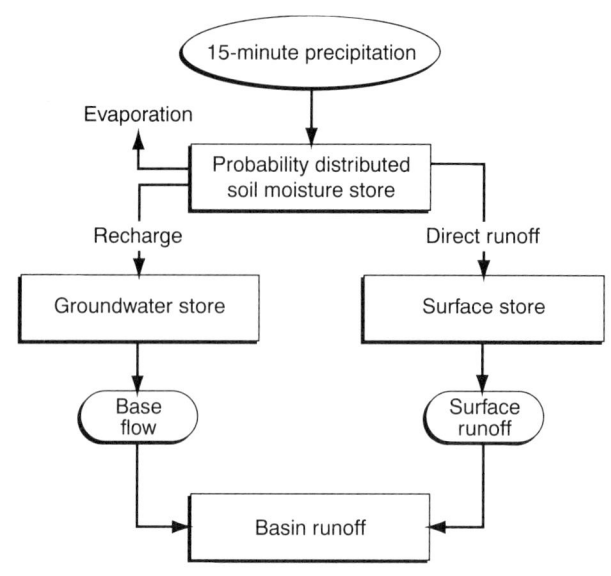

Figure 8.8 Probability Distributed Moisture (PDM) model: main components
Source: Based on a diagram in Moore (1993)

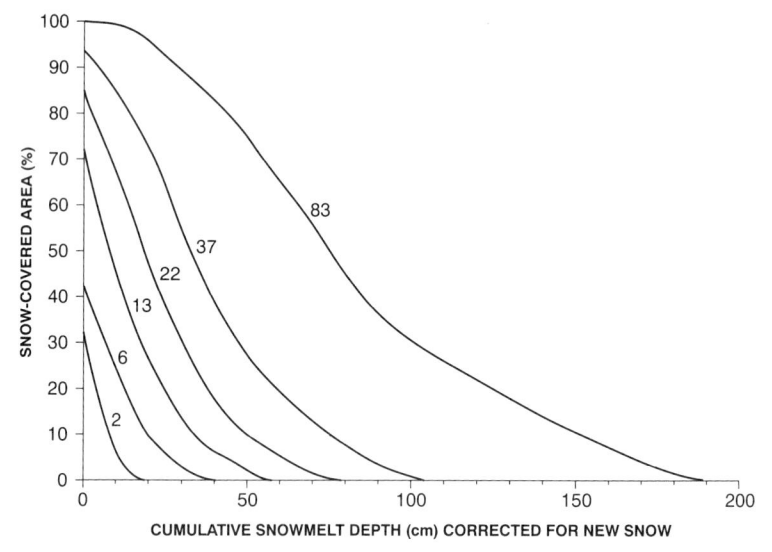

Figure 8.9 Nomograph of modified snow-cover depletion curves indicating the estimated areal average water equivalent of snow (cm) on 1 April in elevation zone B (2925–3353 m a.s.l.) of the Rio Grande basin above Del Norte, Colorado, USA
Source: Based on a diagram in Rango et al. (1990) by permission of IAHS Press

incorporated as one of the process elements in integrated flood forecasting or flood forecasting and warning procedures. Moore (1993) described the PACK snowmelt model which is incorporated into the UK River Flow and Forecasting System. The PACK model (Figure 8.10) subdivides the snowpack into dry and wet stores and a temperature excess melt equation controls the movement of water between them. An areal depletion curve allows for a shallow snowpack to cover only a part of the catchment area. The dry store is added to by snowfall and depleted by melt. The wet store receives water as rainfall and as melt from the dry store and loses water as drainage. Drainage and direct rainfall form an input to either a lumped or a distributed basin runoff model configured to receive snowmelt inputs from elevation zones within the catchment and with a PACK model operating independently within each zone. In the real-time application of these models, measurements of snow depth or water equivalent are continuously assimilated to update and improve forecast performance (Moore et al., 1996).

The break-up and melt of ice on high-altitude or high-latitude water bodies tends to trigger flooding in ways which are largely site-specific. Procedures for forecasting floods in these conditions, although well-developed in areas such as Canada, the USA, Scandinavia and the USSR, are rarely applicable generally. Examples were provided by Zhang and Xie (1991), who showed that forecasting precision was much improved by the use of satellite imagery to monitor the thawing regime of two lakes on the upper Yellow River, China, and by Konovalov (1990) who forecast flood outbursts from Mertzbacher lake on the Inylchek River, USSR, on the basis of air temperature, cloudiness and precipitation data.

8.4.3 Integrated Systems Incorporating Meteorological Models or Forecasts

Most of the catchment models discussed in Sections 8.4.2.1 and 8.4.2.2, although differing in detail, are constructed from essentially similar modules. The same conclusion could be drawn from a much wider selection of examples and indeed Moore (1993) suggested that most 'brand name' catchment models are effectively variations on a common set of components. Recognition of the potential for

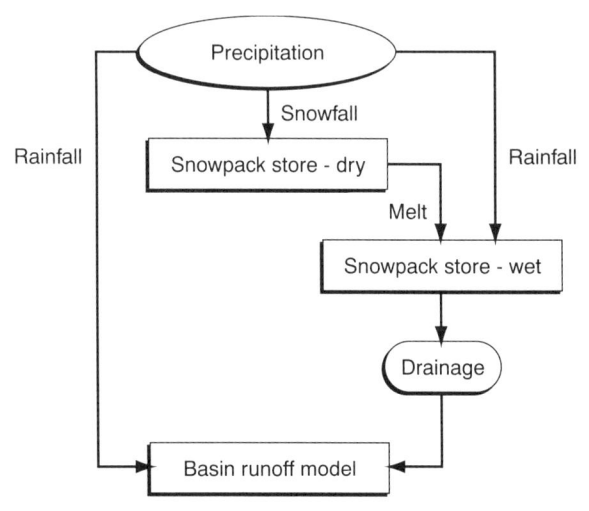

Figure 8.10 The PACK snowmelt model: main components
Source: Based on a diagram in Moore (1993)

interchanging or substituting component modules has been a major factor in the development of more complex and flexible systems for real-time flood forecasting in which the selection of appropriate modules is controlled within a 'shell' or framework. In such systems several rainfall–runoff modules, for example, may be available so that the appropriate one could be chosen for generating the flood hydrograph on each sub-catchment. In the same way the shell might also contain several optional snowmelt or routing modules.

However, another important factor has contributed to the growing interest in developing generic operating shells or frameworks. The progress of technology in weather data collection and processing has been so rapid and so successful that flood forecasting systems for the future must be able to accommodate the fact that an initial alert of possible flooding may be triggered by a wide range of sources. These may include rainfall data, e.g. from raingauges or from weather radar, and rainfall forecasts, e.g. from weather radar, satellite imagery, synoptic analyses, or mesoscale atmospheric models. And as Moore (1993) observed, the detailed information needed for flood warnings to specific localities necessitates

> hydrological and hydrodynamic models of river basins capable of forecasting the rising flood hydrograph at possibly many locations within a river basin, or across a region comprising of several river networks.

To handle the diversity of data inputs, the complexity of flood forecasting and warning requirements, and their application to both simple and complex catchments and river networks, demands the development of fully integrated, automated and transferable forecasting systems (categories 4 and 5 in Table 8.1).

8.4.3.1 River Flow Forecasting System

A good example of such a system, which can be used to illustrate the key issues and development problems, is the UK River Flow Forecasting System (RFFS) described in some detail by Moore (1993) and Moore et al. (1994). The system (see Figure 8.11) can be thought of as comprising a 'kernel' of software for operational forecasting in real time and for off-line model calibration, and a 'shell' environment incorporating the database and the various input and output interfaces. RFFS can accommodate a wide range of models and the Information Control Algorithm (ICA) allows new model formulations to be added, as requirements change or as model improvements are achieved, as well as new telemetered river gauging sites. In addition, not only can the system be configured to any river network without expensive recoding but it is also possible to isolate selected parts of a complex river network in order, for example, to produce flash flood forecasts in headwater catchments without the need to model the tidal lower reaches of the main river system.

The system is not only very flexible; it is also resilient to data loss. For example, at sites within the river model network missing data may simply be replaced by forecasts made by model components upstream, while at sites on the extremities of the network the system relies on the principle of user-defined substitution. Moore et al. (1994) illustrated this in relation to the rainfall data needed for a rainfall–runoff model, where radar data could be used as a first priority and substituted for by raingauge data. If

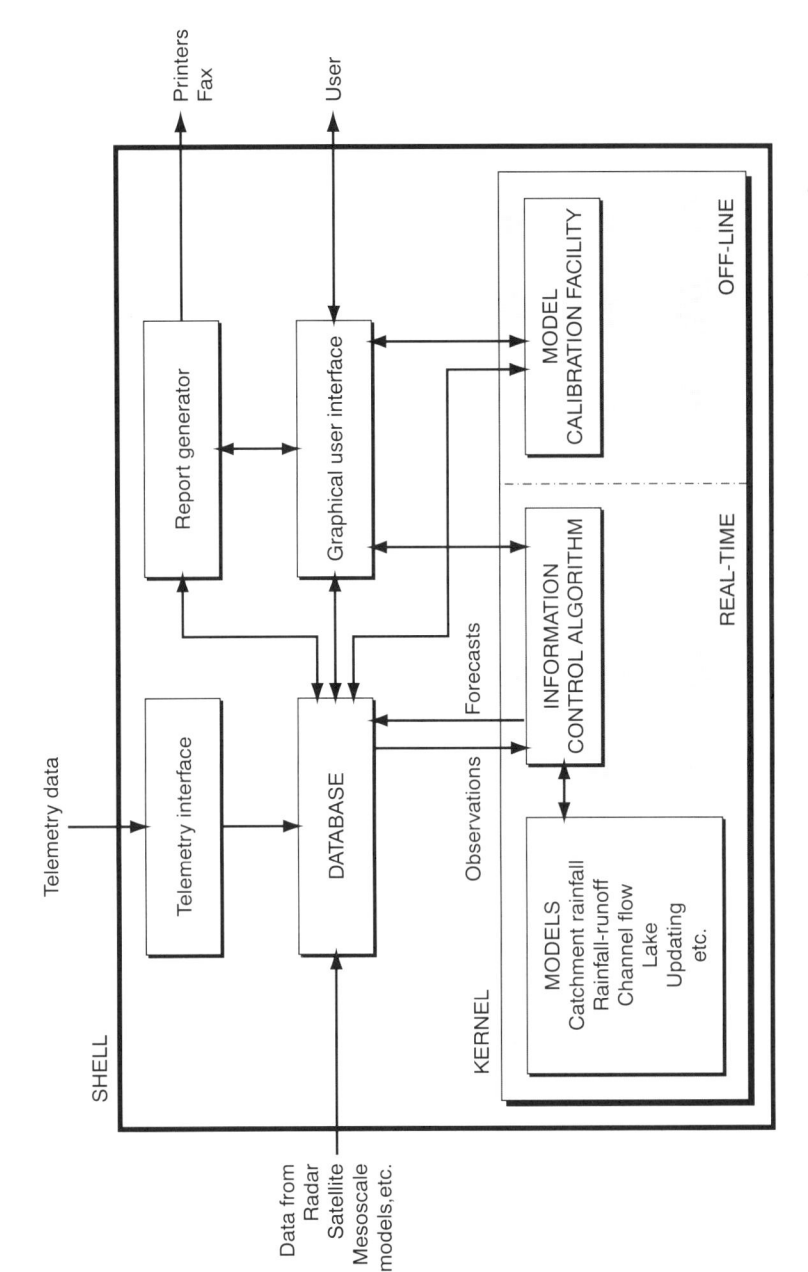

Figure 8.11 Diagrammatic representation of the 'shell' and 'kernel' structure of the RFFS
Source: Based on a diagram in Moore et al. (1994) by permission of the Institute of Hydrology

both were missing then a typical rainfall would be used as a backup, thereby ensuring that catchment rainfall required as input to the model is always present. Telemetered river flow measurements are used to update the interlinked models representing the river network and this controls the propagation of errors downstream.

The first implementation of the RFFS provided forecasts at some 150 locations within the 13 500 km^2 region of Yorkshire, UK, where the system operated on a trial basis from February 1992. By April 1993 the experience gained was sufficient to allow forecasts upstream from the flood-prone city of York to be used for routine flood warning (NERC, 1993).

Because of its flexibility, RFFS can be operated using any of the catchment models discussed in Section 8.4.2.2, i.e. IHCM, TCM or PDM, or indeed many others, and has already been used widely in the UK and in Hong Kong. The growing availability of grid-based spatial data, e.g. from rainfall radar, DTM and GIS, has emphasised the potential value of simple *distributed* rainfall–runoff models for operational flood forecasting. An important development in the UK is that of the IH Grid Model, designed for use in real-time systems such as RFFS, and using inputs from weather radar and a digital terrain model (DTM) of the catchment (Moore and Bell, 1994). The IH Grid Model generates runoff values for weather radar grid-squares using a saturation excess principle to represent the absorption of rainfall by the soil–vegetation assemblage over the grid-square area linked to the average slope within each grid-square. Excess water is assumed to enter either 'fast' or 'slow' channels and is then routed along these. The key linkage function used to define runoff production is a relationship between slope, as measured from the DTM, and aborption capacity (NERC, 1994). It is likely that successive versions of the IH Grid Model will be incorporated into RFFS in order to improve further its flood forecasting capability.

8.4.3.2 Flood Forecasting on Large Rivers

The River Flow Forecasting System described in the previous section serves as a valuable example of what has been achieved to date and indicates some of the likely areas for development in the future. It is a sophisticated integrated flood forecasting and warning system, heavily dependent on current and improving technology, which is both appropriate to and necessary for the small river basins, with very short response times, which characterise the UK. Larger river basins, more typical of major continental areas, have much slower response times and from this point of view may pose fewer problems for the flood forecaster. However, large river basins generate special forecasting problems of their own, especially as many large rivers are either interstate or international in character, may have limited or non-existent rainfall–radar data and in some cases inadequate standard hydrological data, and may have their headwaters in major mountain systems which further complicates the flood forecasting process.

Gutknecht (1991) suggested that some of these problems could be minimised by ensuring that forecasting models for large and international river basins are designed (a) to meet user requirements under various operational constraints; (b) to be capable of adapting to new hydrological situations not previously observed; and (c) to allow for the inadequacies not only of the hydrological data fed into the model but also of the model itself. Not surprisingly, the design of integrated, real-time flood forecasting systems for such rivers has been addressed in a wide variety of ways. Although only a limited number of these can be explored in this section, and then only briefly, it is intended thereby to identify some of the principal issues involved.

The *Mosel River* drains a European catchment of about 28 000 km² in France, Luxemburg, Belgium and Germany. Severe flooding in recent decades has stimulated the development of forecasting procedures which attempt to account for anthropogenic influences associated with changes of climate and land use. Ott et al. (1991) described the ongoing development of a forecasting model for short-term and long-term flow simulation and application for real-time flood forecasting. This distributed model has two main components, the first of which calculates the runoff caused by rainfall on the various subcatchments, and the second of which routes the resulting floodpeaks down the Mosel River and its main tributaries. The model is based on GIS data derived from a digital terrain model, Landsat imagery, and digitised maps. Each catchment is gridded into 30 m by 30 m squares which are aggregated spatially into hydrologically similar units in order to simplify data handling and analysis. Eventually the model will be used to estimate the impact of land-use changes and will also be coupled with an atmospheric general circulation model to enable flood estimates to be made for future climate scenarios.

Another European river, also rising in north-east France, is the *River Meuse* whose catchment area of approximately 33 000 km² is located in five countries. Some of the problems resulting from its international character were discussed by Berger (1991) in relation to flood forecasting in the Netherlands, and Troch et al. (1991) in relation to flood forecasting in Belgium. The chief problems include the technicalities of international data transfer, the contrasting nature of the main river and its flood characteristics in the Netherlands, Belgium and France, and the fact that major flood inputs from tributaries in Belgium significantly affect the flood behaviour of the main river in the Netherlands. Such problems have been exacerbated by inadequate international cooperation, which must be considered inexcusable in view of the political enthusiasm of the Meuse countries for the European Union. Troch et al. (1991), for example, noted that no real-time information on discharge or water level for the French Meuse was available to the Belgian authorities; nor was the equivalent Belgian information available to the Dutch authorities. They concluded, inevitably, that 'international cooperation, at least in data acquisition, would be invaluable for the improvement of real-time flood forecasting systems'.

The flood forecasting model for the Belgian reach of the River Meuse was commissioned by the Belgian Ministry of Public Works in 1986 and successfully completed in 1990 (Troch et al., 1991). The model uses hourly precipitation and discharge data to generate flood flows on each of the main tributaries. Flood hydrographs are routed downstream using a procedure based on the St Venant equations (see Section 8.3). A simplified flow-chart of the model is shown in Figure 8.12.

Such is the size of the country that flood problems in the People's Republic of China are often associated with rivers whose catchments are very large despite being located entirely within national boundaries. The Ch'ang Chiang, for example, has a catchment area of 1 010 000 km² (see Table 1.1). The Pearl River drains an area of 452 600 km² and has three main tributaries of which the *Beijiang* has a catchment area of 46 700 km² and a maximum recorded flood of 18 000 m³ s⁻¹. Gilbert and De Meyer (1994) described the development of a flood forecasting procedure for the Beijiang based on the St Venant

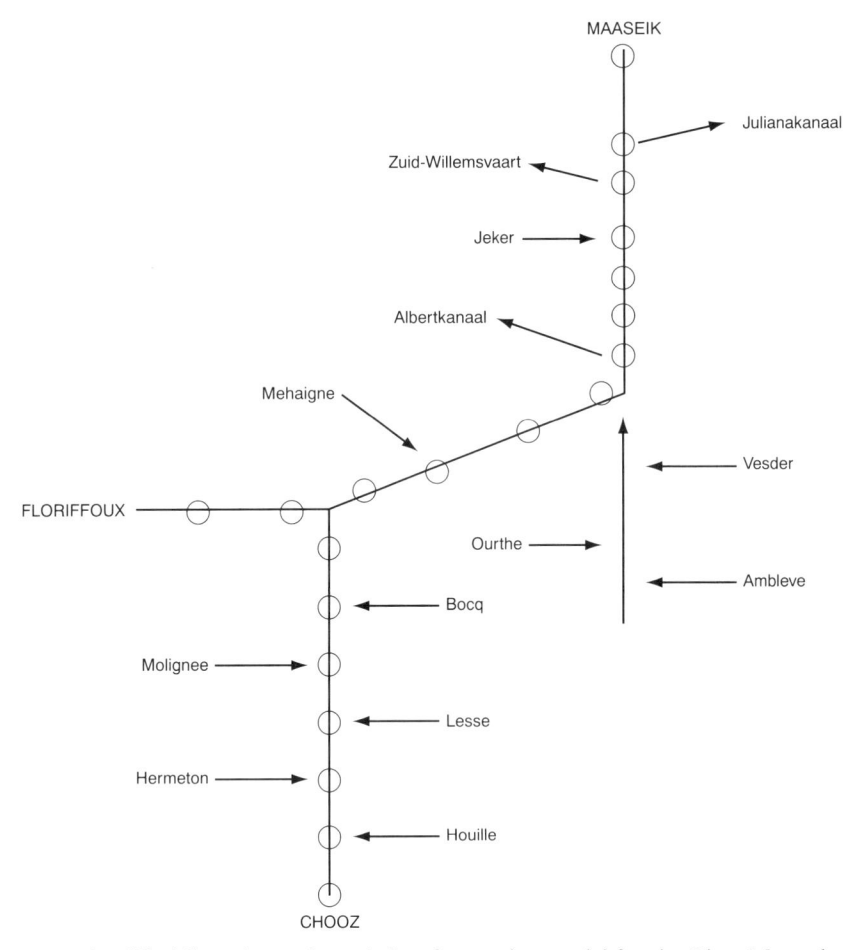

Figure 8.12 Simplified flow-chart of a real-time forecasting model for the River Meuse between Chooz and Maaseik. Arrows entering the main channel represent a rainfall–runoff model for the tributary indicated; arrows leaving the channel represent losses to canals and waterways; circles represent modelling nodes
Source: Based on a diagram in Troch et al. (1991) by permission of IAHS Press

equations at about 250 nodes. The model allows simulation of dike overflooding and breaching and accommodates considerable hysteresis in the stage–discharge relationships. For example, in the downstream reaches of the river, discharges may vary up to 5000 m^3 s^{-1} for the same water level and water levels may vary by up to 1 m for a given discharge. The model, which uses a one-hour time-step, has produced good results and has been integrated into a real-time flood forecasting application.

One of the most familiar of large international rivers is the *River Nile*, whose catchment area of 2 900 000 km^2 is located in parts of Uganda, Ethiopia, Sudan and Egypt and whose floods have been recorded for thousands of years. The Nile system extends over 6000 km and flood peaks take weeks to pass down the full length of the main river and its principal tributaries, thereby appearing to allow adequate lead time for flood warnings. However, Khartoum is a major city situated far upstream and close to the source of flood peaks in the mountainous regions of Ethiopia, resulting in an average flood wave travel time of only five days from the Ethiopian border (Andah and Siccardi, 1991). In addition, both inter- and intra-national communication is difficult and, as the August 1988 floods showed, a lead time of at least one month is required in Sudan in order to allow the evacuation of vulnerable population centres. Andah and Siccardi (1991) argued that the development of adequate flood forecasting and warning procedures for the Nile basin would depend on two main thrusts. First, there is need for an integrated real-time flood forecasting and weather monitoring system covering all the countries involved, together with appropriate technical cooperation. Second, it would be desirable to develop some climatologically mesoscale approaches, such as further analysis of the possible spatial coherence and teleconnections of rainfall anomalies over Africa, in order to further enhance the lead time for forecasts of flooding generated by extreme precipitation events.

8.4.4 Forecasting Flash Floods

The lead times associated with flash floods (see also Section 4.6.2.3) are by definition very brief. This poses particular problems for forecasting and warning operations and underlines the desirability of identifying in real time, or of forecasting, the occurrence of heavy precipitation. Before the advent of radar and satellite imagery, heavy precipitation upstream of areas prone to flash flooding could be identified only from telemetered readings from upstream raingauges or water-level recorders, or by telephone contact from observers in the field. The evolution of reliable rainfall radar and satellite imagery since the 1970s has been largely responsible for significant increases in both the number and viability of flash flood forecasting and warning systems.

Although several forecasting systems were described in a 1974 Symposium on Flash Floods (IAHS, 1974), satellite imagery was used in only one and radar imagery in none of them. However, by 1978 a survey of flash flood forecasting in WMO member countries reported by Hall (1981) showed that 25 of the 45 countries represented provided flash flood forecasts or warnings for at least one catchment. Of these, approximately half used radar and about one-third used satellite imagery to assess areas of heavy rainfall and/or to forecast storm movement.

Flash flood forecasting techniques were considered by Hall (1981) to fall into one of three groups, i.e. meteorological, hydrological, or a combination of the two. *Meteorological techniques* rely essentially on the forecasting or observation and notification of heavy rainfall. Although observation is greatly assisted by the availability of radar imagery, even in mountainous locations (cf. Andrieu et al., 1989), the most useful forecasts, having the longest lead times, usually result from an ability to forecast the location and likely amounts of heavy precipitation. It is shown in Section 4.6.2.3 that, even in the case of thunderstorm precipitation, this increasingly involves the understanding of mesoscale and synoptic atmospheric conditions, often aided by satellite imagery. *Hydrological techniques* mostly involve the elementary downstream routing either of observed upstream increases in river level or discharge, or of discharge peaks estimated from some form of model which may range from a simple 'rule of thumb' to quite sophisticated rainfall–runoff models. The composite hydrological model described by Khavich and Ben-Zvi (1995) produced forecasts with lead times ranging from 0.5 to 3.5 hours, although the longer lead times were associated with relatively large errors of underestimation. This work confirmed that the most effective approach to flash flood forecasting, although still the least commonly used, is likely to be through the use of a *combined meteorological and hydrological method* in which quantitative precipitation forecasts form the crucially important input to runoff and routing models.

Indeed, it is now clear that the most significant factor in successful flash flood forecasting is the ability to make accurate quantitative precipitation forecasts for upstream locations rather than the techniques which are used to 'route' the effect of that precipitation to the downstream forecast location. In this respect satellite imagery is already making an important contribution but, as La Barbera et al. (1993) emphasised, its practical suitability for flash flood forecasting is strongly dependent on the space and time-scale of the 'target' area and on the specific regional climate conditions and morphological configurations.

In major urban areas stormwater flooding during heavy rainfall constitutes a particular case of flash flooding which, because of the high density of population and the high value of the urban infrastructure, is likely to cause considerable damage. As a result the value of weather radar has been investigated in such areas from an early stage of its development. In the USA, for example, the successful Chicago Hydrometeorological Project, completed in 1980, used a combination of weather radar and raingauge data to forecast and monitor heavy storm rainfall which could be extended to other urban areas. The initial success of the project was evaluated by Huff et al. (1981). Similar studies in Europe, including those for Seine-St Denis in France, Bremen in Germany, and the north-west of England, were described by Cluckie and Tyson (1989).

Another special case of flash flooding is represented by dam-break floods. These catastrophic floods result when an artificial or natural dam fails. Prior warning of the failure is rarely available so that the forecasting of dam-break floods is almost always limited to the brief period after failure has been observed and during which the rapidly changing and usually large flood wave moves downstream (Hall, 1981). This means, in effect, that the downstream progress of the flood wave will have been previously modelled so that its likely arrival times at downstream locations are already known. Fread (1989) discussed available models, including **DAMBRK** which is used by the US National Weather Service to develop the outflow hydrograph from a breached dam and hydraulically route the flood through the downstream valley.

FLOOD WARNING

8.5 THE NATURE OF FLOOD WARNING

The practical aim of flood forecasting is to reduce the loss of life and the economic damage caused by floods. However, this aim can only be achieved if an accurate forecast is translated into a reliable warning message which is then disseminated to the people at risk who, in turn, take effective loss-reducing actions before the arrival of the flood. According to Smith et al. (1996), the most significant advance in non-structural flood mitigation measures during the last decade has been in the provision of more accurate flood forecasts. On the other hand, there is a wide recognition that the associated warning procedures have not shown the same degree of improvement. It is important to distinguish between a *flood forecast*, which is a scientific evaluation of an event in real time leading to the issue of a general alert about hazardous conditions, and a *flood warning* which contains additional information, including recommendations or orders for action, such as evacuation or emergency flood proofing, specifically designed to safeguard life or property. In other words, a good flood warning will not only provide advance information on the likely magnitude, location and timing of a flood event but it will also specify the nature of the loss-reducing actions to be taken and will be tailored, in terms of its content and delivery, to achieve an optimal behavioural response from a targeted group of recipients. Therefore, the final outcome of these processes is determined by the actions of a wide range of organisations and individuals.

Differing philosophies and statutory responsibilities between public bodies will determine whether a forecast product or a warning product is issued. Flood forecasting techniques place great emphasis on increasing the reliability of the forward estimation of any potentially hazardous high-water event, whether caused by riverine or coastal processes, and also on extending the time-scale over which this information is available. Flood warnings are based on the stimulus–response model of human behaviour which suggests that, if people at risk receive a warning message with sufficient lead time, they will react in a largely optimal way to reduce their losses. As Nigg (1995) indicated, this model is too simplistic to explain fully the complex performance of flood warning systems and McLuckie (1973) developed a stimulus–actor–response model which emphasised the behavioural characteristics of the warning recipient, such as their perceptions and experience of the flood hazard. But all the models do rely on the assumption that there is a positive relationship between the length of the lead time available for response and the amount of loss reduction which can be achieved. In addition, most emergency managers make the working assumption that a community under threat will react as a single social 'system' to a flood warning or to other stimuli, such as an order to evacuate. In practice, the behavioural response to warnings will differ greatly between individual households due to factors varying from failing to receive the message, the size and composition of the household, past experiences of floods and unwillingness, or inability, to take emergency action. Some responses may even be contradictory. For example, a survey of responses after Hurricane Andrew struck Florida in 1992 showed that, whilst households containing elderly people started preparations earlier than others—perhaps because such people were undistracted by paid employment or had more experience of the hurricane threat—such people were less willing to evacuate their property when eventually ordered to do so (Peacock et al., 1998).

Flood warnings can be viewed as the end part of an integrated sequential system which starts with a river, tidal or storm-surge forecast. In detail, the design of warning systems can be complicated, as indicated by Foster (1980), and alternative views exist with respect to theoretical structures, as shown in Figure 8.13. However, from an operational viewpoint, the most crucial components are the three separate, but linked, procedures of warning preparation, warning dissemination and warning response. These procedures have to occur before the flood although, in some cases, limited loss-reducing responses may continue after the arrival of the flood peak. Like any chain, the system is only as strong as the weakest link and there is some evidence that flood forecasting and warning systems become progressively less efficient as they evolve from warning forecast preparation to warning response.

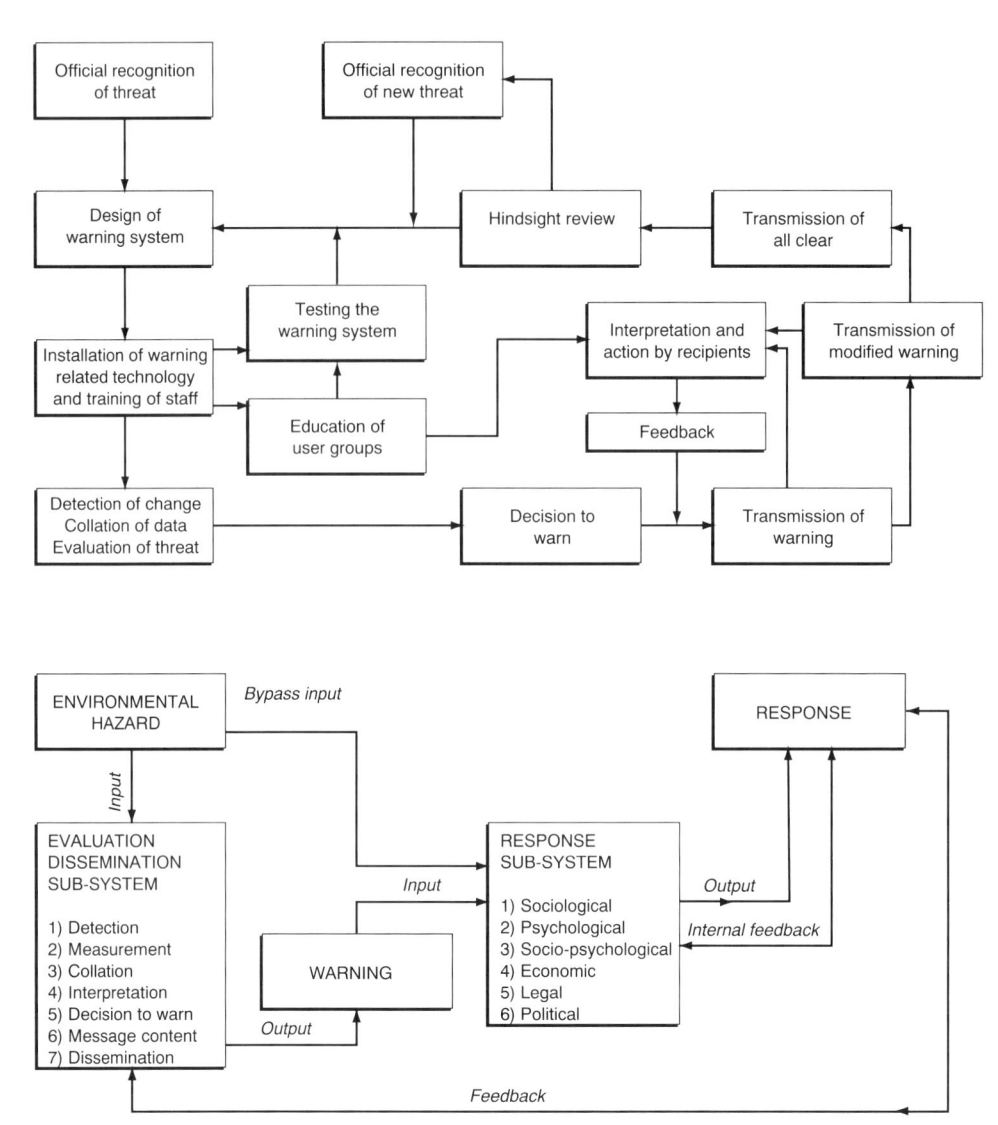

Figure 8.13 Alternative conceptualisations of the design of an integrated flood warning system
Source: After Foster (1980) and White and Haas (1975)

Integrated flood forecasting and warning systems are most effective when they are designed and operated to correspond with the local geographic and socio-economic conditions. In part, this means integration with other flood reduction strategies, such as in the UK where emergency warning systems are complementary with engineered structures and where they comprise the most important non-structural response in England and Wales (Smith and Tobin, 1979). In the USA, flood warnings are used to complement the National Flood Insurance Program. They have been mainly developed to protect urban areas where the highly concentrated risks to life and property offer good opportunities for loss reduction by mass evacuation procedures and favourable cost–benefit ratios. But there are many less favourable situations. For example, in Bangladesh storm-surge warning is necessary to safeguard millions of rural lives along the deltaic coast, although there is a chronic shortage of flood shelters for evacuees.

Flood warning schemes of all types have become much more common in recent decades. This is partly due to the advances in flood forecasting and communication technology (Section 8.4) but it also reflects a growing awareness that warning and evacuation are the best way of ensuring public safety in floods (Stallings, 1991). A recent emphasis has been on warning for very rapid onset events, such as flash floods and dam-break floods. Thus, Gruntfest and Huber (1989) reported that more than 40 automated flash flood warning schemes were operative at the end of the 1980s in the USA. Another recent thrust has been in warning for tropical cyclones. For example, in the USA the Coastal Zone Management Act of 1972, with subsequent amendments, has encouraged several states to fund hurricane evacuation plans, along with other activities. Gross (1991) showed that, despite large increases in the resident coastal population, the loss of life from Atlantic hurricanes has been reduced from over 8000 for the 1900–10 decade to less than 250 in the decades 1970–80 and 1980–90. For Hurricane Hugo in 1989 some 216 000 people were evacuated from coastal areas ranging from the Carolinas to the Caribbean region.

Much of the literature on flood warning is limited to riverine floods. However, we believe that most of this evidence is generic to many natural hazard warning systems, including those for storm surges and tsunamis, always bearing in mind that caution should be exercised when cross-referencing results between different systems. Therefore, the discussion of flood warning in this chapter is intended to be equally applicable to floods in both fluvial and coastal settings. Penning-Rowsell (1986) summarised some important attributes of flood warning systems, such as:

- A relatively low cost of implementation compared to other loss-reducing responses. This is especially so considering the marginal costs for an existing hydrometeorological authority already engaged in weather forecasting or other river management programmes. Low cost is important at a time when reduced capital budgets are available for flood engineering schemes.
- Warning schemes have shown a remarkable ability to benefit from the advances in scientific understanding and new technology associated with flood forecasting. This was documented by Clark (1994) for the USA from the introduction of the Unit-Hydrograph concept in the 1930s, through weather radar and satellite imagery in the 1970s and 1980s, to Quantitative Precipitation Forecasting for periods of less than 24 hours in the 1990s. Such technical innovation is likely to continue.
- Flood warning schemes have either very low, or negligible, environmental impacts compared to other strategies, such as engineered flood defence. This is a significant factor since environmentally sustainable flood reduction is increasingly required.
- Flood warning is often applicable in situations where other methods are not suitable. For example, it can be deployed in intensively developed urban areas where structural works would be either physically impossible or visually intrusive and where land-use controls cannot be used to remove existing buildings.

- Flood warning is likely to have an increasing application in the future. For example, as sea-level rise continues, both managed shoreline retreat and emergency coastal evacuation will become more frequent. Other growing problems exist for which there are few alternative solutions. For example, urban stormwater floods and dam-break floods may well benefit from improved warning schemes in the future.

The remainder of this section examines, in a sequential fashion, the key factors which influence the functioning of flood warning schemes and then provides a critical overview of the performance of such systems.

8.6 PRIOR ASSESSMENT FOR FLOOD WARNINGS

This is the science-based evaluation conducted by national and international agencies to improve the forecast base on which warnings rest. Although reductions in warning time are always desirable, few floods occur without some precursor signs, such as heavy rain or strong winds, on which people can take some early and independent action to reduce losses. It is important, therefore, to ensure that the economic benefits of warning schemes are those which arise through greater forecast accuracy and lead times rather than simply express the difference between the maximum possible and actual recorded damage.

8.6.1 Forecast Accuracy

The effectiveness of flood warning is crucially dependent on the forecast accuracy of certain physical parameters, such as the peak magnitude of the flood, its timing, location and duration. This will not only increase the effectiveness of response to the individual event but will also determine the credibility of, and response attitudes towards, any future flood forecasts. Conversely, warning credibility will be reduced by false warnings and by any failure to warn for damaging flood events. Forecast accuracy is partly determined by the limitations of the data gathering and transmission network. For example, Handmer (1988) reported that, in the Sydney, Australia, flood of 1986, the reporting network which was largely based on manual observers and telephone links failed with the result that many people did not receive warnings. On the other hand, Noonan (1986) described a more secure inland and tidal warning scheme in north-west England which is based on an automated network of rainfall and river flow gauges, plus weather radar, transmitting information via a dedicated microwave radio communications system.

8.6.2 Forecast Lead Time

Many advances in flood forecasting have been geared to speeding up the issue of forecasts in the belief that this will provide more time for loss-reducing responses. This feature is important in the case of public safety because Sorenson (1991) showed that ample time is needed to complete evacuation before the event impact since people are often more vulnerable when on the move than when at home. Such pressures are increasing, as in some coastal states of the USA, where the peak capacity of the road system is insufficient to permit rapid evacuation after a hurricane warning as the population grows. Longer lead times also provide greater opportunities for homeowners and others to protect their property although it should not be assumed that this process follows a linear relationship over long periods of time and that progressively higher investment in the rapid dissemination of warnings will automatically be justified by the savings achieved.

In the UK, severe weather warnings are issued to the emergency services from six hours up to five days ahead (Hunt, 1995). However, many areas at risk of flooding have comparatively little response time. This is especially so for locally generated tsunamis. Thus, whilst the Pacific-wide system processes data and warns coastal populations in about one hour (or >750 km from the earthquake source), various regional warning schemes around the Pacific basin have to warn people within 100–750 km of the tsunami source in about 10 minutes (Bernard et al., 1988). Dam-break floods may also require the issue of warning messages within a matter of minutes. Bossman-Aggrey et al. (1987) drew attention to the presence of 537 large dams in the UK and noted that the short time needed for the passive mode failure of such structures does not accord with the effective response period. For example, after initial site detection, confirmation of probable failure is required by a qualified engineer before the police are alerted. The police then have to decide who is at risk before they can implement warning and evacuation procedures. Flash floods are commonly defined as having less than six hours warning time, a situation faced by many urban communities living near the short rivers of Britain. Table 8.2 shows the improvements in some relevant short-term warnings predicted by the meteorological agencies between 1993 and 2005, which are also expected to show a reduction of 10% in the false alarm rate and an improvement in the confidence level over the next few years (Hunt, 1995).

8.6.3 Threat Evaluation

Flood forecasting authorities often face a difficult decision about when, or whether at all, to issue a forecast. In part this problem stems from dealing with scientific uncertainty but it may also reflect concern about legal liability, which is why forecasts, rather than warnings, are often released by public agencies. Krzyszto-

Table 8.2 Advanced short-term warnings and forecast services

	Existing 1993	Proposed 2005
Flash flood warnings		
Period of warnings (minutes)	15	40
Accuracy of magnitude (%)	55	90
Events with no warning (%)	70	20
Precipitation forecasts		
Period of warning for 25 mm precipitation forecast with same accuracy as a one-day forecast in 1971 (days in advance)	1.8	3.0
Tropical cyclones		
Accuracy of landfall (km) with 24-hour lead time	185	150

Source: After Hunt (1995)

fowicz (1993) worked on producing statistically optimal decisions for issuing warnings based on imperfect forecasts and evaluating system performance for short lead times.

The decision to warn is normally based on some assessment of the likely impact of the event and the probability of its occurrence. However, a public warning stating that an event has a 60% probability of occurrence may be considered as near certainty by some message recipients and as unlikely by others. In turn, such uncertainty will influence the overall confidence level of the warning message and even the nature of the dialogue between the forecast agency and the emergency services. Some public officials may be reluctant to issue warnings at all. This reluctance is likely to grow if the hazard impact is imminent and no previous warnings have been issued because of a fear that people will panic (Nigg, 1995). In practice, these concerns about behaviour are often groundless, although problems of looting may occur in both the MDCs and the LDCs when people leave their homes.

8.7 DISSEMINATION OF FLOOD WARNINGS

The way in which the warning message is composed and issued affects risk communication with both the emergency services and those at risk and also influences the performance of flood warning systems.

8.7.1 Institutional Arrangements

Prior administrative arrangements are necessary to ensure that the functional responsibilities of all those involved in flood warning are well understood in advance of the event, especially when warning times are short. Such arrangements usually seek to attain a balance between general public safety and individual responsibility. The source of the warning message is significant. In developed countries a high credibility generally attaches to official government agencies and the emergency services (Quarantelli, 1984), although the lower socio-economic groups may not be fully reached by large bodies. This is a special problem in many LDCs (Schware and Lippoldt, 1982) or where there are strong language or cultural barriers to effective communication. For example, Patel (1996) highlighted difficulties in flood forecasts from India being passed on to Bangladesh. Some agencies may have only discretionary powers for flood forecasting and warning. In other cases, government bodies may have a statutory responsibility to prepare and issue forecasts, although this responsibility may well not extend to issuing and disseminating the flood warnings (Parker, 1987). Even where an agency has this responsibility, it may not have a follow-up duty to ensure an adequate public response to the warning. The latter parts of the warning process are often the responsibility of voluntary bodies, without explicit statutory functions, who, in turn, may rely on part-time assistance.

Some flood warning schemes are highly international in scope. For example, the Tsunami Warning System in the Pacific is composed of 26 member states. The Pacific Tsunami Warning Center at Ewa Beach, Hawaii, serves as the regional warning centre for Hawaii and as an international warning centre for tsunamis that pose a Pacific-wide threat. The Alaska Tsunami Warning Center in Palmer, Alaska, issues warnings for Alaska, British Columbia, Washington, Oregon and California (Foster and Wuorinen,

1976). Tsunami watch, warning and information bulletins are disseminated to local, state, national and international users, as well as the media. These users then disseminate the information to the public, generally over commercial radio and television channels.

National schemes can also be complex. In the USA the National Weather Service (NWS) is the federal agency with responsibility for issuing weather and river forecasts and warnings to the public. Additionally, the NWS issues forecasts to other agencies such as the Corps of Engineers which also prepares in-house forecasts for operating its major reservoir schemes. In some cases, problems arise because of uncertainties about the exact nature of duties, as in the UK (Handmer et al., 1989), and in Australia where there have been long-standing issues relating to divided responsibility between the Bureau of Meteorology and the state governments in flood forecasting and warning (Downey, 1986).

8.7.2 Message Content

Even when warning messages are received, they have to be understood before action can be taken. Therefore, flood warnings should not simply contain technical information but should be expressed in terms, and relate to conditions, that flood-prone residents can understand and act upon. Confusion is likely if too much information is released. During the 1993 Mid-West floods in the USA, more than 10 000 official forecast statements and warning products were issued from the field offices of the National Weather Service. After the event, it was concluded that many of the end users did not have the expertise necessary to utilise this variety of messages and that the packaging of the relevant information for water control and emergency management bodies could be improved (NOAA, 1994). In particular, it was felt that the media and the public did not fully understand some of the terminology or the procedures involved.

According to Gruntfest (1987), good warning messages should contain a large amount of information yet still retain clarity. Relevant information might well include an estimate of the size of the expected flood, an estimate of the time before impact, a description of the present environmental conditions, an account of the actions of others and an indication of the number of warnings previously issued, whilst conveying a moderate sense of urgency and providing specific instructions for action, such as advising people to stay clear of the hazard zone. Other, more general, attributes include consistency between messages, some statement about the amount of possible damage in order to encourage loss-reducing action and an identification of suitable escape routes for evacuation.

Many authorities issue tiered warnings which increase the level of readiness through Advisory, Watch and Warning stages, e.g. for hurricanes in the USA. The River Tay, Scotland, warning scheme, described by Falconer and Anderson (1993) is typical of UK sequential procedures. The *yellow* warning indicates that minor flooding of upper catchment agricultural land is possible due to above-normal rainfall and river levels. The *amber* warning, which is set at the bankfull level of the mean annual flood, indicates that additional flooding is likely for some roads and high-risk properties. A *red* warning, which corresponds to a 3–5 year return period in the upper reaches and 5–10 years in the lower catchment, signals that serious flooding is likely, including the overtopping of riverbanks and the breaching of other flood defences.

8.7.3 Message Delivery

There is a great variety of warning delivery methods, involving both the physical transmission and the medium employed, which may affect the effectiveness of the response. For example, telephone line communications are often fragile in floods and windstorms and the use of radio links, including the use of tone-alert radios for remote communities, is often preferable. Wherever possible, it is desirable to personalise message delivery at the local level through neighbours and family members. For example, Parker (1991) reported that, if warning delivery is by a siren or loudhailer only, the proportion of residents responding is not likely to exceed 55%. This figure will rise to 65%, and possibly more, where known flood wardens are used or when police visit individual householders. In particular, such highly targeted warning delivery allows groups with special needs, such as the elderly, the very young and the handicapped, to be reached more reliably by the emergency services.

In the UK, flood warnings originate with the Environment Agency in England and Wales (and with the Scottish Environment Protection Agency north of the border) and are then passed to the public by the police or the local authority. The key role of the police is based on their manpower and communications capability together with their level of authority which is important for example in persuading people to evacuate an area. The police may receive their information either directly from telemetered hydrometric stations or, more usually, from the local river forecasting agency. There may be scope for self-help or private flood warning schemes, as described by Wright (1980), but some of the most effective responses are found with well-established structures using teams of local flood wardens to disseminate the message to floodplain residents, as indicated in Figure 8.14 for the River Severn at Shrewsbury (Harding and Parker, 1976).

In some countries, reliance for message delivery is placed on the media, although this is not always satisfactory. For example, the range of television or radio broadcasting may not match the flood-affected area. General flood alerts issued over very large areas are likely to result in some false and withdrawn messages, which may well produce complacency in those still at risk. Even if warning messages are accurately distributed, the continued transmissions in themselves can provoke a false sense of normality. Sometimes journalists can take technical forecast and warning products out of context and allow them to be broadcast in a rather sensational way. In large-scale flood events the media demand for information may overwhelm the capability of the responsible agency and impede their prime activities.

Some studies have shown that the public is highly sceptical of the mass media (Perry, 1985). On the other hand, the high level of local preparation for Hurricane Andrew in 1992, which included loss-reducing activities begun well before the official warnings were issued, has been credited to the extensive, non-stop television coverage of the gathering storm (Peacock et al., 1997). Nearly 85% of households surveyed gave television as their most important source of pre-storm information and 67% rated it as an excellent source. In the USA more specialist media are also deployed and the National Oceanic and Atmospheric Administration Weather Radio service (NWR) is a prime means of dissemination. But the 1993 Mid-West floods showed that issue dates and times were not routinely broadcast with the forecasts, which themselves were not always fully up to date, and that most of the public was unaware

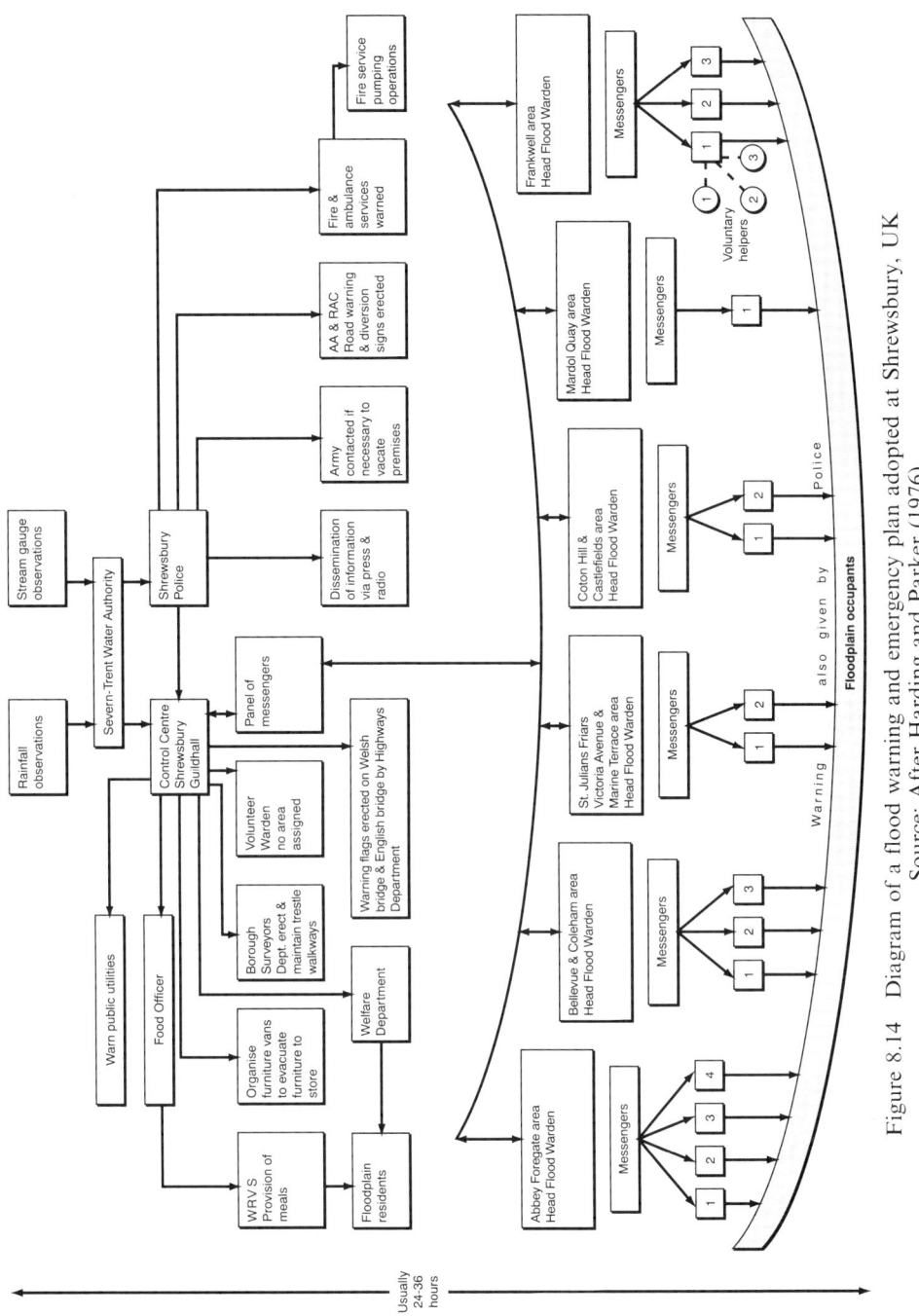

Figure 8.14 Diagram of a flood warning and emergency plan adopted at Shrewsbury, UK
Source: After Harding and Parker (1976)

of the availability of NWR, despite the fact that it broadcasts over most of the United States (NOAA, 1994).

Where the media are involved, it is important that the frequency of warning messages is understood by the public. Wherever possible, messages should be issued at a stated, regular time (e.g. every hour) and there is a clear need for a 24-hour service. It is also necessary for stand-down messages to be issued to confirm that the threat is over. Few media-based warning schemes contain a capability for incorporating feedback from message recipients to the warning source, either during or after an event. In many cases it would be possible for an emergency telephone number, which is often supplied to provide additional information for the public, to be used as a feedback channel of communication. After the emergency, there is also a need for longer-term hindsight reviews of the operation of the systems.

Further advances in information technology are likely to bring future improvements in message delivery and more specialised flood warning. Already automated flood risk mapping, using a combination of remote sensing and GIS technologies, is employed by bodies such as FEMA to detail the lowest-lying areas of floodplains and coasts. If this information is allied to hydrological flood outline data and demographic and socio-economic databases, small areas can be rated for insurance purposes according to elevation and property loss potential. In turn, this can lead to new products, such as an advance warning to insurance companies of their claims exposure in a particular flood. In turn, this warning would allow insurers more time to make funds available to meet their claims liabilities and to target those areas where assistance is needed. In the UK, the Environment Agency has developed a new telephone system that automatically warns people in areas liable to flooding (Utteridge, 1996). Given a combination of weather radar and river, or tidal, height data, a computer works out the areas at risk and can call a database of telephone numbers of properties in the area. More than 1000 residents per hour can be warned with a pre-coded message, after receipt of which the recipient must enter a number into the telephone to record receipt of the warning. The local police can then follow-up such a warning to ensure that evacuation, or other emergency procedures, are followed.

8.8 RESPONSE TO FLOOD WARNINGS

8.8.1 Preparedness

As noted by Gruntfest (1987), the warning response is unlikely to be optimum unless a special publicity campaign is mounted for the potential user groups before the first flood alert. Ideally, a prior education programme should increase the awareness of, and understanding about, flood warning terminology for all at risk using a variety of materials, such as brochures, fact sheets and videos. These awareness programmes should also include rehearsals and planned evacuation measures as appropriate. The nature and effectiveness of such programmes will differ according to the level of economic and social development in the country concerned.

8.8.2 Event Characteristics

The nature of the flood event will itself influence warning response. Thus, Mileti and Beck (1975) found that the time of day and the day of the week were important parameters when warning time was short. Much of the readiness for Hurricane Andrew in the USA has been attributed to the fact that the storm struck at a weekend, when households were not preoccupied with employment duties. Quarantelli (1984) showed that environmental or visual clues, such as heavy rain or a rising river, which people can see and interpret for themselves, may be as important as a formal warning message. In the absence of such clues, some potential victims may delay taking action while seeking to confirm the warning message with neighbours.

8.8.3 Message Recipients

It is not well understood why some people fail to receive flood warnings even though arrangements have been well planned in advance (Mileti, 1994). Given a receipt of the message, the proportion of residents who are able to respond is often dependent on the presence of at least one able-bodied adult in the house. Once alerted, age is a factor in response since elderly people will be less able to undertake emergency flood proofing measures and to evacuate. However, Drabek and Boggs (1968) found that the elderly can respond as well as other groups, if family support is available. Other social attributes such as ethnicity, literacy and gender may all influence the nature of response. Some response will depend on local experience and traditions. For example, if there has been a previous reliance on structural schemes in the area, there might well be a degree of complacency and less of a direct response to warnings. Whilst panic is rare, threat denial may occur and limit response. This is most likely with inexperienced recipients or an unusual type of warning.

On the other hand, previous experience of damaging floods is likely to improve response. Smith and Handmer (1986) indicated that, the more a community becomes experienced with flooding, the more likely it is to develop a reliance on 'informal' warnings. As already indicated, any savings resulting from such arrangements, whilst valuable in themselves, need to be excluded from the benefits credited to any 'formal' system.

8.9 EFFECTIVENESS OF FLOOD WARNING SCHEMES

Attempts to assess the benefits of riverine flood warning schemes began in the USA during the 1960s and have developed in the UK since the late 1970s, mainly as a result of work at the Flood Hazard Research Centre of Middlesex University. According to Parker and Penning-Rowsell (1991), a variety of measures can be used to determine the effectiveness of a warning system. Some of these relate to fairly crude and arbitrary performance criteria, such as the percentage of properties served with warnings on a particular floodplain or the number of people who received a warning message with a pre-determined lead time. However, most assessments to date have relied on either an economic approach, which attempts to assess the amount of tangible economic damage which has been saved, or on a public survey approach, which attempts to reveal the degree of consumer satisfaction with the operation of the scheme. Neither of these approaches fully addresses the question of intangible losses, which are known to be significant for floods.

8.9.1 Economic Assessment

Early attempts to assess damage reduction on the basis of potential flood damages avoided concentrated on the maximum actions that a warning recipient might take. For example, Table 8.3 shows the household items that a typical able-bodied family in the UK could be expected to move to safety given a warning of a major flood for warning periods between 30 minutes and four hours (Chatterton et al., 1979). As can be seen, valuable, non-bulky goods can be removed quickly but it takes longer to remove some semi-fixed items, such as spin driers and freezers, whilst it is assumed that permanent household fixtures, such as central heating systems, cannot be protected. Theoretically, the financial benefit of such actions can then be quantified either for different types of residential property, according to socio-economic status, or for commercial premises. For example, Figure 8.15 shows the expected reduction in gross flood losses for a Pennsylvania, USA, supermarket in relation to the height of the flood and the length of warning period (Day et al., 1969).

As indicated in Figure 8.15, some of this early work was based on a so-called '100% response rate'. This assumed that all the population at risk would receive a warning and that each person would react optimally to reduce their own losses. It is now known that such assumptions are unrealistic. By no means all flood victims receive warnings and, in a study of 13 locations in England and Wales, Penning-Rowsell et al. (1978) found that, of those receiving a warning, about 46% failed to react with any damage-saving actions. This was because the recipients were either too old or infirm, or the flood caught them by surprise (despite the warning), or they were sceptical of the warning message, having received false warnings in the past. Consequently, most schemes are now predicated on a 70–80% response rate for residential properties, meaning that only

Table 8.3 Likely damage-reducing action with different lengths of flood warning

Half-hour warning	2-hour warning	4-hour warning	Not saved
colour television	carpet sweeper	spin drier	dishwasher
monochrome TV	electric food mixer	tumble drier	freezer
portable TV	electric iron	bookcase	electric cooker
hi-fi system	electric fire	dining chair	kitchen utensils
electric kettle	cupboard/cabinet	dining table	washing machine
sewing machine	three-piece suite	occasional chair	central heating boiler
vacuum cleaner	personal effects*	kitchen chairs	fitted electric fire
personal effects*	curtains	sideboard	fitted gas fire
personal computer	car	linen	chest of drawers
		clothing (f)	double bed[†]
		clothing (m)	dressing table[†]
		food stock	piano/organ
		carpets	single bed
		fridge/freezer	wardrobe (fixed)[†]
		refrigerator	wardrobe (free)[†]
		personal effects*	lino/tiles
			rubber flooring

*Progressively more personal effects can be moved with time
[†]In single storey dwelling
Source: After Chatterton et al. (1979)

FLOOD FORECASTING AND WARNING

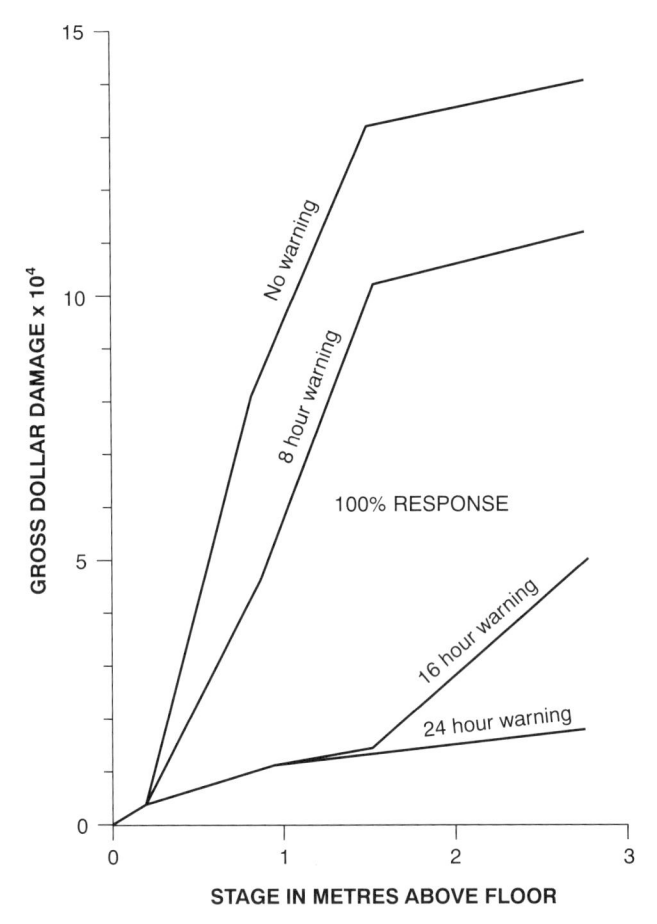

Figure 8.15 The planned mitigation of gross flood losses for a Pennsylvania supermarket in relation to the height of the flood and the length of the flood warning provided
Source: After Day et al. (1969)

about three-quarters of those receiving a flood warning can be expected to achieve the savings targets shown in Table 8.3.

Even when damage-reducing action is taken, it may well be limited to a comparatively short period and may not achieve a very high mitigation of potential losses. Table 8.4 shows the potential flood damage reduction resulting from flood warnings from two to eight hours ahead to residential and sample retail and other business premises at April 1991 prices in the UK. It can be seen that, whilst the proportionate savings increase appreciably with a warning length between two and four hours, given a doubling of the warning period to eight hours the savings are more modest. Generally, savings are less for commercial premises and in no case does the saving exceed 50%. The lower rate of increase in savings for warnings beyond four hours ahead can be attributed to the fact that, whilst many valuables can be moved within the initial four-hour period, the progressive effects of tiredness and lack of storage space for relocated items minimise later efforts. Further substantial improvements are unlikely to be achieved without the deployment of

Table 8.4 Potential flood damage reduction (%) from warnings of different length according to depth of flooding and type of property for those responding

Depth of flooding (m)	Approximate length of warning period (hours)			
	2	2–4	6	8
Residential property				
1.2	25.3	35.7	38.7	40.7
0.9	26.4	37.6	40.6	42.6
0.6	25.5	37.2	40.2	42.2
0.3	30.0	42.1	45.1	47.1
0.1	24.5	32.8	35.8	37.8
Commercial property				
1	20.4	28.9	33.9	38.9
0.6	19.5	27.4	32.4	37.4
0.3	17.2	24.4	29.4	34.4
0.15	1.4	1.4	6.4	11.4

Source: After Parker (1991)

external assistance from volunteer workers. Despite this, Table 8.5 illustrates the substantial estimated financial benefit of flood warning in England and Wales provided that the property owner is given a warning and is able to act on it.

8.9.2 Public Survey Assessment

Comparatively few attempts have been made to determine the level of public satisfaction with flood warning schemes. In a major review, Parker and Penning-Rowsell (1991) drew together the findings from some 1000 interviews with floodplain residents between 1986 and 1990 at several locations in England and Wales. In the Severn–Trent area respondents were asked to scale their level of satisfaction with the flood warning service provided (Table 8.6). In the upper Severn catchment 81.9% of the respondents were either 'completely' or 'quite' satisfied compared with the Avon catchment where 33.3% were found to be either 'not very satisfied' or 'not at all satisfied'. The general satisfaction in the upper Severn basin was attributed to high

Table 8.5 Total annual average savings for effective flood warning in England and Wales

Time to peak	Average annual damage No warning (£M)	Average annual benefit with warning (£M)			
		2 hours	4 hours	6 hours	8 hours
<3 hours	61.18	12.90	18.56	21.91	25.15
3–9 hours	43.97	8.92	12.74	15.18	17.57
>9 hours	55.90	11.33	16.17	19.28	22.32
Total	161.05	33.15	47.47	56.37	65.04

Source: After Heinje et al. (1996)

levels of preparedness and reliance on warnings based on relatively frequent floods and the use of a well-established warning system using local neighbourhood flood wardens (Harding and Parker, 1976). In Avon, on the other hand, liaison and communication problems occurred between the responsible authorities largely due to police jurisdictions crossing more than one water authority division boundary, with different divisions using separate warning procedures.

The evacuation of hazard zones is the ultimate public safety response to flood warnings. Previous studies have found that public compliance with evacuation orders varies greatly, from around 50% overall, according to Drabek (1986), to over 80% for the highest risk zones (i.e. low-lying coastal areas) in major hurricanes (Baker, 1991). In the USA official hurricane warnings, which are not normally issued until 24 hours before an estimated coastal landfall, set evacuation plans in motion. For most hurricane warnings, evacuation is required for coastal islands, areas along the coast and mobile home parks because such zones are susceptible to flooding irrespective of the windspeeds. For Hurricane Andrew in 1992, nearly 500 000 people were under evacuation orders, although it is believed that only 54% of households evacuated entirely (Peacock et al., 1997). This percentage rose to over 70% in the coastal evacuation zones but this means that nearly 30% of households located along the beach did not move away. Despite the comparatively low overall response, it was found that being in an identified evacuation zone was the most important single factor in determining such a response. This pattern is, of course, one on which emergency planners depend and it helps to reduce the problems associated with 'shadow' evacuation. 'Shadow' evacuation occurs when households not under direct threat seek to evacuate and thereby create attendant operational problems for real flood victims, such as additional traffic congestion and strain on the relief agencies at official shelters and refuges. On the other hand, about 250 000 people were successfully evacuated following a major threat of river flooding in the Netherlands in February 1995, comprising the largest evacuation in that country since World War II. Compared to Florida, factors which contributed to this success included plenty of visual warning provided by the waters, which rose slowly over several weeks, and the cultural homogeneity of the region with most people speaking the same language.

Table 8.6 Customer satisfaction with the flood warning service in the Severn–Trent Water Authority area 1986–87

Satisfaction level	Upper Severn (%)	Upper Trent (%)	Avon (%)
Completely	30.4	30.3	24.4
Quite	51.5	20.2	40.0
Not very	8.8	20.3	13.3
Not at all	7.0	21.2	20.0
NA	2.3	0.0	2.3
Reliance (mean)*	2.6	4.0	3.3
Valid cases	384	99	45

*Reliance on the warning system
1 = completely; 2 = considerably; 3 = not very much; 4 = not at all
Source: After Parker and Penning-Rowsell (1991)

8.9.3 Performance-Based Assessment

More recently there has been a move in the UK towards a more comprehensive performance-based assessment of flood warnings on the assumption that there are four principal factors which contribute to an effective flood warning system:

1. Proportion of the population at risk which is warned with sufficient lead time to take action (R).
2. Proportion of residents available to respond to the warning (PRA).
3. Proportion of households able to respond to the warning (PHR).
4. Proportion of households who respond effectively (PHE).

On this basis, the best estimate of present overall performance is only 15% (R $0.45 \times PRA$ $0.55 \times PHR$ $0.85 \times PHE$ $0.70 = 0.15$) (Heinje et al., 1996). This level of performance has been deemed to be unsatisfactory and a performance target of 52% has been suggested as reasonable by the year 2001 based on the following improvement (R $0.80 \times PRA$ $0.80 \times PHR$ $0.95 \times PHE$ $0.85 = 0.52$). This improvement in performance is dependent on a large-scale increase in the public awareness of flood warnings and better methods of message dissemination. Utteridge (1996) indicated that these aims can be achieved by further development of the automatic voice-messaging computer-based alerting system and the further use of local radio, employing a set of nationally consistent flood warning statements, in the dissemination of warnings to the public about the risk of river or coastal flooding.

COASTAL FLOODS

8.10 INTRODUCTION

Coastal flood forecasts may be concerned with high-magnitude waves and tides at extreme points on their magnitude–frequency curves, or with storm-driven tidal surges occurring either separately or coincidentally with high-magnitude waves and tides, or with tsunamis. For a variety of physical and operational reasons, the distinction between forecasting and warning is less meaningful in the case of coastal floods than in the case of river floods. In much of the ensuing discussion, therefore, these two functions are regarded as complementary parts of the same operation.

8.11 THE COMPONENTS OF COASTAL FLOOD FORECASTS

Coastal floods normally result from the interaction of several components. In this section, however, consideration is given briefly to the separate issues related to the forecasting of extreme manifestations of the major components—waves, tides, storms and earthquakes. Subsequent sections then focus on selected examples of forecasting and warning systems which combine information about several of these components.

8.11.1 Waves

Although the detailed mechanisms of wave generation may not yet be fully understood, it may be assumed for forecasting purposes that wave height increases with wind strength and duration and with length of fetch. Very large waves may be generated in large oceans by intense wind systems such as those of tropical and extra-tropical cyclones. For example, the 50-year wave height in the eastern North Atlantic is estimated to be about 30–35 m and exceeds 20 m over much of the northern part of the North Sea (Hardisty, 1990). One of the largest waves ever measured was encountered in the Pacific in February 1933 when typhoon winds, blowing over a fetch of many thousands of kilometres, generated a wave at least 34 m in height (Bascom, 1959). Serious coastal flooding may result from the repeated impact of large waves, for which *significant wave height*, i.e. the average height of the highest one-third of all waves present in a given time period, is a more useful measure. Significant wave heights of 12 m were observed off the Bermudan coast during the approach of Hurricane Felix in 1995. Since, during any three-hour period, the highest individual wave is statistically likely to be twice the significant wave height, it is conceivable that waves of up to 24 m in height were present off Bermuda during Hurricane Felix. Evidence from oil rigs shows that, although rare, waves of this size may also be formed in the North Sea. Even substantially smaller waves, however, can cause flooding of lowland coasts as wind-blown spray and wave runup overtop natural or artifical defences (see Table 8.8).

Although still less than adequate, the worldwide database on wave height, once largely dependent on observations from ships and buoys, is now extensively supplemented by observations from oil rigs and weather satellites, including ERS-1 and ERS-2, which measure wave height using radar altimeters. These more sophisticated data indicate that wave climates and wave heights vary from time to time, although the reasons for this are not yet clear. Wave heights declined by about 0.5 m between the mid-1980s and mid-1990s in the western Mediterranean and central South Atlantic, but between Iberia and Greenland North Atlantic waves almost doubled in size over the four decades preceding the mid-1990s. Satellite observations mean that the maximum wave height likely to threaten a flood-prone coastline can now be estimated more precisely and the real-time analysis of satellite data means that the approach of large waves can now sometimes be forecast with operationally useful lead times.

Wave information is needed in a variety of coastal flood forecasting situations but especially in relation to the forecasting of storm-surge height. In these cases wave height and direction are often derived from models, such as the UK Meteorological Office Wave Forecasting Model (MOWM) which generates forecast wave data for the Storm Tide Warning Service (see Section 8.12.1.1). It is the inshore wave conditions which are directly relevant to flood forecasting and these are derived in the Meteorological Office forecasts from offshore data by the following transformations (Reeve et al., 1996):

Inshore wave height = Offshore wave height × wave height factor
Inshore wave direction = Offshore wave direction + directional shift
Inshore wave period = Offshore wave period

8.11.2 Tides

Coastal flooding is rarely caused by the level of the high tide alone, even in spring tide conditions. However, there are cities such as Hull in eastern England where a significant urbanised area, protected by flood embankments, has been developed below the height of spring high tides which quite commonly cause groundwater flooding. Furthermore, even under average meteorological conditions, tide levels vary considerably with extreme tidal levels being associated with extreme semi-diurnal tidal forces. These occur either when the moon and sun are in line with the earth and at their closest respective distances, i.e. lunar and solar perigee, or when the moon and sun have zero declination, i.e. are in the same equatorial plane. For the sun, this occurs near the March and September equinoxes.

Forecasts for UK coastal waters for the 40 years to 2034 show greater frequencies of extreme tide levels in the years 1997, 2006, 2015, 2024 and 2033 (Alcock and Richards, 1993). Generally, the results show a cycle with high values occurring every four to five years, with the coincidence of the moon's perigee and either of the equinoxes. Overall maxima occur every 18 to 19 years, due to the cycle of the moon's node (the point where its orbit crosses the plane of the earth's orbit). Extreme high levels, at most of the 37 UK sites for which data were analysed, fall within a few days of the maximum equilibrium tidal forcing. However, some extreme levels are shifted by a week, a month or even a year from the theoretical date, indicating the variability of local ocean response to the lunar and solar tidal forces (Alcock and Richards, 1993).

8.11.3 Storms

Since flood-producing surges result largely from the effects of tropical and extra-tropical cyclones, the adequacy of coastal flood forecasting and warning systems is to some extent dependent upon the accuracy with which cyclone characteristics, such as track and intensity, can be forecast. Enormous improvements to forecast lead times have resulted from the use of weather satellites, which now provide continuous global surveillance of the development and movement of major weather systems.

On an even longer time-scale, Gray (1990) argued that patterns of precipitation in West Africa could be used to anticipate hurricane occurrence on the Atlantic and Caribbean coastlines of the USA. Gray showed a strong association between the annual incidence of hurricane activity in the Atlantic and recent multi-decadal periods of high and low rainfall in West Africa (Table 8.7). He also noted long-term evidence of alternating wet and dry multi-decadal periods in West Africa dating back to the seventeenth century and suggested that, if this pattern continued, intense hurricane activity in the Caribbean basin and along the US coastline could be expected to occur during the 1990s and early years of the twenty-first century.

8.11.4 Earthquakes

The prediction and forecasting of the tectonic events which are responsible for triggering tsunamis may be based on either probabilistic or deterministic methods. Where earthquakes are frequent and have been accurately recorded for long periods of time, probabilities may be calculated and, as has been done for New Zealand, return

Table 8.7 Mean annual incidence of Atlantic tropical cyclones in relation to multi-decadal rainfall conditions in the western Sahel

Atlantic hurricanes	Mean annual incidence		
	Wet period 1947–69	Dry period 1970–87	Wet period 1988–89
With central pressure ⩽ 964 hPa	3.30	1.55	2.50
With central pressure < 965 hPa	3.18	3.44	3.50
Hurricane days—all	30.10	15.00	28.00
with central pressure ⩽ 964 hPa	8.53	2.10	9.38

Source: Based on a table in Gray (1990)

periods may be mapped. Such analysis cannot, however, be considered as forecasting other than in the sense of identifying areas of greatest risk.

Forecasting, in the sense of a forewarning of a specific event, must rely on the detection of some physical precursor to an earthquake or submarine slide. On land surfaces tiltmeters and geodetic lasers may identify small but significant movements in the vicinity of active faults which may signify that earthquake activity is imminent. Where such movements are likely to impact upon the adjacent seabed, they may also be used to give warning of an impending tsunami. GPS observations of the 'great crack' on the main island of Hawaii indicate that the whole of the southern part of the island is moving at a rate of about 0.01 m per year. However, although it is known that such movement will eventually result in a major long runout landslide which will trigger a tsunami, there is no way at present to forecast either the timing or the size of that event.

A more promising precursor involves the dilatancy theory of earthquake formation described by Bolt et al. (1975). As crustal rocks deform under the influence of tectonic strain, local cracking causes the volume of rock to increase or dilate. If the cracking occurs rapidly, groundwater may not immediately occupy the additional interstices which, instead, become filled with water vapour. This causes a reduction in pore pressure within the dilated rock which, in turn, could lead to a temporary reduction in the velocity of primary earthquake waves. P-wave velocity would subsequently increase again following the eventual diffusion of groundwater into the interstices. Early evidence to support this dilatancy model came from the former Soviet Union during the 1960s when P-wave velocities appeared to decrease, and then to increase, in advance of significant earthquakes. A similar pattern occurred in California, USA, before the San Fernando earthquake of 1971. It appears that the duration of the precursory phase for tsunami-producing earthquakes, i.e. greater than 7.00 on the Richter scale (see Table 5.4), exceeds six years. This could complicate the precise timing of earthquake/tsunami forecasts and, together with the fact that many earthquakes happen without reliable and observable precursor signals, probably means that for the forseeable future advance warning of tsunamis will not be initiated until the event has been triggered (see Section 8.13).

8.12 STORM-SURGE FORECASTING AND WARNING SYSTEMS

Although the physical factors responsible for the generation of storm surges have been identified (see Section 5.3.1), the accurate forecasting of storm-surge magnitudes and arrival times requires more than a simplified conceptualisation of the ocean response to winds and atmospheric pressure changes. Ideally, forecasting models would include non-linear surface and bottom stresses and would allow for changing depths and boundary conditions as sea levels change through a storm event (Pugh, 1987). The development of such numerical models has been encouraged by improvements in computing facilities and has resulted in a continuing reduction in the computation step and a corresponding increase in forecast lead times. Understandably, earlier forecasting attempts were largely based on empirical relations between meteorological phenomena and observed sea-level changes.

Because storm surges are dependent not only on cyclone characteristics but also on prevailing tidal and bathymetric conditions (see Section 5.3), major flooding from storm surges is less frequent than the occurrence of major cyclones, especially on coastlines having a large tidal range. In addition, the extent of coastal flooding is modified by the effect of breaking waves, which can add 0.5 m or more to water levels inshore of the breaker line, and by land surface topography and roughness which determine the routing of floodwaters inland.

Flood forecasting models for storm-surge conditions are therefore normally hydrodynamically complex. Some are used also for simulation and prediction studies to determine likely frequencies of storm-surge flooding and to identify flood-prone sections of coastline (see also Section 6.6).

Tropical and extra-tropical storm surges are usually modelled in slightly different ways (Pugh, 1987) and in the subsequent sections selected models are used to illustrate some of the basic principles and problems of forecasting both types of surge.

8.12.1 Extra-Tropical Storms

Large-grid, finite-difference models have been developed for surge forecasting on the north-west European continental shelf, including the North Sea, and for some other shelf areas affected by extra-tropical storms. In such shallow-water conditions the main forcing on the ocean is due to the surface-wind stress and the atmospheric pressure. This means that the bulk behaviour of the ocean circulation can be estimated using two-dimensional depth-integrated equations of motion, neglecting baroclinic effects. Hubbert et al. (1990) described a numerical model based upon shallow-water equations for short-range prediction of surges in the Bass Strait between Tasmania and the mainland of Australia. Like the model for the North Sea (see Section 8.12.1.1), that for the Bass Strait is nested so that a relatively coarse grid of about 30 km is used to cover the whole continental shelf and a high-resolution grid of about 10 km can be nested in any sub-region of the coarse-grid domain.

Qin et al. (1994) used a two-dimensional non-linear surge model and a tidal model to simulate storm surges and water levels at Shanghai, China. The model was used mainly for the numerical simulation of the joint effects of surges and tides for 16 tropical cyclones having different kinds of track but was also used for the real-time forecasting of surge and joint tide and surge effects for tropical cyclone Abe in 1990. The resulting errors, though substantial, were considered to result mainly from incorrect forecasting of storm track and central low pressure rather than from inherent model characteristics.

For surge forecasting in smaller, semi-enclosed seas, where the surge takes the form of an oscillating seiche, it is normally necessary to model the boundary and meteorological conditions and attempt a forward extrapolation on the basis of tide-gauge information for a section of the coastline at risk. Such an approach was reported by Labzovskii (1966) for surges in the St Petersburg region of the Gulf of Finland, and by Wróblewski (1994) for five ports on the Baltic coast of Poland. The latter found that the joint application of multiple regression equations of the dynamic system and the empirical orthogonal functions method was effective in producing surge forecasts with a lead time of 24 hours and a forecast step of 3 hours.

The flooding of Venice by storm surges in the Adriatic has been a recurrent and intensifying problem. An early attempt to forecast storm surges (Robinson et al., 1973) used a one-dimensional numerical model to forecast the effect of wind-forcing, astronomical tides and co-tidal oscillations on the magnitude and time of arrival of surges at Venice. Although simple, the model forecasted the surge height with an accuracy of about 0.05 m in sea level and 0.5 hour in arrival time. Later forecasting attempts combined statistical and hydrodynamic models. An improved version, described by Vieira et al. (1993), incorporated a multiple-regression model which extended the forecasting range of the existing model from three hours to 24 hours, with absolute errors of less than 0.1 m for short-term forecasts up to nine hours. The hydrodynamic model, with a grid spacing of 6 km, covered the entire Adriatic Sea and was intended mainly to permit long-term forecasts up to three or four days ahead, with an estimated mean absolute error of 0.2 m.

8.12.1.1 Storm Tide Warning Service

An early and very successful attempt to forecast storm-surge flooding in the North Sea basin was the Storm Tide Warning Service (STWS). This was established in September 1953 in response to the disastrous flooding along the east coast of England in January and February of that year and was based, from the outset, on a network of tidal observations which now comprises 42 tide gauges. The STWS is operated within the Central Forecasting Office (CFO) of the UK Meteorological Office on behalf of the Ministry of Agriculture, Fisheries and Food (MAFF). The forecasting system has been improved and expanded over the years and now covers areas at risk from coastal flooding associated with waves and storm surges around the coastline of England, Wales and parts of Scotland. However, the main 'customers' for the STWS wave and surge forecasts are the former National Rivers Authority regions, which were subsumed in April 1996 into the Environment Agency (EA) in England and Wales, and the former Scottish River Purification Boards, which were similarly subsumed into the Scottish Environment Protection Agency (SEPA) on the same date. The STWS is also involved in the activation of tidal barriers, such as those on the rivers Hull, Colne and Thames. Most of the STWS work is during the winter but there are eight specially trained forecasters in the CFO who maintain a 24-hour continuous coverage for the service.

Storm surges are forecast using a gridded two-dimensional (depth-averaged) finite-difference model, known as CS3 (MAFF, 1995c). Input to the model is wind and barometric pressure data from the UK Meteorological Office atmospheric prediction model which covers Western Europe. CS3 runs twice daily on a Cray supercomputer but if intermediate predictions are required, as when a surge is developing, it can also be run on a dedicated workstation at the CFO, thereby allowing earlier warnings to be issued.

When predicted levels approach established danger levels, water-level forecasts are sent to the Environment Agency and to the police. Regional warning services, which use the STWS surge forecasts, wave forecasts from the Meteorological Office wave model, and other local data, are operated by the Environment Agency regions in England and Wales and the Scottish Environment Protection Agency in Scotland.

Further improvements to the accuracy and timeliness of STWS forecasts are expected to result from three ongoing developments (MAFF, 1995c; Anon., 1996b). First, the more effective incorporation of wave data, including the development of coupled wave–tide–surge models which allow for the effect of the presence of waves on the generation of both surges and waves. Second, for areas of complex nearshore bathymetry, the incorporation of local computer models which have a much finer mesh than the relatively coarse grid of the CS3 model. Already, fine mesh models of the Bristol Channel (4 km grid) and the Severn estuary (1.2 km grid) have added to the quality of the service and there have also been improvements to the models for the English Channel and North Sea. In addition, fine-mesh modelling of storm-surge conditions in the Firth of Clyde, Scotland, using a 0.5 km grid, has confirmed that more than one-third of any surge is generated locally within the area, rather than propagating from the Irish Sea (Collar et al., 1995; Curran, 1995). Third, data assimilation techniques allow real-time tide gauge data to be incorporated in the CS3 model forecasts in order to reduce cumulative errors, and new software at the CFO integrates data from all 42 tide gauges in the STWS network, surge and wave forecasts, and meteorological data within one workstation, thereby automating much of the previously labour-intensive, hand-drawn charting.

Despite these and other possible future improvements, the general STWS system cannot provide detailed forecasts for all flood-risk points along the coast. Accurate *local* forecasts must depend ultimately on the modelling, interpretation and extrapolation to the local coastline and flood defences of the forecasts of water levels and wave heights provided by the STWS for its main reference locations. The tidal forecasting and warning system covering the coastline of the former Yorkshire NRA provides a representative example of such a local model.

Local Coastal Forecasting System, Yorkshire, England The Yorkshire North Sea coast includes not only 150 km of open-sea coastline from just north of Staithes to Spurn Point, but also the major estuary of the River Humber, which penetrates inland some 90 km from Spurn Point to Selby (see Figure 8.16). For this region, the STWS provides tidal forecasts for three locations, i.e. Immingham, Whitby and North Shields on the River Tyne 80 km north-west of Whitby. Wave and wind forecasts are provided for a single location off Flamborough Head.

The local forecasting system, described by Tinnion et al. (1995), became operational in 1992 and considers four significant factors—tide level, wave height and direction, lengths of vulnerable flood defences, and river flows—which can exacerbate tidal flooding close to the tidal limit. Tidal and wave conditions at the STWS reference stations are extrapolated empirically to each of the 13 tidal zones shown in Figure 8.16 and yellow, amber or red warnings issued as appropriate (see Table 8.8).

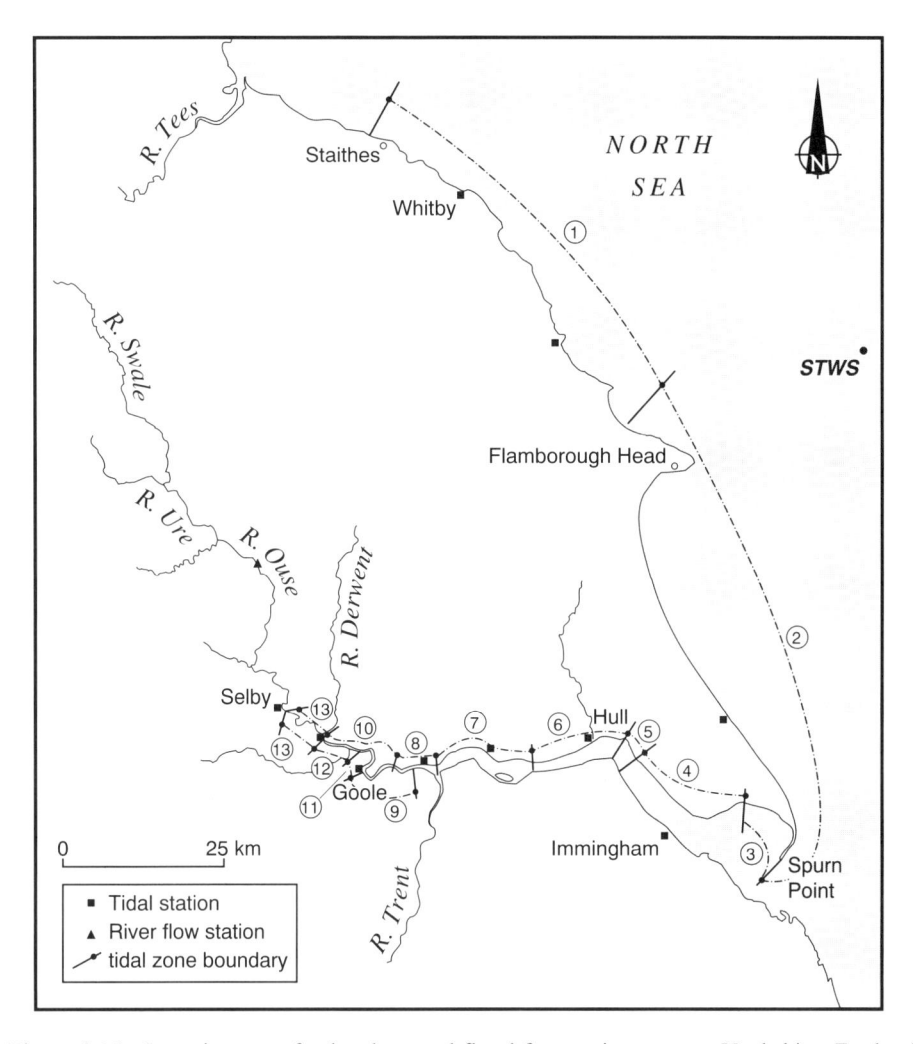

Figure 8.16 Location map for local coastal flood forecasting system, Yorkshire, England
Source: Based on a diagram in Tinnion et al. (1995)

Originally operated manually, the system is now fully computerised in real time, with inputs of weather and wind forecasts, river flow data from the RFFS, and tidal predictions generated by internal software. Coastal flood risk is calculated for each tidal zone and, where appropriate, an alarm is issued to the duty officer. The system is being extended to cover the Tees estuary and later the Northumbrian coast.

8.12.2 Tropical Storms

According to Pugh (1987) the reduced scale of tropical storms requires a smaller grid for their surge effects to be resolved by a shelf model than is the case for extra-tropical storms. For the surge-prone Atlantic and Gulf coastline of the USA (see also Section 5.3.2.3) the National Weather Service developed a depth-integrated finite-difference

Table 8.8 Environment Agency (EA) definition of colour-coded coastal flood warnings used in Yorkshire local coastal flood forecasting system

Colour code	EA definition
Yellow	Wind-blown spray overtopping defences. Flooding of minor roads and agricultural land
Amber	Waves overtopping defences. Flooding of isolated properties and roads
Red	General overtopping of defences with risk of breaches. Flooding of commercial and residential property

Source: Based on a table in Tinnion et al. (1995)

model called SPLASH (Special Program to List Amplitudes of Surges from Hurricanes). This model covers almost 1000 km of coastline from the Mexican border to New England and was designed for real-time surge forecasting. It was subsequently extended for the real-time forecasting of hurricane storm surges on continental shelves, across inland water bodies and along coastlines and for inland routing of water, either from the sea or from inland water bodies. This extended model, known as SLOSH (Sea, Lake and Overland Surges from Hurricanes) is also depth-integrated and uses a polar coordinate grid to allow high resolution in the coastal areas of interest. Six-hourly hurricane data, e.g. central pressure, wind characteristics, storm track, are fed into the model. For 24-hour surge forecasts the reliability of the model is especially vulnerable to inaccurate forecasts of hurricane position (Pugh, 1987) although these have improved greatly with the increased availability of satellite imagery.

Early work on modelling and forecasting surges from tropical cyclones was described by Sobey et al. (1977). A model, SURGE, was applied to 10 coastal locations in Queensland, Australia, to simulate storm-surge levels for a series of nine cyclones having three different intensities and landfalling in three different directions. Modelled water-level contours are shown in Figure 8.17 to illustrate two phases of surge propagation along the Queensland coast between Lucinda and Alva for an approaching cyclone with a central pressure of 933 hPa. Three hours before landfall (Figure 8.17A) local amplification of the surge was already occurring near Alva as the main surge front (0.75 to 1.0 m) was beginning to approach Townsville. Two hours later (Figure 8.17B) levels between Pallarenda and Alva were above 1.5 m, with local amplification to more than 2.25 m as the main surge front moved north of Townsville. And by the time of landfall levels exceeded 1.75 m along much of the coastline.

A basic problem with surge forecasting in many tropical areas is that the archiving of surge data is poor and risk analysis in some of the worst affected areas, such as the Bay of Bengal, is also poorly developed. This is partly because there are so few reliable tide gauges and partly because of inadequate collection of post-surge data which would allow reconstruction of surge profiles and verification of surge-forecasting models (WMO, 1990b). In such areas the World Meteorological Organisation has an important large-scale coordinating role, especially in relation to its basic World Weather Watch programme and projects specifically geared to the goals of the IDNDR, such as the Tropical Cyclone Warning System for the south-west Indian Ocean Region, and the

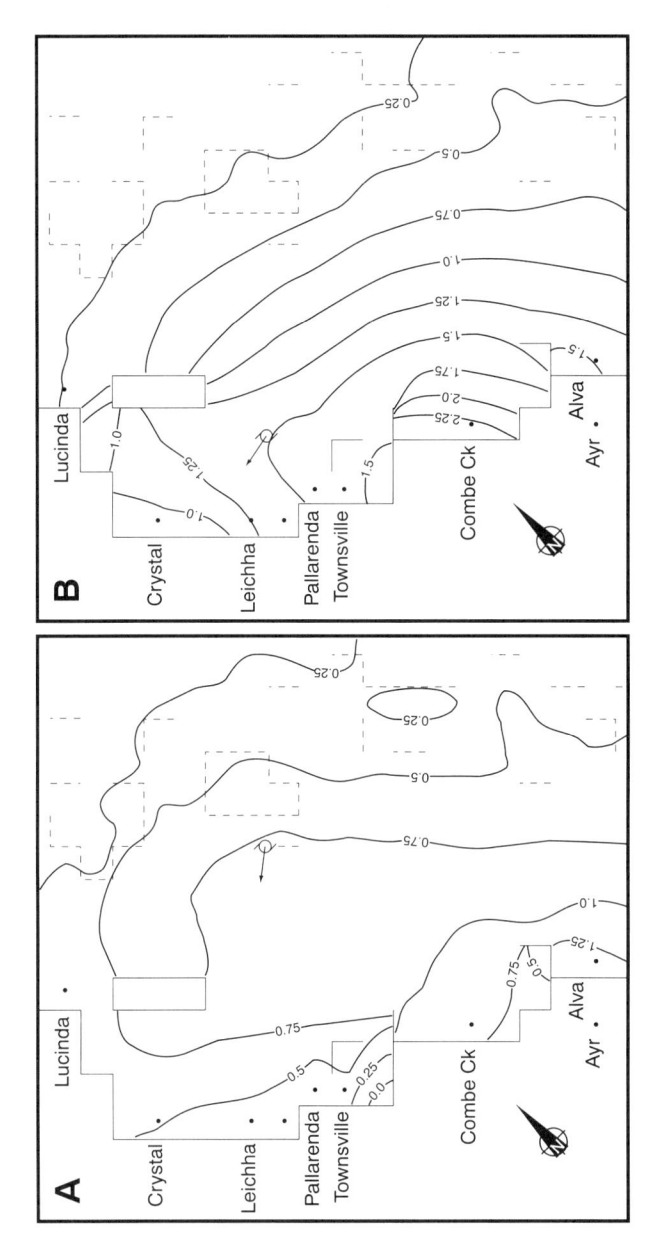

Figure 8.17 Storm-surge contours for a cyclone with a central pressure of 933 hPa as it approaches the coastline just north of Townsville, Australia, at an angle of 75°. Contours produced by SURGE at (A) 3 hours and (B) 1 hour prior to landfall

Source: Based on diagrams in Sobey et al. (1977) by permission of Department of Civil and Environmental Engineering

Comprehensive Risk Assessment project (WMO, 1989; 1990a). Continued improvements to the Tropical Cyclone Warning System are planned both in the specific context of the IDNDR and as part of the long-term programme of the WMO (WMO, 1992b).

Surge models for the Bay of Bengal and for the east coast of India were described respectively by Flather (1994) and Dube et al. (1994). The former combines the use of depth-averaged equations with a numerical scheme involving one-dimensional equations for narrow channels and two-dimensional equations for open sea. In this way it was hoped to simplify the development of realistic models for complex coastal areas which contain open sea, estuarine channels and intertidal banks. Hindcasts developed for known previous cyclone surges in both areas have produced encouraging results.

A potentially promising approach, described by Lai (1992), involves the probabilistic forecasting of tropical storm surges, using a dynamic and statistical algorithm.

8.13 TSUNAMI WARNING SYSTEMS

Tsunamis are triggered by tectonic events such as earthquakes and submarine slumping, and sometimes by submarine volcanic activity. If 'forecasting' is regarded as involving pre-knowledge of the event then, strictly speaking, tsunamis can be forecast only by forecasting the trigger events, several days or even weeks ahead (see Section 8.11). In most cases the imminence of a tsunami becomes known only when the location of a trigger event has been identified by which time tsunami wave trains have already been set in motion. Since these travel at high speeds, crossing the entire width of the Pacific Ocean in less than one day, most tsunami-prone coastlines are afforded a warning lead time of only a few minutes or at most a few hours. In such situations the process of forecasting is subsumed by that of warning. Modern seismological techniques are based on processing the ultra-low-frequency part of the seismic spectrum and should greatly improve real-time recognition of the tsunami-producing potential of earthquakes (Okal, 1994).

Active submarine volcanoes are a potential source of tsunami waves. Just north of Grenada, in the Caribbean, the submarine volcano known as Kick 'em Jenny has erupted 10 times since the late 1970s and has grown in height by some 50 m. Volcanologists believe that the prospect of further eruptions poses a major tsunami threat to surrounding coastlines, especially that of Venezuela. It is hoped that submarine mapping and seismological monitoring may help to clarify this growing threat. Satellite observations of slight surface movement or groundshift, which is now recognised as a precursor to an eruption, may also eventually provide a basis for improved forecasting systems of volcanic activity.

8.13.1 The Pacific Tsunami Warning System

The first effective tsunami warning system was developed for the Pacific Ocean basin in 1948 after the tsunami disaster in Hawaii which resulted from the Aleutian earthquake of 1946 (see Table 5.5). This system, known initially as the Seismic Sea-Wave Warning System (SSWWS), consists of a network of more than 50 tidal stations (planned eventually to increase to 120) and 31 seismographic stations. This Pacific Warning System is centred on Honolulu, Hawaii, where the US National Weather Service operates a coordinating centre, with stations scattered from Alaska to the southern tip of South America and west to Hong Kong (US Coast and Geodetic Survey, 1965; Bolt et al., 1975).

Water movements resulting from normal winds and tides are disregarded but when waves with a critical period are formed, a warning is automatically triggered and transmitted to Honolulu. Simultaneously, the earthquake epicentre is determined from the seismographic information. The water movement and seismographic data can then be combined with information on wave-travel times and ocean depths to enable forecasts of the speed of travel and probable time of arrival of the tsunami at selected locations. The system has been operated by the US National Oceanic and Atmospheric Administration since 1973 and, as in most areas of forecasting, its effectiveness has been greatly increased by the subsequent incorporation of satellite data.

A tsunami generated by an earthquake in the Eastern Aleutian Islands would take five hours to reach Hawaii; one generated by an earthquake off the coast of Ecuador would take just over 12 hours to reach Hawaii and 20 hours to reach Japan. It should be possible, therefore, to provide effective forecast lead times and indeed the system has been very successful in reducing loss of life during tsunami events. However, the system may lead to local under- or over-warning because tsunami waves which impact disastrously in some areas appear minor elsewhere (Hindley, 1978). These local variations, caused by factors such as the morphology of the sea floor and the location and distribution of islands and bays, are still not fully understood. As a result specific disasters have sometimes triggered the establishment of local responses. For example, the Alaska Tsunami Warning Center was set up in 1967, following the disastrous Alaskan earthquake and tsunami of 1964, and in 1982 this regional responsibility was extended to British Columbia, and the Pacific North-Western states of the USA. Improvements to the computer-monitoring and warning-dissemination procedures were described by Sokolowski et al. (1990).

In addition to geographical variations in the effectiveness of a tsunami forecast, there can be a significant difference in the time interval between the receipt of, and an effective response to, this forecast. It has been suggested that only a few of the 22 countries most at risk around the Pacific have standard operating procedures for immediate evacuation or reliable, rapid communication systems capable of receiving real-time warnings from the Pacific Tsunami Warning Centre (UNEP, 1991b).

8.13.2 Other Tsunami Warning Systems

Apart from the Alaskan Tsunami Warning system referred to in the preceding section, there are other regional forecasting and warning systems within the Pacific basin. For example, the Japanese Meteorological Agency has, since 1952, maintained its own system which is designed to issue a warning within 20 minutes of a tsunami-producing earthquake occurring within 600 km of the Japanese coastline.

Also in 1952 the then USSR established a tsunami warning system following the June Kamchatka tsunami disaster. This system was based on a network of seismo-tsunami stations at Petropavlovsk, Kurilsk and Uthno-Schkhalinsk and was initially supervised by the Hydrometeorological Service of the USSR (Bolt et al., 1975). Some problems have arisen because of the remoteness of the region and its proximity to areas of tsunami formation but, as with the Pacific system, the forecasts have significantly reduced the adverse impacts of tsunamis.

Many coastal areas remain more or less unprotected and in the LDCs, in particular, there is a clear need for reliable, low-cost local warning systems (Smith, 1996). Project THRUST (Tsunami Hazards Reduction Utilising Systems Technology) aims to meet that need by combining the advantages of rapid warning via satellite communication links with existing tsunami warning methods. A pilot system was developed for Chile (Bernard and Behn, 1985) and a major upgrade, implemented in 1989, reduced the response time from 88 seconds to 17 seconds and enlarged the broadcast area by 50% (Bernard, 1991).

Tsunami warning systems are now being considered even in comparatively low-risk areas. For example, Chasse et al. (1993) described numerical simulations of tsunami waves for four different earthquake epicentres which were intended to provide a basis for a future tsunami warning system in the St Lawrence estuary, Canada.

CHAPTER 9

Mitigating and Managing Flood Losses

CONTENTS

9.1 Introduction..293
9.2 Disaster aid...295
    9.2.1 Aid within the MDCs...295
    9.2.2 International aid..298
9.3 Insurance...300
    9.3.1 Commercial insurance...302
    9.3.2 Government insurance...304
    9.3.3 The US National Flood Insurance Program 305
9.4 Emergency planning and disaster management 308
9.5 Land-use management..312
9.6 Living with floods...319
    9.6.1 The need for a new approach..319
    9.6.2 Turning theory into practice 320
    9.6.3 Living with floods in Bangladesh..323
Appendix 9.1 Subdivision of Special Flood Hazard Areas......................................326

9.1 INTRODUCTION

A recognition of the problems associated with flood engineering has led to a progressively greater emphasis on non-structural methods. As indicated in Chapter 2, responses based primarily on flood control have been modified by a more integrated approach including measures such as insurance, forecasting, warning and evacuation and land-use planning. This combined approach is intended to reduce human vulnerability to flooding rather than to rely exclusively on a physical confrontation with flood events. The advent of non-structural measures can be viewed within the wider context of the need for all flood-prone communities to develop more hazard-effective and sustainable relationships with their environment in an era of unified floodplain or coastal zone management. Within some LDCs, debate still continues about the relative merits of structural schemes, although there is a growing realisation that they may be inappropriate for long-term development given the relative lack of financial resources coupled with concerns about land degradation. Even in countries like China, with a strong reliance in the past on structural schemes, a survey by Rasid et al. (1996) found that residents in the Yangtze delta expressed overwhelming support for a mix of structural and non-structural measures, including insurance. In some cases, the increasing criticism of engineering projects has led to the emergence of a so-called 'living with floods' philosophy. This can mean the adoption of a sophisticated package of management measures, as in the MDCs, but it also implies a more explicit reliance on the indigenous flood hazard-coping strategies which have been evolved by local communities, especially in the LDCs, over long periods of time.

Non-structural flood mitigation and management methods are varied. The prime intention of *loss-sharing* responses is to provide rapid compensation for flood victims in order that they, their households and their community can get back to normality as quickly as possible. The two main mechanisms for this are *flood aid*, which has proved to be essential in the LDCs, and *flood insurance*, which is still mainly a feature of the MDCs. It is generally accepted that neither of these responses is perfect as a long-term measure. But, in the immediate aftermath of a major flood disaster, the plight of the poorest people, even in the MDCs, requires the supply of emergency aid to save lives and reduce suffering. For wealthier victims, flood insurance provides a more reliable, and perhaps more dignified, way of alleviating the financial hardship.

As originally conceived, aid and insurance do nothing to mitigate and reduce the risk of future disasters. Indeed, such measures may be counter-productive if they continue to encourage settlement in high-risk areas in the expectation that flood victims will be fully compensated through some form of financial subsidy. Because the real costs of flood disasters have continued to rise, the overall effectiveness of aid and insurance has—like flood defence earlier—come under more critical scrutiny. The increasing uses of disaster aid for improving basic infrastructure and the equivalent moves to make insurance reduce future, as well as redistribute present, economic losses have blurred the distinction between short-term flood loss alleviation and longer-term flood mitigation measures. Above all, such developments have highlighted the need for integrated national water planning and coastal zone management policies, which include flood loss reduction, to be adopted by federal, state and local authorities.

These trends have led to more comprehensive floodplain and coastal management strategies as an overall community programme for reducing future flood damage whilst also maintaining the natural values of floodplains and shorelines. The keys to short-term damage reduction are *preparedness and emergency planning* and in the longer term *land-use management*. Apart from their own merits, these approaches have desirable attributes which support other flood hazard responses, such as flood defence and forecasting and warning schemes. For example, they involve public education and encourage community responsibility, at least in the MDCs where individual locational decisions are often important. In the USA there are many organisations concerned with floodplain management, the majority of which are state-wide or regional in geographic scope. The Floodplain Management Association provides education and information at the local level while the larger Association of State Floodplain Managers is more active in influencing national policy and legislation. Floodplain management has sometimes been seen as a local responsibility and, in some cases, has failed to attract necessary levels of support from central government. This situation has tended to accelerate the realisation that all communities at risk have to become more responsible for their own destiny.

Floodplain management concedes that some flood losses will occur. In turn, this has led to an appreciation that a *living with floods* attitude is increasingly important if floodplain communities are to survive into a future where the financial and environmental costs are minimised. The undeveloped parts of floodplains provide natural resource benefits, including wetlands and other zones necessary for a well-functioning ecosystem. In the MDCs, legislation now protects such areas. For example, in the USA the US Army Corps of Engineers Section 404 Permit Program is designed to preserve the integrity of the nation's remaining wetland areas. In the LDCs a growing awareness of the economic and environmental price of flood structures has also brought calls for more sustainable approaches which depend on the rediscovery of traditional community responses to the flood hazard.

The success of non-structural strategies is vitally dependent on the understanding and cooperation of the people at risk. Political wrangles about who should pay for, and who should benefit from, all forms of flood mitigation measures have greatly increased in recent years. Given the shift in attitudes towards longer-term solutions and the belief that, wherever practicable, it is the risk-taker who should bear the costs, it is possible to identify a number of key questions surrounding efficiency and equity in non-structural flood responses such as:

- To what extent does the availability of disaster relief limit the adoption of other hazard reducing measures, including insurance?
- How best to provide affordable insurance cover to those who need it most without penalising the majority of low-, or non-risk, policy holders?
- Can emergency aid be linked effectively with longer-term mitigation strategies?
- Can we ensure that risk-taking for local personal or community profit is not subsidised by national funds from the taxpayer?
- Is it possible to prevent either disaster aid or insurance from encouraging further development in flood-prone areas?
- How can land-use planning for the flood zone cope effectively with market forces?
- Given the strength of the flood defence lobby, can indigenous, low-cost responses to flood hazard be widely adopted?

9.2 DISASTER AID

Disaster aid is largely donated by governments, at either the national or state level, and by charitable Non-Governmental Organisations (NGOs), like the Red Cross and Red Crescent Societies and OXFAM. Even in developed countries, a good deal of assistance is provided by voluntary relief agencies. For example, following the 1993 Mid-West floods, the American Red Cross spent more than US$ 30 million in flood relief, sheltered 14 500 people and served more than 2.5 million meals. In the case of flood disasters in the LDCs, the extent of reliance on foreign aid or the extended family, rather than the national government, may be almost total. For example, Table 9.1, from a local survey by Haque and Zaman (1993) after the 1988 floods in Bangladesh, shows that the principal sources of assistance to villagers were relatives and relief agencies.

9.2.1 Aid Within the MDCs

For smaller-scale flood disasters, voluntary contributions from the public, often coordinated by a relevant local authority, form the bulk of disaster aid. The amount of money collected is often roughly proportional to the degree of devastation and the perceived need, with local events attracting aid from the immediate area and larger

Table 9.1 Sources of assistance (%) received by the flood victims in Sreenagar, Bangladesh, after the 1988 floods

Sources of assistance (multiple response possible)	All households ($n = 280$)
Relatives	78.6
Other villagers	32.9
Local government	7.1
National government	4.3
Relief agencies/other institutions	51.4

Source: After Haque and Zaman (1993)

events stimulating nation-wide contributions. Many appeals raise less than half of the reported losses. If contributions are confined to the disaster area, there is no increase in overall resources for asset replacement and, although wealth redistribution will be of immediate assistance to the disaster victims, this method is neither socially reliable nor financially efficient.

Following a severe flood, it is often left to governments to spread the financial load throughout the whole tax-paying population. This may be achieved either by the advance creation of a designated disaster fund, with legislative arrangements for its distribution, or through *ad hoc* payments from general funds. Typically, disaster arrangements incorporate a formula whereby the national fund contributes at some agreed ratio to local spending once the flood impact has exceeded a threshold figure or the local authority has expended some minimum proportion of its annual budget. Such ratios can be changed in very exceptional circumstances. In Australia, the Natural Disasters Organisation was founded in 1974 within the Department of Defence. Subsequently, financial agreements between the commonwealth government and the states were formalised into the National Disaster Relief Arrangements whereby, if necessary in any year, each state provides a pre-set sum for disaster relief. If payments exceed this threshold, further assistance is jointly funded by the state and the federal government. Not all central government assistance is in the form of direct pay-outs and a substantial proportion may be given as interest-free repayable loans, reconstruction grants and tax write-offs.

In the USA the evolution of federal policy can be traced back through the Disaster Relief Acts of 1950 and 1974. Since 1950, US Presidents have been able to declare disasters and to direct aid to the victims. More than 80% of all Presidentially declared disasters include flooding, and the costs of disaster relief can be high. After the 1993 Mid-West floods, Congress allocated a total of US$ 6.2 billion for disaster relief (Wilkins, 1996) in response to pressure from the public and the media for overt political action. Internal US disaster aid is currently administered through some 12 different federal programmes. Many developed countries retain an arm of government which is charged with the crisis response to major flood disasters. In the USA, the Federal Emergency Management Agency (FEMA) has the lead responsibility for domestic emergency management which includes both internal disaster aid and flood insurance.

The Disaster Relief Fund of the FEMA is the major source of federal disaster recovery assistance to state and local governments. To replenish the fund, FEMA requests annual appropriations from Congress that are based on an average of annual fund expenditures over the previous 10 years. When incurred costs rise above average, FEMA relies on supplementary financial allocations. Flood aid is distributed by FEMA mainly under the Robert T. Stafford Disaster Relief and Emergency Assistance Act of 1974. The basic provision is that, following the declaration of a disaster, 75% of the costs can be met by FEMA with the remaining 25% paid by the appropriate state administration. However, the most typical form of federal disaster assistance is a loan that must be paid back with interest and the average Individual and Family Grant payment is limited to some US$ 2500. In 1988 the Act was amended in various ways. For example, a new category of private non-profit facilities (museums, zoos, libraries, etc.) was recognised as eligible for assistance. More importantly, the Hazard Mitigation Grant Program was authorised with the objective of minimising recurring damage and hardship by providing grants for longer-term mitigation projects, such as the acquisition of flood-damaged properties (buy-outs) and re-building in less hazardous areas. There are also other flood aid programmes in the USA, especially one-off type initiatives. For example, the National Park Service, US Department of the Interior, gave a US$ 5 million grant to the National Trust for Historic Preservation to aid recovery from the 1993 floods.

In the past, most governments have used disaster aid to compensate losses incurred to all uninsured property damaged by a flood. However, the rising cost of disaster relief has prompted criticism of such automatic support given a growing belief that government aid can provide both an inefficient subsidy to risk-takers and can also reduce the willingness of people to make their own preparations for disaster. In the USA an initial debate was triggered by the cost of direct property damage suffered through Hurricane Andrew in 1992 which, at some US$ 18 million, was about three times the equivalent losses in the 1993 floods. But, in 1993, the financial burdens included US$ 1.7 billion allocated in emergency pay-outs by the US Department of Agriculture for agricultural losses. Although these payments provided some immediate financial stability (Zacharias, 1996), they were additional to crop insurance indemnities and it is questionable that such funds should have been released given the participation of only 35% of farmers in the Multiple Peril Crop Insurance Scheme regulated by the USDA as part of its Federal Crop Insurance Corporation. Other critics would go further and regard such disaster aid as a government subsidy to floodplain agriculture which benefits individual farmers and agricultural communities but may not be in the overall national interest. For example, the costs may be high in terms of both taxation levels and the environmental consequences of intensive farming on floodplains. Moreover, such expenditures will continue to be incurred after similar disasters in the future.

Rather like structural measures, disaster aid in the MDCs has come under scrutiny because of the increasing costs, occasionally counter-productive consequences and a growing uneasiness about the transfer of income from the general tax-paying public to a minority of floodplain occupants who suffer from repeated floods. In Canada, for example, a steep rise in flood disaster payments during the early 1970s by the federal government led to the establishment of the National Flood Damage Reduction Program in 1975 (Bruce, 1976; Andrews, 1993). Amongst other things, the Program fostered agreement between the federal and provincial governments that flood disaster assistance would not be provided for any new development in designated flood-risk areas (except for flood-proofed development in the flood fringe). This has led to a move towards more self-reliance, including insurance provision.

Although calls for government-sponsored relief are not uncommon following disastrous floods, these calls are increasingly resisted within the MDCs on the grounds that the taxpayer cannot be expected to fund losses which should have been insured. In the USA, federal disaster assistance is awarded in less than half of all flooding incidents. This attitude is sometimes reinforced by legislation which limits aid, in the form of either grants or loans, to uninsurable losses. The search for alternative strategies has included the possibility of refusing disaster payments to households over certain higher income thresholds and tying assistance more closely to the enforcement of building codes. On the other hand, some targeted schemes have not been successful. For example, after 1987 flooding in suburban Chicago, the Illinois Housing Development Authority made low-interest loans up to US$ 5000 available to finance flood-proofing measures for flooded households with a maximum annual income of US$ 35 000 (Anon., 1993). Some 18 months later only 14 loans, totalling US$ 50 000 out of a designated sum of US$ 500 000, had been allocated, suggesting that the scheme had been inadequately publicised and too restrictive in terms of the size of the loans and the household income limit.

9.2.2 International Aid

Disaster aid is transferred bilaterally from one nation to another by governments and charitable bodies. Often, such transfers occur under the auspices of the UN Disaster Relief Organisation (UNDRO), which exists in Geneva to coordinate the efforts of the many organisations involved in disaster relief. In 1992 a new Department of Humanitarian Affairs (DHA), was created with offices in New York and Geneva. This was in recognition of the need to deal more effectively with the political constraints which hamper the provision of humanitarian assistance in disasters and also to ensure a smoother transition from immediate relief to longer-term rehabilitation and development. Several countries have their own coordinating bodies for overseas disaster reduction. In the USA the Office of Foreign Disaster Assistance (OFDA) operates within the Agency for International Development which coordinates US responses to overseas disasters. This involves the development of early warning systems, technical assistance to strengthen disaster-related institutions in vulnerable countries and the capability to rush life-support goods and services to disaster victims anywhere in the world.

International aid flowing from the MDCs to the LDCs can be unreliable and may not reflect the true need. For example, tropical cyclones, which result in a high ratio of casualties to survivors, usually generate a large donor response irrespective of need. River floods tend to produce a lower response, despite the relatively large numbers of survivors who will be adversely affected and in need of support. However, major disasters prompt significant responses. For example, following floods in Sudan in 1988, there was damage to agriculture, property and social services totalling US$ 1 billion. Some 200 000 homes were either damaged or destroyed and about 2 million people were left homeless. After the emergency relief phase, the World Bank helped the Sudanese government to prepare a US$ 408 million flood reconstruction programme to present to the donor community (Brown and Muhsin, 1991). Sectoral funding was allocated, as shown in Table 9.2, for investment aimed at longer-term rehabilitation. The World Bank coordinated the entire scheme but left the receiving national government with some ownership of the process.

Table 9.2 Funding allocated for reconstruction aid after flooding in the Sudan in 1988

Sector	Local cost	Foreign cost (US$ millions)	Total cost
Agriculture	33.8	63.6	97.4
Rural water	6.6	17.4	24.0
Education	11.9	24.3	36.2
Health	5.9	32.7	38.6
Industry/construction	15.0	35.3	50.3
Power	5.9	29.0	34.9
Telecommunications	3.3	31.1	34.4
Transportation	7.9	25.6	33.5
Urban	31.3	25.0	56.3
Programme coordination and flood prevention	0.6	1.4	2.0
Total	122.2	285.4	407.6

Source: After Brown and Muhsin (1991)

Most of the support is triggered by appeals for help in the emergency relief period immediately following a disaster. After the search and rescue phase, the priority is for medical support (Beinin, 1985). Some flood disasters in the LDCs can create epidemics out of health problems which are already endemic in the country, such as diarrhoeal, respiratory and infectious diseases. In all these cases, the use of local medical teams is preferable to the deployment of expatriate expertise because the former can be mobilised more quickly and are culturally integrated with their patients. This suggests the importance of emergency preparedness training for local medical staff and volunteers.

It is difficult to disentangle disaster and development problems in the LDCs and to make reliable assessments of disaster aid as an adjustment to flood hazard. But, wherever possible, emergency aid should be minimised. Given the fact that some immediate external response will always be necessary in certain cases, attention should be given to optimising this form of relief so that aid can be targeted to improve the situation of the most vulnerable people. The possibility of dependency exists if long-term aid is given to the LDCs although the problem is likely to be greatest for more protracted natural disasters, such as drought, rather than floods.

The more that funds are invested in reconstruction, rather than relief, the greater is the possibility that disaster aid will bring longer-term benefits. Many of the international charitable agencies have changed the emphasis of their work from disaster relief to the reduction of hazard vulnerability, especially amongst the poor, and funding bodies now seek to support more integrated, multi-sectoral development aimed at environmental protection and coordinated flood prevention. For example, Munasinghe et al. (1991) showed how a US$ 393.6 million flood reconstruction package was assembled after floods in the Rio de Janeiro metropolitan area during 1988 left 290 dead and caused damage to the infrastructure and public services. Following the immediate repair of roads and bridges, the emphasis was on strengthening the local authority's institutional and financial ability to promote better urban development and environmental planning. This involved programmes to reverse decades of environmental degradation by improved sanitation and solid waste disposal methods, protection and extension of the forested areas as well as re-settling some 5000 refugee families who lost homes in the flood. Similarly, Kreimer and Preece (1991) cited the case of Taiz, Yemen Arab Republic, where flooding disrupts part of the town up to 10 times per year leaving sediment accumulations up to one metre deep at major road intersections. Apart from the need for localised structures, such as box culverts and sediment traps to protect the most vulnerable areas of the city, it was recognised that more support is required for technical assistance and staff training in order to promote better project management and environmental planning. The flood action plan devised for Vietnam is a further example of comprehensive aid-related planning incorporating 18 interlinked activities ranging from typhoon forecasting to the setting up of a Disaster Relief Management Unit (Department of Humanitarian Affairs, 1994).

9.3 INSURANCE

Like disaster aid, insurance is primarily a method for redistributing flood losses. The difference is that people at risk join forces, in advance, with a large financial organisation to spread the cash burden from one major flood disaster over a number of years through the payment of an annual premium. Assuming that the premiums are set at an appropriate rate, the money received from policyholders can be used to compensate the minority who suffer loss in an individual year. Flood insurance schemes may be underwritten by private companies or national governments. In some cases, a mixed approach is adopted. Because of the high direct costs of flooding, insurance is perceived as an important loss-sharing strategy within the MDCs. According to Blaikie et al. (1994), the general lack of insurance in more traditional societies makes people suffer twice, firstly because they lose goods, such as crops and livestock, and secondly because they lose the time necessary to replace them. As a result, it takes much longer for what passes as normality to be restored. On the other hand, flood losses are often shared amongst an extended family or kinship network. In addition, some LDCs are now moving away from the allocation of disaster relief by central government towards agriculturally based disaster insurance, as in China (Fan, 1991).

Insurance against flood losses is often taken out as part of a wider package of insurance against other natural hazards. As such it can be difficult to isolate flood losses, especially following storms in coastal situations. It is generally accepted that most hurricane damage to property occurs as a result of high winds, rather than flooding. For example, only 10% of the insured losses after Hurricane Hugo in 1989 were flood-related (Anon., 1995).

Flood insurance poses special problems. Individual events can create large losses, as when poorly constructed properties on extensive floodplains or coastal lowlands are damaged. Such disasters impose severe financial burdens because of the unfavourable ratio of claims to premiums. In addition, the tendency for large claims to follow years with few losses makes premium setting difficult and the funding of claims unpredictable. These problems are exacerbated by *adverse selection* which occurs when a large potential loss is spread over a relatively narrow policyholder base. Flood insurance cover can easily be dominated by bad risks because only flood-prone households are likely to seek flood insurance. In the past, individual private companies exposed in this way have become bankrupt, such as the one formed to sell flood

Table 9.3 Effect of insurance on action to prevent water entering residential properties in the UK

	No insurance	Insured
No action	50%	76%
Took action	50%	24%
No. of cases	20	131

Source: After Handmer (1990)

insurance after the Mississippi valley floods of 1895 and 1896 which was quickly wiped out by the flood of 1899 (Hoyt and Langbein, 1955). Therefore, policy underwriters try to ensure that the property they insure is spread over diverse geographical areas so that only a fraction of the total value at risk could be destroyed by a single event.

Other problems are created by the behaviour of the insured people. *Moral hazard* arises when some insured persons deliberately reduce their level of care and thus change the risk probabilities on which the premiums were based. Some people, knowing that they will be compensated in full, may be reluctant to undertake emergency flood proofing, move furniture away from rising floodwater or engage in salvage activities after the event. Table 9.3 shows the effect of insurance on the willingness of some UK flood victims to take action in floods. It has been suggested that insurance should be offered only to those who adopt better standards of building construction. On the other hand, it may be difficult, or prohibitively expensive, to introduce flood proofing to existing structures. Many property owners buy flood insurance policies when they first obtain mortgages but drop them later. Others, especially those located on large rivers served with a reliable flood forecasting system, may even take out insurance cover in the few days before a major flood is expected to arrive. Where disaster aid is known to be available, some floodplain residents will elect either not to take out commercial insurance, or deliberately underinsure their property, and rely on external support.

Even worse, insurance may encourage invasion into flood-prone zones. The very availability of flood insurance guarantees reimbursement and removes the risk of catastrophic financial failure (Lind, 1967). Floodplain land use might be deemed optimum if the long-term benefits exceed the long-term costs but, in order to preserve any financial benefit, flood insurance should cost at least as much as the average annual damages. If the costs are less because insurance premiums are subsidised, or if additional financial benefits are available—such as a river frontage, fishing rights or outstanding coastal views—a property owner is more likely to assume the risk of flood-prone land and floodplain invasion will occur (Smith and Tobin, 1979).

Floodplain invasion may persist even when insurance premiums are geared to risk. Arnell (1987) made a distinction between the financial motives of speculative builders, who develop flood-prone land, and the occupants of residential property. But, unless insurance premiums are raised sufficiently high to deter a potential house purchaser and thereby negate a sale, which is unlikely, both parties have comparatively little to lose in relation to the overall profits and costs involved. The deterrent factor of insurance premiums may be more effective for commercial premises, which are often built by the eventual occupier, but even here premiums would have to be both substantial and compulsory to limit further floodplain invasion. In the areas of highest risk, private insurance may be unobtainable, but is sometimes replaced by government schemes. Since the advent of the US National Flood Insurance Program in 1968, shoreline construction has arguably been made more affordable, i.e. less risky, by the federal government. About 40% of claims on the Program have been from properties flooded at least once before. Half of these repeat claims were from coastal properties involving beach erosion, although others were due to river flooding or stormwater problems (Anon., 1995).

9.3.1 Commercial Insurance

Flood insurance is sold by commercial companies in many countries. Financially, this mechanism is more reliable than disaster relief and appeals to those philosophically committed to the private market. In Britain, comprehensive flood cover for household contents has been a feature for many decades although cover for buildings was typically excluded before 1961. Following river floods in the 1960s, this protection was introduced because of a concern amongst insurance companies that the government might interfere in the industry (Arnell et al., 1984). Since then householders have bought flood cover within comprehensive insurance policies (also covering fire, storm and subsidence) which are separate for buildings and their contents. Insurance companies have tended to charge a flat rate per £1000 of cover for houses, which clearly amounts to a subsidy from the property owners with a low, or even zero, flood risk to those with a high flood risk. It does nothing to encourage those in high-risk areas to protect against future damage and might even provide an incentive for further floodplain invasion. But the advantage for the insurer is that, since buildings insurance is required by mortgage lenders, the policyholder base for flood insurance is very much widened in this way. In addition, individual policies limit the total amount payable under any claim and the company can control the number of policies which it sells in a particular flood-prone area.

Insurance companies also share the risk amongst themselves. This is achieved either by companies joining together to write the primary insurance cover or by passing on part of the risk via *reinsurance*. A typical reinsurance arrangement might be for the primary company to pay the first US$ 5 million in losses from a flood but, for losses in excess of US$ 5 million, the primary company would be entitled to reimbursement of 90–95% of the sum from its commercial partners (AIRAC, 1986). The reinsurance market is international in scope. In this way, risk can be spread from the individual hazard-zone occupant to the world markets, although some companies find it impossible to obtain all the hazard reinsurance they wish to buy.

A major weakness of commercial insurance is that not all flood-prone households elect to take cover and many of those that do will be underinsured. This can happen even when insurance cover is technically required as a condition of obtaining a mortgage. After floods in Texas in 1989, Kunreuther (1993) reported that 79% of owners of damaged property, who were supposed to take out flood cover when granted a mortgage, were uninsured at the time of the disaster. Similarly, only 10% of the buildings damaged in the 1993 Mid-West flood were believed to be covered by flood insurance. Clearly, such low involvement limits the efficacy of insurance as a hazard management tool, although it may occasionally benefit the industry by making claims manageable after a very severe flood. Tenants, pensioners and other lower social status householders are least likely to have adequate cover and are least likely to recover financially after a flood. In the UK, Handmer (1990) cited the 1987 flooding of Strabane, Northern Ireland, a town with a male unemployment rate of over 40% where two-thirds of the flooded households were uninsured at the time.

In summary, there is an absence of a well-functioning flood insurance market in many countries. This inhibits individuals, companies and communities from taking cost-effective steps to reduce future losses. There is no incentive to avoid either new building or occupancy in high-risk areas if losses are to be funded by the general taxpayer. In Britain, flood insurance has not been used as a tool for hazard mitigation. Until very recently, the premium paid by private householders took no account of individual exposure to flood hazards and there was no incentive for policyholders to adopt damage mitigation measures, such as flood proofing.

Private insurers are expected to carry sufficient surplus to keep the probability of insolvency at an acceptably low level. The ratio between the level of surplus and the liability in written cover is known as the 'leverage'. For example, under the Insurance Act of 1973 the Australian insurance industry must adhere to strict rules of financial solvency in order to meet claims arising from all the policies issued. To maintain this obligation the industry would need to rate premiums according to the risks but there is a general lack of information on spatial and temporal flood risk. In the past, the accumulation of a large catastrophe fund has been prevented in some countries by taxation designed to prevent unscrupulous companies manipulating such reserves in order to hide excessive profits (Lockett, 1980). However, government concerns in the UK about the recent rapid growth of insurance claims following floods and other natural disasters may well encourage legislative changes allowing some money, from household and other policy premiums, to be put away tax-free into special catastrophe reserves.

In principle, insurance should provide an equitable distribution of costs and benefits. For active, well-defined floodplains, it may be possible to map areas subject to different flood frequencies but it is sometimes impractical to estimate risk levels for individual properties. A step towards setting premiums more in line with the actual level of risk has recently been taken by a large sector of the UK insurance industry. The advent of Geographical Information Systems (GIS) technology has allowed insurers to accumulate large data-banks of past claims. Spurred on by the recent increase in claims for storm, flood and subsidence damage, several companies now charge householders a premium based on the post-code district, which contains a small group of properties. In some cases premiums have fallen but, if the property has been placed in a higher than average risk band, the annual cost of premiums may have risen by 15–20%. Once properties have been built, it is theoretically possible for insurers to offer lower premiums to policyholders who have taken measures to protect their property. Such measures might include special construction methods and building materials. In extreme cases, insurers could require property owners to retrofit flood-proofing measures before accepting any new policy. Little is known about this potentially important loss-reducing aspect of insurance. Kunreuther (1978) found that those with either flood or earthquake insurance were more likely to have adopted protective measures than those without insurance, although this combination probably reflected a general concern for hazard rather than any premium incentive.

The insurance industry is particularly keen to reduce flood loss claims. Before 1988, the insurance industry worldwide had never faced losses from a single disaster that exceeded US$ 1 billion (Anon., 1995). Since then there have been at least 14 events which have cost more than US$ 1 billion. As a consequence of the large insurance claims following hurricanes Hugo (1989), Iniki (1992) and Andrew (1992), plus the Northridge earthquake (1994), several companies in the USA were forced out of business and the insurance industry became less willing to provide residential cover in what are perceived as flood high-risk areas. For example, companies became reluctant to underwrite further cover in parts of Florida, Hawaii and coastal South Carolina, and state catastrophe funds were required to provide for residents unable to buy policies on the private market. The industry is now much more aware of the need to make better preparations for disaster and adopt new loss-reducing measures (National Association of Independent Insurers, 1994).

The current problems within the US insurance industry have been attributed to several causes (Anon., 1995):

- The coastward shift of population over the last 30 years which has raised the value of property along the Gulf and East coasts at least six-fold over the period. This has greatly increased the potential for catastrophic hurricane losses.
- The rising cost of reinsurance on the world markets which has quadrupled in price since 1989. This has meant that primary insurers have had to assume direct responsibility for a higher proportion of the potential losses.
- The consistent failure of local governments to adopt and enforce stringent building codes. For example, before Hurricane Andrew many buildings in south Florida were poorly designed and constructed and, therefore, very likely to lead to insurance claims.
- The availability of federal disaster assistance for those whose uninsured property has been damaged in any natural disaster, even if they adopt low standards of care. This has led to a culture which tolerates inadequate building construction and maintenance but still believes in automatic compensation following disaster.

9.3.2 Government Insurance

Some of the problems associated with commercial flood insurance can be overcome by government schemes. Such a measure not only greatly widens the policyholder base but can also be used to provide the homeowner with the technical information required for more accurate risk estimation and flood proofing of buildings. This would enable premiums to be related even more sensitively to the potential loss and thus ensure that those living in very hazardous areas paid a realistic contribution towards the cost of their location. On the most risk-prone floodplains and seashores, government would be in a stronger position than private companies to encourage the relocation of existing property and to impose restrictions on the sale of insurance in order to stop further development. For example, national flood insurance programmes often make it a requirement that only properties constructed to certain specified building standards in designated hazard zones may be eligible for insurance. On the other hand, some local authorities may decline to adopt the necessary building codes or zoning requirements and thus make it impossible for residents to participate in the scheme.

Several governments have provided insurance against flood disasters but the arrangements are not always ideal. In New Zealand, the original Earthquake and War Damage Act of 1944 was subsequently extended, through the creation of an Extraordinary Disaster Fund, designed to cover storms, floods and other natural disasters. The scheme is financed by a surcharge on all fire insurance policies of 5 cents per NZ$ 100 of insured value, of which 10% is placed in the Extraordinary Disaster Fund (Falck, 1991). The programme administrators are empowered to set higher premiums in high-risk areas and can refuse to pay for damages to poorly maintained properties. In practice, it has proved difficult to enforce appropriate building codes or to restrict development in high-risk areas because there is political pressure to pay virtually all the claims as presented. A similar scheme, embracing floods, was introduced in France in 1982 as part of a comprehensive natural hazard reduction programme. The insurance is funded by a mandatory 9% surcharge on all property insurance policies.

9.3.3 The US National Flood Insurance Program

In the USA, the National Flood Insurance Program (NFIP) was introduced by the federal government in 1968 because of the reluctance of private industry to continue providing cover. There was also concern about rising federal expenditure on flood aid, plus a desire to exploit the potential of non-structural methods in flood reduction (Arnell, 1984). Thus, the scheme was designed to provide both short-term financial assistance to flood victims and to establish better long-term land-use regulations for floodplains. Such planning regulations aim to restrict development so that no construction is permitted in the floodway and only flood-proofed buildings are allowed on the 1:100 year floodplain (Figure 9.1). This base flood standard was adopted in recognition of the benefits, as well as the costs, associated with floodplain and coastal development.

The NFIP covers most types of flooding resulting from the overflow of rivers and tidal waters, including hurricane activity. Community participation in the NFIP is voluntary, although some states require NFIP participation as part of their own floodplain management programme. If a community does not participate, no federal financial assistance can be provided for the permanent repair or reconstruction of insurable buildings. Insurance cover may be purchased directly from the government programme or through private insurance companies who are authorised to provide policies under the 'Write Your Own' initiative. Flood cover can be purchased at any time but there is normally a 30-day waiting period before the policy is effective to discourage opportunistic participation. Flood insurance claims are paid even if a disaster is not formally 'declared' by the President.

Under the scheme, the Federal Emergency Management Agency (FEMA) initially publishes a Flood Hazard Boundary Map (FHBM) which outlines the approximate area at risk from flooding. In order to join the first phase of the NFIP (the 'Emergency Program'), a community must agree to adopt certain minimum land-use regulations within this area. For example, any new development in the hazard area must be sanctioned by a building permit and be constructed so that it suffers minimal damage through an elevation and dry proofing requirement. Flood insurance is then made available at nation-wide subsidised rates.

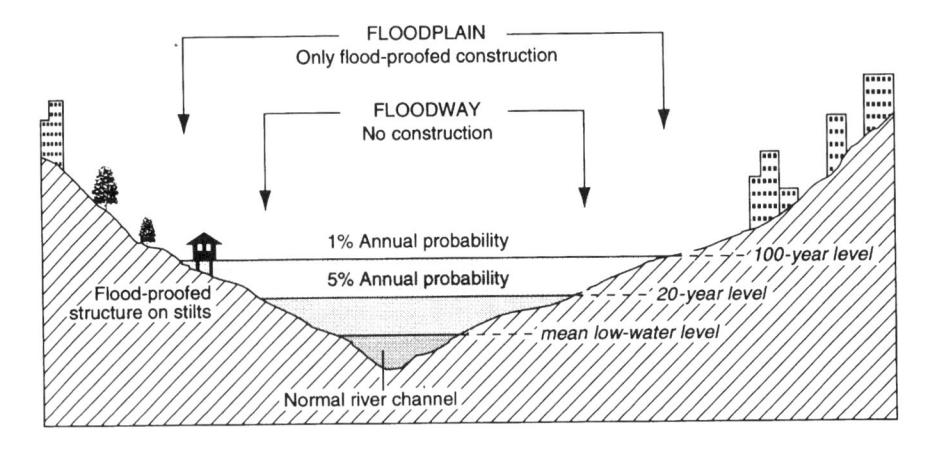

Figure 9.1 The delimitation of land-use planning zones on a hypothetical floodplain

If the hazard is judged to be serious, FEMA provides more detailed maps which identify Special Flood Hazard Areas (SFHAs). About 10% of the nation's 100 million households are located in such designated areas. SFHAs comprise the zone expected to be flooded by a 1:100-year event, defined as the area within which the 100-year flow can be contained without raising the water surface at any point by more than 0.3 m. SFHAs are subdivided into insurance risk rate zones according to the criteria indicated in Appendix 9.1. Areas between the 100-year and 500-year flood boundaries are termed 'moderate flood hazard areas' and areas beyond the 500-year level are termed 'minimal flood hazard areas'. At this stage the community must join the 'Regular Program' of the NFIP and implement more stringent controls. These include prohibitions on further investment in the lowest-lying areas and the elevation of residential property to at least the designated 1:100-year flood level. In addition, non-residential development may be flood proofed.

All new property holders within the designated 1:100-year floodplain must buy insurance at actuarial rates. To facilitate this, the floodplain is subdivided into risk zones on the basis of a Flood Insurance Rate Map (FIRM) which is used to allocate variable insurance premium ratings to individual properties. Accurate maps are essential for the success of the NFIP through the provision of information to planners making land-use decisions, to lenders determining mortgage requirements and to architects and builders designing and placing new structures. In order to improve the use of floodplain maps, the National Flood Insurance Reform Act of 1994 mandated the creation of a new Technical Mapping Advisory Council. The NFIP has encouraged more than 18 400 flood-prone communities in the USA (out of a national total of approximately 21 000) to adopt at least some floodplain regulations and has provided maps of flood hazard areas to nearly 20 000 communities. In 1993 about 2.8 million NFIP policies were in force. Since 1969 over US$ 7 billion has been paid out in claims and the average NFIP claim paid after the Mid-West flood was more than US$ 25 000.

But the scheme has been criticised. Carper (1990) claimed that only some 13% of insurable household units in Special Flood Hazard Areas (100-year floodplain) were actually covered by insurance. Other workers, such as Shilling et al. (1989), have questioned the transfer of national wealth to existing home-owners implicit in the NFIP when the scheme was paying out US$ 3 in claims for every US$ 1 collected in premiums. The NFIP has had only modest success in protecting properties subject to repetitive flooding. About one-third of all claims paid by the NFIP have been for damage in areas designated as 'moderate' or 'minimal' risk, mainly as a result of the inadequacy of local drainage systems, although Duryee (1990) argued that the federal programme was created to guide investment away from flood-prone sites nationally rather than to correct 'nuisance' drainage. Unfortunately, the NFIP has not always met the objectives in discouraging development in key hazard zones, such as regulatory floodways and eroding shorelines. In coastal areas a major criticism has been that FEMA has not mapped ongoing coastal or Great Lakes shoreline erosion and consequently has no knowledge of areas that might flood and undermine property within the next 10 years or so. Insurance cover has, therefore, continued to be available in these threatened areas, for example for properties built as close to the water as possible but raised on stilts to conform to the NFIP elevation requirements, when a better strategy would have been to assist people to relocate further inland. According to Rahn (1994), the biggest defect of FEMA's floodplain policy is the continuing reliance on the 100-year flood because an increasing percentage of the annual flood damage results from very large floods of low probability, such as the Mississippi floods of 1973 and 1993. Therefore, catastrophic flood disasters will continue to affect properties outside the FEMA-designated areas.

During the early 1990s, it became accepted that the NFIP needed reform (Millemann, 1993). Starting 1 October 1991, a move was made to lower insurance rates via the new Community Rating System (CRS) within the Program. Under the CRS, home-owners may receive reductions in their flood insurance premiums when a community demonstrates that its policies have either reduced flood losses or improved the sale of flood insurance beyond the minimum requirements for participation in the NFIP. All participating NFIP communities may apply for classification under the scheme whereby the top rating would provide a 45% premium credit. As a result, it has been estimated that flood insurance premiums will drop 5% for about 400 000 home-owners in 43 states.

More fundamental changes have taken place as a result of the National Flood Insurance Reform Act of 1994 (Anon., 1995). This Act had many detailed provisions, including insurance for agricultural structures and the establishment of the Technical Mapping Advisory Council to advise FEMA on the accuracy and quality of the FIRM. This is important because FEMA is required to review the need to revise and update maps every five years. But the main purpose of the 1994 legislation was to promote flood mitigation opportunities and to ensure better compliance on the part of both mortgage lenders and recipients of federal expenditure. Part of the 1968 National Flood Insurance Act was repealed and replaced by a new National Flood Mitigation Fund into which will be transferred US$ 20 million annually (less during the three-year phase-in period) from the National Flood Insurance Fund. The mitigation fund is intended to provide grants to states and local jurisdictions, on a 75/25% share basis, for urban planning and for the implementation of longer-term mitigation projects such as floor elevation, relocation, property buy-outs, flood proofing, beach nourishment and technical assistance to local governments and individuals.

The 1994 Act also required lending institutions to complete a Standard Flood Hazard Determination Form for every loan secured by improved real estate or a mobile home. The aim was to clarify, for all buildings or mobile homes located within an identified Special Flood Hazard Area, whether flood insurance is required and whether federal flood insurance is available. This should help to ensure that legally required flood insurance is actually purchased. Thus, lenders must notify residential property buyers that they are legally obliged to purchase flood insurance, using information specified in the Act. For existing loans, and where borrowers are not carrying flood insurance, lenders are directed to notify such borrowers that insurance is necessary. If the borrowers do not then purchase cover, the Act states that the lender shall purchase it on their behalf and may charge appropriate premiums and fees. As part of a general tightening up on the allocation of federal funds, FEMA's Federal Insurance Administration (FIA) has implemented new procedures, such as requiring a 30-day waiting period for insurance cover to take effect under the Standard Flood Insurance Policy. This is designed to prevent people from buying insurance as a river crest approaches and then letting it lapse when the danger passes. There are monetary penalties for lenders who fail to meet the flood insurance notice requirements and fail to require mortgagees to purchase flood insurance for structures in the floodplain.

A persistent criticism of the NFIP has been that communities choosing not to join the scheme could still receive substantial disaster assistance from federal sources. The 1994 Act prohibited waiving the requirement to purchase flood insurance as a condition of receiving federal disaster assistance. Further, it directed that no federal disaster relief assistance, including loans, could be provided for a property if the owner was previously required to purchase flood insurance as a condition of receiving disaster assistance and had failed to do so.

As a consequence of these legislative changes, flood mitigation became better integrated with emergency flood response in the USA. In the future, more radical shifts in policy might occur, such as the privatisation of FEMA or the extension of the NFIP to other natural hazards. If federal provision makes it less easy for the private insurance industry to offer home-owner policies, the commercial sector may well wish to reconsider its involvement in this type of business.

9.4 EMERGENCY PLANNING AND DISASTER MANAGEMENT

As indicated by Green (1992), the overall purpose of hazard management is to implement intervention strategies, such as flood defence or flood insurance, which either reduce the likely occurrence of a hazardous event or reduce the negative consequences should such an event arise. Although flood hazard management is—or should be—a continuous process, it cannot eliminate disasters and there is a linked, but separate, need to engage in emergency planning. This process, known as *preparedness*, is designed to lead to rapid response and recovery after disastrous events. Emergency planning for floods is often a relatively specialised activity which takes place within the context of more traditional civil defence arrangements within a country. Many such arrangements still reflect preparations made for war, as in Australia (Handmer, 1992).

Effective preparedness depends on three separate stages:

1. *Organisational arrangements*. Legislation setting up the organisational and financial arrangements is important, together with a clear identification of individual responsibilities. These arrangements often involve a hierarchical structure. For

MITIGATING AND MANAGING FLOOD LOSSES

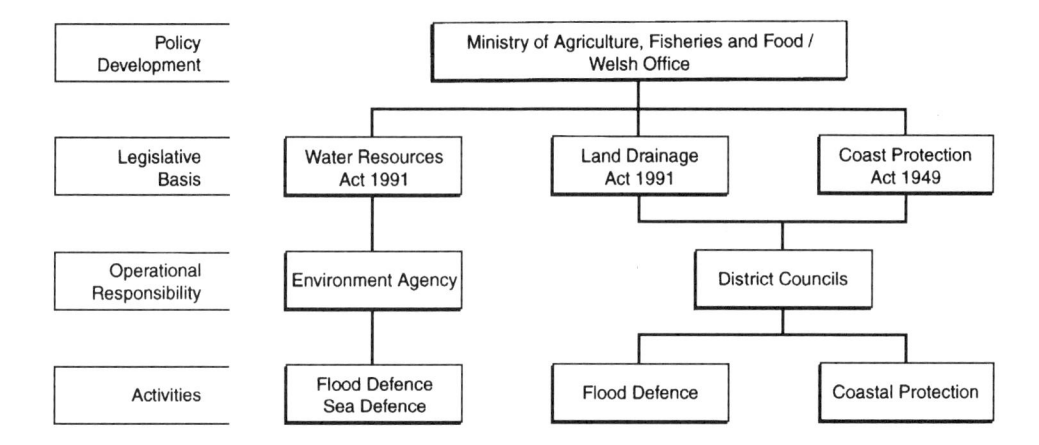

Figure 9.2 The designated agencies involved in flood reduction in England and Wales
Source: Modified after Roberts and Pritchard (1993)

example, in England and Wales the lead government department for flooding is the Ministry of Agriculture, Fisheries and Food in liaison with other national bodies such as the Environment Agency and the Meteorological Office, as shown in Figure 9.2 after Roberts and Pritchard (1993). In practice, the system operates locally through an integrated network of emergency management (Pine, 1995) which reflects the fact that the first responses will rely on the normal local emergency services, e.g. police, fire and ambulance working in association with the local authority. The Environment Agency has the lead role in disseminating flood warning messages with help from the local authorities whilst the police have related roles in the evacuation of people at risk and traffic management. In extreme cases, there are provisions for calling on military support. Thus, no single body is in absolute charge but it is believed that the resultant flexibility of response is a strength. The alternative of a national disaster team on permanent standby is deemed too expensive.

2. *Community preparation.* This involves the provision of contingency plans, including the development of warning systems and the prior publication of evacuation routes. Search and rescue teams have to be established in advance with a capability for emergency medical treatment, shelter and feeding arrangements for the survivors. The material resources for emergency management, such as heavy lifting machinery and drying plant to clean-up properties, have to be stockpiled on a least-cost basis. For example, Egidi (1995) cited equipment deployed after floods in northern Italy which has other uses. Some attempt to assess the costs of the emergency response to floods in the UK has been made by Parker (1988). The advance training of planners, managers and victim counsellors, plus the education of the public, is also vital to ensure an optimum response.

Disaster preparation is a difficult exercise because it requires the allocation of resources for an inherently improbable event that no-one wishes to happen. The authorities responsible for such planning often stress the low likelihood of the event but still expect people to prepare effectively. A key element is risk communication between the planning authority and the public which is often done via brochures distributed through the community. These are cheaper and are believed to be better in terms of community penetration than alternative forms of communication, such as videos. One example is the booklet prepared by the US Army Corps of Engineers (Anon., 1992). Any flood scenarios developed need to be plausible to floodplain residents, which means that they should come from a reliable source. Penning-Rowsell and Winchester (1992) showed how hazard awareness was raised in southeast England with the aid of the Thames flood brochure. Above all, risk communications should contain an 'action outcome' for the target audience, e.g. evacuation. De Vanssay (1995) cited various lessons from the flash flood which claimed at least 37 lives and cost some 980 million FFr in insured losses in southern France during September 1992. A few months earlier, a booklet was issued for the Vaucluse department, including the town of Avignon, designed to reassure residents that a range of hazards, including flooding, was under control and that, in the unlikely event of an emergency, warnings would be issued. This information was designed to pacify rather than prepare. Despite the availability of relevant meteorological information immediately before the flood, the lack of operational preparedness meant that no warning was issued and the loss of telephone lines and blocked access roads contributed significantly to the delayed recovery process.

Preparing for tsunamis poses special problems given the relative rarity and rapid onset of some events. Hazard assessment is usually based on the microzonation of the vulnerable area and the identification of evacuation routes. Figure 9.3 shows such an approach recommended for the industrial port of Salina Cruz, Mexico, where a container terminal, facilities for bulk cargo, fuel storage tanks and the harbour for a 400-vessel fishing fleet lie within the 10 m above sea level line which marks the inland limit of maximum probable tsunami inundation. Some facilities require relocation or protection behind buffer forest and other structures. But a major priority is to provide alternatives to the present evacuation route, which passes through a congested intersection within the urban area, and to implement a public education programme incorporating evacuation plans and exercises.

3. *Disaster management.* The period immediately after the event is critical given the unreliable information, acute time pressures and inadequate communication systems which often prevail at a time when agencies need to liaise effectively. According to Kreps (1992), both formal preparedness and a measure of improvisation are appropriate in post-disaster management. A failure of basic communication systems can occur, either through physical damage or overload, and Green et al. (1985) cited the flood at Uphill, Avon, UK, in 1984 when families seeking advice blocked the telephone network before it was eventually physically damaged by floodwaters. Much of the success of the evacuation in the southern Netherlands in February 1995 was due to the prior placement of communications at the heart of the overall emergency plan for Nijmegen and the surrounding area. Information officials were part of the crisis management team and could transmit formal announcements by radio within two minutes of any new development as well as give regular briefings to the media and others. There is considerable variation in the organisational arrangements for immediate post-disaster management. Several authors, such as Green (1992), have stressed the importance of self-help and Britton (1990) indicated

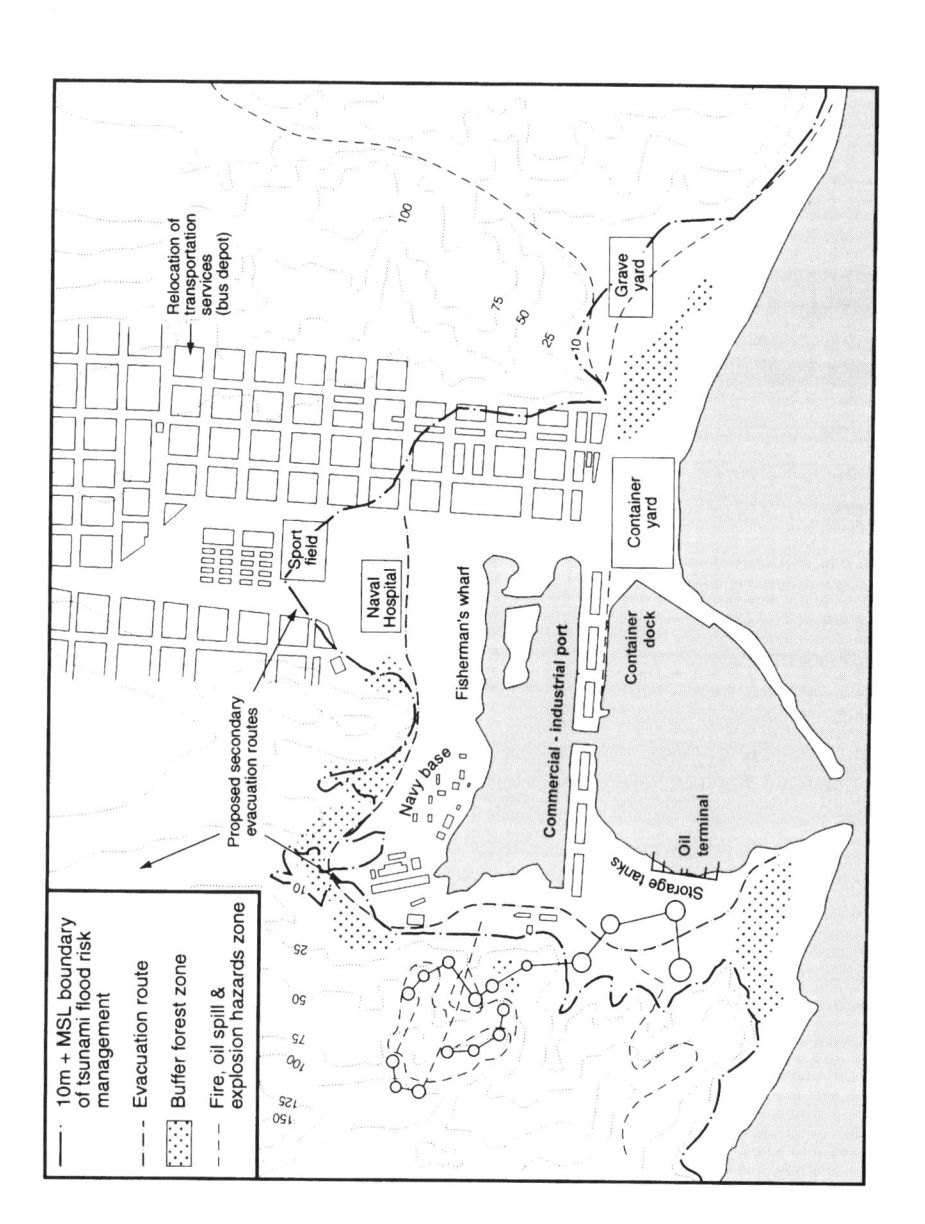

Figure 9.3 Preparedness for the tsunami hazard showing land-use patterns and vulnerability analysis at Salina Cruz, Mexico
Source: After Farreras and Sanchez (1991)

that in Australia about 2.5% of the population are emergency volunteers. In some countries national emergency planning is predicated on survivors coping on their own for up to two days without external support.

In the longer term, it is local government and planning authorities which have to take responsibility for recovery and rehabilitation. Theoretically, in the post-disaster phase, a community has a good opportunity to implement policies which will mitigate future events because the recent flood losses are uppermost in local minds and public pressure can be mobilised to enact some radical land-use changes. On the other hand, there is also a powerful desire to return the community to 'normal' as soon as possible. Commercial and other interests often press for the rapid restoration of facilities at existing sites and, in order to be seen to be making progress and avoid conflict, planning authorities continue to invest in unsuitable locations. Fischer (1989) outlined such missed opportunities along the harbour front at Redondo Beach, California, which is exposed to recurrent flooding from winter storms. In the LDCs, the responsible bodies may well be in an even weaker position to enforce sustainable recovery. Wickramanayake et al. (1995) showed how the local planning agencies were unable to cope with the necessary rehabilitation after severe floods in southern Thailand because of the lack of relevant data and technical expertise in project formulation and suggested that guidelines should be put in place to strengthen local responses after disaster.

9.5 LAND-USE MANAGEMENT

The improved regulation of land use lies at the heart of comprehensive floodplain and coastal zone management. The overall aim is to reduce the risks involved in the present occupation of flood-prone land and to deter further invasion of such areas. Some countries, such as the USA, have adopted an explicit regulatory approach for both floodplain and coastal environments using a combination of hazard zoning and insurance-based incentives. Other countries, such as the UK, rely on essentially voluntary controls. Here, local authorities are generally obliged to consult the appropriate river authority before deciding on planning applications but the information provided is only advisory and can be overruled.

In practice, a very wide variety of measures may be deployed in aid of land-use management, including forecasting and warning, compulsory insurance, zoning and subdivision ordinances, building and health codes, land acquisition, the relocation of structures and a policy of managed retreat from the lowest-lying areas. The ideal would be a pre-planned urban design for street and building alignment which achieved large-scale flood proofing. But this is rarely possible, and many flood-prone areas already have inappropriate development, such as single-storey sheltered housing for the elderly, mobile home parks and recreational camping sites. In all land management schemes, flood risk mapping is the key tool. Flood mapping was first attempted in the USA, where it is now well-developed in the NFIP, but it has been less systematic in Europe where the EU has shown little initiative in setting continent-wide standards (Marco, 1994). For example, just to take the UK, there are no maps relating to dam failure and floodplains have not been routinely mapped to high risk assessment standards (Green, 1992), although this is being remedied (see Section 1.2.3).

To be effective, land-use management has to intervene in the process whereby land is converted from rural to urban uses. As the time for such conversion approaches, the

Table 9.4 Reasons for acquiring floodplain land in the USA

Reason	Percentage
For current income	(72)
Live on property (residential)	35
Use for business purposes	20
Obtain income from renting	13
Use as buffer/addition to adjacent property	7
Use for farm, timber or other production	7
Hold for sentimental reasons	2
For future income	(39)
Profit from future sale	30
Profit from subdivision/development	17

$N = 99$
Source: After Bollens (1990)

proportion of floodplain land likely to be held by land speculators increases (Brown et al., 1981). This triggers powerful profit motives on behalf of private interest groups, such as landowners, builders and estate agents, which may be difficult for public policy to control. In a study of 10 USA cities, Bollens (1990) demonstrated that people bought floodplain land for a variety of financial motives. Although several owners had more than one motive, Table 9.4 shows that use-based, current-income reasons dominated over investment-based future-income potential. Interestingly, the presence of a local floodplain management programme significantly decreased property value expectations and rendered undeveloped floodplain less attractive to all potential investors. This conclusion was broadly endorsed in a separate US study by Burby et al. (1988) who found that the floodplain management policies embedded within the NFIP had been effective in protecting new development from the 1:100 year flood event mainly through influences on the decisions of builders and land developers achieved by heightened flood hazard awareness.

These are important findings given that some structural flood control schemes tend to raise property values and thereby increase floodplain invasion. The deterrent effects were most marked for people looking to buy land for future residential building when plenty of developable land still existed elsewhere in the area. For more general investors, accessibility and infrastructure were perceived to be desirable features of the land whilst, for landowners wishing to live on the property, it was found that hazard zoning was an important deterrent. This would suggest that local authorities should divert roads and other urban infrastructure away from hazardous floodplains and that restrictive zoning ordinances should be implemented wherever appropriate.

For many years, it has been known that the increased physical protection offered by structural flood defence schemes has tended to encourage further floodplain development due to the so-called 'levee' or 'escalator' effect (see also Sections 2.6.2 and 7.2.2). Parker (1995) showed that this occurs because conventional cost–benefit appraisals of engineering schemes fail to take into account likely future developments for which costs are to be borne by the general taxpayer. Detailed studies of several floodplains in England and Wales, summarised in Table 9.5, showed that the increase in development following a structural scheme can be significant. In the case of

Table 9.5 Profile of floodplain development trends in six urban locations in England and Wales

Location	Date of completion of flood defence project	Nature of flood defence project	Flood defence design standard	Estimated number of properties in the 1:100 floodplain Pre-project start (date)	Post-project end (date)
River Thames, Maidenhead	1960–66	Channel improvements to Maidenhead Ditch/Cut	1:10/20	1560 (1947)	3303 (1989)
River Twyi, Carmarthen	1984	Flood wall improvements; new flood walls	1:100	75 (1981)	81 (1989)
Black Brook, Loughborough	1968	Channel realignment; embankments	1:100	0	476
	1979	Embankment/drainage improvement	1:100	476 (1968)	672 (1988)
River Tone, Taunton	1969	Channel improvement; new weirs	1:70/100	795 (1964)	912 (1989)
River Stort, Bishop's Stortford	1979	Channel improvements; new weirs	1:70	129 (1976)	210 (1989)

Source: Modified after Parker (1995)

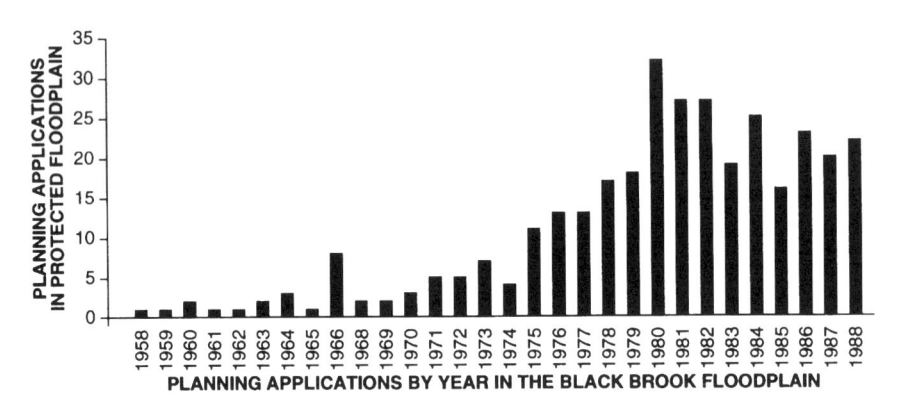

Figure 9.4 The number of planning applications by year in the Black Brook floodplain, Leicestershire, UK. No data for 1967
Source: After Parker (1995)

Loughborough and the Black Brook floodplain, a flood defence scheme was specifically installed to permit the building of 2600 new houses and retail outlets after the land was released for development against the advice of the river authority. A rise in planning applications followed the completion of the scheme in 1968, as shown in Figure 9.4, and only 8% of these applications was refused.

Although the escalator effect has been widely criticised, some alternative locations for new building—as in the Black Brook case—would have developed green belt or

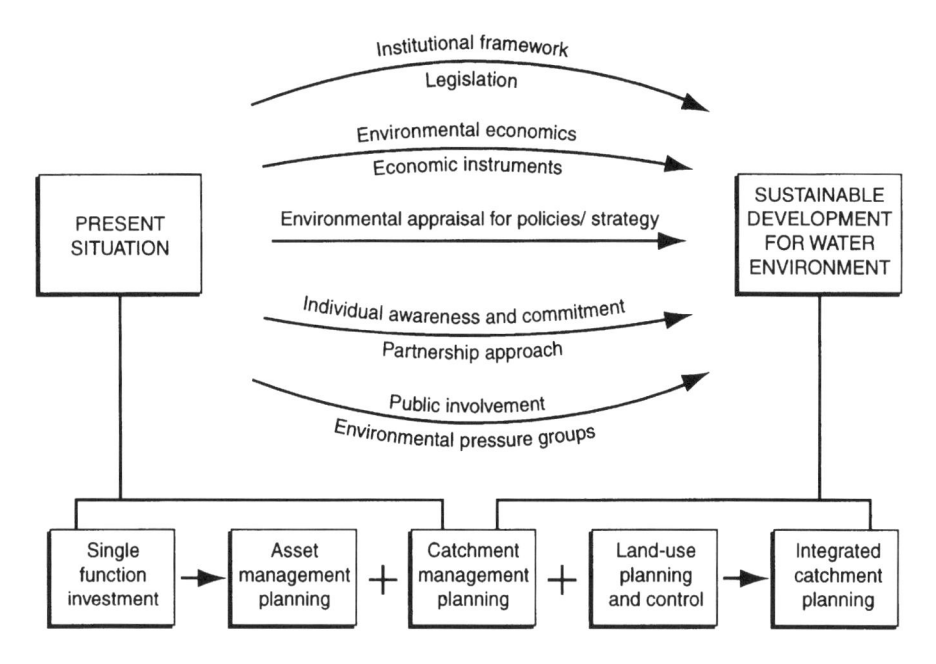

Figure 9.5 Pathways to sustainable development based on integrated catchment planning
Source: After Gardiner (1994)

high-quality agricultural land (Parker, 1995). Where planners have arguably been more at fault is in the many instances when they have failed to resist individual small-scale planning applications on undefended floodplains and have thereby permitted piecemeal development to occur without regard to the cumulative consequences. It is clear that, where land development pressures are high, the present system in the UK does not work well and more strategic alternatives might include a move towards a compulsory insurance-based scheme, as in the USA, or the use of catchment management plans. According to Gardiner (1994), catchment management plans enable flooding to be seen in a basin-wide context and help to ensure that the conservation of the water environment is integrated with development planning and control. This approach has considerable potential but it will take some time to progress to the goals outlined in Figure 9.5.

In the USA, realtors (estate agents) and mortgage lenders have been given an increasingly explicit responsibility to alert prospective home-owners to flood risks. Their statutory role is to inform customers that the property is located within a flood zone and that flood insurance may be required, although the wider purpose is to raise hazard awareness and to discourage further invasion of flood-prone areas. However, this responsibility is contrary to the business motives of such people and the legislation is not always effective. There is also a concern that the very presence of flood hazard insurance may be traded-off against the anxieties of prospective purchasers, especially those specifically attracted to waterfront locations. For example, Cross (1985) investigated the Lower Florida Keys during the mid-1980s when over 5000 dwelling units, plus 1000 mobile homes, were at least seasonally occupied within an area which would be inundated during a 1:100 year storm. Only about half of all new home-owners

Table 9.6 Methods realtors claim to use to inform prospective customers that homes are located in Florida hurricane flood zones

Method employed	Percentage of realtors
Conversation about insurance in general	32
Written disclosure at or before closing	30
Discussion of insurance with financing	26
Comes up in general conversation	26
Conversation while showing homes and area	12
Specific mention of flood levels and hurricanes	10
Contrast of local problems with other areas	4

Note: Responses do not total 100% since some realtors gave several answers to the open-ended question
Source: After Cross (1985)

claimed that they had been warned of the dangers and in most cases the information was communicated in a low-key manner. Table 9.6 shows that only 12% of realtors claimed to raise the issue while initially showing prospective purchasers around the home and the locality. This is despite the fact that discussions about hurricane flooding were twice as likely to make a difference to the choice of a home-buyer if the disclosure took place at this time. Conversely, the written disclosure of risk at or just before the purchase was concluded was likely to have little or no effect on the locational decisions.

Urban redesign has commonly followed major floods. For example, after the June 1972 flood, Rapid City, South Dakota, converted an area four blocks wide and 8 km long into parkland with an 18-hole championship golf course at the centre. More recently, the passing of the Hazard Mitigation and Relocation Act has greatly aided the process. After the Mid-West floods of 1993, FEMA announced the intention to remove some 4300 structures from repeatedly flooded areas in Missouri alone, saving an estimated US$ 200 million in disaster costs over the next 15 years. By October 1994 the federal government had authorised 160 projects for the elevation, acquisition or relocation of some 7500 buildings and it has been predicted that this total will ultimately embrace 8000 buildings in 140 different communities (Wright, 1996). In some cases, whole communities have been relocated. Hanson (1996) described how the three Iowa communities of Nevada, Audubon and Cherokee, with the assistance of various federal agencies, have implemented voluntary buy-outs to create public open space and more affordable housing in neighbourhoods damaged in the 1993 floods. In Cherokee the buy-out involved 187 residential properties on more than 60 acres along the Little Sioux River in order to create leisure trails, river access, wildlife provision, sites for outdoor education, picnic and camping areas and a community garden.

In the emerging era of post-flood hazard reduction, Handmer (1987) drew attention to the growing commitment of public funds for the purchase of flood-prone land and property in North America. The main motive for such acquisition is community safety, but other benefits accrue, such as the creation of parkland, the preservation of wetland habitats, the improvement of waterfront access and the cessation of flood aid pay-outs. Typically, the acquisition areas have a low socio-economic status with low property values. In Australia, where schemes have been voluntary, such acquisition policies have the highest dependence of all flood loss strategies on public cooperation. The process is based on home-owners accepting the estimated market value of their property, as determined by government valuers. Since these values are below the price of equivalent flood-free property, some older residents, in particular, may perceive a major financial loss. Overall, the most important factor in scheme participation has been the residents' belief that they will personally benefit from acquisition, although a high awareness of flood hazard and low attachment to the locality were also relevant. This type of scheme can secure community acceptance because the relocation out of flood-prone land offers families an opportunity to better themselves, despite the inevitable costs for people relocating to higher-value properties. In order to maximise the return on public investment, it has proved necessary in some cases to sell or lease the acquired land for agricultural or recreational use, a function which is still compatible with the overriding need to keep the most hazardous land dedicated to open space use.

Despite this, there are few examples where partial urban relocation has been properly assessed, and costed, as an alternative to more conventional flood mitigation strategies. The most cost-effective relocations are likely to occur when this process is combined with other community goals. One example was the small settlement of Soldiers Grove sited near the Kickapoo River in south-western Wisconsin, USA (David and Mayer, 1984). When the town suffered a series of floods in the 1970s, the Army Corps of Engineers proposed to build two levees, in conjunction with an upstream dam, to protect the central business district, sited within an oxbow on the floodway, and other property (Figure 9.6A). The reservoir project was eventually shelved in 1977 but the residents had been considering other solutions. Following a flood in 1978, they decided that a relocation of the entire business district would yield more benefits than just flood damage reduction for the large cost of the structural scheme. The plan involved public acquisition, evacuation and demolition of all structures in the floodway together with flood proofing of properties in the flood fringe (Figure 9.6B). The community has clearly gained through its non-structural option. Although the levees would have protected the village from most floods, they would not have provided other opportunities. For example, due to compensation payments, businesses could build improved premises. Relocation also gave scope for the use of energy conservation measures at a more attractive site along the major highway into the area.

Property acquisition and relocation schemes can be viewed as part of the wider move towards managed retreat from flood-prone areas. This strategy was originally proposed for coastal zones but it is also relevant to river floodplains where restricting development behind set-back lines can combine a reduced flood risk in association with improved landscape, amenity and habitat values. Managed retreat can take place at all scales including that of the individual building as in the lowest-lying areas of Venice where many residential buildings have blocked doors and windows as ground floors have been abandoned for use. Figure 9.7 shows the most commonly understood options for a coastal zone faced with sea-level rise. Option 2 indicates the traditional policy of maximum shoreline defence whilst option 3 shows a cheaper engineering option, linked to partial inland withdrawal, which still protects the high-value property at risk. The scenario for total retreat (option 4) is the most radical since the coastal zone is left undefended with some loss of nearshore properties.

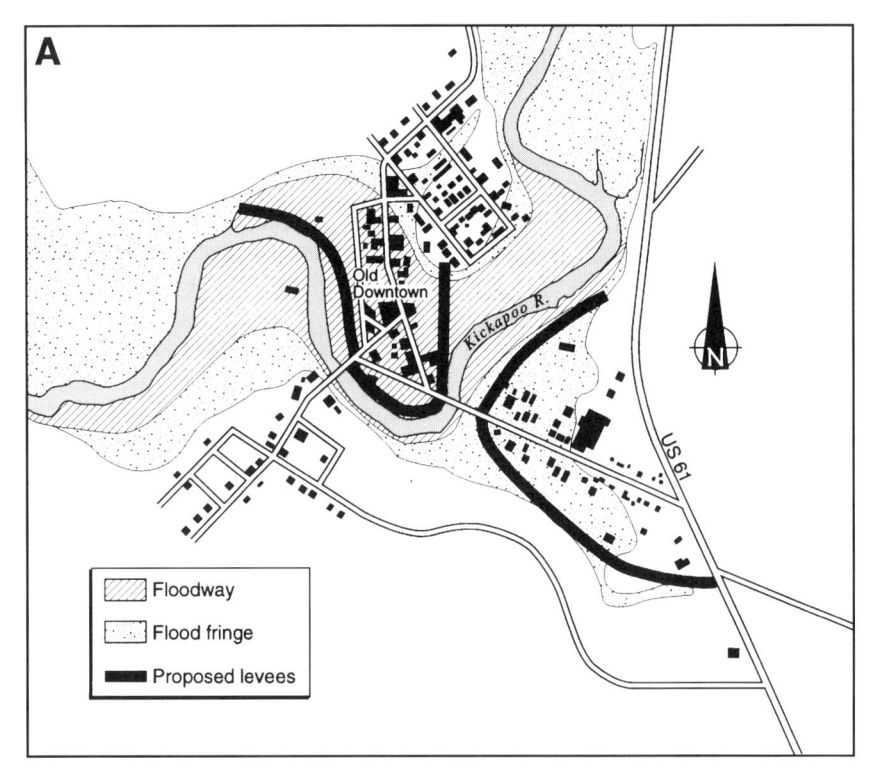

A

- Floodway
- Flood fringe
- Proposed levees

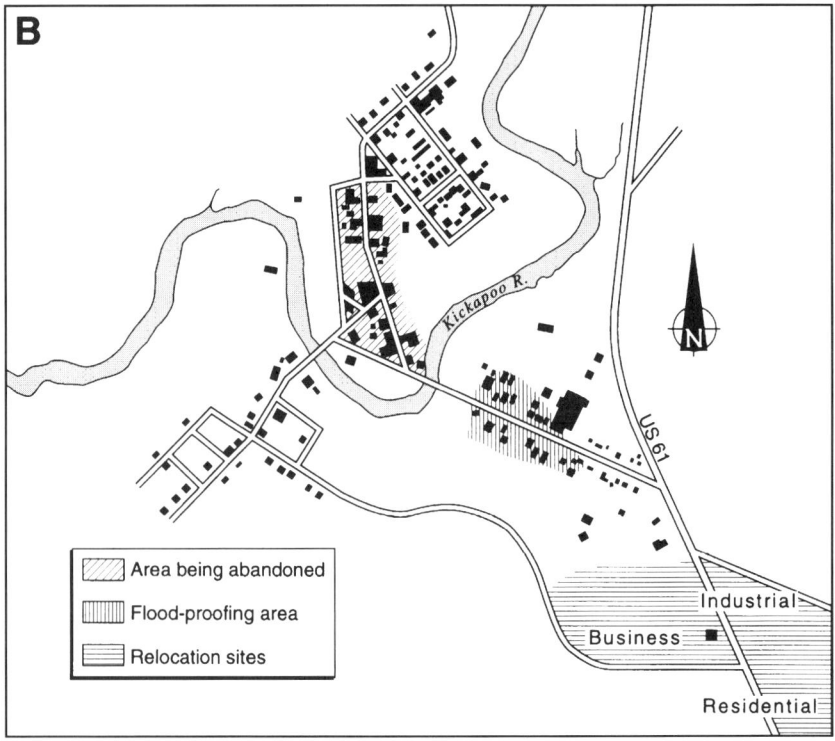

B

- Area being abandoned
- Flood-proofing area
- Relocation sites

MITIGATING AND MANAGING FLOOD LOSSES   319

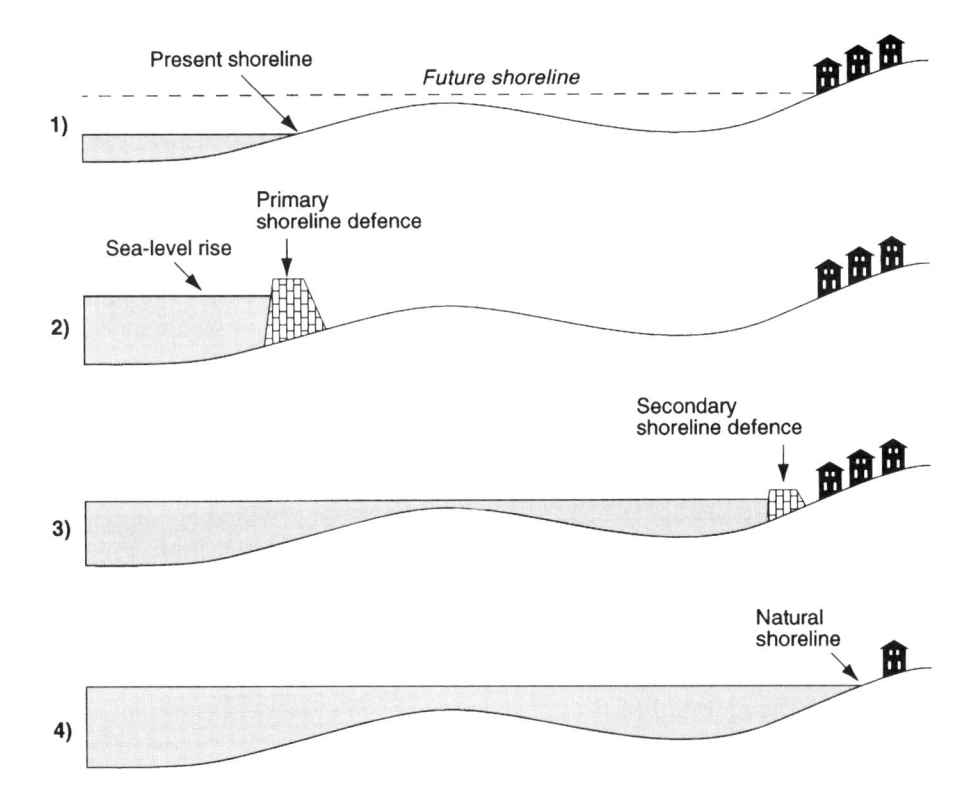

Figure 9.7   The possible adjustments to increased flood hazard in the coastal zone as a result of sea-level rise
Source: Modified after Penning-Rowsell and Peerbolte (1994)

9.6 LIVING WITH FLOODS

9.6.1 The Need for a New Approach

The late twentieth century is characterised by post-normal science in which important environmental decision making has to take place in an uncertain world (Funtowicz and Ravetz, 1995). Issues of great scientific complexity relating to environmental quality are now commonplace and the decision maker often has only a partial knowledge of key elements, such as the consequences of climatic change. Equally, with the increasing dissemination of knowledge, technical solutions which are imposed directly on flood-prone communities are less and less appropriate. Engineered approaches using either structures, or other technical responses such as flood forecasting, are increasingly

---

Figure 9.6 (*opposite*)   Adjustments to the flood hazard at Soldiers Grove, Wisconsin, USA: (A) shows the floodway and the flood fringe, together with the location of the two proposed levees; (B) shows the areas eventually flood proofed and abandoned, together with the relocation sites
Source: After David and Mayer (1984)

questioned. For example, dams, which were formerly seen as an obvious means of exerting control over flood processes, may now be viewed as intrusive, highly expensive and even anti-environmental. Flood embankments in rural areas often take productive land from farmers with small plots of land and may prevent beneficial floods reaching their fields. Thus, there has been a reversal from the traditional acceptance of the application of 'hard' (i.e. well-understood) science in 'soft' (i.e. easy) decisions and today 'soft' science often contrasts strongly with the 'hard' policy decisions which are required, sometimes urgently, to produce flood solutions for a more sustainable future. Flood policy is currently at a cross-roads and, within the last decade, several major disasters have raised the question of how best to reduce flood losses on some of the world's largest rivers in a financially and ecologically enduring fashion.

The underlying policy trend is towards ensuring a more 'self-reliant' attitude by all risk takers in flood-prone environments. This philosophy has widespread appeal because not only does it apply in the wealthiest, market-orientated countries but it is also relevant in the poorest nations which lack the central resources for large schemes. A 'living with floods' strategy is not appropriate in all cases. Where life and property are endangered on a large scale, as in many urban environments, there is little option but to adopt flood defence. But, where smaller numbers or lower-grade land is involved, a process of managed retreat may well be preferable.

9.6.2 Turning Theory Into Practice

In the MDCs, the 1993 Mid-West floods in the USA exposed a variety of failures in conventional flood alleviation policy. Many levees failed along the Mississippi, admittedly often beyond the design limits, only 10% of householders and less than 50% of farms were insured, and organisational muddles occurred over aid distribution. National concern led to the establishment of the Inter-Agency Floodplain Management Review Committee (IFMRC) by the Clinton Administration which suggested some radical reforms. For example, the IFMRC (1994) proposed investment in moving people off the floodplain and reclaiming wetlands which would then act as a natural store for floodwaters along the Mississippi and Missouri Rivers. Such costs would be high. For example, it has been estimated that to expand the existing Wetland Reserve Program, created by the 1990 Farm Act, in the Mississippi basin would cost some US$ 1 billion for each one million acres restored (Tripp, 1994). On the other hand, such costs were only about one-fifth of the cost of federal disaster relief and would be a once-only payment.

Comprehensive floodplain planning is now the declared goal for the Mississippi valley (Rasmussen, 1994). Figure 9.8 shows an idealised pattern for the next century with existing major urban centres protected by a combination of levees and upstream reservoirs and land treatment, whilst frequently flooded built-up areas would become river parks and recreation areas as occupants are relocated to safer sites on higher ground. Key facilities, such as water and waste treatment plants, roads and railways, would be elevated out of harm's way and, like the urban infrastructure, would be fully covered by flood insurance. Some farmers in the lower-lying areas would choose to convert from row crops to grazing or silviculture, although the higher croplands would continue to be very productive. At the upstream end of many levees, federal water

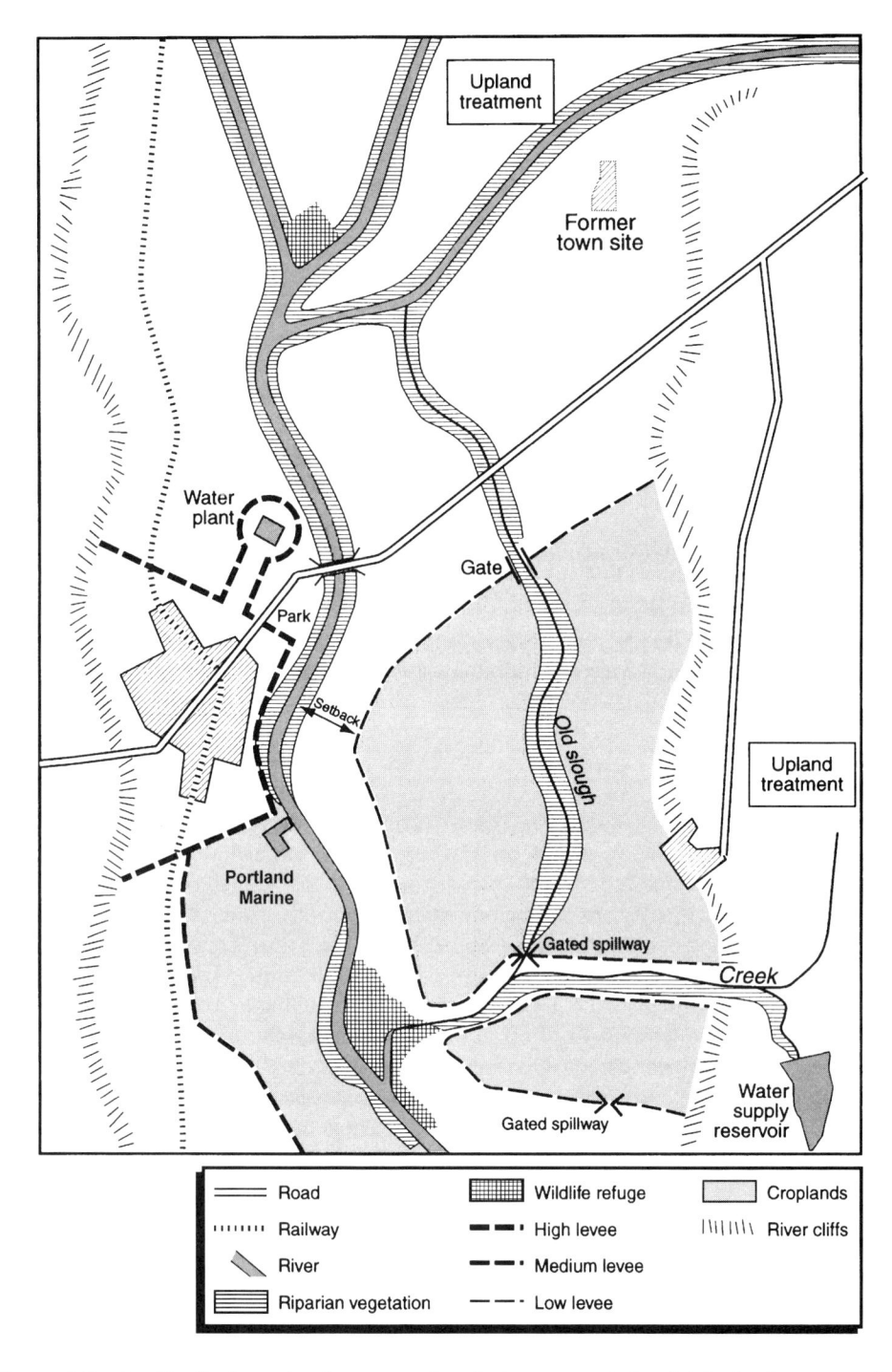

Figure 9.8 An idealised reach of a comprehensively managed floodplain in the twenty-first century
Source: After Rasmussen (1994)

control structures would keep sloughs permanently wet with benefits for wildlife and fisheries.

All this is predicated on a redirection of existing subsidies for traditional activities such as levee construction, aid distribution and even state-sponsored insurance to support a better long-term adaptation to floods. Flood-adapted land uses, such as grazing, floodplain afforestation, biomass crops for energy use and opportunities for wetland recreation are likely to become more prominent while dams will be operated to restore either individual flood events or a 'floodpulse', the regular seasonal rise and fall of the hydrograph, to promote ecological benefits. Already attempts are being made to introduce planned floods to certain river systems and coastal zones in order to rehabilitate some ecosystems (see also Section 3.6.2). For example, water was diverted from an irrigation canal to create a flood on the middle course of the Rio Grande, New Mexico, in order to observe the effects on an accumulation of leaf litter and woody debris resulting from the absence of substantial flooding due to dam construction in the past 50 years (Molles et al., 1995). In northern Canada, river regulation has contributed to the drying-out of the Peace–Athabasca delta—one of the largest freshwater deltas in the world—with adverse consequences for vegetation and wildlife. Prowse et al. (1996) reported the construction of artificial ice jams in an attempt to reproduce the natural periodic meltwater floods of early spring which feed the myriad of channels and lakes across the delta.

A direct 'living with floods' policy is most suitable for river floods with a high degree of seasonality where there is already some positive dependence, either ecologically or economically, on the annual flood pulse. This situation is most commonly found in the rural flood-prone areas of the LDCs. For example, traditional building styles are often flood resistant. Dwellings built on stilts in the wet tropics can provide a significant degree of security, especially when roads are routed parallel to rivers, rather than with many river crossings, as in Thailand. Chan (1995) showed that a significantly higher proportion of people living in Malaysian houses built on stilts more than 1.5 m high were willing to remain in floodplain settings, compared with those with stilts only 1.0 m high or less.

The types of traditional crops can be important, especially if they offer flood-adapted options. For example, in the Mekong delta, Vietnamese farmers routinely reserve a stock of potato seeds to plant after the normal flood recession, using the tailings from water hyacinths deposited by the floodwaters as a mulch (Cuny, 1991). When damaging floods arise, traditional cultivation frequently has benefits. For example, natural varieties of long-grain rice fared much better than the hybrid high-yielding varieties in the 1988 Bangladesh floods. Even when the paddy rice was broken by the force of the floodwaters, the farmers could still collect and market the stunted stems as livestock fodder at advantageous prices because of extensive flood damage to the normal fodder crops. The existence of small boats and rafts owned by individual rural households is often an integral feature of indigenous responses because flood victims, reluctant to abandon their homes, can remain in safety on rafts tethered nearby. In addition, the availability of local water transport can help the distribution of food and other flood relief. In most rural societies the largest part of emergency response is undertaken by the survivors who will not, or cannot, wait for external support before they act.

Living with floods in the LDCs means greater self-reliance on the part of large numbers of less-powerful and even disenfranchised people, such as shanty town dwellers, tenant farmers and subsistence fishermen. For example, Mulwanda (1993) highlighted defects in the official response to floods in Lusaka, Zambia, especially for those living in squatter settlements and has called for a better integration of flood mitigation into longer-term development programmes. Better community participation is the only long-term solution to flood problems in the poorest informal settlements. In parts of South Africa, local residents have worked with local governments and the NGOs to form response groups to carry out early warning, door-to-door alerts and the identification of 'safe' houses as refuges on higher land. In some cases sections of these shanty towns are being relocated above future flood levels on bigger plots using better building material such as galvanised iron rather than plastic sheeting and cardboard.

9.6.3 Living With Floods in Bangladesh

If it is to be credible, the 'living with floods' approach must have some chance of success in countries such as Bangladesh which has a uniquely high level of exposure to both river floods and coastal storm surge. About 80% of the national territory is floodplain land crossed by some 700 rivers with tributaries and distributaries extending to a total length of over 22 000 km (Khalil, 1990). The shallow, alluvial river channels carry an annual sediment load estimated at up to 2.5×10^9 tonnes and they meander widely in an unpredictable fashion during flood events. Most of the plains are at risk from flooding during the monsoon season. At the edge of the low-lying delta, which is the largest in the world, the coastal zone contains about 36 000 km^2 of land which are subject to destructive cyclones from the Bay of Bengal almost every year (Khalil, 1992). In addition to this physical exposure, Bangladesh has one of the world's most vulnerable populations with an average annual per capita income less than US$ 200, illiteracy rates around 75% and an annual growth rate of about 2.5%. Most of the population is engaged in agriculture, although about 80% of the animal protein comes from fish. There is an average population density in excess of 800 people per square kilometre with the most intensively cultivated areas supporting in excess of 12 people per hectare of arable land. Given that well over 100 million people live on the floodplains, and there are a further 20 million in the coastal region, Bangladesh has the highest potential of any country in the world for loss of life from major river floods or storm surges.

Not surprisingly, there is a long history of flood defence in Bangladesh. After floods in 1954, a Master Plan for water resource development was devised and, by 1990, nearly 8000 hydraulic structures and over 7500 km of embankments were protecting some 23% of the total land area (Thompson and Sultana, 1996). These defences were organised into five types which coincide with the chief categories of flooding recognised by Brammer (1990a):

- Embankments along main rivers—this includes some major levees, such as the Brahmaputra Right Embankment some 217 km long, which protect large areas of floodplain during the summer high-flow season.
- Freshwater polders—these are essentially multi-purpose projects on the floodplain designed for flood control with agriculture, fisheries and irrigation as associated functions.
- Submersible embankments—small levees which are used to protect the *boro* paddy rice in the pre-harvest period before they are overtopped by the main monsoon flood pulse.

- Embankments along small rivers—these are similar levee structures which are located principally in the upland areas near the Bangladesh border to protect against flash floods.
- Coastal zone polders—these prevent the intrusion of seawater into the delta area and provide a secondary defence against storm surge. Small inter-tidal channels are closed off but the larger river outlets remain open to support agriculture and shrimp farming.

In the severe floods of 1987 and 1988 this complex of flood defences proved inadequate. In 1987 some 1279 km of embankments failed leading to damages estimated at US$ 0.5 billion and in 1988 the figures increased to 1990 km and US$ 1.3 billion respectively. These losses prompted the national government and the international aid community to develop an ambitious flood action plan (FAP) for the future. The initial five-year FAP started in January 1990 with a series of surveys and pilot projects costing about US$ 150 million and involving 15 donor countries. However, the longer-term direction of the FAP has proved highly controversial given a spectrum of proposals which range from massive structural schemes to a greater emphasis on self-reliant approaches (Boyce, 1990; Brammer, 1990b; Islam, 1990; Rasid and Mallik, 1993). Broadly speaking, many civil engineers and international consultants have favoured further structural intervention. For example, an early French plan, costed provisionally at US$ 5–10 billion with annual maintenance costs of US$ 160–180 million, envisaged main river embankments averaging 4–5 m high rising to a maximum of 7.4 m. Another proposal was for the large-scale protection of the rice crop inside flood control 'compartments' or polders where the flood level would be regulated through a system of drainage channels controlled by sluice gates and culverts. Environmentalists and others have opposed more large-scale engineering works because of concern about the ecological consequences and the financial costs. Although other donors, such as the USA, have advocated some non-structural alternatives, the underlying assumption is that Bangladesh will become committed to more embankment-centred schemes.

The critics of traditional engineering have argued that the floods of 1987 and 1988 merely highlighted some long-standing problems with the existing flood defences, which have frequently performed below expectation. For example, Hoque and Siddique (1995) showed that, whilst the structural schemes implemented by the Bangladesh Water Development Board since its establishment in 1965 had probably reduced the average area of the country which is inundated every year, there were regular seasonal failures of the main embankments. These often hazardous failures are associated with poor planning and design, faulty construction, poor maintenance and river migration. In some instances, the embankments have been deliberately cut, often by landless people living in shelters built on the levees, indicating a need for more involvement of local people as well as more funds and technical expertise. Flood losses have also continued inside the freshwater polder projects, especially when excess water cannot drain out quickly, and there have been conflicts between the needs of rice growers and jute farmers. Thompson and Sultana (1996) claimed that flood losses were actually higher during the 1988 flood inside the project-protected areas than in unprotected control areas and that, in the longer term, flood protection has disadvantaged certain groups, such as fishermen and boatmen.

Given these existing difficulties, any further concentration on so-called megaproject flood engineering poses problems, even if such schemes can be built and maintained over many decades through massive injections of foreign aid. River control engineering will always have severe limitations in a country where 90% of the flood discharge originates outside the national territory, where the volume of water is far too large to be stored upstream, and where perhaps 5% of the floodplain is subject to active riverbank erosion (Hossain, 1993). For example, Khalequzzaman (1994) emphasised that the planned embankments will limit sedimentation on the delta plain, causing riverbed aggradation and the subsequent submergence of large areas by the rising sea. Levees, which rob the country of much-needed agricultural land, are likely to create flood problems downstream and will also interfere further with the open-water fisheries which are important to the rural poor and there is a wider risk to biological diversity on the floodplain.

Even some of the more sensitive proposals are likely to create problems. For example, although some floodplain residents have expressed a preference for the regulation of flood levels within the normal seasonal flood regime range which is suitable for crops and infrastructures (Rasid, 1993), it has been suggested that the proposed compartmentalisation scheme is incompatible with indigenous rice cropping practices because of operational constraints and environmental impacts (Rasid and Mallik, 1995). However, it is widely recognised that some engineering works are essential, especially if they complement other responses. For example, there is a general need to rehabilitate the coastal embankments built in the 1960s and 1970s but this should be done alongside the preservation of the Sunderban mangrove forest which occupies about 4% of Bangladesh and which creates an important buffer against the inland penetration of storm surges (Khalil, 1992; Saenger and Siddiqi, 1993).

Although engineering alone is insufficient, it must also be admitted that the scope for non-structural schemes is also limited in Bangladesh because of the lack of flood-free land, illiteracy, poor communications and social divisions. For example, land zoning in the unprotected areas is largely impractical because of the high population densities. Khan (1991) demonstrated how local elites have influenced the location of flood shelters to their own advantage in northern Bangladesh. But a series of alternative measures was proposed by James (1994) including:

- more raised community flood shelters and health clinics with food, water and sanitation;
- secured storage to preserve capital assets, like cooking equipment and tools, that people need to resume working lives;
- protection of incomes by employing flood victims in the repair of embankments;
- improved design and construction of boats for emergency evacuation, security patrols and temporary housing in floods;
- flood-resistant infrastructure to maintain community services such as water, electricity, health and education during floods;
- more flood-resistant housing built away from areas suffering from severe riverbank erosion.

There are considerable problems associated with the social and political implementation of a 'living with floods' strategy. Critics view these responses as rather vague, and representing an idealistic state of mind, rather than offering clear solutions. Such measures are often not popular with national governments, technical advisers and even some aid agencies because they are less visible and lead to fewer lucrative contracts placed for prestige projects. Small farmers are not powerful politically and have little influence when ranged, for example, against a federal ministry responsible for water power or irrigation which sees engineering as the only way forward. Poverty and a lack of education are major impediments. Mamun (1996) showed that, whilst Bangladeshis are very aware of the damage to homes and farmland from riverbank erosion, it is only the literate who have a positive approach to adjustment. The poorest and illiterate take a fatalistic attitude to the hazard and, although they try to diversify income with less reliance on agriculture and increase fishing and cash employment, it is really the less vulnerable who move away to safer areas. A major challenge is to include the poor, as well as the elites and successful farmers, in decision making.

But 'living with floods' is not a defeatist attitude. The main attraction of this approach is that it seeks to gain benefits from the normal annual floods whilst reducing the risks during abnormal hazardous events. Indigenous strategies are less ambitious than river training projects which imply high capital and maintenance costs, which add to the national debt burden and may never be completed. In particular, they are cost-effective flood reduction schemes, which protect ordinary people rather than urban elites, and, as indicated by Zaman (1991), the plight of the displaced poor commands some priority. A key part of such a strategy is the development of people's self-reliance via small-scale protective devices, which include the greater use of 'floating', quick-growing, flood-resistant varieties of rice, village warnings, the organisation of working parties to repair embankments using locally available materials and labour, the stockpiling of tools and the dissemination of practical information for rapid agricultural recovery after disaster.

Above all, the aim must be to get the country to work together to enhance the collective resilience to major floods.

APPENDIX 9.1   SUBDIVISION OF SPECIAL FLOOD HAZARD AREAS

Zone V: SFHAs along coasts subject to inundation by the 100-year flood with the additional hazards associated with storm waves. Because detailed hydraulic analyses have not been performed, no base flood elevations or depths are shown. Mandatory flood insurance required.

Zones VE and V1-30: SFHAs along coasts subject to inundation by the 100-year flood with additional hazards due to wave action. Base flood elevations derived from detailed hydraulic analyses are shown within these zones. Mandatory flood insurance required.

Zone A: SFHAs subject to inundation by the 100-year flood. Because detailed hydraulic analyses have not been performed, no base flood elevation or depths are shown. Mandatory flood insurance required.

Zones AE and A1-30: SFHAs subject to inundation by the 100-year flood determined in a Flood Insurance Study by detailed methods. Base flood elevations are shown within these zones. Mandatory flood insurance required.

Zone AH: SFHAs subject to inundation by 100-year shallow flooding (ponding) with average depths 1–3 feet (0.3–0.9 m). Base flood elevations derived from detailed hydraulic analyses are shown in this zone. Mandatory flood insurance required.

Zone AO: SFHAs subject to inundation by 100-year shallow flooding (sheetflow) with average depths 1–3 feet (0.3–0.9 m). Average flood depths derived from detailed hydraulic analyses are shown within this zone. Mandatory flood insurance required.

Zone A99: SFHAs subject to inundation by the 100-year flood which will be protected by a federal flood protection system when construction has reached specified statutory progress toward completion. No base flood elevations or depths are shown. Mandatory flood insurance required.

Zones B, C and X: Areas identified with moderate or minimal hazard from the principal source of flood in the area. Buildings could be flooded and flood insurance is available but is not required by regulation in these zones.

Zone D: Unstudied areas where flood hazards are undetermined but flooding is possible. No mandatory flood insurance is required but cover is available in participating communities.

Source: Data obtained from Insurance News Network (1996) site on Internet (LCC\inn@insure.com)

CHAPTER 10

Outlook

CONTENTS

10.1 Introduction 328
10.2 Physical processes: Problems 329
10.3 Physical processes: Some possible solutions 331
10.4 Human impacts: Problems 334
10.5 Human impacts: Some possible solutions 336

10.1 INTRODUCTION

Floods impinge upon individuals and nations in every part of the world. Even as the beginning of the Third Millennium approaches, fluvial and coastal flooding pose complex challenges to scientists, engineers, economists and politicians alike. Through discussion of the physical processes of flooding and their human impacts we have tried in this book to show that, although great progress has been made in understanding the causes of floods and in developing rational responses to the flood hazard, many issues and problems remain unresolved. Some of these may succumb fairly rapidly to further work and investigation; others may remain as intractable problems into the foreseeable future.

What is certain is that the 'flood problem' will always be with us, no matter whether we live in the UK, the USA, Bangladesh or China. In other words, river and coastal floods will continue to happen, economic damage will occur and lives will be lost, but not necessarily at the same scale as at present. Improved understanding of the causes of flooding, better forecasting of flood events, and more sophisticated responses to the flood hazard will, in the long term and on a year-to-year basis, reduce the enormous toll of damage and loss of life to which we have, sadly, become accustomed. But even a basic grasp of statistics is sufficient to confirm that, in terms of the magnitude of individual events, the worst is yet to come. Densely populated floodplains and coastal lowlands will eventually experience a more catastrophic flood than they have ever experienced before, perhaps in the form of a 1000-year flood, or a 2000-year flood or even the Probable Maximum Flood. When that happens flood defences will be overtopped with the direst of consequences.

Rather than dwell upon this unlikely, though of course possible, and gloomy scenario, the purpose of this final chapter is to highlight the main areas in which an understanding of physical processes and their human impacts is still deficient and to discuss ways in which those deficiencies may eventually be overcome and the severity of the flood hazard may thereby be ameliorated.

10.2 PHYSICAL PROCESSES: PROBLEMS

Although, in broad terms, the causes and nature of floods are familiar, there remain frustrating problems and gaps in our understanding of the physical processes of flooding which, quite apart from social, economic and political factors, serve to constrain the effectiveness of our response to the flood hazard. The inadequacy of the floods database is responsible directly for some of these problems and indirectly for our imperfect understanding of some of the underlying hydrological and marine processes. It is increasingly apparent, however, that major flooding may be a response to influences which extend beyond the local or even the regional scale and that, in some cases, an improved explanation of flooding, and therefore an improved ability to forecast and predict floods, will only come about through a better understanding of the large-scale dynamics of the atmosphere and of the implications for flooding of climate change.

The *inadequacy of the floods database* is a recurring theme throughout this book. It cannot be emphasised too often that reliable and comprehensive data on fluvial and coastal flooding are still not universally collected and that in some countries data collection has only begun quite recently. The lack of consistent or appropriate data influences flood studies in different ways. At one extreme, the application of some of the more sophisticated models of catchment hydrology to flood forecasting is limited by their complex data demands; at the other extreme there are many rivers for which flood estimation, by whatever method, is constrained by the sparsity of simple data on annual maximum flood discharge. In other situations, including many LDCs, the information on successive flood catastrophes is neither recorded consistently nor is it archived systematically. To some extent these situations will improve with time, as more data are added year by year or as more effective ways are developed of pooling and archiving data from many different locations.

There are other circumstances, however, in which completely new data will be needed if effective responses to the flood hazard are to be developed. One example is that of accurate long-term weather forecasts, covering periods of several months, which would greatly facilitate flood forecasting and improve responsiveness to the flood hazard in both fluvial and coastal situations. One of the factors which mitigates against such forecasts at present is the lack of detailed monitoring of sea surface temperatures and ocean energy fluxes, although hopefully this will be remedied within the foreseeable future by the World Meteorological Organisation's developing Global Oceanic Observing System (GOOS). Another, potentially less tractable, example concerns the search for a reliable indicator of the tectonic and volcanic activity responsible for triggering tsunamis. Until such indicators have been established and monitored, advance warning of a tsunami will continue to be delayed until after the event has been triggered.

Improved databases are likely to play an important part, but only a part, in the improved *understanding of physical processes* which remains an essential prerequisite for further significant progress in explaining, forecasting and responding to floods. The need for continued scientific advance manifests itself in many aspects of floods, some of them more obvious than others.

One of the most fundamental, which is noted in Chapter 8, is that the derivation of a flow hydrograph from information on the input of rainfall and snowmelt to a river catchment continues to pose one of the greatest challenges in hydrology. This reflects not so much the failure to understand the broad thrust and interrelationships of the hydrological processes at work in the catchment area, but rather the inability to characterise, in a hydrologically meaningful way, the detailed heterogeneity of the catchment geology, slopes, soils and vegetation whose properties determine water storage and the speed and direction of water movement within the catchment.

As a result many conceptually simple tasks are difficult to perform. For example, inferring the flood behaviour of the entire river basin from a knowledge of flood behaviour in its headwater catchments is often a complex and unsatisfactory exercise. Equally frustrating is the frequent inability to relate flood hydrology and environmental change, particularly in a predictive situation. Thus quite apart from the problem of quantifying the likely magnitude of climate change over the next few decades, there is the even more intractable problem of converting that climate change to a change in flood characteristics. Similar gaps and uncertainties diminish our ability to predict the likely impact of land-use change, whether related to forestry or urbanisation or some other process, on flood hydrology.

This is seen very clearly in the context of flood forecasting which, as emphasised in Chapter 8, relies increasingly on the use of conceptual catchment models operating in real-time mode. Many models which work well in hydrologically homogeneous catchments are much less successful when applied to heterogeneous catchments having a mix of, say, urban and rural areas, or of forest and moorland. The success of flood estimation also depends partly on the ability to relate flood hydrology to catchment characteristics, hence the great significance in the UK of the forthcoming *Floods Estimation Handbook*, in which catchment characterisation is the result of a much more sophisticated approach than was possible in the earlier *Flood Studies Report*.

The study of coastal flooding and flood hazard response have also been constrained by an inadequate understanding of physical processes and a lack of basic data, for example on beach and nearshore levels over several decades. Accordingly, research programmes, such as CAMELOT in the UK, have been established in order to improve understanding of the relations between coastal erosion, deposition and flooding. And continued improvements to the Tropical Cyclone Warning System are planned both in relation to the IDNDR and as part of a long-term WMO programme.

In the context of improved understanding of physical processes, it is important to emphasise the unsuitability of some of the mathematical and statistical conventions that have been unthinkingly applied in the probabilistic approach to flood estimation. In the future it must be recognised that extreme flood events result from unusual combinations of circumstances, that they may not therefore fit neatly into any conventional statistical concept of a single frequency distribution and that this has consequential implications for the development of flood estimation methods and particularly for the importance of joint probability analysis.

Two aspects of the need for a better understanding of physical processes which impinge upon the occurrence and nature of flooding are sufficiently important to merit additional emphasis. These relate to the impact of large-scale climate dynamics, and of climate change and variability, on flood production.

It is only recently that the importance of *large-scale climate dynamics* in flood production has begun to emerge. In relation to river floods, for example, it is now clear that flood-producing rainfalls, including thunderstorms, are generated by systems of widely differing scale and complexity, many of which are linked via the macro- and mesoscale elements of the atmospheric circulation, and that a better understanding of such linkages is needed if our understanding of flood hydrology is to progress significantly. Nor is this problem confined to river flooding since, for example, current global climate models do not provide an adequate basis for simulating storm-surge conditions. It is clear then that until the complex interrelationships between hydrological, climatological and ocean circulation processes are better understood and better modelled, it will not be possible to explain more satisfactorily than at present the occurrence, magnitude and distribution of both river and coastal flooding.

It is emphasised several times in this book that the hydrological implications of *climate change and variability* and particularly their implications for flooding constitute some of the most important unanswered questions in flood hydrology. Although it is comparatively easy to show qualitatively that variations of climate, and indeed also changes of land use, may affect hydrological processes, including runoff processes, it is much more difficult to quantify their impact on flood production. In the case of global warming, for example, there continues to be uncertainty about both the magnitude of the event itself and also about its likely impacts on flooding. In the context of river floods this is especially true for the mid-latitudes, where the interaction of tropical and polar airmasses means that future patterns of precipitation are very sensitive to shifts in wind systems. But substantial uncertainties also remain in relation to coastal flooding, despite the fact that increased sea levels are a much surer outcome of global warming than are modified precipitation patterns. For example, the extent to which a rise in mean sea level will be accompanied by an increase in the frequency of coastal storms and by changes in the frequency of extreme sea-level events is far from clear, as also is their impact on coastal flooding.

10.3 PHYSICAL PROCESSES: SOME POSSIBLE SOLUTIONS

Although, undoubtedly, some advancement in our understanding of the physical bases of fluvial and coastal flooding will be brought about by conceptual advances in the relevant hydrological and marine sciences, it is likely that most progress will result from improvements in databases. As indicated in the preceding section, some such improvements will arise from the incremental growth of databases with time. Far more important, however, will be the potential database improvements which are likely to result from the extended application of modern data collection technology. The use of older technology has already resulted in great advances in precipitation measurement and still further improvements will undoubtedly result from the opportunities afforded by radar to improve the robustness of weather prediction models and rainfall–runoff models. More modern approaches, using satellite imagery, global positioning systems, digital terrain models and geographical information systems, have the potential to make an even more effective impact on flood studies.

From early in its development *satellite imagery* has offered considerable scope for both one-off and also repetitive mapping of major flood outlines. Initially the imagery from satellites such as Landsat could be used only for post-event reconstructions. However, modern imagery, such as that from the ERS-1 satellite, can now be processed sufficiently rapidly for use in near-real-time flood forecasting operations and satellite imagery provided evidence of the massive source-area expansion that occurred during the great Mississippi flood of 1993.

A number of flood forecasting systems, such as FRONTIERS and NIMROD in the UK, combine information from radar and weather satellites, and weather satellite data now mean that atmospheric moisture content is a more accessible variable than heretofore.

One of the most intractable areas of data collection pertinent to the generation of river floods has related to quantification of the areal extent of high-altitude and high-latitude snowpacks and to the estimation of their meltwater equivalent. Satellite imagery is now used to map the extent of snow and ice cover in many areas and the resulting data have greatly improved the potential for successful real-time forecasting of floods resulting from snowmelt and icemelt.

Satellite imagery, often in conjunction with radar, is also being used to make improvements in another difficult area, that of flash flood forecasting. Before the advent of radar and satellite imagery, heavy precipitation upstream of areas prone to flash flooding could be identifed only from telemetered readings from upstream raingauges or water-level recorders, or by telephone contact from observers in the field. Technological improvements in both areas have been largely responsible for significant increases in both the number and viability of flash flood forecasting and warning systems, although the practical suitability of satellite imagery for flash flood forecasting is still very dependent on local and regional conditions of climate and topography.

It is perhaps in relation to the understanding and forecasting of coastal flooding, for which relevant observations and data collection have long been difficult, that satellite imagery has made its most successful impact. Data on wave height, for example, were once largely derived from observations on ships and buoys but are now extensively supplemented by weather satellite data, including measurements of wave height by radar altimeters. Such satellite observations mean that the maximum wave height likely to threaten a flood-prone coastline can now be estimated more precisely and the real-time analysis of satellite data means that the approach of large waves can now sometimes be forecast with operationally useful lead times.

Since storm surges and tsunamis may be triggered by very distant events, their accurate and timely forecasting relies heavily on sophisticated technology, such as satellite imagery and radar. In relation to flood-producing surges, for example, which result largely from the effects of cyclones, successful flood forecasting and warning are partly dependent upon the accuracy with which the track and intensity of cyclones can be forecast. Enormous improvements to forecast lead times have resulted from the use of satellite imagery, which now permits continuous global surveillance of the development and movement of major weather systems. Tsunami warning systems, such as that for the Pacific Ocean, have also benefited from the use of satellite imagery, although there is still great potential for yet further improvement. For example, satellite observations of the slight surface movement which is a precursor to a volcanic eruption may eventually provide a basis for improved forecasting of tsunamis resulting from volcanic activity.

A directly related aspect of satellite technology, which also has the potential for improving significantly the understanding and forecasting of floods, is the use of *Global Positioning Systems* (GPSs). Their use has been suggested, for example, for flood outline mapping in the hope that this could lead eventually to the definition of T-year flood outlines, especially for large rivers and for flood-prone coastal areas.

In addition, the recent advent of GPS has made it possible to determine the vertical displacement of tide gauges and thereby to improve greatly the estimation of absolute changes in mean sea level. Again, GPS observations are being used to monitor earth movements, such as those along the 'great crack' on the main island of Hawaii, which are likely to impact upon the adjacent seabed and so trigger a tsunami.

The ability to map flood outlines, model catchment flood hydrology, and develop flood estimation procedures on the basis of relationships between the nature of a catchment and its flood hydrograph, are some of the aspects of flood studies which have been constrained in the past by the sparsity of point information on catchment characteristics. Significant advances have already been made since the comparatively recent development of *Digital Terrain Models* (DTMs) and it is clear that DTMs will play an increasingly important role in flood studies in the future.

For example, definition of the 100-year flood outline in England and Wales became feasible only after the development of the Institute of Hydrology DTM. Also in the UK, one of the most significant improvements offered by the forthcoming *Flood Estimation Handbook* is in the use of catchment characteristics. This was limited in the *Flood Studies Report* to characteristics which could be derived easily from standard maps but in the *FEH* catchment characteristics such as stream length, stream density, and slope gradient can be derived from DTM and other digitised and gridded data.

Some of the greatest benefits of DTM data are likely to be seen in the development of hydrological models for flood simulation and forecasting. Even simple empirical models, such as the Rational Formula, are more successful when applied in situations where DTM data can be used to provide an accurate definition of catchment characteristics. More sophisticated models are likely to benefit to an even greater extent. Thus the growing availability of grid-based spatial data, not just from DTMs but also from rainfall radar and *Geographical Information Systems* (GISs), will undoubtedly enhance the value of *distributed* rainfall–runoff models for operational flood forecasting. An important example in the UK is the IH Grid Model, which is designed for use in real-time systems such as the River Flow Forecasting System and uses inputs from weather radar and DTM. In the longer term, the incorporation of inputs from DTMs and GISs should permit the development of powerful continuous-simulation catchment models which could be applied to a wide range of problems, such as predicting the hydrological effects of land-use change and real-time flow forecasting, as well as to flood estimation. And in mountainous regions information on snow area obtained from satellite imagery will be potentially more valuable when used in conjunction with DTM and GIS data.

In summary, it is important to emphasise that the sort of advances in data collection technology which have been discussed in the preceding paragraphs may not, of themselves and individually, bring about the necessary improvements in our understanding of the physical processes of flooding and in our ability to simulate and forecast flood events. More important are likely to be the ways in which these technologies are used in combination, not only with each other, but also with the continued enhancement of accessible computing power and improved physical models. Evidence of such combined development has been discussed throughout this book. For example, improvements in river flood forecasting and warning have been made possible largely by advances in software design and computing power which enable flexible integrated flood forecasting and warning systems to accommodate the inputs from radar, satellite imagery, digital terrain models, synoptic analyses, mesoscale atmospheric models, and catchment and routing models. In the case of storm surges and tsunamis, which may be triggered by events occurring thousands of kilometres from the location of the subsequent coastal flooding, further improvements in the accuracy and timeliness of forecasts will depend largely on sophisticated telecommunication systems, involving GIS, radar and satellite imagery.

10.4 HUMAN IMPACTS: PROBLEMS

The fundamental problem *appears* to be that of a growing flood hazard in terms of both the number of flood disasters and the associated loss of life and property worldwide. Unfortunately, definite and quantitative statements are difficult to make because the systematic records necessary for monitoring flood impacts over time are unavailable for most countries. Therefore, the *inadequacy of the floods database* is as much a problem when assessing flood impacts as when identifying the geophysical processes which cause floods. Some of the difficulty lies in the definition and archiving of flood disasters which, because they are often part of more complex geophysical events, may be catalogued with other natural hazards. It is also clear that some of the apparent increase in flood hazard impact is due to the improved worldwide reporting of all news items, especially for small events in remote areas, which has followed technical advances in communications and the emergence of continuous 24-hour media coverage in recent decades. In the LDCs, where there is a chronic lack of reliable time-series data, any upward trend in flood disasters will be associated to some extent with the ongoing increase in the world population, most of which is concentrated in the poorest countries, and the consequent growth in the number of people living fragile and vulnerable lives. In the MDCs the progressive increase in personal wealth and the infrastructural assets at risk on floodplains and along shorelines mean that the economic consequences of disaster will increase, even when damages are standardised for cost inflation. Caution is required when interpreting any flood loss statistics. But the available evidence suggests that floods have been under-reported in the past, that the exposure to floods and the potential losses are rising and that most communities are not coping well with the flood hazard, even though conclusive proof for these assertions is often hard to find. Wherever such trends exist, they indicate a lack of sustainability.

The *human causes of flood disaster* are still not well understood and many commentators argue that more attention should be paid to the political, economic, social and demographic aspects of floods. Increasingly the hydrological event is seen as the 'trigger' for a flood disaster which may have its roots in many human-related factors which range potentially from profit-seeking motives or poorly enforced legislation to environmental degradation, colonial expansion or social divisions. In turn, this view has influenced attitudes to flood hazard response where constructive debate has not been helped by the opposing paradigms of hazard represented by the engineering, behavioural and development schools of thought. It is difficult to effect a direct and practical reconciliation between the different interpretations of flood hazard taken by, say, civil engineers and rural sociologists as demonstrated by the ongoing debate regarding solutions to flood problems in Bangladesh. Clearly, any working consensus in a particular area requires a recognition that several disciplines have a contribution to make towards assessing and reducing flood hazards and that the relative importance of these contributions will vary in different settings. It is these implications which make it true to say that less is known about the human than the physical causes of floods.

Especially in the MDCs, floodplain invasion is frequently driven by economic prosperity and land competition, which can be seen by government and others as desirable attributes ranked well above the need to steer new development away from flood-prone sites. Even when the problem of increased hazard exposure is recognised, it is difficult to ensure that the risk-takers assume financial responsibility for their actions. For example, inadequate building regulations or loopholes in insurance legislation often allow local and individual costs to be passed on to the general taxpayer through subsidies or disaster aid payments. Floodplain invasion resulting from rapid economic development and land-use change is nowhere more apparent than in the growth of cities. These create special risks, not simply in terms of the absolute numbers of people crowded onto floodplains and shorelines but also because urbanisation creates largely impermeable surfaces capable of generating local flash flooding. For example, established mid-latitude cities in the MDCs, like London, are already exposed to a threat of summer convectional storms and, even where a comprehensive sewer system exists, the capacity is often underdesigned in relation to the peak flows produced by the higher rates of runoff. Many of the emerging megacities of the world are constructed along low-lying coasts on unconsolidated deposits or landfill. Much of the Japanese population, for example, is located along the Pacific shoreline exposed to storm rainfall and tsunamis. Where large population concentrations do exist they are vulnerable to epidemics and water-borne diseases following floods, especially in the LDCs.

Throughout the LDCs the poor who are at risk in flood-prone environments already pay a high price for their location. In such countries vulnerability, which is poverty dependent, is a key factor and applies at all levels. For example, flood policy pursued by Third World governments may be vulnerable to external influences arising from their burden of foreign debt and a dependency on aid. Regional agencies responsible for flood mitigation may well be seriously underfunded and lacking in well-trained personnel. At the household level vulnerability is reflected in personal circumstances involving factors such as occupation, age, gender, social class, ethnicity and nutritional status. To take age-related factors alone, the very old will have greater mobility problems in responding to flood warnings and the very young will be at extra risk from water-related diseases, like dysentery, which occur after floods. Even in the MDCs floodplains contain a disproportionate number of disadvantaged people, including the elderly, infirm, single parents and others dependent on low incomes. Global environmental change involving continued population growth, the increased per capita use of resources and rising expectations from the public for effective flood mitigation will tend to create more inequality and vulnerability to floods.

If the flood hazard is increasing, part of the reason must be a *failure of existing flood mitigation strategies*. The very number of flood hazard adjustments is itself an indication that none of them is wholly successful. The most fully developed mitigation measures have been structural, often encouraged, as in England and Wales, by grant-aid from central government. It is currently fashionable to blame engineering schemes for underperforming. But, in general, flood defence schemes have worked well within their design limits and saved much economic loss, as demonstrated by the levee schemes during the exceptional Mid-West floods of 1993. Engineered projects will never be able to control all floods and it is not the fault of the dam or levee that planners, developers and others have used the protection available from such structures to accelerate floodplain or coastal invasion and increase the exposed risk. Despite this, flood engineering solutions alone are no longer sufficient and have attracted some valid criticism, not least in terms of their negative environmental effects. At this point it must be recognised that each individual flood response has some flaw; not all those persons who are eligible will take out insurance, aid is short-term and uncertain in its distribution, flood warnings fail to reach all those at risk, and building codes and land-use regulations are sometimes not fully disclosed or enforced. Improving the performance of non-structural measures involves complex problems which go well beyond the often-stated fact that flood forecasts need evolving into an efficient warning and response strategy and will confront some deeply rooted social beliefs and issues. For example, in the MDCs wealth creation and the demand for a home or leisure facilities near water space are powerful personal motives and reflect a belief in individual freedoms to develop flood-prone land. In the LDCs other factors such as poverty, lack of education, ethnic conflict and malnourishment hamper the flood loss reduction process. All these elements are part of the enduring human condition and are not likely to be easily transformed.

10.5 HUMAN IMPACTS: SOME POSSIBLE SOLUTIONS

If the flood hazard is to be reduced in the future, an *improved performance of the individual responses*, both structural and non-structural, is clearly necessary. To some extent this is already taking place with the implementation, at high-risk sites, of more technically and economically sound flood defence measures. Technical soundness will depend, in many cases, on a better understanding of the basic physical processes, including erosion and sedimentation, affecting both fluvial and marine environments. To ensure economic soundness, all new capital schemes should have a benefit–cost ratio of at least unity, with a widening of the scope of economic appraisal to embrace intangible losses and gains such as ill-health and recreational values respectively. Post-project evaluation should be employed in all schemes not just to monitor the degree of success immediately after construction but also to make sure that the 'levee' effect is not permitted as a result of any subsequent development of floodplain or coastal land creating greater property values which, in turn, justify a raising of the structures to ensure a higher standard of flood defence. Such surveillance should, therefore, ensure that structural schemes are environmentally and economically sustainable into the future. Sustainability also depends to some extent on better consultation with the public about structural schemes where the exercise must be conducted as a real partnership between the responsible agencies and the public which they serve.

At the same time, discouraging any further inappropriate development in areas already subject to flooding, or likely to become flood-prone in the future, is also important. There has always been a human dependence on proximity to water but the basic, traditional uses of floodplains and coasts are now accompanied by people pursuing other goals, such as recreational access to a waterfront or the occupation of land with high visual amenity, whilst neglecting the risk of flooding. To counter these trends will probably require a strengthening and a proper enforcement of legislation in order to control the profit-making forces involved in land-use conversion. In some cases, it will be necessary to move people from flood-prone land, especially if the precautionary principle is adopted with respect to global warming. Land and property acquisition in the most threatened areas will then become part of a wider policy of managed retreat including the restoration of wetlands and the release of artificial floods from storages for ecological benefits, in an attempt to re-unite the river channel and the shoreline with the floodplain and the coastal zone respectively.

In general, there is likely to be a continued rise of all types of non-structural measures over the more traditional structural solutions. This means an improved understanding and implementation of alternative 'soft' defensive solutions to flood problems, such as beach management and river corridor development. Some trends will continue such as an increasing reliance on conventional non-structural options, like more efficient and cost-effective flood warning and response schemes if for no other reason than that these schemes are primarily designed to save lives in many areas. The use of new technology, which has already greatly aided the saving of life from hurricanes, will be important partly because it has a low environmental impact and is suitable for a more sustainable future. But, for optimum success, such technology needs to be fully integrated with the warning response and extended to flash floods, for example in the form of automatic dial-up telephone alert systems.

Insurance is destined to become more significant in the future as it spreads from the MDCs to the LDCs. Like aid, insurance can be most effective if it is used to reduce, rather than just redistribute, losses as shown by the land-use regulation function of the NFIP in the USA. This can be done if aid is tied to longer-term development rather than emergency requirements. These are not guaranteed solutions in themselves. Aid can be misdirected by inefficient politicians and agencies, whilst insurance underwriting needs care to ensure that the participation rate is as high as possible amongst those eligible in order to spread the risk and that firm arrangements exist to prevent people who reject insurance from receiving disaster aid.

In addition to improving individual responses, future priorities for research and development must try to achieve a *better integration of flood hazard responses*. It has been known for over 50 years that the choice of any flood response should be made from the widest selection possible and that sole reliance on one approach is rarely successful. Despite this, it has proved difficult to obtain an appropriate synergy between individual methods which are often seen as the preserve of specific organisations representing and safeguarding their own expertise and professional standards. The challenge is to integrate not only between structural and non-structural measures but also within these categories. For example, river channel improvements could be better linked with individual flood-proofing actions, flood forecasting could be associated with more efficient warning and evacuation procedures, and most emergency responses could be better related to long-term planning.

This involves taking a more strategic and holistic approach which is both longer-term and more sustainable. In the first instance, such an approach must address the common lack of political will to tackle flood problems and make sure that clear floodplain management policies are adopted and implemented. Catchment and shoreline management plans, as recently adopted in England and Wales, provide some of the best opportunities for taking such a comprehensive view. This is because catchment management plans take into account the interrelationships between rivers, land use and related developments within drainage basins and address environmental issues which arise in the long term. Similarly, shoreline management plans have gone forward in a practical way and identify the basic building blocks of management units which capture stretches of coast with similar characteristics in terms of natural coastal processes and land use. There remain many problems. For example, coastal defence is not the same as coastal management and there is a need to integrate all the planning activity along the coast within a context of genuine public consultation and the clarification of issues such as that of any legal liability associated with deliberate and managed retreat from the existing shoreline. We must not underestimate these tasks. The complete integration of structural and non-structural responses is an objective which has not yet been achieved in any country. It will be especially difficult to implement in the poorest countries plagued with additional burdens such as low resources and high illiteracy.

Perhaps above all it will be necessary to *change attitudes towards flood hazard*. Unlike most other natural hazards, floods are predictable events within certain physical settings because they are essentially controlled by topography. They create linear outlines and other spatial patterns within which specific settlements, and even individual properties, are known to be most at risk. These certainties have promoted the standard responses to flood hazards through governments and funded agencies working in specialised fields such as flood defence, insurance or land-use planning. Now there is a need to question some of these orthodox mitigation strategies and recognise that other, more radical, options exist. In the future, the exercise of new choices for flood hazard reduction may well become progressively more important and could include some difficult and unpopular decisions, for example, taking central government subsidies away from the protection of intensive floodplain agriculture or the maintenance of some coastal defences. When structures are designed or refurbished to provide a given level of protection it will be necessary to publicise this and to ensure that factors creating greater hazard exposure, such as additional floodplain investment or rising sea levels, do not lead to an automatically heeded call for even more expensive flood control measures.

There is already some evidence that governments are seeking to cut the cost of public expenditure for flood defence and are willing to let some existing works deteriorate. There is also a recognition that the adoption of a design flood standard leaves a community at risk from the rare but possible event above that level and the preparation of plans for managing that emergency, whilst difficult to resource, are still important. Similarly, in all types of political economy, governments are shedding responsibility for routine compensation after disaster. But, ideally, these policies should not be implemented in a haphazard way or by stealth but openly with full public consultation and a wide appreciation of the present and future implications. Part of this process might well be less direct subsidy to flood-prone residents but more technical information, for example on the range of options for flood-proofing properties so that people subsequently moving into such areas can take responsibility for their own futures through a greater degree of self-help. The public has a notoriously short-term memory and flood hazards need to be kept more permanently in the collective conscience.

The increasing reliance on non-structural measures brings with it the explicit need, which has often been ignored, of persuading flood-prone communities to accept some risk and adopt a 'living with floods' attitude. The significance of individual responsibility and self-help goes right through effective hazard adjustment from the immediate emergency response to longer-term planning and public safety issues. The need to prepare for some risk, for example through first-aid training, is vital, although risk communication and education remains an often inefficient process. Given the difficulties which lay persons find in understanding flood return periods and flood hazard maps, flood risk will not be an easy concept to transmit and get accepted at the community level.

The threat of flooding varies widely and there needs to be a better recognition of where the real problems are at the global, regional and local scale so that research, development and investment can be carefully targeted. To some extent this is the role of the *International Decade for Natural Disaster Reduction* which was predicated on easing the plight of the Third World, but not everyone in these areas is equally at risk. In particular the IDNDR recognised the value of transferring best practice from the MDCs to the LDCs but this should not simply be a 'top-down' transfer of technological fixes. Successful flood mitigation is a learning process and there is an equally strong requirement for knowledge and experience to be disseminated between and within the LDCs to ensure that traditional practices are not lost in the face of imported solutions. Flood hazards interact with global environmental changes and the scale of future impacts will be crucially dependent on the balance between vulnerability and resilience of the people at risk. In the newly industrialising countries rising flood problems are a telling indicator of unsustainability. In many of these areas economic development and modernisation threaten to sweep away the traditional lifestyles and the existing use of indigenous responses. Without these responses people will lose their capacity to absorb and recover from a hazardous event. Each country and each community should be encouraged to develop their own capacities and the self-reliance needed to undertake flood hazard reduction measures. By this means national strategies and standards, tailored to a local need, can be developed with investment directed more to the prevention, rather than the cure, of flood disasters.

APPENDIX
Metric Conversion Tables

DISTANCE*

Inches		Millimetres
0.039	1	25.4
0.079	2	50.8
0.118	3	76.2
0.158	4	101.6
0.197	5	127.0
0.236	6	152.4
0.276	7	177.8
0.315	8	203.2
0.354	9	228.6

Feet		Metres
3.281	1	0.305
6.562	2	0.610
9.842	3	0.914
13.123	4	1.219
16.404	5	1.524
19.685	6	1.829
22.966	7	2.134
26.246	8	2.438
29.527	9	2.743

Yards		Metres
1.094	1	0.914
2.187	2	1.829
3.281	3	2.743
4.375	4	3.658
5.468	5	4.572
6.562	6	5.486
7.656	7	6.401
8.750	8	7.316
9.843	9	8.230

Miles		Kilometres
0.621	1	1.609
1.243	2	3.219
1.864	3	4.828
2.486	4	6.437
3.107	5	8.047
3.728	6	9.656
4.350	7	11.265
4.971	8	12.875
5.592	9	14.484

VOLUME*

Cu. feet		Cu. metres
35.315	1	0.028
70.629	2	0.057
105.943	3	0.085
141.258	4	0.113
176.572	5	0.142
211.887	6	0.170
247.201	7	0.198
282.516	8	0.227
317.830	9	0.255

Imp. gallons		Litres
0.220	1	4.544
0.440	2	9.087
0.660	3	13.631
0.880	4	18.174
1.101	5	22.718
1.321	6	27.262
1.541	7	31.805
1.761	8	36.349
1.981	9	40.892

AREA*

Sq. yards		Sq. metres
1.196	1	0.836
2.392	2	1.672
3.588	3	2.508
4.784	4	3.345
5.980	5	4.181
7.176	6	5.016
8.372	7	5.853
9.568	8	6.690
10.764	9	7.526

Sq. miles		Sq. km
0.386	1	2.590
0.772	2	5.180
1.158	3	7.770
1.544	4	10.360
1.931	5	12.950
2.317	6	15.540
2.703	7	18.130
3.089	8	20.720
3.475	9	23.310

Acres		Hectares
2.471	1	0.405
4.942	2	0.809
7.413	3	1.214
9.884	4	1.619
12.355	5	2.023
14.826	6	2.428
17.297	7	2.833
19.768	8	3.237
22.239	9	3.642

*In the Distance, Volume and Area conversion tables the figures in the central columns may be read as either metric or imperial units (e.g. 1 cubic foot = 0.028 cubic metre, or 1 cubic metre = 35.315 cubic feet).

TEMPERATURE
(Celsius to Fahrenheit)

°C	0	1	2	3	4	5	6	7	8	9
40	104.0	105.8	107.6	109.4	111.2	113.0	114.8	116.6	118.4	120.2
30	86.0	87.8	89.6	91.4	93.2	95.0	96.8	98.6	100.4	102.2
20	68.0	69.8	71.6	73.4	75.2	77.0	78.8	80.6	82.4	84.2
10	50.0	51.8	53.6	55.4	57.2	59.0	60.8	62.6	64.4	66.2
0	32.0	33.8	35.6	37.4	39.2	41.0	42.8	44.6	46.4	48.2
0	32.0	30.2	28.4	26.6	24.8	23.0	21.2	19.4	17.6	15.8
−10	14.0	12.2	10.4	8.6	6.8	5.0	3.2	1.4	−0.4	−2.2
−20	−4.0	−5.8	−7.6	−9.4	−11.2	−13.0	−14.8	−16.6	−18.4	−20.2
−30	−22.0	−23.8	−25.6	−27.4	−29.2	−31.0	−32.8	−34.6	−36.4	−38.2
−40	−40.0	−41.8	−43.6	−45.4	−47.2	−49.0	−50.8	−52.6	−54.4	−56.2

References

Abrahams, M.J., J. Price, F.A. Whitlock and G. Williams (1976) The Brisbane floods, January 1974: their impact on health, *Medical J. Australia*, **2**, 936–939.
Acreman, M.C. (1985a) The effects of afforestation on the flood hydrology of the upper Ettrick valley, *Scottish Forestry*, **39**, 89–99.
Acreman, M.C. (1985b) Predicting the mean annual flood from basin characteristics in Scotland, *Hydrol. Sci. J.*, **30**, 37–49.
Acreman, M. and S. Wiltshire (1989) The regions are dead; long live the regions. Methods of identifying and dispensing with regions for flood frequency analysis, in L. Roald, K. Nordseth and K.A. Hassel (eds), *FRIENDS in Hydrology*, IAHS Publ. No. 187, 175–188.
Adams, W.M. (1993) Indigenous use of wetlands and sustainable development in West Africa. *Geogrl. J.*, **159**, 209–218.
Admiralty Tide Tables (1970) Vol. 1.
Agriculture Canada (1975) *Soil Erosion by Water*, Publication 1083, Canada Dept. Agriculture.
Ahmad, M.I., C.D. Sinclair and A. Werritty (1988) Log-logistic flood frequency analysis, *J. Hydrol.*, **98**, 205–224.
Ahuja, A. (1996) Proving Noah's flood, *The Times*, London, 9 December, 14.
AIRAC (1986) *Catastrophic Losses. How the Insurance System would Handle Two 7$-Billion Hurricanes*, All-Industry Research Advisory Council, Oak Brooks, Illinois.
Aitken, A.P. (1975) Catchment models for urban areas, in T.G. Chapman and F.X. Dunin (eds), *Prediction in Catchment Hydrology*, Australian Academy of Science, Canberra, 257–275.
Alam, S.M.N. (1990) Perceptions of flood among Bangladeshi villagers, *Disasters*, **14**, 354–357.
Alcock, G.A. and J.D. Richards (1993) Extreme astronomical tides, *Flood and Coastal Defence*, **4**, p. 2, MAFF, London.
Alexander, D. (1989) Consequences of floods in developing countries: International perspectives for disaster management, *Proc. Internat. Seminar on Bangladesh Floods: Regional and Global Environmental Perspectives*, 4–6 March, 1989, Dhaka.
Ambrus, S., L. Iritz and A. Szöllösi-Nagy (1989) Physically based hydrological models for flood computations, in K. Beven and P. Carling (eds), *Floods: Hydrological, Sedimentological and Geomorphological Implications*, Wiley, Chichester, 47–55.
Amirthanathan, G.E. (1989) Optimal filtering techniques in flood forecasting, in L. Roald, K. Nordseth and K.A. Hassel (eds), *FRIENDS in Hydrology*, IAHS Publ. No. 187, 13–25.
Andah, K. and F. Siccardi (1991) Prediction of hydrometeorological extremes in the Sudanese Nile region: a need for international co-operation, in F.H.M. van de Ven, D. Gutknecht, D.P. Loucks and K.A. Salewicz (eds), *Hydrology for the Water Management of Large River Basins*, IAHS Publ. No. 201, 3–12.
Anderson, J.H., L.E. Ganster, J.P. Scott, et al. (1995) *Mid West Flood: Information on the Performance, Effects and Control of Levees*, Report No. GAO/RCED-95-125. General Accounting Office, Washington, DC.
Andrews, J. (1993) *Flooding: Canada Water Book*, Ecosystem Sciences and Evolution Directorate, Environment Canada, Ottawa.
Andrieu, H., J.D. Creutin, G. Delrieu, J. Leossoff and Y. Pointin (1989) Radar data processing for hydrology in the Cevennes region, in A. Rango (ed.), *Remote Sensing and Large-Scale Global Processes*, IAHS Publ. No. 186, 105–115.

Aneke, D.O. (1985) The effect of changes in catchment characteristics on soil erosion in developing countries (Nigeria), *Agricl. Engg.*, **66**, 131–135.

Anon. (1904) *Relationship of Woods to Domestic Water Supplies*, Leaflet No. 99, Board of Agriculture and Fisheries, London.

Anon. (1970) Forty-year Ice Age is on the way, *The Observer*, London, 2 August.

Anon. (1992) *Explaining Flood Risk*, US Army Corps of Engineers, Washington, DC.

Anon. (1993) *Analysis of the IHDA Floodproofing Loan Program*, Illinois Housing Development Authority, Park Forest, Illinois.

Anon. (1995) Disasters and property insurance: coping with the aftershocks, *Coastal Heritage*, **9**, 3–12.

Anon. (1996a) Nimrod operational, *Outlook*, April/May, p. 17.

Anon. (1996b) Raising the defences, *Outlook*, April/May, p. 17.

Archer, D. (1992) *Land of Singing Waters: Rivers and Great Floods of Northumbria*, Spredden Press, Stocksfield.

Archer, D. (1994) Walls of water, *Circulation, British Hydrol. Soc. Newsletter*, **44**, 1–3.

Archer, D.R. (1989) Flood wave attenuation due to channel and floodplain storage and effects on flood frequency, in K. Beven and P. Carling (eds), *Floods: Hydrological, Sedimentological and Geomorphological Implications*, Wiley, Chichester, 37–46.

Arnell, N.W. (1984) Flood hazard management in the United States and the National Flood Insurance Program, *Geoforum*, **15**, 525–542.

Arnell, N.W. (1987) Regional institutions and floodplain management in England and Wales, in R.H. Platt (ed.), *Regional Management of Metropolitan Floodplains*, Monograph No. 45, Institute of Behavioral Science, University of Colorado, 193–222.

Arnell, N.W. (1989) Changing frequency of extreme hydrological events in northern and western Europe, in L. Roald, K. Nordseth and K.A. Hassel (eds), *FRIENDS in Hydrology*, IAHS Publ. No. 187, 237–249.

Arnell, N.W. (1992) Factors controlling the effects of climate change on river flow regimes in a humid temperate environment, *J. Hydrol.*, **132**, 321–342.

Arnell, N.W. (1994) Variations over time in European hydrological behaviour: a spatial perspective, in P. Seuna, A. Gustard, N.W. Arnell and G.A. Cole (eds), *FRIEND: Flow Regimes from International Experimental and Network Data*, IAHS Publ. No. 221, 179–184.

Arnell, N.W., M.J. Clark and A.M. Gurnell (1984) Flood insurance and extreme events: the role of crisis in prompting changes in British institutional response to flood hazard, *Appl. Geog.*, **4**, 167–181.

Arnell, N.W., R.P.C. Brown and N.S. Reynard (1990) *Impact of Climatic Variability and Change on River Flow Regimes in the UK*, Institute of Hydrology, Report No. 107.

ASCE (1972) *Re-evaluation of the Adequacy of Spillways of Existing Dams*, Report of the Task Committee of the Hydrometeorology Committee of the American Society of Civil Engineers.

Aschwanden, H., R. Weingartner and H. Duster (1993) The requirement and advantage of the application of GIS at a national level, *Hydrol. Sci. J.*, **38**, 529–537.

Ashkenazi, V., G. Beamson, R. Bingley, C.C. Chang, A. Dodson, T. Moore and T. Baker (1995) Measuring changes in mean sea level to millimetres by GPS, Section 7.1 in *Proc. 30th MAFF Conference of River and Coastal Engineers*, Keele, 5–7 July, Ministry of Agriculture, Fisheries and Food, London.

Askew, A.J. (1992) Flooding—Contributions from the engineering profession and an international perspective, Paper presented at *Hazards Forum Seminar*, Inst. Civ. Engrs, London, January 1992.

Atwood, G. (1994) Geomorphology applied to flooding problems of closed basin lakes...specifically Great Salt Lake, Utah, *Geomorphology*, **10**, 197–219.

Austin, B.N., I.D. Cluckie, C.G. Collier and P.J. Hardaker (1995) *Radar-based Estimation of Probable Maximum Precipitation and Flood*, The Meteorological Office, Bracknell.

Austin, R.M. and R.J. Moore (1996) Evaluation of radar rainfall forecasts in real-time flood forecasting models, in M. Borga and R. Casale (eds), *Integrating Radar Estimates of Rainfall in Real Time Flood Forecasting*, Special Issue of *Quaderni di Idronomia Montana*, **16**, 19–28.

AWRC (1972) *Hydrology of Smooth Plainlands of Arid Australia*, Hydrol. Series No. 6, Australian Water Resources Council, Canberra.

REFERENCES

AWRC (1978) *Variability of Runoff in Australia*, Hydrol. Series No. 11, Australian Water Resources Council, Canberra.

Bailey, J.F. and J.L. Patterson (1975) *Hurricane Agnes Rainfall and Floods, June–July 1972*, USGS Professional Paper, 924.

Bailey, R.A. and C. Dobson (1981) Forecasting for floods in the River Severn catchment, *J. Inst. Wat. Engrs. Sci.*, **35**, 168–178.

Baird, A.W. (1884) Report on the tidal disturbances caused by the volcanic eruptions at Java, August 27th–28th, 1883, *Proc. Roy. Soc.*, **36**, 248–253.

Baker, E.L. (1991) Hurricane evacuation behaviour, *Internat. J. Mass Emergencies and Disasters*, **9**, 287–310.

Baker, V.R. (1989) Magnitude and frequency of palaeofloods, in K. Beven and P. Carling (eds) *Floods: Hydrological, Sedimentological and Geomorphological Implications*, Wiley, Chichester, 171–183.

Baker, V.R. (1994) Geomorphological understanding of floods, *Geomorphology*, **10**, 139–156.

Baker, V.R., R.C. Kochel and P.C. Patton (1979) Long-term flood frequency analysis using geological data, in *The Hydrology of Areas of Low Precipitation*, IAHS Publ. No. 128, 3–9.

Baker, V.R., R.C. Kochel and P.C. Patton (eds) (1988), *Flood Geomorphology*, Wiley, New York.

Bandarin, F. (1994) The Venice project: a challenge for modern engineering, *Proc. Inst. Civil Engrs.*, **102**, 163–174.

Bardsley, W.E. (1994) Against objective statistical analysis of hydrological extremes, *J. Hydrol.*, **162**, 429–431.

Bari, M.F. (1994) Flood proofing potentials in Bangladesh flood management, in W.R. White and J. Watts (eds), *River Flood Hydraulics*, Wiley, Chichester, 579–593.

Barnett, T., N. Graham, M. Cane, S. Zebiak, S. Dolan, J. O'Brien, and D. Legler (1988a) On the prediction of the El Niño of 1986–1987, *Science*, **241**, 192–195.

Barnett, T.P., L. Dümenil, U. Schlese and E. Roeckner (1988b), The effect of the Eurasian snow cover on global climate, *Science*, **239**, 504–507.

Baron, B.C., D.H. Pilgrim and I. Cordery (1980) *Hydrological Relationships Between Small and Large Catchments*, Tech. Paper No. 54, Australian Water Resources Council, Canberra.

Barry, R.G. and R.J. Chorley (1982) *Atmosphere, Weather & Climate*, 4th edition, Methuen, London.

Bascom, W. (1959) Ocean waves, *Scientific Amer.*, Aug., 2–12.

Bates, C.G. (1921) First results in the stream-flow experiment: Wagon Wheel Gap, Colorado, *J. Forestry*, 402–408.

Bates, C.G. and A.J. Henry (1928) Forest and streamflow experiment at Wagon Wheel Gap, Colorado, *Monthly Weather Rev., Supplement 30*.

Bathurst, J.C., L. Hubbard, G.J.L. Leeks, M.D. Newson and C.R. Thorne (1990) Sediment yield in the aftermath of a dambreak flood in a mountain stream, in R.O. Sinniger and M. Monbaron (eds), *Hydrology in Mountainous Regions, II Artificial Reservoirs, Water and Slopes*, IAHS Publ. No. 194, 287–294.

Bayliss, A.C. and R.C. Jones (1993) *Peaks-over-threshold Flood Database: Summary Statistics and Seasonality*, Institute of Hydrology, Rept. No. 121.

Bayliss, A.C. and R.M.J. Scarrott (1996) Catchment characteristics for flood estimation: Indexing catchment urbanization using gridded spatial data, *Flood Estimation Handbook, Note 24*, Institute of Hydrology, Wallingford.

Beard, L.R. (1971) Closing the technology gap, *Computer Applications in Hydrology*, US Army Corps of Engineers, Hydrologic Engineering Center, pp. 1–7.

Beaumont, P. (1978) Man's impact on river systems: a world-wide view, *Area*, **10**, 38–41.

Becchi, I., E. Caporali and E. Palmisano (1994) Hydrological response to radar rainfall maps through a distributed model, *Natural Hazards*, **9**, 95–108.

Beckinsale, R.P. (1969) River regimes, in R.J. Chorley (ed.), *Water, Earth and Man*, Methuen, London, 455–471.

Beinin, L. (1985) *Medical Consequences of Natural Disasters*, Springer, New York.

Bell, G.D. and J.E. Janowiak (1995) Atmospheric circulation associated with the Midwest floods of 1993, *Bull. Amer. Met. Soc.*, **76**, 681–695.

Belt, C.B. (1975) The 1973 flood and man's constriction of the Mississippi River, *Science*, **189**, 681–684.
Bennet, G. (1970) Bristol floods 1968: controlled survey of effects on health of local community disaster, *British Medical J.*, **3**, 454–458.
Benson, M.A. (1950) Use of historical data in flood-frequency analysis, *Trans. Amer. Geophys. Union*, **31**, 419–424.
Berger, H.E.J. (1991) Flood forecasting for the River Meuse, in F.H.M. van de Ven, D. Gutknecht, D.P. Loucks and K.A. Salewicz (eds), *Hydrology for the Water Management of Large River Basins*, IAHS Publ. No. 201, 317–328.
Bergeron, L. (1996) Will El Niño become El Hombre?, *New Scientist*, 20 January.
Bergmann, H., G. Richtig and B. Sackl (1990) A distributed model describing the interaction between flood hydrographs and basin parameters, in M.A. Beran, M. Brilly, A. Becker and O. Bonacci (eds), *Regionalization in Hydrology*, IAHS Publ. No. 191, 91–102.
Bernard, E.N. (1991) Assessment of Project THRUST: Past, present, future, *Natural Hazards*, **4**(2/3), 285–292.
Bernard, E.N. and R.R. Behn (1985) Regional tsunami warning system (THRUST), *Ocean Engineering and the Environment*, Proceedings, Oceans '85, San Diego, California, 12–14 Nov., **1**, 215–219.
Bernard, E.N., R.R. Behn, G.T. Hebenstreit, et al. (1988) On mitigating rapid onset natural disasters: Project THRUST, *EOS Trans. Amer. Geophys. Union*, **69**, 649–661.
Bernstein, J. (1954) Tsunamis, *Scientific Amer.*, Aug., 3–6.
Berris, S.N. and R.D. Harr (1987) Comparative snow accumulation and melt during rainfall in forested and clearcut plots in the western cascades of Oregon, *Water Resources Res.*, **23**, 135–142.
Beven, K. (1993) Riverine flooding in a warmer Britain, *Geogrl. J.*, **159**, 157–161.
Beven, K. and P. Carling (eds) (1989) *Floods: Hydrological, Sedimentological and Geomorphological Implications*, Wiley, Chichester.
Bhowmik, N.G., A.G. Buck, S.A. Changnon, et al. (14 others) (1994) *The 1993 Flood on the Mississippi River in Illinois*, Miscellaneous Publication No. 151, Illinois State Water Survey, Champaign-Urbana.
Biswas, A.K. (1970) *History of Hydrology*, North-Holland, Amsterdam–London.
Biswas, A.K. and S. Chatterjee (1971) Dam disasters—An assessment, *Eng. J.*, **54**(3), 3–8.
Black, P.E. (1968) Streamflow increases following farm abandonment on eastern New York watershed, *Water Resources Res.*, **4**, 1171–1178.
Blaikie, P., T. Cannon, I. Davis and B. Wisner, (1994) *At Risk: Natural Hazards, People's Vulnerability and Disasters*, Routledge, London and New York.
Bleasdale, A. and C.K.M. Douglas (1952) Storm over Exmoor on August 15, 1952, *Met. Mag.*, **81**, 353–367.
Blyth, K. and D.S. Biggin (1993) Monitoring floodwater inundation with ERS-1 SAR, *Earth Observation Quarterly*, **42**, 6–8.
Boardman, J. (1995) Damage to property by runoff from agricultural land, South Downs, southern England, 1976–93, *Geogrl. J.*, **161**, 177–191.
Boardman, J., L. Ligneau, A. de Roo and K. Vandaele (1994) Flooding of property by runoff from agricultural land in northwest Europe, *Geomorphology*, **10**, 183–196.
Bollens, S.A. (1990) Public policy and land conversion: lessening urban growth pressure in river corridors, *Growth and Change*, **21**, 40–58.
Bolt, B.A., W.L. Horn, G.A. Macdonald and R.F. Scott (1975) *Geological Hazards*, Springer-Verlag, Berlin.
Bonell, M. and J. Williams (1986) The generation and redistribution of overland flow on a massive oxic soil in a eucalypt woodland within the semi-arid tropics of north Australia, *Hydrol. Processes*, **1**, 31–46.
Bonell, M., D.A. Gilmour and D.S. Cassells (1983) Runoff generation in tropical rainforests of northeast Queensland, Australia, and the implications for land use management, in R. Keller (ed.), *Hydrology of Humid Tropical Regions*, IAHS Publ. No. 140, 287–297.
Boorman, L.A., J.D. Goss-Custard and S. McGrorty (1989) *Climatic Change, Rising Sea Level and the British Coast*, NERC, ITE Res. Publ. No. 1, HMSO, London.

Bossmann-Aggrey, P., C.H. Green and D.J. Parker (1987) *Dam Safety Management in the United Kingdom*, Geography and Planning Paper No. 21, School of Geog. and Planning, Middlesex Polytechnic, Enfield.

Boughton, W.C. (1980) A frequency distribution for annual floods, *Water Resources Res.*, **16**, 347–354.

Boyce, J.K. (1990) Birth of a megaproject: political economy of flood control in Bangladesh, *Environmental Management*, **14**, 419–428.

Bramley, E. (1987) The River Foss flood alleviation scheme, *Circulation, Newsl. Brit. Hydrol. Soc.*, **16**, 6–7.

Brammer, H. (1990a) Floods in Bangladesh I. Geographical background to the 1987 and 1988 floods, *Geogrl. J.*, **156**, 12–22.

Brammer, H. (1990b) Floods in Bangladesh II. Flood mitigation and environmental aspects, *Geogrl. J.*, **156**, 158–165.

Brammer, H. (1993) Geographical complexities of detailed impact assessment for the Ganges–Brahmaputra–Megna delta of Bangladesh, in R.A. Warrick, E.M. Barrow and T.M.L. Wigley (eds), *Climate and Sea Level Change*, Cambridge University Press, Cambridge, 246–262.

Branson, F.A. (1956) Range forage production changes on a water spreader in S E Montana, *J. Range Management*, **9**, 187–191.

Bras, R.L. (1990) *Hydrology: An Introduction to Hydrologic Science*, Addison Wesley, Reading, Massachusetts.

Bree, T., J. Curran and C. Cunnane (1989) Applications of regional flood frequency procedures in Ireland, in L. Roald, K. Nordseth and K.A. Hassel (eds), *FRIENDS in Hydrology*, IAHS Publ. No. 187, 189–196.

Bretschneider, C.L. (1967) Storm surges, *Adv. Hydroscience*, **4**, 341–418.

Breusers, H.N.C. and M. Vis (1994) Policy analysis of river-dike improvement in the Netherlands, in W.R. White and J. Watts (eds), *River Flood Hydraulics*, Wiley, Chichester, 193–199.

Britton, N. (1990) Disaster volunteers—What do we know about them? in the *Newsletter of the International Hazard Panel*, Flood Hazard Research Centre, Middlesex Polytechnic, London.

Broadus, J.M. (1993) Possible impacts of, and adjustments to, sea level rise: the cases of Bangladesh and Egypt, in R.A. Warrick, E.M. Barrow and T.M.L. Wigley (eds), *Climate and Sea Level Change*, Cambridge University Press, Cambridge, 263–275.

Brooke, J. (1995) Cost implications of dealing with environmental issues in river and coastal engineering schemes, *J. Inst. of Water Engrs. and Managers*, **9**, 1–6.

Brookes, A. (1985) River channelisation: traditional engineering methods, physical consequences and alternative practices, *Progress in Physical Geog.*, **9**, 1–15.

Brookes, A. (1988) *Channelised Rivers: Perspectives for Environmental Management*, Wiley, Chichester.

Brooks, C.E.P. and J. Glasspoole (1928) *British Floods and Droughts*, Benn, London.

Brown, A.G., K.J. Gregory and E.J. Milton (1987) The use of Landsat multispectral scanner data for the analysis and management of flooding on the River Severn, England, *Environmental Management*, **11**, 695–701.

Brown, H.J., R.S. Phillips and N.A. Roberts (1981) Land markets at the urban fringe: new insights for policyholders, *J. Amer. Planning Assoc.*, **47**, 131–144.

Brown, J. and M. Muhsin (1991) Case study: Sudan emergency flood reconstruction program, in A. Kreimer and M. Munasinghe (eds), *Managing Natural Disasters and the Environment*, Environment Department, The World Bank, Washington, DC, 157–162.

Browning, K.A., R.B. Bussell and J.A. Cole (1977) Radar for rain forecasting and river management, *Water Power and Dam Construction*, **29**, 38–42.

Bruce, J.P. (1976) The national flood damage reduction program, *Canadian Water Resources J.*, **1**, 5–14.

Brugge, R. (1994) The floods of October and November 1894 in southern Britain, *Weather*, **49**, 383–390.

Bruijnzeel, L.A. (1983) Evaluation of runoff sources in a forested basin in a wet monsoonal environment: a combined hydrological and hydrochemical approach, in R. Keller (ed.), *Hydrology of Humid Tropical Regions*, IAHS Publ. No. 140, 165–174.

Bruijnzeel, L.A. (1990) *Hydrology of Moist Tropical Forests and Effects of Conversion: A State of Knowledge Review*, Free University, Amsterdam.

Bruijnzeel, L.A. and C.N. Bremer (1989) *Highland–Lowland Interactions in the Ganges Brahmaputra River Basin: A Review of the Published Literature*, Occ. Pap. No. 11, International Centre for Integrated Mountain Development, Kathmandu.

Bruton, M.J. (1980) Public participation, local planning and conflicts of interest, *Policy and Politics*, **8**, 423–442.

Burby, R.J., S.A. Bollens, J.M. Holloway, E.J. Kaiser, D. Mullan and J.R. Sheaffer (1988) *Cities Under Water*, Monograph No. 47, Institute of Behavioral Science, University of Colorado, Boulder.

Burgess, K. and D. Reeve (1994) The development of a method for the assessment of sea defences and risk of flooding, Section 5.3 in *A Strategic Approach, Proc. 1994 MAFF Conference of River and Coastal Engineers*, Loughborough, 4–6 July, Ministry of Agriculture, Fisheries and Food, London.

Burkham, D.E. (1978) Accuracy of flood mapping, *J. Res. US Geol. Survey*, **6**, 515–527.

Burroughs, W. (1995) A climate of confusion, *The Times*, London, 27 March, p. 16.

Burt, T.P. (1989) Storm runoff generation in small catchments in relation to the flood response of large basins, in K. Beven and P. Carling (eds), *Floods: Hydrological, Sedimentological and Geomorphological Implications*, Wiley, Chichester, 11–35.

Burton, I. (1989) Natural environmental hazards, in J.G. Henry and G.W. Heinke (eds), *Environmental Science and Engineering*, Prentice-Hall, Englewood Cliffs, 86–113.

Burton, I. and R.W. Kates (1964) The floodplain and the seashore: a comparative analysis of hazard-zone occupance, *Geogrl. Rev.*, **54**, 366–385.

Burton, I., R.W. Kates and G.F. White (1968) *The Human Ecology of Extreme Events*, Natural Hazard Research Working Paper No. 1, Dept. Geog., University of Toronto.

Burton, I., R.W. Kates and R.E. Snead (1969) *The Human Ecology of Coastal Flood Hazard in Megalopolis*, Univ. of Chicago, Dept. of Geogr. Res. Paper, No. 115, 195.

Caine, N. (1995) Snowpack influences on geomorphic processes in Green Lakes valley, Colorado Front Range, *Geogrl. J.*, **161**, 55–68.

Caissie, D. and N. El-Jabi (1993) Characterisation of floods in Canada, in Z.W. Kundzewicz, D. Rosbjerg, S.P. Simonovic and K. Takeuchi (eds), *Extreme Hydrological Events: Precipitation, Floods and Droughts*, IAHS Publ. No. 213, 325–332.

Cane, M.A. (1983) Oceanographic events during El Niño, *Science*, **222**, 1189–1194.

Carper, T.R. (1990) The National Flood Insurance Program: beating a retreat, *Natural Hazards Observer*, **14**, 1–2.

Carter, R.W.G. (1988) *Coastal Environments*, Academic Press, London.

Cavadias, G.S. (1990) The canonical correlation approach to regional flood estimation, in M.A. Beran, M. Brilly, A. Becker and O. Bonacci (eds), *Regionalization in Hydrology*, IAHS Publ. No. 191, 171–178.

CCIRG (1991) *The Potential Effects of Climate Change in the United Kingdom*, Climate Change Impacts Review Group, 1st Report, HMSO, London.

CCIRG (1996) *Review of the Potential Effects of Climate Change in the United Kingdom*, Climate Change Impacts Review Group, 2nd Report, HMSO, London.

Cedeno, J.E.M. (1986) Rainfall and flooding in the Guayas river basin and its effects on the incidence of malaria 1982–1985, *Disasters*, **10**, 107–111.

Chan, N.W. (1995) Choice and constraints in floodplain occupation: the influence of structural factors on residential location in Peninsular Malaysia, *Disasters*, **19**, 287–307.

Chan, N.W. and D.J. Parker (1996) Response to dynamic flood hazard factors in peninsular Malaysia, *Geogrl. J.*, **162**, 313–325.

Chandler, R. (1995) HYREX—the Hydrological Rainfall Experiment, Circulation, *Newsl. Brit. Hydrol. Soc.*, **47**, 12–15, British Hydrological Society, London.

Changnon, S.A. (1987) Climate fluctuations and record-high levels of Lake Michigan, *Bull. Amer. Met. Soc.*, **68**, 1394–1402.

Changnon, S.A. (1996) The lessons from the flood, in S.A. Changnon (ed.), *The Great Flood of 1993 Causes, Impacts and Responses*, Westview Press, Boulder and Oxford, 300–319.

Chasse, J., M.I. El-Sabh and T.S. Murty (1993) A numerical model for water level oscillations in the St Lawrence estuary, Canada. Pt II: tsunamis, *Marine Geodesy*, **16**(2), 125–148.

Chatterton, J.B., J. Pirt and T.R. Wood (1979) The benefits of flood forecasting, *J. Inst. Water Engrs. and Scientists*, **33**, 237–252.

Chau, K.W. (1990) Application of the Preissmann scheme on flood propagation in river systems in difficult terrain, in H. Lang and A. Musy (eds), *Hydrology in Mountainous Regions I Hydrological Measurements, The Water Cycle*, IAHS Publ. No. 193, 535–543.

Chettri, R. and B. Bowonder (1983) Siltation in Nizamsagar reservoir: environmental management issues, *Appl. Geog.*, **3**, 193–204.

Chow, V.T. (1956) *Hydrologic Studies of Floods in the United States*, Internat. Assoc. Sci. Hydrol. Publ. No. 42, 134–170.

Chowdhury, A.M.R. (1988) The 1987 flood in Bangladesh: An estimate of damage in twelve villages, *Disasters*, **12**, 249–300.

Chowdhury, A.M.R., A.V. Bhuiya, A.Y. Choudhury and R. Sen (1993) The Bangladesh cyclone of 1991: why so many people died, *Disasters*, **17**, 291–302.

Church, J.S. (1974) The Buffalo Creek disaster: extent and range of emotional and/or behavioral problems, *Omega*, **5**, 61–63.

Church, M. (1988) Floods in cold climates, in V.R. Baker, R.C. Kochel and P.C. Patton (eds), *Flood Geomorphology*, Wiley, New York, 205–229.

Clark, G.M., R.B. Jacobson, J.S. Kite and R.C. Linton (1987) Storm-induced catastrophic flooding in Virginia and West Virginia, November, 1985, in L. Mayer and D. Nash (eds), *Catastrophic Flooding*, Unwin Hyman, London, 355–379.

Clark, M.J. (1988) Periglacial hydrology, in M.J. Clark (ed.), *Advances in Periglacial Geomorphology*, Wiley, Chichester, 415–462.

Clark, R.A. (1994) Evolution of the national flood forecasting system in the USA, in G. Rossi, N. Harmancioglu and V. Yevjevich (eds), *Coping with Floods*, Kluwer Academic Publishers, Dordrecht, 437–444.

Cluckie, I.D. and C.G. Collier (eds) (1991) *Hydrological Applications of Weather Radar*, Ellis Horwood, Chichester.

Cluckie, I.D. and J.M. Tyson (1989) Weather radar and urban drainage systems, in *Weather Radar and the Water Industry*, British Hydrological Society, Occasional Paper No. 2, 67–76.

Cohen, O. and A. Ben-Zvi (1979) Regional analysis of peak discharges in the Negev, in *The Hydrology of Areas of Low Precipitation*, IAHS Publ. No. 128, 23–31.

Colbeck, S.C. (1979) Water flow through heterogeneous snow, *Cold Regions Science Technology*, **3**, 37–45.

Colbeck, S.C., E.A. Anderson, V.C. Bissell, A.G. Crook, D.H. Male, C.W. Slaughter and D.R. Wiesnet (1979) Snow accumulation, distribution, melt and runoff, *EOS*, **60**, 464–471.

Collar, R., J. Townson, Y. Kaya, J. Wark and J. Curran (1995) Coastal flood warning and surge modelling for the Firth of Clyde, Section 1.3 in *Proc. 30th MAFF Conference of River and Coastal Engineers*, Keele, 5–7 July, Ministry of Agriculture, Fisheries and Food, London.

Collier, C.G. (1992) International radar networking, *Meteorological Mag.*, **121**, 221–241.

Conner, W.C., R.H. Kraft and D.L. Harris (1957) Empirical methods for forecasting the maximum storm tide due to hurricanes and other tropical storms, *Mon. Wea. Rev.*, **85**, 113–116.

Cordery, I., D.H. Pilgrim and D.G. Doran (1983) Some hydrological characteristics of arid western New South Wales, *Hydrology and Water Resources Symposium 1983*, Inst. Engrs. Austral., Publ. No. 83/13, 287–292.

Corkan, R.H. (1948) Storm surges, *Dock and Harbour Authority*, Feb., 3–19.

Corkan, R.H. (1950) The levels of the North Sea associated with the storm disturbance of 8th January, 1949, *Phil. Trans. Royal Soc.*, A, **242**, 493–525.

Cornelius, S.C., D.A. Sear, S.J. Carver and D.I. Heywood (1994) GPS, GIS and geomorphological field work, *Earth Surf. Proc. and Landforms (Technical and Software Bull., No. 3)*, **19**(9), 777–787.

Corps of Engineers (1960) *Water Resources Activities in the US: Future Needs for Navigation*, Committee Print No. 11 of the US Senate Select Committee on National Water Resources, US Government Printing Office, Washington, DC.

Cosandey, C. and J.F. Didon-Lescot (1990) Etude des crues cévenoles: Conditions d'apparition dans un petit bassin forestier sur le versant sud du Mont Lozère, France, in M.A. Beran, M. Brilly, A. Becker and O. Bonacci (eds), *Regionalization in Hydrology*, IAHS Publ. No. 191, 103–115.

Costa, J.E. (1988) Floods from dam failures, in V.R. Baker, R.C. Kochel and P.C. Patton (eds), *Flood Geomorphology*, Wiley, New York, 439–463.

Coxon, P., C.E. Coxon and R.H. Thorn (1989) The Yellow River (County Leitrim Ireland) flash flood of June 1986, in K. Beven and P. Carling (eds), *Floods: Hydrological, Sedimentological and Geomorphological Implications*, Wiley, Chichester, 199–217.

Craig, R.G. (1987) Dynamics of a Missoula flood, in L. Mayer and D. Nash (eds), *Catastrophic Flooding*, Unwin Hyman, London, 305–332.

Cross, J.A. (1985) *Flood Hazard Information Disclosure by Realtors*, Working Paper No. 52, Natural Hazards and Applications Information Center, Institute of Behavioral Science, University of Colorado, Boulder.

Cullingford, R.A., C.J. Caseldine and P.E. Gotts (1989) Evidence of early Flandrian tidal surges in Lower Strathearn, Scotland, *J. Quat. Sci.*, **4**(1), 51–60.

Cunge, J.A. (1969) On the subject of a flood propagation method, *J. Hydraulics Res. IAHR*, **7**, 205–230.

Cunnane, C. (1985) Factors affecting choice of distribution for flood series, *Hydrol. Sci. J.*, **30**, 25–36.

Cunnane, C. (1989) *Statistical Distributions for Flood Frequency Analysis*, WMO Operational Hydrology Rept. No. 33, WMO, Geneva.

Cuny, F.C. (1991) Living with floods: alternatives for riverine flood mitigation, in A. Kreimer and M. Munasinghe (eds), *Managing Natural Disasters and the Environment*, The World Bank, Washington, DC, 62–73.

Curran, J.C. (1994) It never rains but it pours, *Circulation, Newsl. Brit. Hydrol. Soc.*, **42**, 13.

Curran, J.C. (1995) Coastal flooding in the Clyde estuary, *Circulation, Newsl. Brit. Hydrol. Soc.*, **45**, 17.

David, E. and J. Mayer (1984) Comparing costs of alternative flood hazard mitigation plans, *J. Amer. Planning Assoc.*, **50**, 22–35.

Davidson, D.D. and B.L. McCartney (1975) Water waves generated by landslides into reservoirs, *J. Hydraul. Div.*, ASCE, **101**, HY 12, 1489–1501.

Davies, A.M. and J.E. Jones (1993) On improving the bed stress formulation in storm surge models, *J. Geophys. Res. C. Oceans*, **98**, 7023–7038.

Dawson, A.G. (1994) Geomorphological effects of tsunami run-up and backwash, *Geomorphology*, **10**, 83–94.

Dawson, A.G., D.E. Smith and H. Nichols (1994) Making waves, *The Times*, London, 18 October.

Day, H.J., G. Bugliarello, P.H.P. Ho and V.T. Houghton (1969) Evaluation of benefits of a flood warning system, *Water Resources Res.*, **5**, 937–946.

Department of Humanitarian Affairs (1994) *Strategy and Action Plan for Mitigating Water Disasters in Vietnam*, United Nations Development Programme, New York and Geneva.

de Ronde, J.G. (1993) What will happen to the Netherlands if sea level rise accelerates? in R.A. Warrick, E.M. Barrow and T.M.L. Wigley (eds), *Climate and Sea Level Change*, Cambridge University Press, Cambridge, 322–335.

de Vanssay, B. (1995) Assessing social vulnerability through risk representation: the flash-floods in the Vaucluse on 22 September 1992, in T. Horlick-Jones, A. Amendolla and R. Casale (eds), *Natural Risk and Civil Protection*, E & FN Spon, London, 424–434.

de Vries, J. (1985) Analysis of historical climate–society interaction, in R.W. Kates, J.H. Ausubel and M. Berberian (eds), *Climate Impact Assessment*, Wiley, New York, 277–291.

Dhar, O.N. and P.R. Rakhecha (1979) Incidence of heavy rainfall in the Indian desert region, in *The Hydrology of Areas of Low Precipitation*, IAHS Publ. No. 128, 33–42.

Díaz Arenas, A. (1983) Tropical storms in Central America and the Caribbean: characteristic rainfall and forecasting of flash floods, in R. Keller (ed.), *Hydrology of Humid Tropical Regions*, IAHS Publ. No. 140, 39–51.

Dickinson, W.T. and J.R. Douglas (1972) *A Conceptual Runoff Model for the Cam Catchment*, Report No. 17, Institute of Hydrology, Wallingford.

Ding, Y. and J. Liu (1992) Glacier lake outburst flood disasters in China, *Annals of Glaciology*, **16**, 180–184.

Di Silvio, G. (1994) Floods and sediment dynamics in mountain rivers, in G. Rossi, N. Harmancioglu and V. Yejevich (eds), *Coping with Floods*, Kluwer, 375–392.

Dobbie, C.H. and P.O. Wolf (1953) The Lynmouth flood of August 1952, *Proc. Instn. Civ. Engrs.*, **2**, 522–588.

Dobson, C. and R.C. Cross (1994) Optimising a conceptual catchment model in real time, in W.R. White and J. Watts (eds), *River Flood Hydraulics*, Wiley, Chichester, 49–58.

Dolan, R. and P. Godfrey (1973) Effects of hurricane Ginger on the barrier islands of North Carolina, *Bull. Geol. Soc. America*, **84**, 1329–1334.

Douglas, I. (1980) Flooding in Australia: A review, in R.L. Heathcote and B.G. Thom (eds), *Natural Hazards in Australia*, Australian Academy of Science, Canberra, 143–163.

Douglas, J.R. and C. Dobson (1987) Real-time forecasting in diverse drainage basins, in V.K. Collinge and C. Kirby (eds), *Weather Radar and Forecasting*, Wiley, Chichester, 153–169.

Downey, W.K. (1986) Commonwealth–State relations in flood forecasting and warning: a perspective of the future role of the Commonwealth, in D.I. Smith and J.W. Handmer (eds), *Flood Warning in Australia*, Centre for Environmental Studies, Canberra, 39–62.

Dozier, E.F. and T.N. Yancey (1993) *Floodproofing Options for Virginia Homeowners*, US Army Corps of Engineers and Commonwealth of Virginia, Norfolk, Virginia.

Drabek, T.E. (1986) *Human System Response to Disaster: an Inventory of Sociological Findings*, Springer-Verlag, New York.

Drabek, T.E. and K.S. Boggs (1968) Families in disaster: reaction and relatives, *J. Marriage and the Family*, **30**, 443–451.

Dragoun, F.J. (1969) Effects of cultivation and grass on surface runoff, *Water Resources Res.*, **5**, 1078–1083.

Driever, S.L. and D.M. Vaughn (1988) Flood hazard in Kansas City since 1880, *Geogrl. Rev.*, **78**, 1–19.

Dube, S.K., A.D. Rao, P.C. Sinha and P. Chittibabu (1994) A real time storm surge prediction system: An application to the east coast of India, *Proc. Indian Natl. Sci. Acad., A. Phys. Sci.*, **60**(1), 157–170.

Dubief, J. (1953) *Essai sur l'hydrographie superficielle au Sahara*, Vol. 1, Direction du Service de la Colonisation et de l'Hydraulique, Direction des Études Scientifiques, Birmandreis, Algiers.

Dudgeon, D. (1995) River regulation in Southern China: ecological implications, conservation and environmental management, *Regulated Rivers: Res. and Managt.*, **11**, 35–54.

Dunn, G.E. (1962) The tropical cyclone problem in East Pakistan, *Mon. Wea. Rev.*, **91**, 83–86.

Duryee, H.T. (1990) What the NFIP should (and should not) do, *Natural Hazards Observer*, **14**, 6–7.

Ebel, U. and H. Engel (1994) *The 'Christmas Floods' in Germany 1993/94*, Bayerische Rückversicherung, Special Issue 16, Munich.

Edge, R. (1996) Measuring absolute gravity for the fixing of tide gauge bench marks, *Flood and Coastal Defence*, **9**, 5.

Edmonds, R.L. (1991) The Sanxia (Three Gorges) project: the environmental argument surrounding China's super dam, *Global Ecology and Biogeography Letters*, **1**, 105–125.

Edwards, K.C. (1953a) The storm floods of 1st February, 1953: III. A note on the River Trent, *Geography*, **38**, 161–164.

Edwards, K.C. (1953b) The storm floods of 1st February, 1953: VIII. The Netherlands floods: some further aspects and consequences, *Geography*, **38**, 182–187.

Edwards, R. (1996) Chernobyl floods put millions at risk, *New Scientist*, **149**, 4.

Egidi, D. (1995) The Emilia–Romagna approach to civil protection, in T. Horlick-Jones, A. Amendolla and R. Casale (eds), *Natural Risk and Civil Protection*, E & FN Spon, London, 435–444.

Ely, L.L., Y. Enzel and D.R. Cayan (1994) Anomalous North Pacific atmospheric circulation and large winter floods in the southwestern United States, *J. of Climate*, **7**, 977–987.

Engel, H. (1994) The Rhine flood 1993/94 (Christmas flood), Section 7.4 in *A Strategic Approach, Proc. 1994 MAFF Conference of River and Coastal Engineers*, Loughborough, 4–6 July, Ministry of Agriculture, Fisheries and Food, London.

Engel, H., N. Busch, K. Wilke, P. Krahe, H.G. Mendel, H. Giebel and C. Zieger (1994), *The 1993/94 Flood in the Rhine Basin*, Rept. No. 833, Federal Institute of Hydrology, Koblenz.

Engler, A. (1919) Einfluss des Waldes auf der Stand der Gewasser, *Mitt. Schweiz Aust. Forstl. Versachsw*, 12, 626. Reported in H.L. Penman (1963) *Vegetation and Hydrology*, Technical Communication No. 53, Commonwealth Agricultural Bureaux, Farnham Royal.

Ericksen, N.J. (1975) A tale of two cities: flood history and the prophetic past of Rapid City, South Dakota, *Economic Geog.*, **51**, 305–320.

Ericksen, N.J. (1986) *Creating Flood Disasters? New Zealand's Need for a New Approach to Urban Flood Hazard*, National Water and Soil Conservation Authority, Wellington.

Erikson, K.T. (1976) *Everything in its Path*, Simon and Schuster, New York.

Erskine, W.D. (1992) Channel response to large-scale river training works: Hunter River, Australia, *Regulated Rivers: Res. and Managt.*, **7**, 261–278.

Etcheverry, B.A. (1931) *Land Drainage and Flood Protection*, McGraw-Hill Book Company Inc., New York.

Evans, R. (1994) Run river run, *Geographical*, July, 17–20.

Falck, L.B. (1991) Disaster insurance in New Zealand, in A. Kreimer and M. Munasinghe (eds), *Managing Natural Disasters and the Environment*, Environment Department, The World Bank, Washington, DC, 120–125.

Falconer, R.H. and J.L. Anderson (1993) Assessment of the February 1990 flooding in the river Tay and subsequent implementation of a flood-warning system, *J. Inst. Water Engrs. and Managers*, **7**, 134–148.

Fan, Y. (1991) Disaster relief in China, *Disasters*, **15**, 379–381.

Farreras, S.F. and A.J. Sanchez (1991) The tsunami threat on the Mexican west coast: a historical analysis and recommendations for hazard mitigation, *Natural Hazards*, **4**, 301–316.

Fatorelli, S., M. Borga and D. Da Ros (1995) Integrated systems for real-time flood forecasting, in T. Horlick-Jones, A. Amendola and R. Casale (eds), *Natural Risk and Civil Protection*, R & FN Spon, London, 191–212.

Feldman, A.D. (1994) Assessment of forecast technology for flood control operation, in G. Rossi, N. Harmancioglu and V. Yejevich (eds), *Coping with Floods*, Kluwer Academic Publishers, Dordrecht, 445–458.

Fernández, P., J. Maza and A.V. Aranibar (1994) Prediction of floods from a mountain river with glacierized and snow covered areas, in W.R. White and J. Watts (eds), *River Flood Hydraulics*, Wiley, Chichester, 27–35.

Ferrari, E., S. Gabriele & P. Villani (1993) Combined regional frequency analysis of extreme rainfalls and floods, in Z.W. Kundzewicz, D. Rosbjerg, S.P. Simonovic and K. Takeuchi (eds), *Extreme Hydrological Events: Precipitation, Floods and Droughts*, IAHS Publ. No. 213, 333–346.

Finch, C.R. (1972) Some heavy rainfalls in Great Britain, 1956–1971, *Weather*, **17**, 364–377.

Fischer, D.W. (1989) Response to coastal storm hazard: short-term recovery versus long-term planning, *Ocean and Shoreline Management*, **12**, 295–308.

Flather, R.A. (1994) A storm surge prediction model for the northern Bay of Bengal with application to the cyclone disaster in April 1991, *J. Phys. Oceanogr.*, **24**(1), 172–190.

Flather, R.A. and H. Khandker (1993) The storm surge problem and possible effects of sea level changes on coastal flooding in the Bay of Bengal, in R.A. Warrick, E.M. Barrow and T.M.L. Wigley (eds), *Climate and Sea Level Change*, Cambridge University Press, Cambridge, 229–245.

Fordham, M., S. Tunstall and E.C. Penning-Rowsell (1991) Choice and preference in the Thames floodplain: the beginnings of a participatory approach?, *Landcape and Urban Planning*, **20**, 183–187.

Foster, H.D. (1980) *Disaster Planning: The Preservation of Life and Property*, Springer-Verlag, Berlin.

Foster, H.D. and V. Wuorinen (1976) British Columbia's tsunami warning system: an evaluation, *Syesis*, **9**, 113–122.

Fowler, A.M. and K.J. Hennessy (1995) Potential impacts of global warming on the frequency and magnitude of heavy precipitation, *Natural Hazards*, **11**, 283–303.

REFERENCES

Fox, W.E. (1965) Methods of river forecasting, *Proc. Conf. Hydrologic Activities in the South Carolina Region*, Clemson Univ., Clemson, South Carolina.

Francou, J. and J.A. Rodier (1969) Essai de classification des crues maximales, *Floods and their Computation*, IAHS/UNESCO/WMO, 518–527.

Frank, N.L. and S.A. Husain (1971) The deadliest tropical cyclone in history?, *Bull. Amer. Met. Soc.*, **52**(6), 438–444.

Fread, D.L. (1980) *DAMBRK—The NWS Dam-Break Flood Forecasting Model*, National Weather Service, Office of Hydrology, Silver Spring, Maryland.

Fread, D.L. (1985) Channel routing, in M.G. Anderson and T.P. Burt (eds), *Hydrological Forecasting*, Wiley, Chichester, 437–504.

Fread, D.L. (1989) National Weather Service models to forecast dam-breach floods, in Ö. Starosolszky and O.M. Melder (eds), *Hydrology of Disasters*, WMO, Geneva, 192–211.

Fruget, J.F. (1992) Ecology of the lower Rhone after 200 years of human influence: a review, *Regulated Rivers: Res. and Managt.*, **7**, 233–246.

Fukushima, Y., O. Watanabe and K. Higuchi (1991) Estimation of streamflow change by global warming in a glacier-covered high mountain area of the Nepal Himalaya, in H. Bergmann, H. Lang, W. Frey, D. Issler and B. Salm (eds), *Snow, Hydrology and Forests in High Alpine Areas*, IAHS Publ. No. 205, 181–188.

Funtowicz, S.O. and J.R. Ravetz (1995) Planning and decision-making in an uncertain world: the challenge of post-normal science, in T. Horlick-Jones, A. Amendolla and R. Casale (eds), *Natural Risk and Civil Protection*, E & FN Spon, London, 415–423.

Ganoulis, J. (1994) Flood retention basins in the Mediterranean urban areas, in G. Rossi, N. Harmancioglu and V. Yevjevich (eds), *Coping with Floods*, Kluwer Academic Publishers, Dordrecht, 759–765.

Gardiner, J.L. (1994) Sustainable development for river catchments, *J. Inst. Water and Environmental Management*, **8**, 308–320.

Georgiadi, A.G. (1993) Historical high water marks as a basis of estimation of spring discharges of Russian plain rivers, in Z.W. Kundzewicz, D. Rosbjerg, S.P. Simonovic and K. Takeuchi (eds), *Extreme Hydrological Events: Precipitation, Floods and Droughts*, IAHS Publ. No. 213, 207–210.

Gerard, R. and E.W. Karpuk (1979) Probability analysis of historical flood data, *J. Hydraul. Div. Am. Soc. Civ. Eng.*, **105**, 1153–1166.

Gilbert, M. and J. De Meyer (1994) Flood forecasting for Beijiang River (People's Republic of China), in W.R. White and J. Watts (eds), *River Flood Hydraulics*, Wiley, Chichester, 59–63.

Gill, M.A. (1986) Unified theory for flood and pollution routing (Discussion), *J. Hydraul. Engg.*, **112**, 981–983.

Gilvear, D.J., J.R. Davies and S.J. Winterbottom (1994) Mechanisms of floodbank failure during large flood events on the rivers Tay and Earn, Scotland, *Quart. J. Engg. Geology*, **27**, 319–332.

Giuli, D., L. Baldini and L. Facheris (1994) Simulation and modeling of rainfall radar measurements for hydrological applications, *Natural Hazards*, **9**, 109–122.

Gladwell, J.S. and K.S. Low (1993) *Tropical Cities: Managing their Water*, IHP Humid Tropics Programme Series No. 4, UNESCO, Paris.

Glickman, T.S., D. Golding and E.D. Silverman (1992) *Acts of God and Acts of Man. Recent Trends in Natural Disasters and Major Industrial Accidents*, Discussion paper CRM 92-02, Resources for the Future, Washington, DC.

Goddard, J.E. (1976) The nation's increasing vulnerability to flood catastrophe, *J. Soil and Water Conservation*, **31**, 48–52.

Gordon, D.M. (1988) Disturbance to mangroves in tropical-arid Western Australia: hypersalinity and restricted tidal exchange as factors leading to mortality, *J. Arid Environments*, **15**, 117–145.

Graftdijk, K. (1960) *Holland Rides the Sea*, World's Window Ltd, Baarn.

Gray, W.M. (1990) Strong association between West African rainfall and U.S. landfall of intense hurricanes, *Science*, **249**, 1251–1256.

Green, C.H. (1983) *Evaluating Road Traffic Disruption from Flooding*, Geography and Planning Paper No. 11, Flood Hazard Research Centre, Middlesex Polytechnic, London.

Green, C.H. (1992) Enabling effective hazard management by the public, in D.J. Parker and J.W. Handmer (eds), *Hazard Management and Emergency Planning*, James and James, London, 175–193.

Green, C.H., D.J. Parker, P. Thompson and E.C. Penning-Rowsell (1983) *Indirect Losses from Urban Flooding: An Analytical Framework*, Geography and Planning Papers 6, Flood Hazard Research Centre, Middlesex Polytechnic, London.

Green, C.H., P.J. Emery, E.C. Penning-Rowsell and D.J. Parker (1985) *The Health Effects of Flooding: A Survey at Uphill, Avon*, Flood Hazard Research Centre, Middlesex Polytechnic, London.

Green, C.H., S.M. Tunstall and M.H. Fordham (1991) The risks from flooding: which risks and whose perception?, *Disasters*, **15**, 227–236.

Greswell, R.B., J.W. Lloyd and D.N. Lerner (1994) Rising groundwater in the Birmingham area, in W.B. Wilkinson (ed.), *Groundwater Problems in Urban Areas*, Thomas Telford, London, 330–341.

Grieve, H. (1959) *The Great Tide*, County Council of Essex, Chelmsford.

Grigg, N.S. and O.J. Helweg (1975) State-of-the-art estimating flood damage in urban areas, *Water Resources Bull.*, **11**, 379–390.

Groen, P. and G.W. Groves (1962) Surges, in M.N. Hill (ed.), *The Sea*, Interscience, New York, 611–646.

Gross, E.M. (1991) The hurricane dilemma in the United States, *Episodes*, **14**, 36–45.

Gruntfest, E. (1987) Warning dissemination and response with short lead times, in J. Handmer (ed.), *Flood Hazard Management: British and International Perspectives*, Geo Books, Norwich, 191–202.

Gruntfest, E. (1994) Flood disaster relief, rehabilitation and reconstruction, in G. Rossi, N. Harmancioglu and V. Yevjvich (eds), *Coping with Floods*, Kluwer Academic Publishers, Dordrecht, 723–731.

Gruntfest, E. and C. Huber (1989) Status report on flood warning systems in the United States, *Environmental Management*, **13**, 279–286.

Gueri, M., C. Gonzalez and V. Morin (1986) The effect of the floods caused by 'El Niño' on health, *Disasters*, **10**, 118–124.

Guilcher, A. (1965) *Précis d'Hydrologie, marine et continentale*, Masson, Paris.

Guillot, P. (1993) The arguments of the gradex method: a logical support to assess extreme floods, in Z.W. Kundzewicz, D. Rosbjerg, S.P. Simonovic and K. Takeuchi (eds), *Extreme Hydrological Events: Precipitation, Floods and Droughts*, IAHS Publ. No. 213, 287–298.

Gunn, J.P., H.J. Todd and K. Mason (1930) The Shyok flood 1929, *Himalayan J.*, **2**, 35–47.

Gutknecht, D.K. (1991) On the development of 'applicable' models for flood forecasting, in F.H.M. van de Ven, D. Gutknecht, D.P. Loucks and K.A. Salewicz (eds), *Hydrology for the Water Management of Large River Basins*, IAHS Publ. No. 201, 337–345.

Hagen, V.K. (1982) Re-evaluation of design floods and dam safety. Paper presented at *14th International Commission on Large Dams Congress*, Rio de Janeiro.

Haigh, M.J., J.S. Rawat and H.S. Bisht (1990) Hydrological impact of deforestation in the central Himalaya, in L. Molnar (ed.), *Hydrology of Mountainous Areas*, IAHS Publ. No. 190, 419–433.

Hall, A.J. (1981) *Flash Flood Forecasting*, Operational Hydrology Rept. No. 18, WMO, Geneva.

Hall, M.J. (1984) *Urban Hydrology*, Elsevier, London.

Hamid, J. and K. Amaning (1991) River engineering, in J.L. Gardiner (ed.), *River Projects and Conservation: A Manual for Holistic Appraisal*, Wiley, Chichester, 95–101.

Hamilton, L.S. (1987) What are the impacts of Himalayan deforestation on the Ganges–Brahmaputra lowlands and delta? Assumptions and facts, *Mountain Res. and Developt.*, **7**(5), 256–363.

Handmer, J.W. (1987) Guidelines for floodplain acquisition, *Appl. Geog.*, **7**, 203–221.

Handmer, J.W. (1988) The performance of the Sydney flood warning system, August 1986, *Disasters*, **12**, 37–48.

Handmer, J.W. (1990) *Flood Insurance and Relief in the US and Britain*, Working Paper No. 68, Natural Hazards Research and Applications Center, Institute of Behavioral Science, University of Colorado, Boulder.

Handmer, J.W. (1992) Emergency management in Australia: concepts and characteristics, in D.J. Parker and J.W. Handmer (eds), *Hazard Management and Energy Planning*, James and James, London, 227–241.

Handmer, J.W. and D.I. Smith (1983) Health hazards of floods: hospital admissions for Lismore, *Australian Geogrl. Studies*, **21**, 221–230.

Handmer, J.W., D.J. Parker and J. Neal (1989) British authorities liable for flood warning failure, *Macedon Digest*, **4**, 4–6.

Hanna, J.E. and D.N. Wilcock (1984) The prediction of mean annual flood in Northern Ireland, *Proc. Instn. Civ. Engrs., Part 2*, **77**, 429–444.

Hanson, K. (1996) Building partnerships to restore floodplain open space, *Natural Hazards Observer*, **20**, 12–13.

Hanwell, J.D. and M.D. Newson (1970) *The Great Storms and Floods of July 1968 on Mendip*, Wessex Cave Club, Wells.

Haque, C.E. and D. Blair (1992) Vulnerability to tropical cyclones: evidence from the April 1991 cyclone in coastal Bangladesh, *Disasters*, **16**, 217–229.

Haque, C.E. and M.Q. Zaman (1989) Coping with riverbank erosion hazard and displacement in Bangladesh: survival strategies and adjustments, *Disasters*, **13**, 300–314.

Haque, C.E. and M.Q. Zaman (1993) Human responses to riverine hazards in Bangladesh—a proposal for sustainable floodplain development, *World Development*, **21**, 93–107.

Hardaker, P. (1996) Can weather radar forecast rainfall at lead times of 1 hour up to 2.5×10^5 years?, *Circulation, Newsl. Brit. Hydrol. Soc.*, **50**, 5–6.

Harding, D.M. and D.J. Parker (1976) Flood loss reduction: a case study, *Water Services*, **80**, 24–28.

Hardisty, J. (1990) *The British Seas*, Routledge, London.

Harlin, J., G. Lindström and S. Bergström (1993) New guidelines for spillway design floods in Sweden, in Z.W. Kundzewicz, D. Rosbjerg, S.P. Simonovic and K. Takeuchi (eds), *Extreme Hydrological Events: Precipitation, Floods and Droughts*, IAHS Publ. No. 213, 237–244.

Harmancioglu, N.B. (1994) Flood control by reservoirs, in G. Rossi, N. Harmancioglu and V. Yevjevich (eds), *Coping with Floods*, Kluwer Academic Publishers, Dordrecht, 637–652.

Harris, D.L. (1967) A critical survey of the storm surge protection problem, *11th Pacific Science Congress. Sympos. on Tsunami and Storm Surges*, 47–65.

Haslam, S.M. (1978) *River Plants*, Cambridge University Press, Cambridge.

Hassan, M.A. (1990) Observations of desert flood bores, *Earth Surf. Proc. and Landforms*, **15**, 481–485.

Hatton, R. (1994) Weather radar for flood forecasting—NRA research, Section 9.2 in *A Strategic Approach, Proc. 1994 MAFF Conference of River and Coastal Engineers*, Loughborough, 4–6 July, Ministry of Agriculture, Fisheries and Food, London.

Hayden, B.P. (1988) Flood climates, in V.R. Baker, R.C. Kochel and P.C. Patton (eds), *Flood Geomorphology*, Wiley, New York, 13–26.

HEC (1975) *Techniques for Real-time Operation of Flood Control Reservoirs in the Merrimack River Basin*, Tech. Paper No. 45, Hydrologic Engineering Center, US Army Corps of Engineers, Davis, California.

Heck, N.H. (1947) List of seismic sea waves, *Bull. Seism. Soc. Amer.*, **37**, 4.

Hederra, R. (1987) Environmental sanitation and water supply during floods in Ecuador (1982–1983), *Disasters*, **11**, 297–309.

Heijne, I., M. Butts and J. Chatterton (1996) Flood forecasting, warning and response systems: technologies, costs and benefits, *Proc. 31st Conference of River and Coastal Engrs.*, MAFF, London, 4.2.1.–4.2.12.

Hendriks, M.R. (1990) *Regionalisation of Hydrological Data*, Geografisch Instituut, Rijksuniversiteit, Utrecht.

Henry, R.F. and T.S. Murty (1990) Relevance of data dossiers for storm-surge prediction, *Natural Hazards*, **3**, 413–417.

Heras, R. (1974) Les crues brutales en Espagne, in *Flash Floods*, Proc. Paris Symposium, September 1974, IAHS Publ. No. 112, 93–99.

Hershfield, D.M. (1961) Estimating the probable maximum precipitation, *J. Hydraul. Div., ASCE*, **87**, 5.

Hershfield, D.M. (1965) Method for estimating probable maximum rainfall, *J. Amer. Waterworks Assoc.*, **57**, 965–972.

Hewitt, K. (1982) Natural dams and outburst floods of the Karakoram Himalaya, in J.W. Glen (ed.), *Hydrological Aspects of Alpine and High Mountain Areas*, IAHS Publ. No. 138, 259–269.

Hewitt, K. and I. Burton (1971) *The Hazardousness of a Place: A Regional Ecology of Damaging Events*, Department of Geography, University of Toronto, Toronto.

Hewlett, J.D. (1961) Watershed management, in *Report for 1961 Southeastern Forest Experiment Station*, US Forest Service, Ashville, North Carolina.

Hewlett, J.D. (1982) *Principles of Forest Hydrology*, University of Georgia Press, Athens.

Hey, R.D., G.L. Heritage and M. Patteson (1994) Impact of flood alleviation schemes on aquatic macrophytes, *Regulated Rivers: Res. and Managt.*, **9**, 103–119.

Hibbert, A.R. (1967) Forest treatment effects on water yield, in W.E. Sopper and H.W. Lull (eds), *Forest Hydrology*, Pergamon, Oxford, 527–543.

Higgins, R.J. and D.J. Robinson (1981) *An Economic Comparison of Different Flood Mitigation Strategies in Australia: A Case Study*, Technical Paper No. 65, Australian Water Resources Council, Australian Government Publishing Service, Canberra.

Hill, H.C. (1994) Environmentally-led engineering: the Maidenhead, Windsor and Eton flood alleviation scheme, in W.R. White and J. Watts (eds), *River Flood Hydraulics*, Wiley, Chichester, 513–524.

Hillen, R. and R.E. Jorissen (1995) River flooding and flood management in the Netherlands, Section 5.2 in *Proc. 30th MAFF Conference of River and Coastal Engineers*, Keele, 5–7 July, Ministry of Agriculture, Fisheries and Food, London.

Hindley, K. (1978) Beware the big wave, *New Scientist*, **77**, 346–347.

Hirschboeck, K.K. (1987) Catastrophic flooding and atmospheric circulation anomalies, in L. Mayer and D. Nash (eds), *Catastrophic Flooding*, Unwin Hyman, London, 23–56.

Hirschboeck, K.K. (1988) Flood hydroclimatology, in V.R. Baker, R.C. Kochel and P.C. Patton (eds), *Flood Geomorphology*, Wiley, New York, 27–49.

Hobbs, J.E. and S. Lawson (1982) The tropical cyclone threat to the Queensland Gold Coast, *Appl. Geog.*, **2**, 207–219.

Hofer, T. (1993) Himalayan deforestation, changing river discharge, and increasing floods—Myth or reality, *Mountain Res. and Development*, **13**, 213–233.

Hollinrake, P.G. and P. Samuels (1994) *Flood Discharge Assessment—An Engineering Guide*, Report SR 379, HR Wallingford.

Holmes, C.G. (1995) The West Sussex floods of December 1993 and January 1994, *Weather*, **50**, 2–6.

Hoque, B.A., S.R.A. Huttly, K.M.A. Aziz, Z. Hasan and M.Y. Patwary (1989) Effects of floods on the use and condition of pit latrines in rural Bangladesh, *Disasters*, **13**, 315–321.

Hoque, B.A., R. Bradley-Sack, M. Siddiqui, A.M. Jahangir, N. Hazera and A. Nahid (1993) Environmental health and the 1991 Bangladesh cyclone, *Disasters*, **17**, 144–152.

Hoque, M.M. (1994) Evaluating design characteristics for floods in Bangladesh, in W.R. White and J. Watts (eds), *River Flood Hydraulics*, Wiley, Chichester, 15–26.

Hoque, M.M. and M.A.B. Siddique (1995) Flood control projects in Bangladesh: reasons for failure and recommendations for improvement, *Disasters*, **19**, 260–263.

Horner, R.W. (1981) Flood prevention works with specific reference to the Thames barrier, in D.H. Peregrine (ed.), *Floods Due to High Winds and Tides*, Academic Press, London, 95–106.

Horton, R.E. (1933) The role of infiltration in the hydrologic cycle, *Trans. Amer. Geophys. Union*, **14**, 446–460.

Hossain, M. (1993) Economic effects of riverbank erosion: some evidence from Bangladesh, *Disasters*, **17**, 25–32.

Houghton, J.T., G.J. Jenkins and J.J. Ephraums (eds) (1990) *Climate Change: The IPCC Scientific Assessment*, Cambridge University Press.

Houghton, J.T., B.A. Callander and S.K. Varney (1992) *Climate Change 1992: The Supplementary Report to the IPCC Scientific Assessment*, Cambridge University Press.

Hoyt, W.G. and W.B. Langbein (1955) *Floods*, Princeton University Press, Princeton, New Jersey.

Hubbert, G.D., L.M. Leslie and M.J. Manton (1990) A storm surge model for the Australian region, *Quart. J. Roy. Met. Soc.*, **116**, 1005–1020.

Huff, F.A., J.L. Vogel and S.A. Changnon (1981) Real-time rainfall monitoring—prediction system and urban hydrologic operations, *J. Water Resources Planning and Management Div., Proc. ASCE*, **107**, No. WR2, 419–435.

Hughes, R. (1982) The effects of flooding upon buildings in developing countries, *Disasters*, **6**, 183–194.

Hunt, J.C.R. (1995) Forecasts and warnings of natural disasters: the roles of national and international agencies, *Meteorological Applications*, **2**, 53–64.

Hurni, H. (1983) Soil erosion and soil formation in agricultural ecosystems, Ethiopia and Northern Thailand, *Mountain Res. and Development*, **3**, 131–142.

Hutchison, J. (1994) Shoreline Management Plans, Section 4.2 in *A Strategic Approach, Proc. 1994 MAFF Conference of River and Coastal Engineers*, Loughborough, 4–6 July, Ministry of Agriculture, Fisheries and Food, London.

IAHS (1974) *Flash Floods*, Proc. Paris Symposium, September 1974, IAHS Publ. No. 112.

ICE (1933) *Floods in Relation to Reservoir Practice*, Interim Report of Committee on Floods in Relation to Reservoir Practice, The Institution of Civil Engineers, London.

ICE (1967) *Flood Studies for the United Kingdom*, Report of the Committee on Floods in the United Kingdom, Institution of Civil Engineers, London.

IH (1986a) *Hydrological Data UK, 1983 Yearbook*, Institute of Hydrology/British Geological Survey.

IH (1986b) *Hydrological Data UK, 1984 Yearbook*, Institute of Hydrology/British Geological Survey.

IH (1993a) Flood routing in mixed catchments, *Flood & Coastal Defence*, **4**, 6–7, MAFF, London.

IH (1993b) *Report of the Institute of Hydrology 1992/1993*, Natural Environment Research Council, 41–42.

IH (1994a) *Hydrological Data UK, 1993 Yearbook*, Institute of Hydrology/British Geological Survey.

IH (1994b) *The Implications of Climate Change for the National Rivers Authority*, R&D Report 12, HMSO, London.

IH (1994c) *Annual Report 1993–1994*, Institute of Hydrology, Natural Environment Research Council.

Inter-agency Floodplain Management Review Committee (1994) *Sharing the Challenge: Floodplain Management into the 21st Century*, Washington, DC.

International Commission on Large Dams (1973) *Lessons From Dam Incidents*, abridged edition, US Commission on Large Dams, Boston, Massachusetts.

Iritz, L., B. Johansson and L. Lundin (1994) Impacts of forest drainage on floods, *Hydrol. Sci. J.*, **39**, 637–661.

Islam, N. (1990) Let the Delta be a Delta: an essay in dissent on the flood problem of Bangladesh, *J. Social Studies*, **48**, 21–22.

Ives, J.D. and B. Messerli (1989) *The Himalayan Dilemma: Reconciling Development and Conservation*, Routledge, London.

Jabbar, M.A. (1990) Floods and livestock in Bangladesh, *Disasters*, **14**, 358–365.

Jakeman, A.J., I.G. Littlewood and P.G. Whitehead (1993) An assessment of the dynamic response characteristics of streamflow in the Balquhidder catchments, *J. Hydrol.*, **145**, 337–355.

Jakob, D. (1997) Proposing regions for flood frequency analysis, *Flood Estimation Handbook, Note 35*, Institute of Hydrology, Wallingford.

James, L.D. (1994) Flood action: an opportunity for Bangladesh, *Water Internat.*, **19**, 61–69.

Jansen, R.B. (1980) *Dams and Public Safety*, US Department of the Interior, Bureau of Reclamation, Denver, Colorado.

Jarrett, R.D. (1990) Hydrologic and hydraulic research in mountain rivers, in L. Molnar (ed.), *Hydrology of Mountainous Areas*, IAHS Publ. No. 190, 107–117.

Jelesnianski, C.P. (1967) Numerical computations of storm surges with bottom stress, *Mon. Wea. Rev.*, **95**, 740–756.

Jelgersma, S., M.J.F. Stive and L. van der Valk (1995) Holocene storm surge signatures in the coastal dunes of the western Netherlands, *Marine Geology*, **125**, 95–110.

Jensen, H.A.P. (1953) Tidal inundations past and present, *Weather*, **8**, 85–89 and 108–112.

Johnson, E.P. (1967) Example of radar as a tool in forecasting tidal flooding, *US Wea. Bur., Eastern Region Tech. Mem.*, WBTM-ER-24.

Johnston, H.T., N.N.J. Higginson and T. McCully (1994) A review of the hydraulic and environmental performance of river schemes in N. Ireland, in W.R. White and J. Watts (eds), *River Flood Hydraulics*, Wiley, Chichester, 217–229.

Joly, F. (1953) Quelques phénomènes d'ecoulement sur la bordure du Sahara dans les confins algéro-marocains et leurs conséquences morphologiques, *Compt. Rend. XIX Congrès Géol. Intern.*, Algiers, 1952, Pt. VII, Déserts actuels et anciens, 135–146.

Junk, W.J., P.B. Bayley and R.E. Sparks (1989) The flood pulse concept in river–floodplain systems, *Fishery and Aquatic Sci. (Special Publication)*, **106**, 110–127.

Kadoya, M., H. Chikamori and T. Ichioka (1993) Some characteristics of heavy rainfalls in the Yamato river basin found by the principal component and cluster analyses, in Z.W. Kundzewicz, D. Rosbjerg, S.P. Simonovic and K. Takeuchi (eds), *Extreme Hydrological Events: Precipitation, Floods and Droughts*, IAHS Publ. No. 213, 75–85.

Kale, V.S., L.L. Ely, Y. Enzel and V.R. Baker (1994) Geomorphic and hydrologic aspects of monsoon floods on the Narmada and Tapi Rivers in central India, *Geomorphology*, **10**, 157–168.

Kapotov, A.A. (1989) Permeability of frozen and thawed soils and sub-soils during spring snowmelt flood, in Roald, L., K. Nordseth and K.A. Hassel (eds), *FRIENDS in Hydrology*, IAHS Publ. No. 187, 27–34.

Karim, M.A. and J.U. Chowdhury (1995) A comparison of four distributions used in flood frequency analysis in Bangladesh, *Hydrol. Sci. J.*, **40**(1), 55–66.

Karl, T.R., R.W. Knight and N. Plummer (1995) Trends in high-frequency climate variability in the twentieth century, *Nature*, **377**, 217–220.

Kattelmann, R. (1990a) Conflicts and cooperation over floods in the Himalaya–Ganges region, *Water Internat.*, **15**, 189–194.

Kattelmann, R. (1990b) Floods in the high Sierra Nevada, California, USA, in R.O. Sinniger and M. Monbaron (eds), *Hydrology in Mountainous Regions II. Artificial Reservoirs, Water and Slopes*, IAHS Publ. No. 194, 311–317.

Kattelmann, R. (1991a) Peak flows from snowmelt runoff in the Sierra Nevada, USA, in H. Bergmann, H. Lang, W. Frey, D. Issler and B. Salm (eds), *Snow, Hydrology and Forests in High Alpine Areas*, IAHS Publ. No. 205, 203–211.

Kattelmann, R. (1991b) Hydrologic regime of the Sapt Kosi basin, Nepal, in F.H.M. van de Ven, D. Gutknecht, D.P. Loucks and K.A. Salewicz (eds), *Hydrology for the Water Management of Large River Basins*, IAHS Publ. No. 201, 139–147.

Kay, R., A. Carman-Brown and G. King (1995) Western Australian experiences in preparing and implementing coastal management plans: Some possible implications for shoreline management planning in England and Wales, Section 8.1 in *Proc. 30th MAFF Conference of River and Coastal Engineers*, Keele, 5–7 July, Ministry of Agriculture, Fisheries and Food, London.

Khairulmaini, O.S. (1994) Perception of and adaptation to flood hazard—a preliminary study, *Malaysian J. Tropical Geog.*, **25**, 99–106.

Khalequzzaman, M. (1994) Recent floods in Bangladesh: Possible causes and solutions, *Natural Hazards*, **9**, 65–80.

Khalil, G.M. (1990) Floods in Bangladesh: A question of disciplining the rivers, *Natural Hazards*, **3**, 379–401.

Khalil, G.M. (1992) Cyclones and storm surges in Bangladesh: some mitigative measures, *Natural Hazards*, **6**, 11–24.

Khan, M. (1969) Influence of Upper Indus basin on the elements of the flood hydrograph at Tarbela-Attock, in *Floods and their Computation*, IASH/UNESCO/WMO, 918–925.

Khan, Md.M.I. (1991) The impact of local elites on disaster preparedness planning: the location of flood shelters in northern Bangladesh, *Disasters*, **15**, 340–354.

Khavich, V. and A. Ben-Zvi (1995) Flash flood forecasting model for the Ayalon stream, Israel, *Hydrol. Sci. J.*, **40**, 599–613.

Kidson, C. (1953) The Exmoor storm and the Lynmouth floods, *Geography*, **38**, 1–9.
Kimmage, K. and W.M. Adams (1992) Wetland agricultural production and river basin development in the Hadejia–Jama'are valley, Nigeria, *Geogrl. J.*, **158**, 1–12.
Klemes, V. (1993) Probability of extreme hydrometeorological events—a different approach, in Z.W. Kundzewicz, D. Rosbjerg, S.P. Simonovic and K. Takeuchi (eds), *Extreme Hydrological Events: Precipitation, Floods and Droughts*, IAHS Publ. No. 213, 167–176.
Knox, J.C. (1984) Fluvial responses to small scale climatic changes, in J.E. Costa and P.J. Fleisher (eds), *Developments and Applications of Geomorphology*, Springer-Verlag, Berlin Heidelberg, 318–342.
Knox, J.C. (1993) Large increases in flood magnitude in response to modest changes in climate, *Nature*, **361**(6411), 430–432.
Koellner, W.H. (1996) The flood's hydrology, Ch 4 in S.A. Changnon (ed.), *The Great Flood of 1993*, Westview Press, Boulder and Oxford.
Können, G.P. (1995) Storms and surges at the coast of the Netherlands: probabilities of flooding, in T. Horlick-Jones, A. Amendola and R. Casale (eds), *Natural Risk and Civil Protection*, E & FN Spon, London, 178–179.
Konovalov, V.G. (1990) Methods for the computations of onset date and daily hydrograph of the outburst from the Mertzbacher Lake, Tien-shan, in H. Lang and A. Musy (eds), *Hydrology in Mountainous Regions. I. Hydrological Measurements, The Water Cycle*, IAHS Publ. No. 193, 181–187.
Koopmans, B.N., C. Pohl and Y. Wang (1995) The 1995 flooding of the Rhine, Waal and Maas rivers in The Netherlands, *Earth Observation Quarterly*, **47**, 11–12.
Koussis, A.D. (1983) Unified theory for flood and pollution routing, *J. Hydraul. Engg.*, **109**, 1652–1664.
Koussis, A.D. (1986) Unified theory for flood and pollution routing, *J. Hydraul. Engg.*, **112**, 983–985.
Kovács, Z.P. (1980) *Maximum Flood Peak Discharges in South Africa: An Empirical Approach*, Tech. Rept. TR 105, RSA Department of Environment Affairs, Pretoria.
Kramer, L.A. and A.T. Hjelmfelt (1989) *Watershed Erosion for Ridge-Till and Conventional-Till Corn*, Paper No. 89-2511. American Society of Agricultural Engineers, St Joseph, Missouri.
Krasovskaia, L., N.W. Arnell and L. Gottschalk (1994) Flow regimes in northern and western Europe: development and application of procedures for classifying flow regimes, in P. Seuna, A. Gustard, N.W. Arnell and G.A. Cole (eds), *FRIEND: Flow Regimes from International Experimental and Network Data*, IAHS Publ. No. 221, 185–192.
Kreimer, A. and M. Preece (1991) Case study: Taiz flood disaster prevention and municipal development project, in A. Kreimer and M. Munasinghe (eds), *Managing Natural Disasters and the Environment*, Environment Department, The World Bank, Washington, DC, 186–190.
Kreps, G. (1992) Foundations and principles of emergency planning and management, in D.J. Parker and J.W. Handmer (eds), *Hazard Management and Emergency Planning*, James and James, London, 159–174.
Krzysztofowicz, R. (1993) A theory of flood warning systems, *Water Resources Res.*, **29**, 3981–3994.
Kuchment, L.S., V.N. Demidov, Y.G. Motovilov, N.A. Nazarov and V.Y. Smakhtin (1993) Estimation of disastrous floods risk via physically based models of river runoff generation, in Z.W. Kundzewicz, D. Rosbjerg, S.P. Simonovic and K. Takeuchi (eds), *Extreme Hydrological Events: Precipitation, Floods and Droughts*, IAHS Publ. No. 213, 177–182.
Kuichling, E. (1889) The relation between the rainfall and the discharge of sewers in populous districts, *Trans. Amer. Soc. Civil Engrs.*, **20**, 1–56.
Kuittinen, R. (1989) Determination of snow water equivalents by using NOAA-satellite images, gamma ray spectrometry and field measurements, in A. Rango (ed.), *Remote Sensing and Large-Scale Global Processes*, IAHS Publ. No. 186, 151–159.
Kunreuther, H. (1978) *Disaster Insurance Protection: Public Policy Lessons*, Wiley, New York.
Kunreuther, H. (1993) Combining insurance with hazard mitigation to reduce disaster losses, *Natural Hazards Observer*, **27**, 1–3.
Kuzmin, P.P. (1961) *Melting of Snowcover*, Hydrometeorological Institute, Leningrad (Israel Programme for Scientific Translations, Jerusalem, 1972).

La Barbera, P., L. Lanza and F. Siccardi (1993) Flash flood forecasting based on multisensor information, in Z.W. Kundzewicz, D. Rosbjerg, S.P. Simonovic and K. Takeuchi (eds), *Extreme Hydrological Events: Precipitation, Floods and Droughts*, IAHS Publ. No. 213, 21–32.

Labzovskii, N.A. (1966) (Sea floods in the estuary of the Neva River) Morkskie navodneniia v ust'e r. Neva, *Moscow. Gosudarstvennyi Okeanograficheskii Institut, Trudy*, **79**, 3–40.

Lai, C.C.A. (1992) Probabilistic forecast of tropical cyclone-generated storm surge with a dynamic-statistical approach, *J. Mar. Technol. Soc.*, **26**(2), 33–43.

Lal, R. (1983) Soil erosion in the humid tropics with particular reference to agricultural land development and soil management, in R. Keller (ed.), *Hydrology of Humid Tropical Regions*, IAHS Publ. No. 140, 221–239.

Lambert, W.G. and A.R. Millard (1969) *Atra-hasís—The Babylonian Story of the Flood*, Oxford University Press, London.

Langbein, W.B. (1949) Annual floods and partial-duration series, *Trans. Amer. Geophys. Union*, **30**, 879–881.

Larson, L.W. (1995) The great USA flood of 1993, Paper presented at US–Italy Research Workshop, Water Resources Research and Documentation Center (WARREDOC), Perugia, Italy, 13–17 November, 1995.

Laska, S.B. (1991) *Floodproof Retrofitting: Homeowner Self-Protective Behavior*, Monograph No. 49, Institute of Behavioral Science, University of Colorado.

Law, F. (1957) Measurement of rainfall, interception and evaporation losses in a plantation of sitka spruce, *Toronto General Assembly*, IAHS Publ. 43, 397–411.

Lawler, D.M. (1987) Climate change over the last millennium in central Britain, in K.J. Gregory, J. Lewin and J.B. Thornes (eds), *Palaeohydrology in Practice*, Wiley, Chichester, 99–129.

Lawrence, P.L. and J.G. Nelson (1994) Flooding and erosion hazards on the Ontario Great Lakes shoreline: a human ecological approach to planning and management, *J. Environmental Planning and Management*, **37**, 289–304.

Lee, M. and B. Marker (1995) Earth science information in coastal planning, Section 6.2 in *Proc. 30th MAFF Conference of River and Coastal Engineers*, Keele, 5–7 July, Ministry of Agriculture, Fisheries and Food, London.

Legome, E., A. Robins and D.A. Rund (1995) Injuries associated with floods: the need for an international reporting scheme, *Disasters*, **19**, 50–54.

Lerner, D.N. and M.H. Barrett (1996) Urban groundwater issues in the United Kingdom, *J. Hydrogeology*, **4**(1), 80–89.

Lewin, J. (1989) Floods in fluvial geomorphology, in K. Beven and P. Carling (eds), *Floods: Hydrological, Sedimentological and Geomorphological Implications*, Wiley, Chichester, 265–284.

Lewin, J. and A. Scotti (1990) The flood-prevention scheme of Venice: experimental module, *J. Inst. Water and Environmental Management*, **4**, 70–77.

Lind, R.C. (1967) Flood control alternatives and the economics of flood protection, *Water Resources Res.*, **3**, 345–357.

Linsley, R.K. and J.B. Franzini (1964) *Water Resource Engineering*, McGraw-Hill, New York.

Lloyd-Davis, D.E. (1906) The elimination of storm water from sewerage systems, *Min. Proc. Inst. Civil Engrs.*, **164**, 41–67.

Lockett, J.E. (1980) Catastrophes and catastrophe insurance, *J. Institute of Actuaries Students' Soc.*, **24**, 91–134.

Lott, J.N. (1994) The US summer of 1993: A sharp contrast in weather extremes, *Weather*, **49**, 370–383.

Lowing, M.J. (1995) *Linkage of Flood Frequency Curve with Maximum Flood Estimate*, Report FR/D 0023, Published for the Department of the Environment by the Foundation for Water Research, Marlow.

Lucas, R.M., A.R. Harrison and E.C. Barrett (1989) A multispectral snow area algorithm for operational 7-day snow cover monitoring, in A. Rango (ed.), *Remote Sensing and Large-Scale Global Processes*, IAHS Publ. No. 186, 161–166.

Lvovitch, M.I. (1973) The global water balance, United States, International Hydrological Decade Bulletin 23, *Trans. Amer. Geophys. Union*, **54**, 28–42.

McCarthy, G.T. (1938) The unit hydrograph and flood routing. Unpublished paper presented at Conference of North Atlantic Div., US Army Corps of Engineers, New London, Conn., 24 June 1938.

McEwen, L.J. and A. Werritty (1988) The hydrology and long-term geomorphic significance of a flash flood in the Cairngorm mountains, Scotland, *Catena*, **15**, 361–377.

McLuckie, B.J. (1973) *The Warning System: A Social Science Perspective*, US National Weather Service, Washington, DC.

McMahon, T.A. (1979) Hydrological characteristics of arid zones, in *The Hydrology of Areas of Low Precipitation*, IAHS Publ. No. 128, 105–123.

Maddox, R.A. (1980) Mesoscale convective complexes, *Bull. Amer. Met. Soc.*, **61**, 1374–1387.

Maddox, R.A. (1983) Large scale meteorological conditions associated with mid-latitude, mesoscale convective complexes, *Mon. Weather Rev.*, **111**, 1475–1493.

Maddox, R.A., C.F. Chappell and L.R. Hoxit (1979) Synoptic and meso-α scale aspects of flash flood events, *Bull. Amer. Met. Soc.*, **60**, 115–123.

Maddox, R.A., F. Canova and L.R. Hoxit (1980) Meteorological characteristics of flash flood events over the western United States, *Monthly Weather Rev.*, **108**, 1866–1877.

Maddrell, R., A. Toft, N. Stevens, D. Leggett, M. Dixon and J. Morgan (1995) Saltmarsh management and set back research, Section 7.3 in *Proc. 30th. MAFF Conference of River and Coastal Engineers*, Keele, 5–7 July, Ministry of Agriculture, Fisheries and Food, London.

MAFF (1992) Saltmarsh erosion and accretion processes, *Coastal Defence*, **1**, 2–3.

MAFF (1993) *Strategy for Flood and Coastal Defence in England and Wales*, Ministry of Agriculture, Fisheries and Food and the Welsh Office.

MAFF (1994) Floods in large basins, *Flood and Coastal Defence*, **5**, 4.

MAFF (1995a) Holderness experiment: the observational phase, *Flood and Coastal Defence*, **7**, 1–2.

MAFF (1995b) Flood risk map of England and Wales, *Flood and Coastal Defence*, **7**, 2.

MAFF (1995c) Storm Tide Warning Service workshop, *Flood and Coastal Defence*, **7**, 6.

MAFF (1995d) Tollesbury Breach experiment, *Flood and Coastal Defence*, **8**, 1–2.

MAFF (1996) *Developments in Flood Estimation and Flood Warning*, MAFF R&D Project Sheet, June 1996.

Maheshwari, B.L., K.F. Walker and T.A. McMahon (1995) Effects of regulation on the flow regime of the river Murray, Australia, *Regulated Rivers: Res. and Managt.*, **10**, 15–38.

Maizels, J. (1989) Sedimentology and palaeohydrology of Holocene flood deposits in front of a jökulhlaup glacier, south Iceland, in K. Beven and P. Carling (eds), *Floods: Hydrological, Sedimentological and Geomorphological Implications*, Wiley, Chichester, 239–251.

Mamun, M.Z. (1996) Awareness, preparedness and adjustment measures of river-bank erosion-prone people: a case study, *Disasters*, **20**, 68–74.

Mann, R.H.K. (1988) Fish and fisheries of regulated rivers in the UK, *Regulated Rivers: Res. and Managt.*, **2**, 411–424.

Marco, J.B. (1994) Flood risk mapping, in G. Rossi, N. Harmancioglu and V. Yevjvich (eds), *Coping with Floods*, Kluwer Academic Publishers, Dordrecht, 353–373.

Marco, J.B. and A. Cayuela (1994) Urban flooding: the flood-planned city concept, in G. Rossi, N. Harmancioglu and V Yevjvich (eds), *Coping with Floods*, Kluwer Academic Publishers, Dordrecht, 705–721.

Marino, M.J. (1992) Implications of climatic change on the Ebro delta, in L. Jeftic, J.D. Milliman and G. Sestini (eds), *Climatic Change and the Mediterranean*, Edward Arnold, London, 304–327.

Marsh, G.P. (1864) *Man and Nature; or Physical Geography as Modified by Human Action*, Charles Scribner, New York.

Marsh, P. and M.K. Woo (1985) Meltwater movement in natural heterogeneous snow covers, *Water Resources Res.*, **21**, 1710–1716.

Marshall, W.A.L. (1952) The Lynmouth floods, *Weather*, **7**, 338–342.

Martin, D.J. (1994) The impact of conservation issues on a sea-defence scheme at Pennington, *J. Inst. Water Engrs. and Managers*, **8**, 567–575.

Martinec, J. (1985) Time in hydrology, in J.C. Rodda (ed.), *Facets of Hydrology, II*, Wiley, Chichester, 249–290.

Martinec, J. and A. Rango (1989) Effects of climate change on snowmelt runoff patterns, in A. Rango (ed.), *Remote Sensing and Large-Scale Global Processes*, IAHS Publ. No. 186, 31–38.

Mastenbroek, C., G. Burgers and P.A.E.M. Janssen (1993) The dynamical coupling of a wave model and a storm surge model through the atmospheric boundary layer, *J. Phys. Oceanogr.*, **23**, 1856–1866.

Meier, M.F. (1964) Ice and glaciers, in V.T. Chow (ed.), *Handbook of Applied Hydrology*, McGraw-Hill, New York, 16.1–32.

Melentijevich, M. (1969) Estimation of flood flows using mathematical statistics, *Floods and their Computation*, IASH/UNESCO/WMO, Wallingford, 164–170.

Mercado, A., J.D. Thompson and J.C. Evans (1993) Requirements for modelling of future storm surge and ocean circulation, chapter 4 in G.A. Maul (ed.), *Climatic Change in the Intra-Americas Sea*, Edward Arnold, London, 75–84.

Midgley, P. (1995) Groundwater flooding: a time to run?, *Circulation*, **45**, 13–14.

Mikolajewicz, U., B. Santer and E. Maier-Reimer (1990) *Ocean Response to Global Warming*, Max-Planck-Institut für Meteorologie, Report 49.

Mileti, D.S. (1994) Public response to flood warnings, in G. Rossi, N. Harmancioglu and V. Yejevich (eds), *Coping with Floods*, Kluwer Academic Publishers, Dordrecht, 549–563.

Mileti, D.S. and E.M. Beck (1975) Communication in crisis: Explaining evacuation symbolically, *Communication Res.*, **2**, 24–49.

Miljukov, P.I. (1972) Status and trends in hydrological forecasts, *Status and Trends of Research in Hydrology 1965–74*, UNESCO, Paris, 85–100.

Millemann, B. (1993) The national flood insurance program, *Oceanus*, **36**, 6–8.

Miller, A.A. (1950) *Climatology*, 7th edition, Methuen, London.

Miller, D.J. (1960) The Alaska earthquake of July 10, 1958: giant wave in Lituya Bay, *Bull. Seism. Soc. Amer.*, **50**, 253–266.

Miller, J.F. (1973) Probable maximum precipitation—the concept, current procedures and outlook, in E.F. Schulz, V.A. Koelzer and K. Mahmood (eds), *Floods and Droughts*, Water Resources Publications, Fort Collins, Colorado, 50–61.

Miller, R.F., I.S. McQueen, F.A. Branson, L.M. Shown and W. Buller (1969) An evaluation of range floodwater spreaders, *J. Range Management*, **22**, 246–257.

Milliman, J.D., J.M. Broadus and F. Gable (1989) Environmental and economic implications of rising sea-level and subsiding deltas: The Nile and Bengal examples, *Ambio*, **6**, 340–345.

Milliman, J.W. (1983) An agenda for economic research on flood hazard mitigation, in S.A. Changnon et al. (eds), *A Plan for Research on Floods and their Mitigation in the United States*, Illinois State Water Survey, Champaign, Illinois, 83–104.

Milne, A. (1986) *Flood Shock*, Sutton, Gloucester.

Mirtskhoulava, T.E. (1994) Flood control in the former USSR, in G. Rossi, N. Harmancioglu and V. Yejevich (eds), *Coping with Floods*, Kluwer Academic Publishers, Dordrecht, 751–758.

Mitchell, J.F.B., S. Manabe, V. Meleshko and T. Tokioka (1990) Equilbrium climate change—and its implications for the future, in J. T. Houghton, B.A. Callender and S. K. Varney (eds) *Climate Change 1992*, Cambridge University Press, Chapter 5.

Miyazaki, M. (1967) Storm surges along the Japanese coast and their prediction, *11th Pacific Science Congress, Sympos. on Tsunami and Storm Surges*, 74.

Mo, K.C., J. Noguespaegle and J. Paegle (1995) Physical mechanisms of the 1993 summer floods, *J. Atmospheric Sci.*, **52**, 879–895.

Mokadem, A-I., J.O. Motta and S. Dautrebande (1989) Characterization of watersheds by the integration of remote sensing and cartographic digital data, in L. Roald, K. Nordseth and K.A. Hassel (eds), *FRIENDS in Hydrology*, IAHS Publ. No. 187, 147–152.

Molinaroli, R. (1965) Flood plain studies, *Proc. Conf. Hydrologic Activities in the South Carolina Region*, Clemson Univ., Clemson, 1–6.

Molles, M.C., C.S. Crawford and L.M. Ellis (1995) Effects of an experimental flood on litter dynamics in the middle Rio Grande riparian ecosystem, *Regulated Rivers: Res. and Managt.*, **11**, 275–281.

Montz, B.E. (1992) The effects of flooding on residential property values in three New Zealand communities, *Disasters*, **16**, 283–298.

Montz, B.E. and G.A. Tobin (1988) The spatial and temporal variability of residential real estate values in response to flooding, *Disasters*, **12**, 345–355.

Moore, J.W and D.P. Moore (1989) *The Army Corps of Engineers and the Evolution of Federal Floodplain Management Policy*, Special Publication No. 20, Natural Hazards Research and Applications Center, Institute of Behavioral Science, University of Colorado, Boulder.

Moore, R.J. (1985) The probability-distributed principle and runoff production at point and basin scales, *Hydrol. Sci. J.*, **30**(2), 273–297.

Moore, R.J. (1986) Advances in real-time flood forecasting practice, in *Symposium on Flood Warning Systems*, unpublished paper presented at winter meeting of the River Engineering Section, Inst. Water Engineers and Scientists.

Moore, R.J. (1993) Real-time flood forecasting systems: Perspectives and prospects, *UK–Hungarian Workshop on Flood Defence*, Budapest, 6–10 September 1993.

Moore, R. J. (1995) Rainfall and flow forecasting using weather radar, unpublished paper presented to British Hydrological Society National Meeting on *Hydrological Uses of Weather Radar*, 9 January 1995, Institution of Civil Engineers, London.

Moore, R.J. and V.A. Bell (1994) A grid-square rainfall–runoff model for use with weather radar data, in M.E. Almeida-Teixeira, R. Fantechi, R. Moore and V.M. Silva (eds), *Advances in Radar Hydrology*, EUR 14334 EN, Directorate-General Science, Research and Development, European Commission, Brussels.

Moore, R.J. and V.A. Bell (1996) A grid-based flood forecasting model using weather radar, digital terrain and Landsat data, in M. Borga and R. Casale (eds), *Integrating Radar Estimates of Rainfall in Real Time Flood Forecasting*, Special Issue of *Quaderni di Idronomia Montana*, **16**, 97–105.

Moore, R.J., D.A. Jones, K.B. Black, R.M. Austin, D.S. Carrington, M. Tinnion and A. Akhondi (1994) RFFS and HYRAD: Integrated systems for rainfall and river flow forecasting in real-time and their application in Yorkshire, in *Analytical Techniques for the Development and Operations Planning of Water Resource and Supply Systems*, British Hydrological Society, Occasional Paper No. 4.

Moore, R.J., R.J. Harding, R.M. Austin, V.A. Bell and D.R. Lewis (1996) *Development of Improved Methods for Snowmelt Forecasting*, R&D Note 402, Institute of Hydrology/National Rivers Authority, Research Contract for The Foundation for Water Research, Marlow, 192pp.

Morris, D.G. and R.W. Flavin (1995) *Flood Risk Map for England and Wales*, IH Report to MAFF, Institute of Hydrology, Wallingford.

Morrisette, P.M. (1988) The stability bias and adjustment to climatic variability: the case of the rising level of the Great Salt Lake, *Appl. Geog.*, **8**, 171–189.

Mulwanda, M. (1993) The need for new approaches to disaster management: the 1989 floods in Lusaka, Zambia, *Environment and Urbanization*, **5**, 67–77.

Munasinghe, M., B. Menezes and M. Preece (1991) Case study: Rio flood reconstruction and prevention project, in A. Kreimer and M. Munasinghe (eds), *Managing Natural Disasters and the Environment*, Environment Department, The World Bank, Washington, DC, 28–31.

Munk, W.H. (1962) Long ocean waves, in M.N. Hill (ed.), *The Sea*, Interscience, New York, 647–663.

Munk, W.H., F. Snodgrass and G. Carrier (1956) Edge waves on the continental shelf, *Science*, **123**, 127–132.

Munkvold, G.P. and X.B. Yang (1995) Crop damage and epidemics associated with 1993 floods in Iowa, *Plant Disease*, **79**, 95–101.

Murray, I. (1995) Downpour overwhelms decrepit sewer system, *The Times*, London, 14 February.

Murty, T.S. and M.I. El-Sabh (1992) Mitigating the effects of storm surges generated by tropical cyclones: a proposal, *Natural Hazards*, **6**(3), 251–273.

Murty, T.S. and V.R. Neralla (1992) On the recurvature of tropical cyclones and the storm surge problem in Bangladesh, *Natural Hazards*, **6**(3), 275–279.

Naden, P.S. (1992) Spatial variability in flood estimation for large catchments: the exploitation of channel network structure, *Hydrol. Sci. J.*, **37**, 53–71.

Naden, P., S. Crooks and P. Broadhurst (1996) Impact of climate and land use change on the flood response of large catchments, Section 2.1 in *Proc. 31st MAFF Conference of River and Coastal Engineers*, Keele, 3–5 July, Ministry of Agriculture, Fisheries and Food, London.

Naef, F. and G.R. Bezzola (1990) Hydrology and morphological consequences of the 1987 flood event in the upper Reuss valley, in R.O. Sinniger and M. Monbaron (eds), *Hydrology in Mountainous Regions, II Artificial Reservoirs, Water and Slopes*, IAHS Publ. No. 194, 339–346.

Namias, J. (1973) Hurricane Agnes—an event shaped by large-scale air–sea systems generated during antecedent months, *Quart. J. Roy. Met. Soc.*, **99**, 506–519.

Nanson, G.C. (1986) Episodes of vertical accretion and catastrophic stripping: a model of disequilibrium floodplain development, *Bull. Geol. Soc. America*, **97**, 1467–1475.

NAS (1987) *Confronting Natural Disasters: An International Decade for Natural Disaster Reduction*, US National Academy of Sciences, Washington, DC.

National Association of Independent Insurers (1994) *Mitigating Catastrophic Property Insurance Losses*, National Association of Independent Insurers, Des Plaines, Illinois.

NERC (1975) *Flood Studies Report*, Natural Environment Research Council, 5 vols.

NERC (1993) *Institute of Hydrology 1992-1993 Report*, Natural Environment Research Council.

NERC (1994) *Institute of Hydrology 1993-1994 Report*, Natural Environment Research Council.

Newman, C.J. (1976) Children of disaster: clinical observations at Buffalo Creek, *Amer. J. Psychiatry*, **133**, 306–312.

Newman, M. (1995) River Lavant flood investigation, Section 4.3 in *Proc. 30th MAFF Conference of River and Coastal Engineers*, Keele, 5–7 July, Ministry of Agriculture, Fisheries and Food, London.

Newsome, D.H. (1992) *Weather Radar Networking, COST 73 Project/Final Report*, Kluwer, Dordrecht.

Newson, M. (1992) *Land, Water and Development: River Basin Systems and their Sustainable Management*, Routledge, London.

Newson, M. (1994) *Hydrology and the River Environment*, Oxford University Press.

Newson, M. and J. Lewin (1991) Climatic change, river flow extremes and fluvial erosion—scenarios for England and Wales, *Progress in Physical Geog.*, **15**, 1–17.

Newson, M.D. (1989) Flood effectiveness in river basins: Progress in Britain in a decade of drought, in K. Beven and P. Carling (eds), *Floods: Hydrological, Sedimentological and Geomorphological Implications*, Wiley, Chichester, 151–169.

Ngan, P. and S.O. Russell (1986) Example of flow forecasting with Kalman filter, *J. Hydraul. Engg.*, **112**(9), 818–832.

Nigg, J. (1995) Risk communication and warning systems, in T. Horlick-Jones, A. Amendolaz and R. Casale (eds), *Natural Risk and Civil Protection*, E & FN Spon, London, 368–382.

Nimmo, W.H.R. (1947) *Summary of the Hydrology of Cooper's Creek*, Technical Bull. No. 1, Queensland Bureau of Investigation, 134–148.

N'Jai, A., S.M. Tapsell, D. Taylor, P.M. Thompson and R.C. Witts (1990) *Flood Loss Assessment Information Report (FLAIR)*, Flood Hazard Research Centre, Middlesex Polytechnic, London.

NOAA (1994) *The Great Flood of 1993*, Natural Disaster Survey Report, US National Weather Service, Office of Hydrology.

Noonan, G.A. (1986) An operational flood warning system, *J. Inst. Wat. Engrs. Sci.*, **40**, 437–453.

Odemerho, F.O. (1993) Flood control failures in a Third World city: Benin City, Nigeria—some environmental factors and policy issues, *Geojournal*, **29**, 371–376.

OFDA (1996) *Disaster History: Significant Data on Major Disasters Worldwide 1900–1995*, Office of US Foreign Disaster Assistance, U.S. Agency for International Development, Washington, DC.

Okal, E.A. (1994) Tsunami warning: beating the waves to death and destruction, *Endeavour*, **18**(1), 38–43.

Oliver, J.E. (1981) *Climatology: Selected Applications*, Edward Arnold, London.

Onesti, L.J. (1985) Meteorological conditions that initiate slushflows in the central Brooks Range, Alaska, *Annals of Glaciology*, **6**, 23–25.

Ott, M., Z. Su, A.H. Schumann and G.A. Schultz (1991) Development of a distributed hydrological model for flood forecasting and impact assessment of land-use change in the international Mosel river basin, in F.H.M. van de Ven, D. Gutknecht, D.P. Loucks and K.A. Salewicz (eds), *Hydrology for the Water Management of Large River Basins*, IAHS Publ. No. 201, 183–194.

PAHO (1981) *Emergency Health Management after Natural Disaster*, Pan American Health Organisation, Washington, DC.
Pardé, M. (1955) *Fleuves et Rivières*, 3rd edition, Armand Colin, Paris.
Pardé, M. (1964) Floods, floods, floods!, *UNESCO Courier*, 17, July–August, 54–59.
Parker, D.J. (1985) British structural flood mitigation experience: A critical look at response to coastal flood disasters, *Internat. Symposium on Housing and Urban Redevelopment after Natural Disasters: Mitigating Future Losses*, Bal Harbour, Miami.
Parker, D.J. (1987) Flood warning dissemination: the British experience, in J. Handmer (ed.), *Flood Hazard Management: British and International Perspectives*, Geo Books, Norwich, 169–190.
Parker, D.J. (1988) Emergency service response and costs of British floods, *Disasters*, 12, 50–69.
Parker, D.J. (1991) *The Damage-Reducing Effects of Flood Warnings*, Publication No. 197, Flood Hazard Research Centre, Middlesex Polytechnic, London.
Parker, D.J. (1992) The mismanagement of hazards, in D.J. Parker and J.W. Handmer (eds), *Hazard Management and Emergency Planning*, James and James, London, 3–21.
Parker, D.J. (1995) Floodplain development policy in England and Wales, *Appl. Geog.*, 15, 341–363.
Parker, D.J. and E.C. Penning-Rowsell (1972) *Problems and Methods of Flood Damage Assessment*, Progress Report No.3, Flood Hazard Research Centre, Middlesex Polytechnic, London.
Parker, D.J. and E.C. Penning-Rowsell (1983a) Flood hazard research in Britain, *Progress in Human Geog.*, 7, 182–202.
Parker, D.J. and E.C. Penning-Rowsell (1983b) Flood risk in the urban environment, in D.H. Herbert and R.J. Johnston (eds), *Geography and the Urban Environment*, Wiley, Chichester, 201–239.
Parker, D.J. and E.C. Penning-Rowsell (1991) *Institutional Design for Effective Hazard Management: Flood Warnings and Emergency Response*, Paper presented to a ESRC/LSE seminar, 1 May, Middlesex Polytechnic, London.
Parker, D.J., C.H. Green and P.M. Thompson (1987) *Urban Flood Protection Benefits: A Project Appraisal Guide*, Gower Technical Press, Aldershot.
Parrett, C., N.B. Melcher and R.W. James (1993) *Flood Discharges in the Upper Mississippi River Basin 1993*, US Geol. Surv., Circular 1120-A.
Patel, T. (1996) Indian red tape holds back vital flood warnings, *New Scientist*, 23 November, p. 10.
Patt, H. (1994) Rehabilitation of an urban river, in W.R. White and J. Watts (eds), *River Flood Hydraulics*, Wiley, Chichester, 483–489.
Paul, B.K. (1984) Perception of and agricultural adjustment to floods in the Jamuna floodplain, Bangladesh, *Human Ecology*, 12, 3–19.
Peacock, W.G., B.H. Morrow and H. Gladwin (eds) (1998) *Hurricane Andrew and the Reshaping of Miami: Ethnicity, Gender and Socio-Political Ecology of Disasters*, Routledge, London (in press).
Pearce, F. (1988) Cool oceans caused floods in Bangladesh and Sudan, *New Scientist*, 8 September, p. 31.
Pellymounter, D. (1994) The Yorkshire Humber estuary tidal defence strategy, in W.R. White and J. Watts (eds), *River Flood Hydraulics*, Wiley, Chichester, 15–26.
Penning-Rowsell, E.C. (1976) The effect of flood damage on land use planning, *Geographica Polonica*, 34, 139–153.
Penning-Rowsell, E.C. (1986) The development of integrated flood warning systems, in D.I. Smith and J.W. Handmer (eds), *Flood Warning in Australia*, Centre for Resource and Environmental Studies, Canberra, 15–36.
Penning-Rowsell, E.C. and J.B. Chatterton (1977) *The Benefits of Flood Alleviation: A Manual of Assessment Techniques*, Saxon House, Farnborough.
Penning-Rowsell, E.C. and D.J. Parker (1987) The indirect effects of floods and benefits of flood alleviation: evaluating the Chesil sea defence scheme, *Appl. Geog.*, 7, 263–288.
Penning-Rowsell, E.C. and B. Peerbolte (1994) Concepts, policies and research, in E.C. Penning-Rowsell and M. Fordham (eds), *Floods Across Europe: Flood Hazard Assessment, Modelling and Management*, Middlesex University Press, London, 1–17.

Penning-Rowsell, E.C. and D.I. Smith (1987) Self-help hazard mitigation: the economics of house raising in Lismore, NSW, Australia, *Tijdschrift Economic and Social Geog.*, **78**, 176–189.

Penning-Rowsell, E.C. and P. Winchester (1992) Scenario construction for risk communication in emergency planning: six 'golden rules', in D.J. Parker and J.W. Handmer (eds), *Hazard Management and Emergency Planning*, James and James, London, 203–217.

Penning-Rowsell, E.C., J.B. Chatterton and D.J. Parker (1978) *The Effect of Flood Warning on Flood Damage Reduction*, Report for the Central Water Planning Unit, Flood Hazard Research Centre, Middlesex Polytechnic, London.

Penning-Rowsell, E.C., J.W. Handmer and D.I. Smith (1986) Australian flood warnings: policy changes for the next decade, in D.I. Smith and J.W. Handmer (eds), *Flood Warnings in Australia*, Australian National University, Canberra, 293–310.

Penning-Rowsell, E.C., C.H. Green, P.M. Thompson et al. (1992) *The Economics of Coastal Management*, Belhaven Press, London.

Pereira, H.C. (1973) *Land Use and Water Resources*, Cambridge University Press, London.

Perry, A.H. (1981) *Environmental Hazards in the British Isles*, George Allen & Unwin, London.

Perry, C.A. (1994) *Effects of Reservoirs on Flood Discharges in the Kansas and Missouri River Basins 1993*, Circular 120-E, US Geological Survey, Denver, 20 pp.

Perry, R.W. (1985) *Comprehensive Emergency Management: Evacuating Threatened Populations*, Jai Press, London.

Perumal, M. (1993) Comparison of two variable parameter Muskingum methods, in Z.W. Kundzewicz, D. Rosbjerg, S.P. Simonovic and K. Takeuchi (eds), *Extreme Hydrological Events: Precipitation, Floods and Droughts*, IAHS Publ. No. 213, 129–138.

Peters, J.C. and P.B. Ely (1985) Flood-runoff forecasting with HEC-1F, *Water Resources Bull.*, **21**(1), 7–13.

Petts, G.E. (1988) Regulated rivers in the United Kingdom, *Regulated Rivers: Res. and Managt.*, **2**, 201–220.

Pierson, T.C. (1989) Hazardous hydrologic consequences of volcanic eruptions and goals for mitigative action: an overview, in Ö. Starosolszky and O.M. Melder (eds), *Hydrology of Disasters*, Proceedings of a WMO Technical Conference, James and James, London, 220–236.

Pilgrim, D.H. (ed.) (1987) *Australian Rainfall and Runoff: A Guide to Flood Estimation*, Institution of Engineers Australia, Canberra.

Pilgrim, D.H. and D.G. Doran (1993) Practical criteria for the choice of method for estimating extreme design floods, in Z.W. Kundzewicz, D. Rosbjerg, S.P. Simonovic and K. Takeuchi (eds), *Extreme Hydrological Events: Precipitation, Floods and Droughts*, IAHS Publ. No. 213, 227–235.

Pilgrim, D.H., T.G. Chapman and D.G. Doran (1988) Problems of rainfall–runoff modelling in arid and semiarid regions, *Hydrol. Sci. J.*, **33**, 379–400.

Pine, T. (1995) Integrated emergency management: hierarchy or network?, in T. Horlick-Jones, A. Amendolla and R. Casale (eds), *Natural Risk and Civil Protection*, E & FN Spon, London, 494–501.

Pircher, W. (1990) The contribution of hydropower reservoirs to flood control in the Austrian Alps, in R.O. Sinniger and M. Monbaron, *Hydrology in Mountainous Regions. II Artificial Reservoirs, Water and Slopes*, IAHS Publ. No. 194, 3–10.

Platt, R.H. (1987) (ed.) *Regional Management of Metropolitan Floodplains*, Monograph No. 45, Institute of Behavioral Science, University of Colorado, Boulder.

Platt, R.H. and G.M. McMullen (1980) *Post-flood Recovery and Hazard Mitigation: Lessons from the Massachusetts Coast, February, 1978*, Publ. No. 115, Water Resources Research Center, University of Massachusetts, Amherst.

Platt, R.H., T. Beatley and H.C. Miller (1991) The folly at Folly Beach and other failings of U.S. coastal erosion policy, *Environment*, **33**, 7–9, 25–32.

Ponte Ramirez, R.R. and E.M. Shaw (1983) Simulating flood hydrographs from storm rainfalls in Venezuela, in R. Keller (ed.), *Hydrology of Humid Tropical Regions*, IAHS Publ. No. 140, 435–445.

Pope, J. (1994) Replacing the SPM: The Coastal Engineering Manual, Section 2.5 in *A Strategic Approach, Proc. 1994 MAFF Conference of River and Coastal Engineers*, Loughborough, 4–6 July, Ministry of Agriculture, Fisheries and Food, London.

Prasuhn, A.L. (1987) *Fundamentals of Hydraulic Engineering*, Holt, Rinehart and Winston, New York.

Prestegaard, K.L., A.M. Matherne, B. Shane, K. Houghton, M. O'Connell and N. Katyl (1994) Spatial variability in the magnitude of the 1993 floods, Raccoon River basin, Iowa, *Geomorphology*, **10**, 169–182.

Price, A.G., L.K. Hendrie and T. Dunne (1979) Controls on the production of snowmelt runoff, in S.C. Colbeck and M. Ray (eds), *Proceedings, Modeling of Snow Cover Runoff*, US Army Corps of Engineers, Cold Regions Research and Engineering Lab., Hanover, 257–268.

Price, R.K. (1978) *FLOUT—A River Catchment Flood Model*, Hydraulics Research Station Report No. IT 168, Wallingford.

Prince, H. (1995) Floods in the upper Mississippi River basin, 1993: newspapers, official views and forgotten farmlands, *Area*, **27**(2), 118–126.

Prowse, T.D., B. Aitken, M.N. Demuth and M. Peterson (1996) Strategies for restoring spring flooding to a drying northern delta, *Regulated Rivers: Res. and Managt.*, **12**, 237–250.

Pugh, D.T. (1987) *Tides, Surges and Mean Sea-Level*, Wiley, Chichester.

Qian, W. (1983) Effects of deforestation on flood characteristics with particular reference to Hainan Island, China, in R. Keller (ed.), *Hydrology of Humid Tropical Regions*, IAHS Publ. No. 140, 249–257.

Qin Zenghao, Duan Yihong, Wang Yinong, Shen Zhengfen and Xu Kuanren (1994) Numerical simulation and prediction of storm surges and water levels in Shanghai harbour and its vicinity, *Natural Hazards*, **9**, 167–188.

Quarantelli, E.L. (1984) Perceptions and reactions to emergency warnings of sudden hazards, *Ekistics*, **309**, 511–515.

Rahn, P.H. (1994) Flood plains, *Bull. Assoc. Engg. Geologists*, **331**, 171–181.

Ramachandran, R. and S.C. Thakur (1974) India and the Ganges floodplains, in G.F. White (ed.), *Natural Hazards: Local, Natural, Global*, Oxford University Press, New York, 36–43.

Ramamoorthi, A.S. and H. Haefner (1991) Runoff modelling and forecasting of river basins, and Himalayan Snow-cover Information System (HIMSIS), in F.H.M. van de Ven, D. Gutknecht, D.P. Loucks and K.A. Salewicz (eds), *Hydrology for the Water Management of Large River Basins*, IAHS Publ. No. 201, 347–355.

Rangachari, R., R.S. Prasad, T.K. Mukhopadhyay and I. Rishiraj (1989) Conceptual river routing model 'CWCFF', in Ö. Starosolszky and O.M. Melder (eds), *Hydrology of Disasters*, Proceedings of a WMO Technical Conference, James and James, London, 123–133.

Rango, A. and V. van Katwijk (1990) Development and testing of a snowmelt-runoff forecasting technique, *Water Resources Bull.*, **26**(1), 135–144.

Rango, A., V. van Katwijk and J. Martinec (1990) Snowmelt runoff forecasts in Colorado with remote sensing, in H. Lang and A. Musy (eds), *Hydrology in Mountainous Regions I Hydrological Measurements, The Water Cycle*, IAHS Publ. No. 193, 627–634.

Rappaport, E.N. (1994) Hurricane Andrew, *Weather*, **49**(2), 51–60.

Rasid, H. (1993) Preventing flooding or regulating flood levels? Case studies on perception of flood alleviation in Bangladesh, *Natural Hazards*, **8**, 39–57.

Rasid, H. and A. Mallik (1993) Poldering vs compartmentalisation: the choice of flood control techniques in Bangladesh, *Environmental Management*, **17**, 59–71.

Rasid, H. and A. Mallik (1995) Flood adaptations in Bangladesh, *Appl. Geog.*, **15**, 3–17.

Rasid, H., S. Sun, X. Yu and C. Zhang (1996) Structural vs non-structural flood alleviation measures in the Yangtze delta: a pilot survey of floodplain residents' preferences, *Disasters*, **20**, 93–110.

Rasmussen, J.L. (1994) Floodplain management into the 21st century: a blueprint for change—sharing the challenge. *Water Internat.*, **19**, 166–176.

Raymo, M.E. and W.F. Ruddiman (1992) Tectonic forcing of late Cenozoic climate, *Nature*, **359**, 117–122.

Redfield, A.C. and A.R. Miller (1957) *Water Levels Accompanying Atlantic Coast Hurricanes*, Amer. Met. Soc., Met. Monog., 2, 1–23.

Reed, D.W. (1994a) Some notes on generalized methods of flood estimation in the United Kingdom, in G. Rossi, N. Harmancioglu and V. Yejevich (eds), *Coping with Floods*, Kluwer Academic Publishers, Dordrecht, 185–191.

Reed, D.W. (1994b) Plans for the *Flood Estimation Handbook*, Section 8.3 in *Proc. 29th MAFF Conference of River and Coastal Engineers*, Keele, 4–6 July, Ministry of Agriculture, Fisheries and Food, London.

Reed, D.W. (1997) Regional frequency analysis and the *Flood Estimation Handbook*, Presentation to Pennines Hydrological Group, 19 February.

Reeve, D., B. Li and C. Fleming (1996) Validation of storm wave forecasting, Section 4.3 in *Proc. 31st MAFF Conference of River and Coastal Engineers*, Keele, 3–5 July, Ministry of Agriculture, Fisheries and Food, London.

Reich, B.M. (1973) *Effects of Agnes floods on Annual Series in Pennsylvania*, Pennsylvania State University, Inst. for Res. on Land and Water Resources, Res. Publ. 74.

Reid, I. and L.E. Frostick (1987) Flow dynamics and suspended sediment properties in arid zone flash floods, *Hydrol. Processes*, **1**, 239–253.

Reid, I., D.M. Powell, J.B. Laronne and C. Garcia (1994) Flash floods in desert rivers: Studying the unexpected, *EOS, Trans. Amer. Geophys. Union*, **75**, 452–453.

Renard, K.G. and R.V. Keppel (1966) Hydrographs of ephemeral streams in the Southwest, *J. Hydraul. Div., ASCE*, **92** (No. HY2), 33–52.

Riehl, H. (1965) *Introduction to the Atmosphere*, McGraw-Hill, New York.

Roald, L.A. (1989) Application of regional flood frequency analysis to basins in northwest Europe, in L. Roald, K. Nordseth and K.A. Hassel (eds), *FRIENDS in Hydrology*, IAHS Publ. No. 187, 163–173.

Roberts, A. (1996) HYREX conference, *Circulation, Newsl. Brit. Hydrol. Soc.*, **52**, 9, British Hydrological Society, London.

Roberts, A. and V. Pritchard (1993) Coastal engineering in the future: is coastal management the answer?, *Proc. Inst. Civil Engineers and Municipal Engineers*, **98**, 161–168.

Roberts, C.R. (1989) Flood frequency and urban-induced channel change: Some British examples, in K. Beven and P. Carling (eds), *Floods: Hydrological, Sedimentological and Geomorphological Implications*, Wiley, Chichester, 57–82.

Robinson, A.H.W. (1953) The storm floods of 1st February, 1953: I. The storm surge of 31st January–1st February, 1953, *Geography*, **38**, 134–141.

Robinson, A.H.W. (1961) The Pacific tsunami of May 22nd, 1960, *Geography*, **46**, 18–24.

Robinson, A.R., A. Tomasin and A. Artegiani (1973) Flooding of Venice: phenomenology and prediction of the Adriatic storm surge, *Quart. J. Roy. Met. Soc.*, **99**, 688–692.

Robinson, M. (1989) Small catchment studies of man's impact on flood flows: Agricultural drainage and plantation forestry, in L. Roald, K. Nordseth and K.A. Hassel (eds), *FRIENDS in Hydrology*, IAHS Publ. No. 187, 299–308.

Robinson, M. (1990) *Impact of Improved Land Drainage on River Flows*, IH Report No. 113, Institute of Hydrology, Wallingford.

Rockwood, D.M. (1969) Application of streamflow synthesis and reservoir regulation— 'SSARR'—program to the Lower Mekong River, in *The Use of Analog and Digital Computers in Hydrology*, IASH–UNESCO, **1**, 329–344.

Rodda, J.C. (1970) Rainfall excesses in the United Kingdom, *Trans. Inst. Brit. Geog.*, **49**, 49–60.

Rodier, J.A. (1983) Aspects scientifiques et techniques de l'hydrologie des zones humides de l'Afrique centrale, in R. Keller (ed.), *Hydrology of Humid Tropical Regions*, IAHS Publ. No. 140, 105–126.

Rodier, J.A. (1985) Aspects of arid zone hydrology, Chapter 8 in J.C. Rodda (ed.), *Facets of Hydrology II*, Wiley, Chichester, 205–247.

Rodier, J.A. and M. Roche (1984) *World Catalogue of Maximum Observed Floods*, Internat. Assoc. Hydrol. Sci. Publ. No. 143.

Rogers, P., P. Lydon and D. Seckler (1989) *Eastern Waters Study: Strategies to Manage Flood and Drought in the Ganges–Brahmaputra Basin*, Irrigation Support Project for Asia and the Near East, US Agency for International Development, Arlington.

Ross, S.M., J.B. Thorne and S. Nortcliffe (1990) Soil hydrology, nutrient and erosional response to the clearance of terra firme forest, Maraca Island, Roraima, northern Brazil, *Geogrl. J.*, **156**(3), 267–282.

Rossi, F. and P. Villani (1994) Regional flood estimation methods, in G. Rossi, N. Harmancioglu and V. Yevjevich (eds), *Coping with Floods*, Kluwer Academic Publishers, Dordrecht, 135–169.

Rossiter, J.R. (1954) The great tidal surge of 1953, *The Listener*, 8 July, 55–56.

Rostvedt, J.O. et al. (1968) *Summary of Floods in the United States during 1963*, USGS Wat. Sup. Pap., 1830-B.

Rumsby, B.T. and M.G. Macklin (1994) Channel and floodplain responses to recent abrupt climate change: The Tyne basin, Northern England, *Earth Surf. Proc. and Landforms*, **19**(6), 499–515.

Russac, P.A. (1986) Epidemiological surveillance: malaria epidemic following the Niño phenomenon, *Disasters*, **10**, 112–117.

Russell, A.P.G. and P.H. von Lany (1994) A risk methodology for determining an appropriate set-back distance for a flood embankment on the right bank of the Brahmaputra river, in W.R. White and J. Watts (eds), *River Flood Hydraulics*, Wiley, Chichester, 561–578.

Rutter, E.J. and L.R. Engstrom (1964) Hydrology of flow control: Part III—reservoir regulation, in V.T. Chow (ed.), *Handbook of Applied Hydrology*, McGraw-Hill, New York, section 25-III.

Saenger, P. and N.A. Siddiqi (1993) Land from the sea: the mangrove afforestation program of Bangladesh, *Ocean and Coastal Management*, **20**, 23–39.

Salas, J.D., E.E. Wohl and R.D. Jarrett (1994) Determination of flood characteristics using systematic historical and paleoflood data, in G. Rossi, N. Harmancioglu and V. Yejevich (eds), *Coping with Floods*, Kluwer Academic Publishers, Dordrecht, 111–134.

Samuels, P. and A. Brampton (1996) The effects of climate change on flood and coastal defence in the UK, Section 2.2 in *Proc. 31st MAFF Conference of River and Coastal Engineers*, Keele, 3–5 July, Ministry of Agriculture, Fisheries and Food, London.

Samuels, P. and P.G. Hollinrake (1995) Flood discharge assessment, Section 3.5 in *Proc. 30th MAFF Conference of River and Coastal Engineers*, Keele, 5–7 July, Ministry of Agriculture, Fisheries and Food, London.

Sapsted, D. (1996) Flood gives new life to Grand Canyon, *Daily Telegraph*, London, 10 October.

Sargent, R.J. (1984) The Haddington flood warning scheme, *J. Inst. Wat. Engrs. Sci.*, **38**, 108–118.

Sastry, N.S.R. (1994) Managing the livestock sector during floods and cyclones, *J. Rural Development*, **13**, 583–592.

Schermerhorn, V.P. and D.W. Kuehl (1969) Operational streamflow forecasting with the SSARR model, in *The Use of Analog and Digital Computers in Hydrology*, IASH–UNESCO, **1**, 317–328.

Schick, A.P. (1971) A desert flood: Physical characteristics; effects on man, geomorphic significance, human adaptation, *Jerusalem Studies in Geography*, **2**, 91–155.

Schick, A.P. and J. Lekach (1987) A high magnitude flood in the Sinai Desert, in L. Mayer and D. Nash (eds), *Catastrophic Flooding*, Unwin Hyman, London, 381–410.

Schiller, H. (1994) Flood frequency analysis with consideration of representativity and homogeneity of the flood series, in P. Seuna, A. Gustard, N.W. Arnell and G.A. Cole (eds), *FRIEND: Flow Regimes from International Experimental and Network Data*, IAHS Publ. No. 221, 257–266.

Schneider, W.J. and J.E. Goddard (1974) Extent and development of urban floodplains, *US Geological Survey Circular*, **601-J**, Washington, DC.

Schulze, O., R. Roth and O. Pieper (1994) Probable maximum precipitation in the Upper Harz Mountains, in P. Seuna, A. Gustard, N.W. Arnell and G.A. Cole (eds), *FRIEND: Flow Regimes from International Experimental and Network Data*, IAHS Publ. No. 221, 315–321.

Schulze, R.E. (1980) *Potential Flood Producing Rainfall of Medium and Long Duration in Southern Africa*, Water Research Commission, Pretoria, 37 pp.

Schumm, S.A. (1965) Quaternary palaeohydrology, in H.E. Wright and D.G. Frey (eds), *Quaternary of the United States*, Princeton University Press, Princeton, 783–794.

Schwab, G.O., D.D. Fangmeier, W.J. Elliot and R.K. Frevert (1993) *Soil and Water Conservation Engineering*, Wiley, New York.

Schware, R. and P. Lippoldt (1982) An examination of community flood warning systems: are we providing the right assistance?, *Disasters*, **6**, 195–201.

Scientific Software Group (1997) Hydrogeology, Hydrology, Geology, *Best Buys Software Bull.*, Washington, DC.

Scotti, A. (1993) Progettazione delle opere di defesa dell acque alte: I, *Quaderni Trimestrali, Consorzio Venezia Nuova*, **1**, 9–29.

Sempere, C.M., M. del R. Vidal-Abarca and M.L. Suárez (1994) Floods in arid south-east Spanish areas: a historical and environmental review, in G. Rossi, N. Harmancioglu and V. Yevjevich (eds), *Coping with Floods*, Kluwer Academic Publishers, Dordrecht, 271–278.

Sestini, G. (1992) Implications of climatic changes for the Po delta and Venice lagoon, in L. Jeftic, J.D. Milliman and G. Sestini (eds), *Climate Change and the Mediterranean*, Edward Arnold, London, 428–494.

Sewell, W.R.D. (1969) Human response to floods, in R.J. Chorley (ed.), *Water, Earth and Man*, Methuen, London, 431–451.

Shan, F. (1996) The impact of 1995 floods on Mainland China grain production, *Issues and Studies*, **32**, 52–64.

Sharon, D. (1980) The distribution of hydrologically effective rainfall incident on sloping ground, *J. Hydrol.*, **46**, 165–188.

Sharon, D. (1981) The distribution in space of local rainfall in the Namib Desert, *J. Climatol.*, **1**, 69–75.

Sheaffer, J.R. (1961) Flood-to-peak interval, in G.F. White (ed.), *Papers on Flood Problems*, Univ. of Chicago, Dept. of Geog. Res. Pap. No. 70, 95–113.

Sheaffer, J.R. et al. (1976) *Flood Hazard Mitigation Through Safe Land Use Practices*, Kiefer and Associates, Chicago.

Shennan, I. (1989) Holocene crustal movements and sea-level changes in Great Britain, *J. Quat. Sci.*, **4**, 77–89.

Shennan, I. (1993) Sea-level changes and the threat of coastal inundation, *Geog. J.*, **159**, 148–156.

Shepard, F.P. (1948) *Submarine Geology*, Harper, New York.

Sherman, L.K. (1932) Streamflow from rainfall by the unit-graph method, *Engg. News Record*, **108**, 501–505.

Shilling, J.D., C.F. Sirmans and J.D. Benjamin (1989) Flood insurance, wealth distribution and urban property values, *J. Urban Economics*, **26**, 43–53.

Siah, S.J. and W.D. Lasch (1989) A synthetic simulation model of storm surge history and inland drainage in coastal zone areas, in M.A. Ports (ed.), *Hydraulic Engineering*, Proc. 1989 National Conference on Hydraulic Engineering, 780–783.

Sikander, A.S. (1983) Floods and families in Pakistan—a survey, *Disasters*, **7**, 101–106.

Simon, A. (1990) *Gradation Processes and Channel Evolution in Modified West Tennessee Streams: Process, Response and Form*, US Geological Survey Professional Paper 1470.

Sinniger, R.O. and M. Monbaron (eds) (1990) *Hydrology in Mountainous Regions, II Artificial Reservoirs, Water and Slopes*, IAHS Publ. No. 194.

Slatyer, R.O. and J.A. Mabbutt (1964) Hydrology of arid and semiarid regions, in V.T. Chow (ed.), *Handbook of Applied Hydrology*, McGraw-Hill, New York, Section 24.

Smith D.I. (1993) Greenhouse climatic change and flood damages, the implications, *Climatic Change*, **25**, 319–333.

Smith, D.I. (1994) Flood damage estimation—a review of urban stage–damage curves and loss functions, *Water SA*, **20**, 231–238.

Smith, D.I. and M.A. Greenaway (1980) *Computer Flood Damage Maps of Lismore*, Centre for Resource and Environmental Studies, Australian National University, Canberra.

Smith, D.I. and J.W. Handmer (eds) (1986) *Flood Warning in Australia*, Centre for Resource and Environmental Studies, Australian National University, Canberra.

Smith, D.I. and R.G. Munro (1980) *Richmond River Valley, Flood Damages and Social Attitudes: A Summary*, Centre for Resource and Environmental Studies, Australian National University, Canberra.

Smith, D.I., J.W. Handmer, J.M. McKay, M.A.D. Switzer and D.J. Williams (1996) Non-destructive measures in Australian urban floodplain management: barriers and recommendations, *Australian J. Emergency Management*, **11**, 51–56.

Smith, D.K. (1989) *Natural Disaster Reduction: How Meteorological and Hydrological Services Can Help*, Publication No. 772, World Metorological Organisation, Geneva.

Smith, K. (1975) *Principles of Applied Climatology*, McGraw-Hill, London.

Smith, K. (1995) Precipitation over Scotland, 1757–1992: Some aspects of temporal variability, *Internat. J. of Climatology*, **15**, 543–556.

Smith, K. (1996) *Environmental Hazards: Assessing Risk and Reducing Disaster*, 2nd edition, Routledge, London.

Smith, K. and A.M. Bennett (1994) Recently increased river discharge in Scotland: effects on flow hydrology and some implications for water management, *Appl. Geog.*, **14**, 123–133.

Smith, K. and G.A. Tobin (1979) *Human Adjustment to the Flood Hazard*, Longman, London.

Sobey, R.J., B.A. Harper, and K.P. Stark (1977) *Numerical Simulation of Tropical Cyclone Storm Surge*, Res. Bull., CS14, Dept. Civil and Systems Engg., James Cook University of North Queensland, Townsville.

Sokolov, A.A. (1969) Methods of snowmelt maximum discharge computation in cases of absence or insufficiency of hydrometric data, *Floods and their Computation*, IAHS/UNESCO/WMO, Wallingford, 671–680.

Sokolowski, T.J., P.M. Whitmore and W.J. Jorgensen (1990) Alaska Tsunami Warning Center's automatic and interactive computer processing system, *Pure and Appl. Geophysics*, **134**(2), 163–174.

Solín, L. and S. Polácik (1994) Identification of homogeneous hydrological regional types of basins, in P. Seuna, A. Gustard, N.W. Arnell and G.A. Cole (eds), *FRIEND: Flow Regimes from International Experimental and Network Data*, IAHS Publ. No. 221, 467–473.

Sorenson, J.H. (1991) When shall we leave? Factors affecting the timing of evacuation departing, *Internat. J. Mass Emergencies and Disasters*, **9**, 183–200.

Soulsby, R., H. Southgate, P. Thorne and R. Flather (1994) Predicting long-term coastal morphology, Section 2.4 in *A Strategic Approach, Proc. 1994 MAFF Conference of River and Coastal Engineers*, Loughborough, 4–6 July, Ministry of Agriculture, Fisheries and Food, London.

Spencer, T. and J. French (1993) Coastal flooding: transient and permanent, *Geography*, **78**, 179–182.

Stallings, R.A. (1991) Ending evacuations, *Internat. J. Mass Emergencies and Disasters*, **9**, 183–200.

Starosolszky, O. (1994) Flood control by levees, in G. Rossi, N. Harmancioglu and V. Yevjevich (eds), *Coping with Floods*, Kluwer Academic Publishers, Dordrecht, 617–635.

Steers, J.A. (1953) The east coast floods, January 31–February 1, 1953, *Geogr. J.*, **119**, 280–298.

Stern, C.M. (1976) Disaster at Buffalo Creek: from chaos to responsibility, *Amer. J. Psychiatry*, **133**, 300–301.

Strahler, A.N. (1964) Quantitative geomorphology of drainage basins and channel networks, in V.T. Chow (ed.), *Handbook of Applied Hydrology*, McGraw-Hill, New York, section 4-11.

Strahler, A.N. and A.H. Strahler (1983) *Modern Physical Geography*, 2nd edition, Wiley, New York.

Stringfield, V.T. and H.E. LeGrand (1969) Hydrology of carbonate rock terranes—a review, with special reference to the United States (continuation), *J. Hydrol.*, **8**(4), 442–450.

Stromquist, L. (1992) Environmental-impact assessment of natural disasters, a case study of the recent Lake Babati floods in northern Tanzania, *Geografisk Annaler, Series A*, **74**, 81–91.

Sturm, M., J. Beget and C. Benson (1987) Observations of jökulhlaups from ice-dammed Strandline Lake, Alaska: Implications for paleohydrology, in L. Mayer and D. Nash (eds), *Catastrophic Flooding*, Unwin Hyman, London, 79–94.

Suleman, M.S., A. N'Jai, C.H. Green and E.C. Penning-Rowsell (1988) *Potential Flood Damage Data: A Major Update*, Flood Hazard Research Centre, Middlesex Polytechnic, London.

Svoboda, A. (1991) Change in flood regime of large rivers as a consequence of a significant water diversion: example of the hydropower plant Gabcikovo on the River Danube, in F.H.M. van de Ven, D. Gutknecht, D.P. Loucks and K.A. Salewicz (eds), *Hydrology for the Water Management of Large River Basins*, IAHS Publ. No. 201, 203–208.

Takács, A. (1989) Monitoring of weather conditions on mesoscale to improve a precipitation warning system for hydrological purposes, in A. Rango (ed.), *Remote Sensing and Large-Scale Global Processes*, IAHS Publ. No. 186, 117–123.

Tayag, J.C. and R.S. Punongbayan (1994) Volcanic disaster mitigation in the Philippines: experience from Mt Pinatubo, *Disasters*, **18**, 1–15.

Taylor, C. (1995) Splash out on a view, *The Times (Weekend)*, London, Saturday, March 4, p. 13.

Telleria, A.V. (1986) Health consequences of the floods in Bolivia in 1982, *Disasters*, **10**, 88–106.

Thomas, C.A. and R.D. Lamke (1962) Floods of February 1962 in southern Idaho and northeastern Nevada, *US Geol. Surv. Circ.*, **467**, 30.

Thompson, G. and C. Frith (1995) The use of joint probability analysis for the design of flood defences, Section 3.2 in *Proc. 30th MAFF Conference of River and Coastal Engineers*, Keele, 5–7 July, Ministry of Agriculture, Fisheries and Food, London.

Thompson, J. (1989) *Case Studies in Drainage and Levee District Formation and Development on the Floodplain of the Lower Illinois River 1890s to 1930s*, Special Report 17, Water Resources Center, University of Illinois, Urbana.

Thompson, P.M. and P. Sultana (1996) Distribution and social impacts of flood control in Bangladesh, *Geogrl. J.*, **162**, 1–13.

Thompson, S.M. (1993) Estimation of probable maximum floods from the southern Alps, New Zealand, in Z.W. Kundzewicz, D. Rosbjerg, S.P. Simonovic and K. Takeuchi (eds), *Extreme Hydrological Events: Precipitation, Floods and Droughts*, IAHS Publ. No. 213, 299–305.

Thórarinsson, S. (1953) Some new aspects of the Grimsvötn problem, *J. Glaciol.*, **2**, 267–275.

Thorn R.G. and A.G. Roberts (1981) *Sea Defence and Coast Protection Works*, Thomas Telford Ltd, London.

Thorne, C.R., A.P.G. Russell and M.K. Alam (1993) Planform pattern and channel evolution of the Brahmaputra River, Bangladesh, in J.L. Best and C.S. Bristow (eds), *Braided Rivers*, Geological Society, Special Publ. No. 75, 257–276.

Tiefenbrun, A.J. (1965) Bank stabilisation of Mississippi River between the Ohio and Missouri Rivers, *Proc. the Federal Inter-Agency Sedimentation Conference 1963*, USDA Publication 970, 387–399.

Tilford, K.A. (ed.) (1995) *Hydrological Uses of Weather Radar*, Proc. BHS National Meeting, 9 January, British Hydrological Society, Occasional Paper No. 5.

Tilford, K.A. and I.D. Cluckie (1990) The hydrological utilization of weather radar data in the UK, *Workshop on the Remote Sensing of Precipitation with Hydrological Applications*, 5–7 March, University of São Paulo, Brazil (Preprint).

Tinnion, M., D. Pelleymounter and P. Parle (1995) The National Rivers Authority Northumbria and Yorkshire Region's tidal forecasting system, Section 1.2 in *Proc. 30th MAFF Conference of River and Coastal Engineers*, Keele, 5–7 July, Ministry of Agriculture, Fisheries and Food, London.

Tobin, G.A. and B.E. Montz (1990) Response of the real estate market to frequent flooding: the case of Des Plaines, Illinois, *Bull. Illinois Geogrl. Soc.*, **32**, 11–21.

Tollerud, E.I. and R.S. Collander (1993) Mesoscale convective systems and extreme rainfall in the central United States, in Z.W. Kundzewicz, D. Rosbjerg, S.P. Simonovic and K. Takeuchi (eds), *Extreme Hydrological Events: Precipitation, Floods and Droughts*, IAHS Publ. No. 213, 11–19.

Torry, W.I. (1979) Hazards, hazes and holes: a critique of *The Environment as Hazard* and general reflections on disaster research, *Canadian Geographer*, **23**, 383.

Toth, S. (1994) Flood control in Hungary and piping failures at river levees, in W.R. White and J. Watts (eds), *River Flood Hydraulics*, Wiley, Chichester, 249–255.

Trimble, S.W. (1995) Catchment sediment budgets and change, in A. Gurnell and G. Petts (eds), *Changing River Channels*, Wiley, Chichester, 201–215.

Tripp, J.T.B. (1994) Flooding: who is to blame?, *USA Today*, 32–33, Valley Stream, New York.

Troch, P.A., F.P. de Troch and D. van Erdechem (1991) Operational flood forecasting on the River Meuse using on-line identification, in F.H.M. van de Ven, D. Gutknecht, D.P. Loucks and K.A. Salewicz (eds), *Hydrology for the Water Management of Large River Basins*, IAHS Publ. No. 201, 379–389.

Truitt, R.V. (1967) *High Winds... High Tides, a Chronicle of Maryland's Coastal Hurricanes*, Natural Resources Inst., Univ. of Maryland, Educational Series, No. 77.
Tufnell, L. (1984) *Glacier Hazards*, Longman, London.
UN Department of Humanitarian Affairs (1994) *Strategy and Action Plan for Mitigating Water Disasters in Viet Nam*, United Nations, New York and Geneva.
UNEP (1991a) *Environmental Data Report, 1991/92*, United Nations Environment Programme, Blackwell, Oxford.
UNEP (1991b) Tsunamis, *Environmental Data Report, 1991/92*, United Nations Environment Programme, Blackwell, Oxford, 363.
UNESCO (1969) *Discharge of Selected Rivers of the World*, Studies and Reports in Hydrology, UNESCO, 3 vols, Paris.
UNESCO (1991) *The Disappearing Tropical Forests*, IHP Humid Tropics Programme Series No. 1, UNESCO, Paris.
United Nations (1993) *Landslide La Josefina on the Paute River, Cuenca, Ecuador*, Report on Disaster Management, DHA/93/135, United Nations Department for Humanitarian Affairs, Geneva.
US Army Corps of Engineers (1972) *SSARR—Streamflow Synthesis and Reservoir Regulation*, Program Description and User Manual, North Pacific Division, Portland, Oregon.
US Army Corps of Engineers (1973) *Flood Plain Information: Onion Creek, Austin, Texas*, US Army Corps of Engineers, Galveston, Texas.
US Coast and Geodetic Survey (1965) *Tsunami! The story of the Seismic Sea-Wave Warning System*, USGPO, Washington, DC.
US Department of Commerce (1973) *Final Report of the Disaster Team on the Events of Agnes*, National Oceanic and Atmospheric Administration (NOAA) Natural Disaster Survey Report 73-1, 45 pp.
USGS (1944) *Surface Water Supply of the United States 1942, Part 2. South Atlantic Slope and Eastern Gulf of Mexico Basins*, US Geological Survey Water-Supply Paper 952, USGPO.
US Soil Conservation Service (1981) *Simplified Dam-Breach Routing Procedure*, Tech. Release No. 66, US Dept. of Agriculture, Washington, DC.
US Weather Bureau (1959) *National Hurricane Research Project*, Rept. No. 32, Washington, DC.
Utteridge, B. (1996) Flood warning into the next decade, *Proc. 31st Conference of River and Coastal Engineers*, MAFF, London, 4.1.1.–4.1.11.
van de Griend, A.A. (1981) *A Weather-type Hydrologic Approach to Runoff Phenomena*, Free University of Amsterdam.
Van Overeem, J. and E.B. Peerbolte (1994) Evaluation of coastal defence strategies for the Dutch island of Texel, Section 7.3 in *A Strategic Approach, Proc. 1994 MAFF Conference of River and Coastal Engineers*, Loughborough, 4–6 July, Ministry of Agriculture, Fisheries and Food, London.
Van Ufford, H.A.Q. (1953) The disastrous storm surge of 1st February, *Weather*, **8**, 116–121.
Verry, E.S., J.R. Lewis and K.N. Brooks (1983) Aspen clearcutting increases snowmelt and storm flow peaks in north-central Minnesota, *Water Resources Bull.*, **19**, 59–67.
Vieira, J., J. Foens and G. Cecconi (1993) Statistical and hydrodynamic models for the operational forecasting of floods in the Venice Lagoon, *Coast. Eng.*, **21**(4), 301–331.
Viles, H. and T. Spencer (1995) *Coastal Problems: Geomorphology, Ecology and Society at the Coast*, Edward Arnold, London.
Volker, A. (1983) Rivers of southeast Asia: their regime, utilization and regulation, in R. Keller (ed.), *Hydrology of Humid Tropical Regions*, IAHS Publ. No. 140, 127–138.
Volker, M. (1953) La marée de tempête du 1er février 1953 et ses conséquences pour les Pays-Bas, *La Houille Blanche*, **2**, 207–216.
Wagner, M.J. (1994) ERS-1 images of the Christmas flood over Europe, *Earth Observation Quarterly*, **43**, 12–13.
Wain, A.S. (1994) Diurnal river flow variations and development planning in the tropics, *Geogrl. J.*, **160**(3), 295–306.
Walling, D.E. (1971) Streamflow from instrumented catchments in S.E. Devon, in K.J. Gregory and W.L.D. Ravenhill (eds), *Exeter Essays in Geography*, Exeter, 77.
Walsh, J.E. (1977) Exploitation of mangal, in V.J. Chapman (ed.), *Wet Coastal Ecosystems*, Elsevier, Amsterdam, 347–362.

Walsh, R.P.D., H.R.J. Davies, and S.B. Musa (1994) Flood frequency and impacts at Khartoum since the early nineteenth century, *Geogrl. J.*, **160**(3), 266–279.

Wang, Y. (1993) Solar activity and maximum floods in the world, in Z.W. Kundzewicz, D. Rosbjerg, S.P. Simonovic and K. Takeuchi (eds), *Extreme Hydrological Events: Precipitation, Floods and Droughts*, IAHS Publ. No. 213, 121–127.

Wanielista, M. (1990) *Hydrology and Water Quantity Control*, Wiley, New York.

Ward, R.C. (1971) *Small Watershed Experiments: An Appraisal of Concepts and Research Developments*, Occ. Papers in Geog., No. 18, University of Hull.

Ward, R.C. (1978) *Floods: A Geographical Perspective*, Macmillan, London.

Ward, R.C. (1981) River systems and river regimes, in J. Lewin (ed.), *British Rivers*, Allen & Unwin, London, 1–33.

Ward, R.C. and M. Robinson (1990) *Principles of Hydrology*, 3rd edition, McGraw-Hill, Maidenhead.

Warrick, R. and J. Oerlemans (1990) Sea level rise, in J.T. Houghton, G.J. Jenkins and J.J. Ephraums (eds) *Climate Change: The IPCC Scientific Assessment*, Cambridge University Press, 257–281.

Watanabe, M. (1989) Some problems in flood disaster prevention in developing countries, in Ö. Starosolszky and O.M. Melder (eds), *Hydrology of Disasters*, Proceedings of a WMO Technical Conference, James and James, London.

Waylen, P.R. and C.N. Caviedes (1987) El Niño and annual floods in coastal Peru, in L. Mayer and D. Nash (eds), *Catastrophic Flooding*, Unwin Hyman, 57–77.

Webb, C. (1996) A sea view is worth its salt, *The Times* (*Weekend*), London, Saturday, 24 August, p. 9.

White, E.L. and B.M. Reich (1970) Behavior of annual floods in limestone basins in Pennsylvania, *J. Hydrol.*, **10**(2), 193–198.

White, G.F. (1945) *Human Adjustment to Floods: a Geographical Approach to the Flood Problem in the United States*, Res. Paper No. 29, Dept. Geog., University of Chicago.

White, G.F. (1964) *Choice of Adjustment to Floods*, Res. Paper No. 93, Dept. Geog., University of Chicago.

White, G.F. (1975) *Flood Hazard in the United States: A Research Assessment*, Monograph No. 6, Institute of Behavioral Science, University of Colorado, Boulder.

White, G.F. and J.E. Haas (1975) *Assessment of Research on Natural Hazards*, MIT Press, Cambridge, Massachusetts.

White, W.R. and J. Watts (eds) (1994) *River Flood Hydraulics*, Wiley, Chichester.

White House (1993) *Protecting America's Wetlands: A Fair, Flexible and Effective Approach*, Office on Environmental Policy, 24 August, GPO Washington, DC.

Whitmore, P.M. (1993) Expected tsunami amplitudes and currents along the North American coast for Cascadia subduction zone earthquakes, *Natural Hazards*, **8**(1), 59–73.

Wickramanayake, E. (1994) Flood mitigation problems in Vietnam, *Disasters*, **18**, 81–86.

Wickramanayake, E., G.A. Shook and T. Rojnkureesatien (1995) Rehabilitation planning for flood affected areas of Thailand: experience from Phipun District, *Disasters*, **19**, 348–355.

Wiesner, C.J. (1970) *Hydrometeorology*, Chapman & Hall, London.

Wijkman, A. and L. Timberlake (1984) *Natural Disasters: Acts of God or Acts of Man?*, Earthscan, London.

Wilke, K. and F. Barth (1991) Operational river-flood forecasting by Wiener and Kalman filtering, in F.H.M. van de Ven, D. Gutknecht, D.P. Loucks and K.A. Salewicz (eds), *Hydrology for the Water Management of Large River Basins*, IAHS Publ. No. 201, 391–400.

Wilkins, L. (1996) Living with the flood: human and governmental response to real and symbolic risk, in S.A. Changnon (ed.), *The Great Flood of 1993 Causes, Impacts and Responses*, Westview Press, Boulder and Oxford, 218–224.

Williams, J.R., A.D. Nicks and J.G. Arnold (1985) Simulator for water resources in rural basins, *J. Hydraul. Engg.*, **111**, 970–986.

Wilson, E.M. (1983) *Engineering Hydrology*, 3rd edition, Macmillan, London.

Wiltshire, S.E. (1986) Regional flood frequency analysis II: Multivariate classification of drainage basins in Britain, *Hydrol. Sci. J.*, **31**, 335–346.

REFERENCES

Winchester, P. (1992) *Power, Choice and Vulnerability: A Case Study in Disaster Management in South India, 1977–88*, James and James, London.
Wind, H.G. and E.B. Peerbolte (1993) Sea level rise: assessing the problems, in R.A. Warrick, E.M. Barrow and T.M.L. Wigley (eds), *Climate and Sea Level Change*, Cambridge University Press, 297–309.
Winkley, B.R. (1971) Practical aspects of river regulation and control, in H.W. Shen (ed.), *River Mechanics*, Vol. 1. Fort Collins, Colorado, 19.1–19.79.
Wisler, C.O. and E.F. Brater (1959) *Hydrology*, 2nd edition, Wiley, New York.
WMO (1983) Operational hydrology in the humid tropical regions, in R. Keller (ed.), *Hydrology of Humid Tropical Regions*, IAHS Publ. No. 140, 3–26.
WMO (1986) *Manual for Estimation of Probable Maximum Precipitation*, World Meteorological Organization, Operational Hydrology Report No. 1, 2nd edition, WMO, Geneva.
WMO (1989) *Natural Disaster Reduction: How Meteorological and Hydrological Services Can Help*, WMO 722, World Meteorological Organization, Geneva.
WMO (1990a) *The Role of the World Meteorological Organization in the International Decade for Natural Disaster Reduction*, WMO 745, World Meteorological Organization, Geneva.
WMO (1990b) *Tropical Cyclone Warning Systems*, Tropical Cyclone Programme Rept. No. TCP-26, World Meteorological Organization, Geneva.
WMO (1992a) The worst cyclone in 20 years, in *Annual Report 1991*, WMO No. 774, World Meteorological Organization, Geneva, p. 7.
WMO (1992b) *The World Weather Watch Programme 1992–2001, Third WMO Long-Term Plan, Part II*, WMO 761, World Meteorological Organization, Geneva.
Wolf, P.O. (1952) Forecast and record of floods in Glen Cannich in 1947, *J. Inst. Wat. Engrs. and Scientists*, **6**, 298–324.
Wolman, M.G. and R. Gerson (1978) Relative scales of time and effectiveness of climate in watershed geomorphology, *Earth Surf. Proc.*, **3**, 189–208.
Woodruff, B. A. et al. (1990) Disease surveillance and control after a flood: Khartoum, Sudan, 1988, *Disasters*, **14**, 151–163.
Woodworth, P. (1990) Measuring and predicting long term sea level changes, *NERC News*, October, 22–25.
Woodworth, P.L., W.E. Spencer and G. Alcock (1990) On the availability of European mean sea level data, *Internat. Hydrographic Rev.*, **67**, 131–146.
Woolley, R.R. (1946) *Cloudburst Floods in Utah, 1850–1938*, US Geol. Surv. Water-Supply Paper, 994.
WRC (1981) *Guidelines for Determining Floods Flow Frequency*, Bull. 17B, Hydrol. Comm. Washington, DC.
Wright, C. and A. Mella (1963) Modification to the soil patterns of south central Chile resulting from seismic and associated phenomena during the period May to August 1960, *Bull. Seismol. Soc. Am.*, **53**, 1367–1402.
Wright, J.M. (1996) Effects of the flood on national policy: some achievements, major challenges remain, in S.A. Changnon (ed.), *The Great Flood of 1993 Causes, Impacts and Responses*, Westview Press, Boulder and Oxford, 245–275.
Wright, S.K. (1980) Benefits of self-help flood warning systems, Paper presented at the *2nd Conference on Flash Floods*, American Meteorological Society, Atlanta, Georgia (Quoted by Parker, 1987).
Wróblewski, A. (1994) Empirical Orthogonal Functions (EOF) method in determining and forecasting storm floods in the coastal zones of the sea, in G. Rossi, N. Harmancioglu and V. Yevjevich (eds), *Coping with Floods*, Kluwer Academic Publishers, Dordrecht, 503–512.
Wu, Q. (1989) The protection of China's ancient cities from flood damage, *Disasters*, **13**, 193–227.
Yair, A. (1990) Runoff generation in a sandy area—The Nizzana Sands, western Negev, Israel, *Earth Surf. Proc. and Landforms*, **15**, 597–609.
Yair, A. and H. Lavee (1985) Runoff generation in arid and semi-arid zones, in M.G. Anderson and T.P. Burt (eds) *Hydrological Forecasting*, Wiley, Chichester, 183–220.
Yang Yicheng (1989) Flood monitoring from space in China, in A. Rango (ed.), *Remote Sensing and Large-Scale Global Processes*, Internat. Assoc. Hydrol. Sci. Publ. No. 186, 183–185.

Yevjevich, V. (1979) Extraction of full information on flood peaks in arid areas, in *The Hydrology of Areas of Low Precipitation*, IAHS Publ. No. 128, 223–232.

Yevjevich, V. (1994) Classification and description of flood mitigation measures, in G. Rossi, N. Harmancioglu and V. Yevjevich (eds), *Coping with Floods*, Kluwer Academic Publishers, Dordrecht, 573–584.

Yoshimoto, T. and T. Suetsugi (1990) Comprehensive flood disaster prevention measures in Japan, in H. Massing, J. Packman and F.C. Zuidema (eds), *Hydrological Processes and Water Management in Urban Areas*, IAHS Publ. No. 198, 175–181.

Young, G.J. and K. Hewitt (1990) Hydrology research in the upper Indus basin, Karakoram Himalaya, Pakistan, in L. Molnar (ed.), *Hydrology of Mountainous Areas*, IAHS Publ. No. 190, 139–152.

Young, S.W. and J.D. Pos (1995) Objective setting within Shoreline Management Plans, Section 8.2 in *Proc. 30th MAFF Conference of River and Coastal Engineers*, Keele, 5–7 July, Ministry of Agriculture, Fisheries and Food, London.

Zacharias, T.P. (1996) Impacts on agricultural production: huge financial losses lead to new policies, in S.A. Changnon (ed.), *The Great Flood of 1993*, Westview Press, Boulder, Colorado, 163–182.

Zaman, M.Q. (1991) The displaced poor and resettlement policies in Bangladesh, *Disasters*, **15**, 117–125.

Zaman, M.Q. (1993) Rivers of life: living with floods in Bangladesh, *Asian Survey*, **33**, 985–996.

Zhang Shunying and Xie Zichu (1991) Analysis of the relationship between lakes' melting regimes and stream runoff, in H. Lang, W. Frey, D. Issler and B. Salm (eds), *Snow, Hydrology and Forests in High Alpine Areas*, IAHS Publ. No. 205, 245–250.

Zon, R. (1927) Do forests prevent floods?, *Amer. Forests and Forest Life*, **33**, 387–392 and 432.

Index

abatement of floods 206, 207–208, 220–225
absolute flood magnitude 6
adverse selection 300–301
afforestation in flood control 84–85, 221, 224–225
aggradation of river sediments 86–87, 118, 218
agriculture
　effect on floods of 79, 85–86, 221–224
　impact of floods on 22, 31–32, 45–46, 51, 53–54, 220–221, 322
aid following floods 294, 295–300
Alaska Tsunami Warning Center 291
annual flood series 189
Ash Wednesday storm (1962) 162
atmospheric pressure in cyclones 150–152, 157–158, 164, 283, 288
atmospheric processes 75, 90–91, 94, 103, 104, 111–112, 119, 121, 127–128, 133, 135–135, 140, 151–152, 153–155, 157–158, 162, 164, 182–183

Bakerganj cyclone (1876) 158
band of tolerance 20–21
Bangladesh floods (1987) and (1988) 37, 45, 47, 49, 116–119, 120, 295, 296, 324
barrages 227–230
barrier islands 27, 36, 230–231
barsha flood 37
beaches 175–176, 232
Beijiang river: flood forecasting on 261–262
benefits from floods 35
　agricultural 36–38
　ecological 36
best estimate of mean annual flood (BESMAF) 97–98
Blue Manual 43
bonna flood 37
building codes 298, 301, 304, 305, 335, 336

causes of floods
　dam failure 50–51
　on coasts and in estuaries 10, 145–148
　icejams 109–111
　melting ice 65, 76–79, 95
　rainfall 75–76, 120
　in rivers 10
　storm surges 146
　tidal surges 146
　tsunamis 165–168
catchment characteristics 12–15, 99, 101–102, 194, 197
channel characteristics 86–87
channel improvements 209–210, 218–219, 220–221
channel networks 13–14, 99, 101, 128–129
channel obstruction 104, 108–111
chars 36, 54
check dams 222–224
'Chicago school' 55
climate change 4–5, 45, 89, 92–95, 170–175, 191, 200, 319, 330, 331
climate partitioning 89–90
climatic variability 82, 88–95
climatic zones 103–104
Coastal Area Modelling for Engineering in the Long-Term (CAMELOT) 177, 330
coastal floods 10, 11–12, 51, 118, 119, 143–177, 175, 280–292
Coastal Zone Management Act (1972) 267
Coast Protection Act (1949) 309
coasts
　configuration of 152–153, 154, 157, 159, 168, 175
　defence of 144–145, 226–231
　erosion of 28, 176–177, 226
　human invasion of 22–23, 27–30, 143, 304
computer simulation 180, 196–197, 239
contour ploughing 221–222

dams
　advantages of 214
　disadvantages of 216–217
　for flood control 206–207, 213–219
　floods from failures of 50–51, 264
　ice jams 109–111
　operation of 215–217
　reservoir 142
・warning of failure of 269

databases
  inadequacy of 3–4, 23, 92, 93, 170, 189–192, 329, 334
  methods of extending 192–194, 200, 201
deaths
  in flood disasters 4–5, 23, 25, 31–32, 47, 50, 126–127, 130–131, 151, 153, 157, 158, 159, 169–170, 214, 299, 310
  from tropical cyclones 31–32, 151, 153, 159, 267
deforestation 84–85, 114–116, 207, 220, 224–225, 230
degradation of river sediments 218
deliberate flooding 87, 232
deltas 146, 148, 233
depth–damage relationships *see also* stage–damage curves
design flood 17, 178–179, 198–202, 207, 215, 227, 233, 338
  deterministic estimation of 179–188
  estimation on coasts 199–201
  probabilistic estimation of 188–191, 201
development/underdevelopment and floods 299–300
digital terrain models (DTM) 18, 244, 260
dikes *see* levees and sea walls
direct losses from floods 34–35, 38, 51, 277
Disaster Relief Acts (1950) and (1974) 296
drainage 84, 137
Dutch Delta Plan 228, 231
dysentery 31, 335

earthquake activity and floods 118, 131, 164–170, 282–283, 290–292
Earthquake and War Damage Act (1944) 305
economic losses from floods 4, 23, 25, 29, 45, 47, 48, 50–51, 52–53
  in agriculture 45–46, 297
  calculation of 39–45
  mapping of 39–40
  from tsunamis 169
El Niño Southern Oscillation (ENSO) 48, 88, 89, 90–91, 133, 136
engineering approaches to floods 205–207, 293, 320, 335–336
  on coasts 226–230
  on rivers 208–220
Environment Agency 208, 272, 274, 288, 309
environmental impact of flood defence 208, 217–220, 230–231, 324å325
epidemics 299, 335
erosion
  and agriculture 220–221
  on coasts 28, 175–177, 226, 230–231, 232, 302, 307
  on floodplains 53–54
  on hillslopes 207
  of riverbanks 53–54, 210, 213, 325, 326
'escalator effect' *see* 'levee effect'
estuaries
  causes of floods in 145–148
  flood estimation in 201–202
  sediment in 227–228
  walls on 226–227
estuarine floods *see* coastal floods
evacuation 4, 267, 268, 270, 274, 279, 309, 310, 311

Federal Emergency Management Agency (FEMA) 274, 296–297, 305–308
Federal Insurance Administration 43, 308
flash floods 124–131
  on alluvial fans 124, 129–130
  causes of 67, 75, 94, 116, 335
  and dam failure 130–131
  deaths from 25, 126–127, 130–131
  definition of 15–16, 125
  forecasting of 263–264, 332
  losses from 47
  warnings for 269
Flood Action Plan 324–325
flooding
  causes of 5, 9–14
  definition of 8–9
  paradigms of 54–57, 334–335
  types of 10,
floods in carbonate terrain 65–66
Flood Control Act (1936) 8, 54
Flood Control Act (1960) 17
flood defence 32–33, 205–237, 320, 323–325, 335–336
flood depth 41, 46, 50
flood disasters
  definition of 23
  early legends of 6–8, 168
  history of 5–8
  trends in 21–25
flood duration 41, 43, 46, 67, 70, 124
flood estimation 178–202
Flood Estimation Handbook 8, 102, 330, 333
flood fringe 317–318
flood hydrograph 12–16, 65–71, 81, 83, 97, 106, 110–111, 115, 129, 134, 138, 141, 195, 329–330
flood outline 17–18, 28, 117, 156, 158, 332–333
floodplain(s)
  acquisition of 297, 312–317
  attributes of 36
  ecology of 218–219
  human invasion of 22, 26–30, 301–302, 312–317, 335–337

management of 294–295
optimum use of 140, 301
flood proofing 233–237, 303, 322, 338
flood pulse 21, 35, 37, 322
Flood Studies Report 8, 74, 99, 100, 128, 179, 181, 182, 184, 187, 192, 194, 195, 197, 199, 330, 333
floodway 305–306
Forecasting Rain Optimised Using New Techniques of Interactively Enhanced Radar and Satellite (FRONTIERS) 242–243, 332
forecasting schemes 241–264, 280–292
  in the Bay of Bengal 288, 290
  on the Beijiang river 261–262
  on coasts 280–292
  derivation of flood hydrograph 245–247
  for flash floods 263–264
  history of 239–240
  methodology of 241–245
  on the Meuse river 261, 262
  models of 249–251
  on the Mosel river 261
  on the Nile river 262–263
  in the North Sea 286–287
  role of computer simulation in 239
  for snow melt floods 254–257
  for storm surge 283–290
  for tropical cyclones 282, 283, 287–290
  for tsunamis 269, 270–271
forests *see* afforestation/deforestation
frozen ground and flood effects 65–66

gastroenteritis 48–49, 299
Geographical Information Systems (GIS) 244, 303, 333
geomorphology and floods 86–88
Global Positioning System (GPS) 19, 170, 283, 332–333
global warming *see* climate change and climate variability
Great Atlantic Hurricane (1944) 159–161
greenbelt 27–28, 314

housing
  effects of flooding on prices 46–47
  effects of flooding on structures 234, 236–237, 322
  floodproofing of 233–237
  relocation of 234, 297, 304, 307, 316–319, 320
  value on floodplains of 26–27, 313, 317
human adjustments to floods 32–33
human vulnerability to floods 19–22, 23, 25, 30–32, 45, 47, 50, 56–57, 275, 276, 294, 322–323, 324–326, 335
hurricane(s) *see* tropical cyclones

Hydrologic Engineering Center (HEC) 251, 252, 253
Hydrological Radar (HYRAD) 242–243

ice dams/jams 109–111
ice melt 65, 76–79, 95, 257
ill health after floods 44, 47–51
indirect losses from floods 34–35, 44, 51–54
industry, impact of floods on 51–52, 227
infiltration 12–13, 80, 84, 123–124, 181, 225
infiltration capacity 62–64
Institute of Hydrology 18, 189, 254
Institute of Hydrology Digital Terrain Model (IHDTM) 18, 180, 182
insurance, against floods 294, 300–308, 337
intangible loss(es) from floods 34–35, 52–54, 298
Intermediate Regional Flood 17–18
internal government aid 295–298, 308
international aid 298–300
International Decade for Natural Disaster Reduction (IDNDR) 288, 290, 330, 339
invasion of flood-prone land 22–23, 27–30, 143, 301–302, 304, 312–317, 335–337

jhils 38
jökulhlaups 65, 104, 110, 198

lahars 88
lake levels 91–92
Land Drainage Act (1991) 309
landslides 11, 165, 167
land use change 5
land use planning for floods 30–31, 230, 305–306, 310, 311, 312–319
La Niña 90
levee(s)
  failures of 137, 139, 212–213, 324
  for flood control 206–207, 210–213, 218–219, 321, 323–325
'levee effect' *see also* invasion of flood-prone land 55, 211, 313–315, 336
living with floods 56–57, 293–295, 319–326, 338–339

magnitude–frequency relationships 17, 20, 76, 86, 93
malaria 48–49
managed retreat
  from coasts 145, 174–175, 232–233, 268, 312, 317–319, 320, 337, 338
  from floodplains 312, 317–319, 320
management plans
  for coastal flooding 176–177, 208, 232–233, 337–338
  for river flooding 208, 315, 337–338

mapping
  of flood damage 39–40
  of flood hazard zones 17–19, 274, 303, 306, 312
maximum rainfall *see* Probable Maximum Precipitation
mean annual flood 99, 101
media coverage of floods 271, 272, 274, 310, 334
melt rates, calculation of 76–77
meltwater flooding 64–65, 88, 94, 105–111, 198–199
mental health after floods 47, 49–51
meso-scale convective complexes 75
Meteorological Office 77, 93, 155, 243, 254, 281, 285, 286, 309
Meteorological Office Rainfall and Evaporation Calculation System (MORECS) 246, 254
Meuse river, flood forecasting on 261, 262
mid-latitude storms 72, 75, 93–94, 136, 153, 173, 183–186, 284–285
Mid-West floods (1993) 24, 45, 133–139, 213, 215–216, 271, 272–274, 295, 296, 302, 306, 316, 320, 332, 335–336
Ministry of Agriculture, Fisheries and Food (MAFF) 18, 177, 208, 232, 233, 285, 309
models
  of catchment hydrology 193, 198, 246–247, 251–257, 330
  in flood forecasting 249–251
  rainfall-runoff 191, 196–197, 258, 259, 333
  of runoff 187
  of storm surge 201, 284–290
monsoon flooding/rains 72, 74, 85, 90, 95, 112, 116–119, 120
moorland gripping 86
moral hazard 301
Mosel river, flood forecasting on 261
mudflows *see* lahars
Muskingum method 248, 249, 252, 253, 254

National Flood Insurance Program 266, 302, 305–308, 313, 337
natural hazards, floods as 19–21
Nile river, flood forecasting on 262–263
Noah 6–8, 168
North Sea floods (1953) 149, 153–158, 240

overland flow 62–64, 114–116, 121, 123–124
Oxford Committee for Famine Relief (OXFAM) 295

Pacific Tsunami Warning System 290–291, 332
paleohydrology 193

peak discharge 11, 16, 72–73, 92, 94, 97, 106, 108, 109, 110–111, 112, 114, 118, 124–126, 128, 133, 139, 142, 198–199, 209, 216
perception of floods 4, 19, 47, 56, 220, 326, 338
Permanent Service for Mean Sea Level (PSMSL) 171
permeability *see* infiltration
persistence 72, 89, 92
pollution 88
population and floods
  density of 29–30, 323, 325
  growth of 27, 80
poverty
  and disaster vulnerability 30–32, 45, 56–57, 294, 322–323, 324–326, 334, 336
  and under-insurance 303
precipitation *see* rainfall
preparedness for floods 294, 308–312, 339
  success of 278–279
prism storage 248–249
Probable Maximum Flood (PMF) 41, 42, 182, 186–188, 199
Probable Maximum Precipitation (PMP) 182–188

quickflow 14–15, 62–64, 80, 84, 98, 128, 181

radar, role in flood forecasting 242–244, 332
rainfall
  as a cause of floods 75–76, 120
  duration of 71, 74
  extremes of 71–72, 75, 121, 128, 136, 182–186
  intensity of 71, 82, 112, 114, 140
  persistence of 72, 89, 92
rainprints of storms 64, 65, 67–68, 74, 121
rain-on-snow floods 70–71, 76, 107–108
rainfall–runoff relationships 191, 196–197, 258, 259
rainwater flooding 10, 118
Rational Flood Formula 180
real estate agents 315–316
Red Manual 44
regionalisation of floods 97–102, 192–193
reinsurance 302, 304
relocation of flood-prone properties 234, 297, 304, 307, 316–319, 320
remote sensing for snow cover 244–245, 254–255, 257
representativeness of small drainage basins 97
reservoirs for flood control 142
retrofitting 234, 236–237
return periods
  of floods 17, 102, 107, 119, 122–123, 127, 133, 138, 139, 142, 145, 173–174, 188–194, 202, 206, 211, 213
  of rainfall 74

Rhine river floods (1993–94)   4, 139–142, 310
risk communication   309–310
riverbank erosion   53–54, 210, 213
river corridors   36, 218, 219, 316, 337
river floods   10–12
River Flow Forecasting System (RFFS)   254, 258–260, 333
river restoration   219
Robert T Stafford Disaster Relief and Emergency Assistance Act (1974)   297
routing of floods   247–249, 251

Saffir/Simpson scale   152, 164
St Venant equations   247, 261
saltmarshes   226, 227, 230, 231–233
sand dunes, protective role of   231–232
satellite imagery   62, 93, 240, 242–244
Scottish Environment Protection Agency (SEPA)   272
Sea Lake and Overland Surges for Hurricanes (SLOSH)   201, 288
sea level change   170–175
sea level rise   118
   future estimates of   171–172, 200, 231
   observations of   146–148, 171, 200
seasonal floods   16–17, 37–38, 46, 52
sea surface temperature   90–91
sea walls/dikes   17, 226–227, 233
sediment
   on coasts   143–144, 175–176, 231
   in deltas   148
   deposition of   30–31, 300
   in estuaries   227–228
   in rivers   86–88, 213, 217, 218, 323
sewer and stormwater systems, failures in flooding   10, 30–31, 48–49, 80–83, 210, 236–237, 264, 307, 335
shanty housing   30
slush avalanches   9, 79
snow melt   64–65, 76–79, 94, 105–108, 140, 199, 198–199, 225
   snow cover equivalent of   244–245
snow melt floods, forecasting of   254–257
snowpack
   energy balance   76–77
   water equivalent of   64–65, 332
Soil Conservation and Rivers Control Act (1941)   54–55
Special Flood Hazard Areas (SFHAs)   306, 308, 326–327
specific discharge   67, 69–70, 76, 97, 98, 124–125
specific flood magnitude   6–7
stage–damage curves   39, 41–44
stage of rivers   16, 20, 29, 137–139, 142
Standard Project Flood   17–18

stationarity   191
storm surge
   on Atlantic coast of USA   159–164
   in the Bay of Bengal   31–32, 157–160
   as a cause of floods   146, 148–153, 158, 160–164
   deaths from   151, 153, 157, 158
   effect of storm intensity on   173
   flooding on English coasts   175
   forecasting and warning for   283–290, 332
   in the North Sea   153–158, 228
   models of   201, 284–285
   prediction of   199–201
   from tropical cyclones   148–153
Storm Tide Warning Service (STWS)   228, 240, 281, 285–287
strip cropping   222
sub-surface flooding   9, 82
SURGE   201, 288
Susquehanna river floods (1972)   132–133
sustainability and floods   56–57, 231, 293, 295, 312, 315, 320, 334, 336, 339
synthetic streamflow records   193

tangible losses from floods   34–35, 45–47, 51–52
terracing   221–222
Thames tidal barrier   228
thermal expansion of oceans   171–172
tidal range   145–146, 148–149, 153, 158, 162, 173–174, 200
transport, effects of floods on   44, 51–52
tropical cyclones see also hurricanes and typhoons
   areas of vulnerability to   113
   as a cause of floods   67, 72, 74, 88, 90, 93–94, 112–114, 132–133
   deaths from   31–32, 151, 153, 159, 267
   forecasting and warning of   282, 283, 287–290
   storm surge from   148–151, 157–164, 287–290
tsunamis
   causes of   165–168
   deaths from   169–170
   defence against   229–230
   economic losses from   169
   forecasting for   269, 270–271
   land use planning for   230, 310, 311, 332
   magnitude of   165–167
   travel time of   164, 166–170, 291
   warning systems for   290–292
   wave characteristics of   12, 164–168
typhoid   48
typhoon(s) see hurricanes and tropical cyclones

under-insurance 302–303
UK Geological Survey 19
United Nations Disaster Relief Organisation (UNDRO) 298
unit hydrograph 181, 187, 194, 195, 246
urbanisation and flooding 18, 80–83, 225, 335
US Army Corps of Engineers 17, 26, 54, 176, 207, 208, 213, 219, 248, 252, 253, 295, 310, 317
US Department of Commerce 25
US Office of Foreign Disaster Assistance (OFDA) 23–24, 298
US Geological Survey 17, 139
US National Weather Service 251, 264, 271, 287, 290

variable source areas 62, 116
velocity of water 129, 130
volcanic activity and floods 88, 92–93, 110, 164, 167, 290

warning systems for floods 264–280
  assessment of 269–270, 278–280
  attributes of 267–268
  content of 271
  delivery of 272–274
  design of 265–269
  dissemination of 270–274
  lead time of 268–269, 275–278, 290–292
  for tsunamis 290–292
washlands *see also* floodplains 5, 10, 17
Water Act (1973) 17
Watershed Protection and Flood Prevention Act (1954) 207
Water Resources Act (1991) 208, 309
water spreading 222–223
wetlands 36, 137, 207, 236, 295, 320–322, 337
'wild' rivers 36
World Meteorological Organisation (WMO) 240, 288, 290, 329

yellow fever 48
Yellow Manual 44